INFORMATION TECHNOLOGY FOR BALANCED MANUFACTURING SYSTEMS

IFIP – The International Federation for Information Processing

IFIP was founded in 1960 under the auspices of UNESCO, following the First World Computer Congress held in Paris the previous year. An umbrella organization for societies working in information processing, IFIP's aim is two-fold: to support information processing within its member countries and to encourage technology transfer to developing nations. As its mission statement clearly states,

> *IFIP's mission is to be the leading, truly international, apolitical organization which encourages and assists in the development, exploitation and application of information technology for the benefit of all people.*

IFIP is a non-profitmaking organization, run almost solely by 2500 volunteers. It operates through a number of technical committees, which organize events and publications. IFIP's events range from an international congress to local seminars, but the most important are:

- The IFIP World Computer Congress, held every second year;
- Open conferences;
- Working conferences.

The flagship event is the IFIP World Computer Congress, at which both invited and contributed papers are presented. Contributed papers are rigorously refereed and the rejection rate is high.

As with the Congress, participation in the open conferences is open to all and papers may be invited or submitted. Again, submitted papers are stringently refereed.

The working conferences are structured differently. They are usually run by a working group and attendance is small and by invitation only. Their purpose is to create an atmosphere conducive to innovation and development. Refereeing is less rigorous and papers are subjected to extensive group discussion.

Publications arising from IFIP events vary. The papers presented at the IFIP World Computer Congress and at open conferences are published as conference proceedings, while the results of the working conferences are often published as collections of selected and edited papers.

Any national society whose primary activity is in information may apply to become a full member of IFIP, although full membership is restricted to one society per country. Full members are entitled to vote at the annual General Assembly, National societies preferring a less committed involvement may apply for associate or corresponding membership. Associate members enjoy the same benefits as full members, but without voting rights. Corresponding members are not represented in IFIP bodies. Affiliated membership is open to non-national societies, and individual and honorary membership schemes are also offered.

INFORMATION TECHNOLOGY FOR BALANCED MANUFACTURING SYSTEMS

IFIP TC5, WG 5.5 Seventh International Conference on Information Technology for Balanced Automation Systems in Manufacturing and Services, Niagara Falls, Ontario, Canada, September 4-6, 2006

Edited by

Weiming Shen

Springer

Library of Congress Control Number: 2006929211

Information Technology for Balanced Manufacturing Systems

Edited by W. Shen

p. cm. (IFIP International Federation for Information Processing, a Springer Series in Computer Science)

ISSN: 1571-5736 / 1861-2288 (Internet)
ISBN: 10: 0-387-36590-7
ISBN: 13: 9780-3870-36590-7
eISBN: 10: 0-387-36594-X
Printed on acid-free paper

Printed in the United States of America.

9 8 7 6 5 4 3 2 1
springer.com

TABLE OF CONTENTS

SPONSORS

ifip IFIP – The International Federation for Information Processing

COVE

IFIP WG 5.5 COVE
Co-Operation infrastructure for Virtual Enterprise and electronic business

CSCWD International Working Group on Computer Supported Cooperative Work in Design

SoCol net

Society of Collaborative Networks

Western University of Western Ontario

UNIVERSITY OF WINDSOR University of Windsor

COMMITTEES

Steering Committee

L. M. Camarinha-Matos (PT) **[Chair]**
H. Afsarmanesh (NL)
V. Marik (CZ)
H. Erbe (DE)

Program Committee Chairs

Weiming Shen, National Research Council / University of Western Ontario, Canada
Robert Brennan, University of Calgary, Canada
Luis M. Camarinha-Matos, New University of Lisboa, Portugal

Track Chairs

Track A: Vladimir Marik, Czech Technical University, Czech Republic
Track B: Hamideh Afsarmanesh, University of Amsterdam, The Netherlands
Track C: Mauro Onori, Royal Institute of Technology, Sweden
Track D: Paulo E. Miyagi ,University of Sao Paulo, Brazil

Organizing Committee Chairs

Xiaobu Yuan, University of Windsor, Canada
Qi Hao, National Research Council, Canada

Program Committee Members

Track A: Multi-Agent and Holonic Systems in Manufacturing

- J. Barata (PT)
- R. W. Brennan (CA)
- Y. Fan (CN)
- S. Feng (US)
- M. Fletcher (UK)
- H. Ghenniwa (CA)
- W. Gruver (CA)
- K. Hall (US)
- P. Kopacek (AT)
- V. Marik (CZ) **[Chair]**
- F. Maturana (US)
- E. Oliveira (P)
- M. Pechoucek (CZ)
- W. Shen (CA)
- L. Sheremetov (MEX)
- P. Sousa (PT)
- S. Tamura (J)
- P. Valckenaers (BE)
- L. Wang (CA)

Track B: Networked Enterprises

- H. Afsarmanesh (NL) **[Chair]**
- P. Bernus (AU)
- F.W. Bruns (DE)
- L.M. Camarinha-Matos (PT)
- W. Cellary (PL)
- B. Chaib-draa (CA)
- S. D'Amours (CA)
- S. Dudstar (AT)
- H. Erbe (DE)
- C. Garita (CR)
- T. Goranson (US)
- H.S. Jagdev (UK)
- T. Kaihara (JP)
- K. Kosanke (DE)
- G. Kovács (HU)
- A. Molina (MX)
- L. Nemes (AU)
- G. Olling (US)
- M. Ollus (FI)
- J. Pinho Sousa (PT)
- G. Putnik (PT)
- R. Rabelo (BR)

Track C: Integrated Design and Assembly

- T. Arai (JP)
- R. Bernhardt (DE)
- J. Camelio (US)
- Q. Hao (CA)
- J. Jacot (CH)
- S. Lang (CA)
- R. Molfino (IT)
- M. Onori (SW) **[Chair]**
- S. Ratchev (UK)
- M. Santocchi (IT)
- G. Schreck (DE)
- R. Tuokko (FI)
- S. Yang (CA)
- X. Yuan (CA)
- Y. Zeng (CA)
- D. Zhang (CA)

Track D: Monitoring and Control

- M. Barata (PT)
- E. Bordatchev (CA)
- S. Feng (CA)
- A. Giua (IT)
- L. Gomes (PT)
- G.K. Knopf (CA)
- P. Li u (CA)
- P.E. Miyagi **[Chair]**
- P. Orban (CA)
- L. Seabra Lopes (PT)
- J. Shi (US)
- X. Tu (CA)
- R. Valette (FR)
- E. Villani (BR)
- J. Zaytoon (FR)

PREFACE

The manufacturing sector has been facing major challenges as it undergoes revolutionary changes fuelled by new and sophisticated demands from customers, global competition, distribution of manufacturing and marketing activities, and technological advances. In order to address these challenges, manufacturing enterprises need to change the way they do business and adopt innovative technologies and solutions to increase their responsiveness and production efficiency. Information technology plays an essential role in this process.

Current manufacturing systems are collections of complex systems or subsystems operating in distributed collaborative environments involving software, hardware, humans, and organizations. It is crucial to keep a balance between the technical aspects of automation and the human and social facets when applying information technology in industrial applications, particularly with the rapid advancements in information and communication technologies and the wide deployment of automated manufacturing systems. However, in order to create appropriate frameworks for exploring the best synergies between humans and automated systems, there are still numerous issues in terms of processes characterization, modeling, and development of adequate support tools. BASYS conferences have been developed and organized to promote the development of balanced automation systems in an attempt to address these issues.

The first BASYS conference was successfully launched in Victoria, Brazil (1995), and then the following conferences were held in Lisbon, Portugal (1996), Prague, Czech Republic (1998), Berlin, Germany (2000), Cancun, Mexico (2002), and Vienna, Austria (2004).

BASYS'06 is the 7th edition in its series. It keeps the same objective and style of previous BASYS conferences. While keeping the three tracks the same as BASYS'04, BASYS'06 changes the fourth track (Track D) from "Machine Learning and Data Mining in Industry" to "Monitoring and Control", which currently has 17 accepted papers among the total of 49 regular papers. From another point of view, BASYS'06 has a smooth shift of the focus on the integration and balanced automation from the high levels (enterprise and virtual enterprise levels) to the low levels (shop floor and machine levels).

This book contains three invited keynote papers and forty-nine regular papers accepted for presentation at the conference.

Three keynote papers are presented by internationally recognized experts in the related fields:

- Prof. Soundar Kumara provides an overview of Agent-Based Manufacturing. An extended abstract is included in this book with a reference to a full survey paper co-authored by Monostori, Váncza, Kumara for CIRP - College

International pour la Recherche en Productique (The International Academy for Production Engineering).

- Prof. Mo A. Elbestawi presents a comprehensive review of tool monitoring systems, techniques, their components, and particularly the Multiple Principle Component fuzzy neural network for tool condition monitoring.
- Prof. Sophie D'Amours proposes an architecture integrating agent technology and operational research with the objective to enable the development of advanced planning systems for the forest products industry.

Forty-nine regular papers are organized in four main tracks:

- *Track A: Multi-Agent and Holonic Systems in Manufacturing*
- *Track B: Networked Enterprises*
- *Track C: Integrated Design and Assembly*
- *Track D: Monitoring and Control*

All together, the papers will make significant contributions to the literature of Intelligent Technology for Balanced Manufacturing Systems. However, it seems that significant efforts are still required to develop practical IT solutions for balanced manufacturing systems where humans and intelligent machines are in perfect harmony.

Special thanks to all the authors for their contributions to the book and to the BASYS'06 Program Committee members (particularly Program Committee Co-Chairs and Track Chairs) for their efforts in promoting the conference and reviewing / selecting submitted papers.

The editor,
Weiming Shen
National Research Council Canada/
University of Western Ontario
weiming.shen@nrc.gc.ca
wshen@ieee.org

KEYNOTE PAPERS

1 AGENT BASED MANUFACTURING[1]

L. Monostori, J. Váncza
Computer and Automation Research Institute,
Hungarian Academy of Sciences, Budapest, Hungary;
Department of Production Informatics, Management and Control,
Budapest University of Technology and Economics, Budapest, Hungary
laszlo.monostori@sztaki.hu, vancza@sztaki.hu

Soundar Kumara
Department of Industrial and Manufacturing Engineering,
The Pennsylvania State University, University Park, PA 16802, USA
skumara@psu.edu

Agent based manufacturing is an important area from a research and implementation points of view. In the past decade agent based manufacturing has become popular and many applications are built. In this keynote paper we survey the field with special emphasis on The International Institution of Production Research's activities in this area. The original extended paper is referenced at the end of this paper.

EXTENDED ABSTRACT

The emerging paradigm of agent-based computation has revolutionized the building of intelligent and decentralized systems. The new technologies met well the requirements in all domains of manufacturing where problems of uncertainty and temporal dynamics, information sharing and distributed operation, or coordination and cooperation of autonomous entities had to be tackled. In the paper software agents and multi-agent systems are introduced and through a comprehensive survey, their potential manufacturing applications are outlined. Special emphasis is laid on methodological issues and deployed industrial systems. After discussing open issues and strategic research directions, we conclude that the evolution of agent technologies and manufacturing will probably proceed hand in hand. The former can receive real challenges from the latter, which, in turn, will have more and more benefits in applying agent technologies, presumably together with well-established or emerging approaches of other disciplines.

The past decade has witnessed an explosive growth in computer, communication, and information technologies. High-performance computing, the world-wide web, universal access and connectivity, virtual reality, and enterprise integration are but a sampling of this revolution's many facets. At the same time, organizations and

---

Please use the following format when citing this chapter:

Monostori, L., Váncza, J., Kumara, S., 2006, in IFIP International Federation for Information Processing, Volume 220, Information Technology for Balanced Manufacturing Systems, ed. Shen, W., (Boston: Springer), pp. 3–4.

markets have also changed dramatically, represented by developments such as virtual organizations, business process integration, customer-centric supply chains, and electronic commerce. And, although industry strategists and academics continue to debate the precise future trajectory of changes in technologies and organizations, they agree that **information** - its availability, and the ability to exchange it seamlessly and process it quickly - lies at the core of organizations' abilities of meeting escalating customer expectations in global markets.

Agent-based computation is a new paradigm of information and communication technology that largely shapes and, at the same time, provides supporting technology to the above trends. Agent theories and applications have appeared in many scientific and engineering disciplines. Agents address autonomy and complexity: they are adaptive to changes and disruptions, exhibit intelligence and are distributed in nature. In this setting computation is a kind of social activity. Agents can help in self-recovery, and react to real-time perturbations. Agents are vital in the globalization context, as globalization refers to an inherently autonomous and distributed world both from geographical and information processing perspectives.

Agents – and similar concepts – were welcome in **manufacturing** because they helped to realize important properties as autonomy, responsiveness, redundancy, distributedness, and openness. Agents could be designed to work with uncertain and/or incomplete information and knowledge. Hence, many tasks related to manufacturing - from engineering design to supply chain management - could be conducted by agents, small and large, simple and sophisticated, fine- and coarse-grained that were enabled and empowered to communicate and cooperate with each other.

---

[1] The original version of this paper can be found in Annals of CIRP, Vol. 2, 2006.

2 TOOL CONDITION MONITORING IN MACHINING – NEURAL NETWORKS

Mo A. Elbestawi
McMaster University
elbestaw@mcmaster.ca

Mihaela Dumitrescu
McMaster University
dumitrm@mcmaster.ca

Condition monitoring and diagnosis systems capable of identifying machining system defects and their location are essential for unmanned machining Unattended (or minimally manned) machining would result in increased capital equipment utilization, thus substantially reducing the manufacturing costs. A review of tool monitoring systems and techniques and their components and the Multiple Principle Component fuzzy neural network for tool condition monitoring machining are presented.

1. INTRODUCTION

Increased demands for even higher product quality, reliability, and manufacturing efficiency levels have imposed stringent requirements on automated product measurement and evaluation. Manufactured products of the modern day command ever-higher precision and accuracy, therefore automated process monitoring becomes crucial in successfully maintaining high quality production at low cost.

The automated tool condition monitoring processes imply the identification of cutting tool condition without interrupting the manufacturing process operation, under minimum human supervision. Unattended or minimally manned machining leads to increased capital equipment utilization, thus substantially reducing the manufacturing costs. Both these situations require intelligent sensor systems.

An "Intelligent Sensor System" was defined by (Dornfeld, 1986)[1] as an integrated system consisting of sensing elements, signal conditioning devices, signal processing algorithms, and signal interpretation and decision making procedures. In the absence of a human operator, the system should sense signals indicating the process status and its changes, interpret incoming sensed information, and decide on the appropriate control action.

A system could be defined as *Automated/Intelligent Monitoring System* if sensing, analyzing, knowledge learning, and error correction abilities, essential to machining tool condition monitoring, are incorporated.

Please use the following format when citing this chapter:

Elbestawi, M. A., Dumitrescu, M., 2006, in IFIP International Federation for Information Processing, Volume 220, Information Technology for Balanced Manufacturing Systems, ed. Shen, W., (Boston: Springer), pp. 5–16.

An automated/intelligent machining process and tool condition monitoring system should be able to emulate as closely as possible the abilities of human operators. Thus, the following four essential components have to be included in any automated tool condition monitoring system to emulate the human monitoring action:

Sensing Technique Systems: Typically, indirect sensing techniques such as cutting forces, vibrations, and acoustic emission are used. Different types of sensors and sensory data from different locations are combined to yield maximum useful information.

Feature Extraction Systems: Ideally, sensory signals contain the necessary information required to discriminate between different process and tool conditions. However, these signals are usually noisy and require further processing to yield useful features, highly sensitive to the tool conditions but insensitive to external noises.

Decision Making Systems: Decision making strategies process incoming signal features and perform a pattern association task, mapping the signal feature to a proper class (tool condition). This processing task can be done sequentially or in parallel depending on the monitoring system architecture.

Knowledge Learning Systems: In order to make a correct decision, learning algorithms have to be provided. Such algorithms tune system parameters by observing the sample features corresponding to different tool conditions. Like human operators, automated monitoring systems should have the ability to learn from their experiences (past work) as well as from the new information generated from the machining process.

Key concerns with both signal processing and decision-making algorithms, jointly known as monitoring methods, include reliable and fast identification or response to an abnormal event occurring at normal process conditions (Du, 1995, [2].)

2. RESEARCH ISSUES

The major research goals for tool condition monitoring are to develop self-adjusting and integrated monitoring systems able to function under various working conditions with minimum operator supervision.

The purpose of automated tool condition monitoring in machining is to relate the process signals to the tool conditions, and detect or predict tool failure. Automated tool condition monitoring implies identifying the characteristic changes of the machining process based on the evaluation of process signatures without interrupting normal operations. Basically, a monitoring process has two parts:

- *Sensing-* obtaining cutting process signals from sensors. Appropriate signals used for tool condition monitoring are force, torque, vibration, temperature, acoustic emission, electric current, etc...)
- *Monitoring-* composed of *signal processing* and *decision making,* can be divided into model-based and feature-based methods. Both methods use sensor signals from the cutting process for the system input.

Any automated machining process and tool condition monitoring systems should include:

- *"Multi-Sensor Systems"* - more than one sensor should be used for monitoring machining processes and tool conditions, yielding an extended survey of sensitive features.
- *"Automated Feature Extracting Systems"* - automatically generate monitoring features through learning. The signals sensed from multiple sensors are analyzed, compacted, and selected by the system to yield the most sensitive features to the monitoring subjects. The extracted features are also further refined or reselected by the monitoring system.
- *"Learning and Decision Making Systems"* - build up flexible and comprehensive monitoring strategies and automatically generate control parameters. The concentrated information from the learning procedure is stored in the system for classification purposes and can be modified by knowledge updating procedures. With increasing experience, the system will become more and more reliable and promote the monitoring/control functions. These strategies should be robust and valid for a reasonable range of cutting conditions.

Significant research work performed in this research field focused on analytical forecast, dynamic structure identification, monitoring techniques, and adaptive control approaches; Thonshoff et.al.,1998,[3]; Tlusty and Andrews, 1983,[4]; Isserman, 1984,[5]; and Dornfeld, 1990,[6] published research papers on the development of modern monitoring techniques for machining.

Critical reviews on sensors for machining monitoring were published by Tlusty and Andrews, 1983 and Dornfeld, 1992,[7]. Applications include geometric corrections, machine diagnosis, surface finish controls, tool condition monitoring, and machining process monitoring.

Thonshoff et.al., 1998, identified five monitoring tasks: machine, tool, process, tool condition, and workpiece; the author described the monitored conditions in machining processes and classifies the monitored functions into two groups: time critical and non-time critical. The former requires a system response within a range of milliseconds while the later may take seconds or even minutes.

2.1 Sensing Techniques

Metal cutting is a dynamic process; the sensor signals can be considered as the output of the dynamic system in a form of time series. Consequently, process and tool condition monitoring can be conducted based on system modeling and model evaluation. One of the most used models is linear time-invariant system, such as state space models, input-output transfer function models, Auto-Regressive (AR) models, Auto-Regressive and Moving Average (ARMA) models, and Dynamic Data Systems (DDS) methodology. When a suitable model is identified, monitoring can be performed by detecting the changes of the model parameters and/or the changes of expected system responses.

Current sensing techniques can be divided into two basic types:

- *Direct Techniques* – the most accurate measures for determining tooling failure; however the trade-off is production stoppage. With these methods, a direct analysis of the tool or workpiece surface is performed at the end of the machining cycle. Basic analysis methods include optical measurements, surface finish using contact probes (profilometers), chip size measurements, etc. The main disadvantages of this methods is that any significant deterioration occurring in between measurements goes unnoticed, allowing for potential damage to the "machine tool- tooling-workpiece" (MTW) system.

- *Indirect Techniques* - use correlated variables with process signals to monitor for a specific signature of tool failure. These techniques can be applied continuously while machining, and therefore can be used in an online monitoring algorithm. The most common methods able to correlate a variable with tool condition are:
  - Spindle Current Monitoring
  - Acoustic Emissions
  - Vibration Signature Analysis

Sensor Fusion

In most cases, signals coming from only one sensor are typically insufficient to give enough information for machining and tool condition monitoring.

Using several sensors at different locations simultaneously was proposed for data acquisition by Ruokangas et.al., 1986[8]; McClelland, 1988[9]; Chryssolouris et.al., 1998[10], 1989[11] and 1992[12]; and Dornfeld, 1990. Signals from different sources are integrated to provide the maximum information needed for monitoring and control tasks. A schematic diagram of using multiple sensors in monitoring systems is shown in Figure **1**.

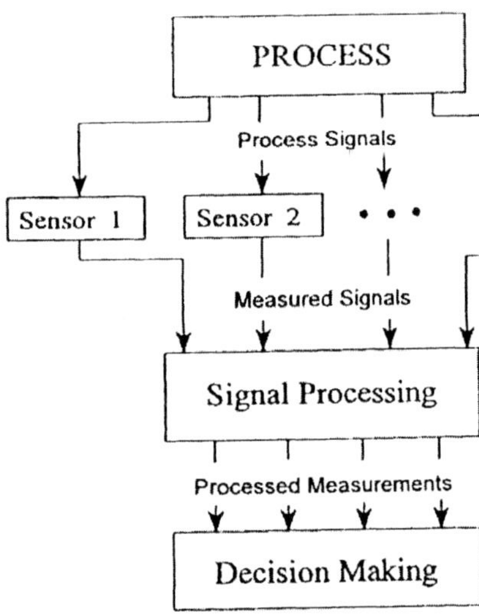

Figure 1 - Multiple Sensors in Monitoring

Sensor fusion generally covers all the issues of linking sensors of different types together in one underlying system architecture (McClelland, 1988, [9].) The strategy of integrating the information from a variety of sensors will increase the accuracy and resolve ambiguities in the knowledge about the environment. The most significant advantageous aspect of sensor fusion is its enriched information for feature extraction and decision making strategy, and its ability to accommodate changes in the operating characteristics of the individual sensors due to calibration, drift, failure, etc... The type and number of sensors used in tool conditioning monitoring are chosen according to the type of monitoring tasks. The most common types of sensors used in monitoring systems were identified by Tlusty, 1983.

The use of multiple sensors for machining process and tool condition monitoring gives extended information about the process. As most process variables have an influence on one another, more than one model is typically needed to analyze the sensor signals.

Some models considered are: dynamic structure models for cutting force, such as empirical cutting force model (Endres et.al, 1990,[13]), dynamic models for tool wear such as diffusion wear models, adhesive wear models, empirical models (Koren et.al., 1991,[14]), linear steady-state models for tool wear and cutting forces (Koren, 1978,[15]; Koren et.al, 1987,[16] and Matsumoto et.al, 1988,[17]), and parametric models including Auto-Regressive methods (AR) for chatter (Yang et.al, 1982,[18] and Tsai et.al, 1983,[19]), AR for tool wear (Liang et.al, 1989,[20])], and AR for tool breakage (Takata and Sata, 1986,[21].)

Emphasis on high reliability and fast response is placed on tool condition monitoring systems to ensure the manufacturing of high quality parts in an efficient manner. Early detection of tool deterioration improves process productivity and reliability. Therefore, monitoring systems must be developed with the above criteria in mind. Computing time and adaptive learning are also both important factors to consider in developing monitoring systems (Du et.al, 1995,[22].)

2.2 Feature Extraction Methods

According to Du,1995, monitoring methods can be sub-divided into two basic types: model-based and feature-based methods:

Model-based methods involve finding a model that fits the process and monitoring specific parameters in that model to detect changes. Changes in the expected system response can be interpreted as changes in the process, signifying an abnormal cycle. Model based monitoring methods are also referred to as failure detection methods by Isserman et.al, 1993,[23] and Lee et.al;, 1998,[24].

Feature-based monitoring methods use mapping processes to identify the process and tool conditions and relate the tool conditions to the sensor features. Such techniques include pattern recognitions, expert systems, neural networks, and fuzzy classifications. The feature-based methods consist of two phases: *learning* and *classification.*

Learning, also called training, is the procedure of establishing the system structure and classification rules. The knowledge for decision making is obtained from the learning samples as well as from instructions. Knowledge *updating* or *continuous learning* implies that the system is retrained with new available information.

Monitoring tasks are done in *classification phase.* The structure and knowledge base built in the learning phase are used for the decision making in monitoring.

2.3 Decision-Making Methods

Decision making strategies are one of the major issues in the development of automated machining process and tool condition monitoring. They range from simple threshold limit values being exceeded and triggering abnormal conditions, to systems requiring substantial training and learning to determine process characteristics of each cutting condition and using these to detect out of specification conditions. A decision making process in monitoring is based on the relationship between the process/tool conditions and the feature-bearing signals (monitoring indices).

Currently used feature-based monitoring methods include: pattern recognition, fuzzy systems; decision trees, expert systems and neural networks – which will be discussed further.

Among the large number of decision making methods that have been developed, statistical pattern recognition, neural networks and fuzzy classification are very interesting aspects in the development of automated/intelligent tool condition monitoring in machining. They have been applied successfully in many cases of monitoring tasks in turning, milling, drilling, and other metal cutting processes.

The pattern recognition technique has been applied to recognize the cutting states and to monitor the tool conditions in machining for decades. The most simple and popular algorithms use linear classifiers.

A linear model was used in applications of linear classifiers by Zhang et.al, 1982,[25] ; Marks and Elbestawi, 1988, [26]; and Liu et.al, 1988,[27].

The features for classifying the cutting states included cutting speed, feed and the power spectrum in different frequency bands.

Features usually used for tool condition monitoring are: feed rate, depth of cut, cutting force, cutting torque, and the sum of the magnitudes of spectral components at certain frequencies. Experiments showed that the number of features and the different combinations of features had great effects on the correct classification rates. The success rate of classification with these cases is 77% or higher.

Other pattern recognition algorithms for tool condition monitoring in machining included the class-mean scatter criterion, the class variance criterion, and Fisher's weighted criterion (Emmel et.al. 1987,[28], and 1988,[29].) The class-mean scatter criterion maximizes class separation and minimizes within-class variance. The class variance criterion maximizes the difference between the within-class variance of each class. Fisher's weight criterion maximizes class separation and minimizes the within-class variance between each pair of classes. This methodology was applied in order to detect tool wear and breakage in turning operations using acoustic emission spectral information under fixed cutting conditions. The tool wear sensing results had performances ranging from 84 to 94%.

3. NEURAL NETWORKS FOR TOOL CONDITION MONITORING

Neural networks are computing systems made up of a number of simple, highly interconnected processing elements that provide the system with the capability of self-learning. Using neural networks, simple classification algorithms can be used and the system parameters are easily modified. One major characteristic of building neural networks is the training time. Training times are typically longer when complex decision regions are required and when networks have more hidden layers. As with other classifiers, the training time is reduced and the performance improved if the size of a network is optimally tailored. The tasks of an automated tool condition monitoring system involve the ability to recognize the tool condition by analyzing measured cutting process parameters such as forces and vibrations. This ability is based on the accumulation of useful information from related laws of physics and operators' experiences. In building automated/intelligent tool condition monitoring systems, some basic functions have to be considered:

1. Fusion of multiple sensors;
2. Learning or training strategies for the monitoring system;
3. Knowledge updating techniques; and
4. Description of the imprecision in tool conditions for various cutting conditions.

With the increasing needs for effective and robust automated machining process and tool condition monitoring, a significant amount of research work has been performed to find decision making strategies. The principal constituents of soft computation include fuzzy logic for imprecision in the acquired data, neural networks for learning, and probability reasoning for uncertainty. These three components are usually overlapped. The "soft computation" is easily implemented by fuzzy neural networks.

3.1. Multiple Principal Component (MPC) Fuzzy Neural Networks Structure

Learning refers to the processes which build the monitoring system in a given structure with information from the learning data. In addition, some logic rules are also created, which determine the data processing and govern the relationship between the processing elements. During the learning phase, a limited amount of data is used to adjust the parameters of the monitoring system. The trained monitoring system uses the stored knowledge to classify the data regarding the successive tool conditions.

If the sampled data for training the system are labeled with the class to which a sample belongs, the decision making is performed with a priori knowledge. This is called pattern classification, or supervised classification, common in automated tool condition monitoring in machining. When the training samples are collected, the tool conditions related to each training sample are provided to give the necessary information.

Knowledge updating, or self-learning, essential for an automated tool condition monitoring system, refers to processes in which the structure and the parameters of a monitoring system are modified according to the new information about the classification. Classification results should be checked on-line to ensure the system gives correct results. If the results are not correct, the system should be retrained or modified.

3.2 Construction of the MPC Fuzzy Neural Networks

The Multiple Principal Component (MPC) Fuzzy Neural Networks are constructed based on the idea of "soft computation." Neural networks, fuzzy logic and statistical reasoning are employed. Simple classification procedures can be implemented at individual processing elements (neurons). The interconnections between neurons in the network communicate the information and make it possible to solve complex classification problems. Statistical reasoning is used in the learning procedure for the feature extraction and the partition strategies.

For conventional neural networks, each of the processing elements (for input, output, and hidden layers) is always connected to every single processing element in the neighboring layers. Decision tree classifiers are hyperplane classifiers that can be regarded as a type of partially connected neural networks since each node in the tree is connected to only its "father" and "sons", requiring comparatively less classification computations and can be implemented using parallelism from decision region by performing simple, easily understood operations on the neural network. In more sophisticated implementations, multi-layered neural networks, consisting of nonlinear connections between the inputs and the outputs are employed.

As an alternative to conventional neural networks, a partially connected, fuzzy neural network approach can be used for automated tool condition monitoring in machining. Different from matrix-type decision making, a tree structure is used for reducing unnecessary connections between elements in the input and the output layer. The fuzzy classifications are used in the neural networks to provide a comprehensive solution for certain complex problems. The neural network that utilizes fuzzy classification is shown in Figure 2.

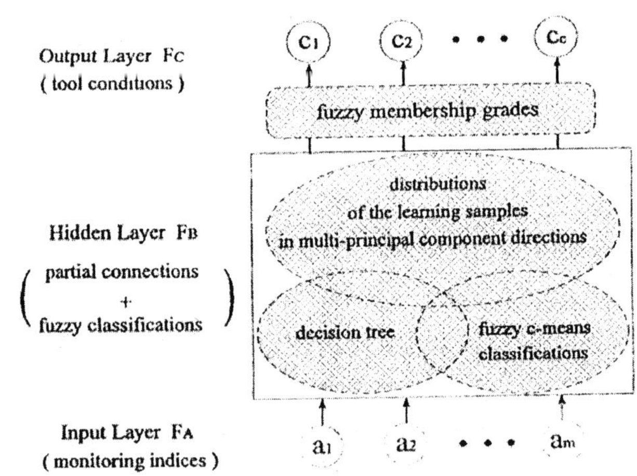

Figure 2 - The Multiple Principal Component (MPC) fuzzy Neural Network

The input layer, FA = (a_1, a_2, ..., a_m), has m processing elements, one for each of the m dimensions of the input pattern x_m. The hidden layer of the network, FB, consists of the neurons that use the fuzzy classification to separately address the subsets of the original data set while invoking necessary information from other neurons. The probability distribution and the membership function are used for interconnections within the hidden layer and the connections to the output layer. The neurons of the output layer, FC, represent the degrees to which the input pattern xk fits within the each class. There are two possible ways that the outputs of FC class nodes can be utilized. For a soft decision, outputs are defined with the fuzzy grades. For a hard decision, the FC nodes with the highest membership degree are located (a "winner-take-all" response). The connections within the hidden layer are not from one element to every one in the neighboring layers. The structure depends on the training data and is created through the training process. These partial connections result in the simpler and faster training and classification.

3.3 Evaluation of MPC Fuzzy Neural Networks

A pattern classifier should possess several properties: on-line adaptation, nonlinear separability, dealing with overlapping classes, short training time, soft and hard decisions, verification and validation, tuning parameters, and nonparametric classification (Simpson, 1992,[30].) The MPC fuzzy neural networks for automated tool condition monitoring has been developed in consideration of these requirements. The training and classification algorithms are based on the theories of neural networks, fuzzy logic, and probability reasoning.

With the application of fuzzy classification, the neural networks are effective in dealing with nonlinear separable and/or class-overlapping classification problems, which are common in the case of tool condition monitoring in machining, especially for the monitoring with varying cutting conditions. The partial interconnections within the fuzzy neural networks make the training time very short compared to that of fully connected networks such as the back-propagation neural networks. The calculations necessary for the classification are also significantly reduced since not all the neurons in the hidden layer are used while a sample is being processed. Soft and hard decisions are optional for different applications. The maximum partition algorithm is based on the distributions of the learning samples and no parameter estimations are needed.

This method functions similarly, in the partition of training samples, to the linear fuzzy equation algorithm proposed by Du and Elbestawi, 1992,[31]. The linear fuzzy equation method uses a matrix to describe the relationship between the monitoring indices and the tool conditions. The proposed MPC fuzzy neural networks use a tree structure similar to that in the fuzzy decision tree described by Li et.al.,1992, [32]. Because the decision tree is more flexible than a matrix approach, it has better performance in the case of tool condition monitoring in machining. In constructing the fuzzy decision tree, the maximum partition generates nodes holding the samples from only one tool condition. The other samples are put into other nodes. This means each partition leads to a final decision at a leaf node of the tree. The maximum partition in the MPC fuzzy neural networks chooses a better partition so that a new-born neuron can hold samples from different tool conditions. A neuron can lead to other neurons in the hidden layer as well as neurons in the output layer.

The consequence of this structure is simplicity in the interconnection and the short routines in the classification. Experimental tests by using the same set of data showed that the MPC fuzzy neural networks gave a better success rate than the fuzzy decision tree algorithm (Li and Elbestawi, 1994,[33].)

In the consideration of on-line adaptation (on-line learning) and the tuning parameters, a classifier should have as few parameters to tune in the system as possible. Both the back-propagation neural networks and the proposed MPC fuzzy neural networks have very few tuning parameters. The structure of the MPC fuzzy neural network is, however, easily modified with new learning samples. Unlike the back-propagation neural networks that require a complete retraining of the system with both the old and the new learning data, the MPC fuzzy neural networks need only to change partially their neurons and the connections when the new learning information is added. Both supervised and unsupervised classification algorithms are easily implemented with the available learning samples.

To insure a good distribution of the learning data, the training samples have to be representative of the whole span of the feature space. In tool condition monitoring, all applicable tool and cutting conditions have to be considered during the training phase. On the other hand, if a poor distribution is encountered, then a modified feature extraction procedure has to be implemented. Information about new phenomena can be added to the monitoring system by knowledge updating.

3.4 Fuzzy Classification and Uncertainties in Tool Condition Monitoring

During machining, cutting conditions *(e.g.,* cutting speed, feed, depth of cut) as well as tool conditions *(e.g.,* tool wear) significantly affect the process parameters such as cutting forces and vibrations, which are usually used as the input signals to a monitoring system. Deterministic models which attempt to describe the relationship between the tool conditions and the various measured parameters are typically valid for a limited range of cutting conditions. The fuzzy classification can be used to describe the uncertainties and the overlapped relationship of the tool conditions and the monitoring indices. Briefly, the fuzzy expression of a tool condition, A, can be defined by:

$$A = \{x \mid \mu_A(x)\} \tag{1}$$

where x is the value of A, and $\mu_A(x)$ is a fuzzy measure, also known as the membership function. $\mu_A(x)$ is a monotonous function, and $0 \le \mu_A(x) \le 1$. The function increases with respect to the decrease of the uncertainty of A. If B is also a fuzzy set and is more uncertain than A, then:

$$\mu_A(x) > \mu_B(x) \tag{2}$$

This might be interpreted as "the membership grade of small tool wear is greater than that of large tool wear." The fuzzy representation of the tool conditions in machining has its advantages. The concept of fuzzy decision making in machining tool condition monitoring is illustrated in Figure 3, where, A_H and B_H are categories classified by the hard decision, while A_F and B_F are classified by the fuzzy decision.

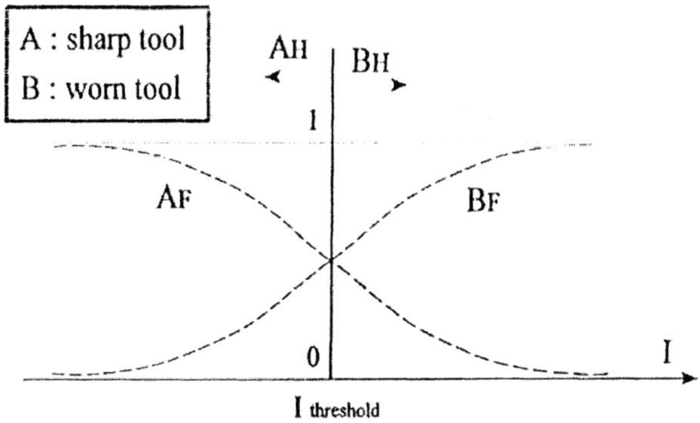

Figure 3 - Soft Boundaries in Fuzzy Classification

4. CONCLUSIONS

Three major components of the "soft computation" are involved in the construction of theMPC Fuzzy Neural Networks. The combination of fuzzy logic with neural networks has a sound technical basis because these two techniques approach the design of intelligent machines from different angles. Fuzzy neural networks employ the advantages of both neural networks and fuzzy logic. Neural networks offer good performance in dealing with sensor information in parallel at a low computational level. The high interconnection within the networks gives the capabilities of exchanging the information sufficiently and managing nonlinearity. Fuzzy logic gives a means for representing, manipulating, and utilizing the data and the information that possess non-statistical uncertainties.

5. REFERENCES

1. Dornfeld, D. A., 1986, "Acoustic Emission Monitoring for Untended Manufacturing," Proceedings. of Japan/USA Symposium. on Flexible Automation, JAACE.
2. Du, R., Elbestawi, M. A. and Wu, S. M., 1995, "Automated Monitoring of Manufacturing Processes, Part 1: Monitoring Methods," ASME J. of Eng. for Industry, Vol.117, pp. 121-132.
3. Tonshoff, H. K. and Wulfsberg, J. P., 1988, "Developments and Trends in Monitoring and Control of Machining Processes," Annals of the CIRP, Vol.37, No.2, pp. 611-622.
4. Tlusty, J. and Andrews, G. C., 1983, "A Critical Review of Sensors for Unmanned Machining," Annals of the CIRP, V 01.32, No.2, pp. 563-572.
5. Iserrmann, R., 1984, "Process Fault Detection Based on Modeling and Estimation Methods -A Survey," Automatica, Vol.20, No.4, pp. 387-404.
6. Dornfeld, D. A., 1990, "Neural Network Sensor Fusion for Tool Condition Monitoring," Annals of the CIRP, Vol.39, No.1, pp. 101-105.
7. Dornfeld, D. A., 1992, "Monitoring of Machining Process -Literature Review," Presented at CIRP STC"C" Meeting, Paris, France.
8. Ruokangas, C. C., Blank, M. S., Matin, J. F. and Schoenwald, J. S., 1986, "Integration of Multiple Sensors to Provide Flexible Control Strategies," Proceedings of the 1986 IEEE International Conference on Robotics and Automation, pp. 947-953.
9. McClelland, S., 1988, "Tacking the Problem of Sensor Fusion," Sensor Review, Vol.8/No.2, pp. 89-91.

10. Chryssolouris, G. and Domroese, M., 1988, "Sensor Integration for Tool Wear Estimation in Machining," Sensors and Control for Manufacturing, ASME PED- Vol. 33, pp. 115-123.
11. Chryssolouris, G. and Dornroese, M., 1989, "An Experimental Study of Strategies for Integrating Sensor Information in Machining," Annals of the CIRP, Vol. 38, No. I, pp. 425- 428.
12. Chryssolouris, G., Domroese, M. and Beaulieu, P., 1992, "Sensor Synthesis for Control of Manufacturing Process," ASME J. of Engineering for Industry, Vol.114, pp. 158-174.
13. Endres, W. J., Sutherland, J. W., DeVor, R. E. and Kapoor, S. G., 1990, "A Dynamic Model of the Cutting Force System in the Turning Process," Monitoring and Control for Manufacturing Process, ASME PED. Vol.44, pp. 193-212.
14. Koren, Y., Ko, T. R., Ulsoy, A. G. and Danai, K., 1991, "Flank Wear EstimationUnder Varying Cutting Conditions," ASME J.of Engineering for Industry, Vol.113, pp. 300-307.
15. Koren, Y., 1978, "Flank Wear Model of Cutting Tools Using Control Theory," ASME Journal of Engineering for Industry, paper No. 77-WA/PROD-20.
16. Koren, Y., Danai, K., Ulsoy, A. G. and Ko, T. R., 1987, "Monitoring Tool Wear Through Force Measurement," Proceedings of the 15th NAMRI Conference, pp. 463-468.
17. Matsumoto, Y., Tjinag, N. and Foote, B., 1988, "Tool Wear Monitoring Using A Linear Extrapolation," Sensors and Control for Manufacturing, ASME-PED Vol. 33, pp.83-88.
18. Yang, W. Q., Hsieh, S. H. and Wu, S. M., 1982, "Adaptive Modelling and Characterization for Machining Chatter," Measurement and Control for Batch Manufacturing, ASME, pp. 135-144.
19. Tsai, S. Y., Eman, K. F. and Wu, S. M., 1983, "Chatter Suppression in Turning," Proceedings of the 11th NAMRC, pp. 399-402.
20. Liang, S. Y. and Dornfeld, D., 1989, "Tool Wear Detection Using Time Series Analysis of Acoustic Emission," ASME J Journal of Engineering for Industry, Vol.111,.pp. 199-205.
21. Takata S. and Sata, T., 1986, "Model Referenced Monitoring and Diagnosis Application to the Manufacturing System," Computers in Industry, Vol. 7, pp. 31-43
22. Du, R., M.A. Elbestawi, and S. Wu (1995, May). Automated Monitoring of Manufacturing Processes, Part 2: Applications. Journal of Engineering for Industry, Vol. 117, pp.133-141
23. Isermann, R., M. Ayoubi, H. Konrad, and T. Reib (1993). Model Based Detection of Tool Wear and Breakage for Machine Tools. Proceedings of the IEEE International Conference System Manufacturing Cybernetics Vol. 3, pp72-77
24. Lee, B.Y., H.S. Liu, and Y.S. Tarng (1998). Modeling and Optimization of Drilling Process. Journal of Material Processing Technology Vol.74, pp.149-157
25. Zhang, Y. Z., Liu, Z. F, Pan, L. X., Liu, Y. J. and Yang, W. B., 1982, "Recognition of the Cutting States for the Difficult-to-Cut Materials Application of the Pattern Recognition Technique," Annals of the CIRP, Vol.31, No.1, pp. 97-101
26. Marks, J. and Elbestawi, M. A., 1988, "Development of Machine Tool Condition Monitoring System Using Pattern Recognition," Sensors and Controls for Manufacturing, ASME-PED- Vol.33, pp. 89-98.
27. Liu, T. I. and Wu, S. M., 1988, "On-Line Drill Wear Monitoring," Sensors and Control for Manufacturing, ASME-PED-Vol. 33, pp. 99-104.
28. Emmell, E. and Kannatey-Asibu, E. Jr., 1987, "Tool Failure Monitoring in Turning by Pattern Recognition Analysis of AE Signals," Sensors for Manufacturing, ASME-PED. Vol.3D, pp. 39-57
29. Emmell, E. and Kannatey-Asibu, E. Jr., 1988, "Tool Failure Monitoring in Turning by Pattern Recognition Analysis of AE Signals," ASME J. of Engineering for Industry, Vol.112, pp. 137-145
30. Simpson, P. K., 1992, "Fuzzy Min-Max Neural Networks -Part 1: Classification," IEEE Transactions on Neural Networks, Vol. 3, No.5, pp 776-786.
31. Du, R. X., Elbestawi, M. A. and Li, S., 1992, "Tool Condition Monitoring in Turning Using Fuzzy Set Theory," International Journal of Machine Tool and Manufacturing, Vol. 32, No.6, pp. 781-796.
32. Li, S., Elbestawi. M. A. and Du, R. X., 1992, "A Fuzzy Logic Approach for Multi- Sensor Process Monitoring in Machining," Sensors and Signal Processing for Manufacturing, ASME PED-Vol. 55, pp. 1-16.
33. Li, S. and Elbestawi, M. A., 1994, "Tool Condition Monitoring in Machining by Fuzzy Neural Networks," Dynamic Systems and Control, ASME-DSC. Vol. 55-2, pp. 1019- 1034.

3

AGENT-BASED SUPPLY CHAIN PLANNING IN THE FOREST PRODUCTS INDUSTRY

Sophie D'Amours[1,2], Jean-Marc Frayret[1], Alain Rousseau[1],
Steve Harvey[1], Pascale Plamondon[1], Pascal Forget[1]
[1]*Research Consortium in e-Business in the Forest Products Industry (FOR@C)*
[2]*Network Organization Technology Research Center (CENTOR)*
Université Laval, Quebec, Canada
Sophie.Damours@forac.ulaval.ca

Due to new economical challenges and recent trends regarding international trade and globalization, many companies from the Canadian forest products industry are now facing the need to reengineer their organizational processes and business practices with their partners. This paper proposes an architecture which aims to enable the development of advanced planning systems for the forest products industry. This architecture combines agent technology with operational research, in order to take advantage of the ability of agent-based technology to integrate distributed decision problems, and the ability of operation research to solve complex decision problems This paper describes how this architecture has been configured into an advanced planning and scheduling tool for the lumber industry, and how it is being validated.

1. INTRODUCTION

Due to new economical challenges and recent trends regarding international trade and globalization, many companies from the Canadian forest products industry have reached the point where profit improvement cannot be reaped without the involvement of their entire organization (e.g., their distributed facilities and offices spread across North America and the world) as well as their business partners (e.g., raw material suppliers, industrial clients and distributors). If companies from other sectors have faced these challenges a few years ago (e.g., the semi-conductor and automotive industries), companies of the Canadian forest products industry are now facing the need to reengineer their organizational processes and business practices with their partners and adopt supply chain management best practices.

Supply chains are global networks of organizations where material and information flow in many directions within and across organizational boundaries through complex business networks of suppliers, manufacturers, and distributors, to the final customers. Organizations of the supply chain cooperate to improve the flows of materials and information in order to maximize customer satisfaction at the lowest possible cost and highest possible speed.

Please use the following format when citing this chapter:

D'Amours, S., Frayret, J.-M., Rousseau, A., Harvey, S., Plamondon, P., Forget, P., 2006, in IFIP International Federation for Information Processing, Volume 220, Information Technology for Balanced Manufacturing Systems, ed. Shen, W., (Boston: Springer), pp. 17–26.

In that regard, the forest product supply chain is similar to other industries: forest product material flow from forest contractors, to lumber or paper or panel facilities, to value-added mills (referred to as secondary transformation), and through many channels of distributors and wholesalers to finally reach the markets. However, unlike the traditional manufacturing supply chain, which has a convergent product structure (i.e., assembly), the forest product industry needs to master industry specific production processes due to its divergent product structure (i.e., trees are broken down into many products) and the highly heterogeneous nature of its raw material, which make planning and control a difficult task. Because of this inherent complexity, the forest products industry needs to compensate its lack of control over these various stochastic elements by the ability to (1) exchange promptly information throughout its supply chain regarding supply availability and quality, production output and demand, and (2) quickly react in a coordinated manner with supply chain members to correct any deviances or disturbances to the plan. Here lies a need for reactive and specific information and decision support systems to address both the need to produce feasible operation plans and to quickly adapt these plans when perturbing events occur. The experimental planning platform presented in this paper attempts to address these two issues. In particular, an agent-based architecture is proposed in order to develop an experimental environment to design and test various configurations of planning systems for the forest products industry. The remaining of this paper is organized as follows. First, a review of the literature dealing with Advanced Planning and Scheduling systems (i.e., APS) and with agent-based manufacturing and supply chain management systems is presented. Next, the different elements of the proposed generic architecture are introduced. Then, the functional and technological aspects of a specific implementation of the proposed approached are illustrated. Finally, the validation process for correction and improvements is detailed.

2. LITERATURE REVIEW

Operation planning within supply chains is a complex issue. Companies usually face this by implementing and using information and decision support systems, which address various planning tasks. Some companies also adopt just-in-time approaches control the pace of production and replenishment using Kanban. This paper focuses on computer supported planning systems in a distributed planning environment.

2.1 Advanced Planning and Scheduling

Advanced Planning and Scheduling (APS) systems are considered by many as the state of the art of manufacturing and supply chain planning and scheduling practices. The reader is referred to Stadtler and Kilger (2000) and Stadtler (2005) for a thorough description of APS. These systems usually exploit operations research (OR) technologies or constraint programming in order to carry out in an integrated manner true finite capacity planning and scheduling optimization at the long, mid and short term levels of decision making (Fleischmann and Meyr (2003)). For a specific description of some of the many planning problems and optimization applications in the forest product industry, the reader is referred to Rönnqvist

(2003), Epstein, Morales, Seron, and Weintraub (1999) and Frayret, Boston, D'Amours, and Lebel (2005).

2.2 Agent-based manufacturing and supply chain management systems

In order to address the planning and scheduling of manufacturing and supply chain systems, academics have initiated in the middle of the 1980s a new body of approaches and distributed computing techniques drifting away from traditional OR-based solutions. Some of these approaches are referred to as agent-based manufacturing and supply chain management systems. They are designed to tackle the need for reactive, reliable, and (re)configurable operation management systems. These approaches are rooted in the multi-agent technologies which are related to distributed artificial intelligence (Weiss (1999)).

An agent-based manufacturing system may be defined as a planning and control system made of interdependent software agents designed to (1) individually handle a part of a manufacturing planning and control problem, such as planning a single order or allocating tasks to resources, and (2) collectively carry out specific higher functionalities such as planning an entire manufacturing system. Software agents generally exhibit characteristics that allow them to individually behave and interact with each other in such a manner that they collectively fulfill the purpose of the entire system. In their book, Shen, Norrie, and Barthes (2001) identify 12 desirable characteristics of an agent: network-centric, communicative, semi-autonomous, reactive, deliberative, collaborative, pro-active, predictive, adaptive, flexible, persistent, mobile.

Many agent-based approaches for the planning and control of supply chains and manufacturing activities have been proposed in the literature. For a review of these approaches, the reader is referred to Shen and Norrie (1999) and Parunak (1999). The reader is also referred to Caridi and Cavalieri (2004) who provide a recent critical analysis of multi-agent technology applied to manufacturing. Their analysis notably reveals the lack of real world applications and the low maturity level of agent-based manufacturing technology. The reader is referred to Tharumarajah (2001) and Frayret, D'Amours, and Montreuil (2004) who analyze some of the important issues related to the design of such distributed systems, including the distribution of decisions and the coordination of manufacturing activities. These recent developments provide many new concepts and ideas to design solutions to support interdependent supply chain members to make their planning and control decisions so as to be collectively more efficient than if they were not coordinated.

3. GENERIC ARCHITECTURE

The architecture that is introduced in this paper aims at addressing two particular issues relevant to provide the forest products industry with specific normative decision and planning tools. The first issue concerns the ability to plan and coordinate operations throughout the supply chain, while the second issue concerns the ability to analyze through simulation the dynamics and performance of various supply chain scenarios. In order to address these challenges, an experimental platform was developed by the FOR@C Research Consortium, which is a joint industry and university research initiative focused on supply chain planning within the forest product industry. The design of this platform exploits both multi-agent technology and OR techniques. Figure 1 presents the different functions of this platform.

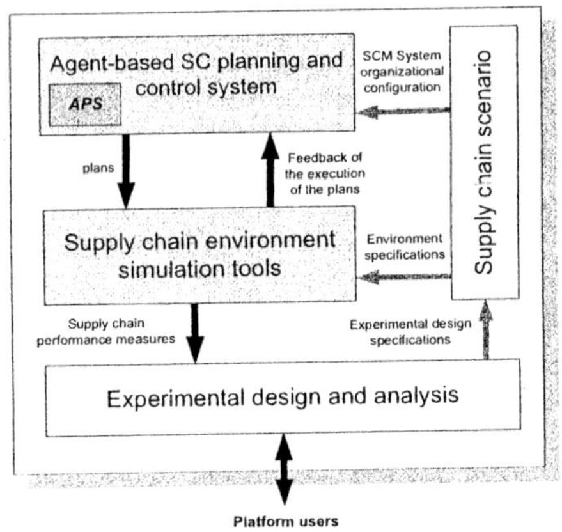

Figure 1 – General overview of the experimental platform

3.1 Architecture description

In the context of this paper, the architecture is defined through its generic principles, mechanisms and software components used to build particular implementations of agent-based supply chain planning systems. The description of this architecture can be made from three different perspectives: the business level, the planning level and the implementation level.

The business level of the architecture encompasses the organizational principles that are used to design a supply chain business configuration from which a detailed configuration of an agent-based supply chain planning system is based on. These principles, inspired by previous work (Montreuil, Frayret, and D' Amours (2000), Frayret (2002)), propose to model supply chains as networks of autonomous and interdependent business units responsible for fulfilling the agreements and commitments they have with their partners. From a company's point of view, a business unit can also be external entities that represent a business partner such as a customer, a supplier or a contractor.

The planning level of the architecture consists in all the generic structures and mechanisms required to implement agent-based supply chain planning systems for the forest products industry. First, there is a direct structural relationship between a business units' supply chain configuration and its corresponding distributed planning configuration. In other words, there is a direct correspondence between business units and their planning level counterpart, referred to as a Planning Unit (PU). Consequently, PUs' relationships are specified through a supply chain business configuration. Because of the flexibility of the business level principles, the scope of a PU can vary greatly. For example, a PU can be set to support the operation planning of a department, a complete facility, or an entire multi-facility division. Second, the internal configuration of a PU is composed of specialized agents capable of maintaining complex communication and collaborative workflows with each other (whether they are in the same PU or not) through structured conversations based on specifically designed conversation protocols, commonly used in multi-agent systems. These configurations are built and customized in order to implement the particular planning support models of their corresponding PU's decision processes. Figure 2 presents a specific implementation of a supply chain configuration, including different PUs.

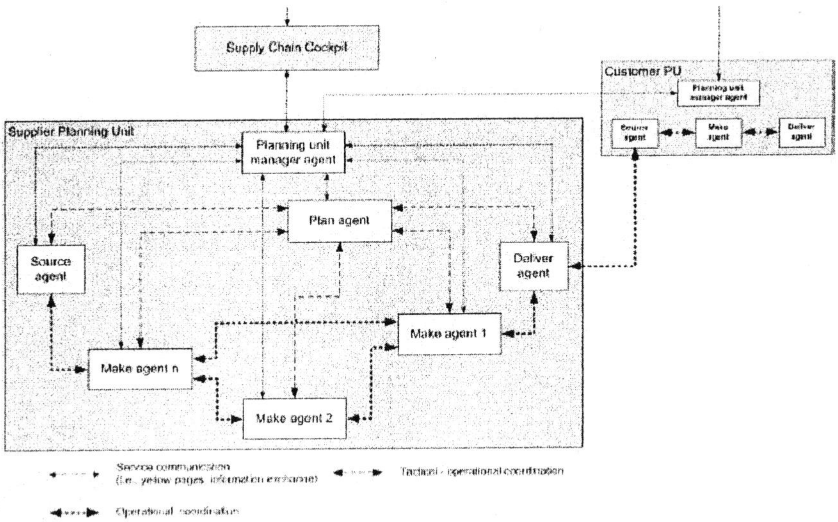

Figure 2 – Components of a specific implementation of a supply chain configuration

The configuration of an agent-based supply chain planning system specifies the PU's internal configurations and its relationships with other PUs. Its design and implementation, which includes the customization of the PU, is made through a software component called Supply Chain Modeler. Supply Chain Modeler enables system designers to customize and deploy PUs through a network of computer.

3.2 Planning unit configuration

Within a PU, planning activities are carried out by various agents responsible for the many planning functions of the planning process model. This functional distribution is inspired by the SCOR model defined by the Supply Chain Council (Stephens (2000)) and the agent-based supply chain management application presented in Fox, Barbuceanu, and Teigen (2000). Each PU has a deliver agent which function is to manage all relationships with the PU's customers. On the other side of the functional spectrum, each PU has also a source agent which role is to manage relationships with all the PU's suppliers. One function of the Source Agent is to forward to the right suppliers the needs of the PU that cannot be fulfilled internally or for which it is more economical to outsource the fulfillment. If the PU is responsible for carrying out production activities (i.e., it is not a warehouse or a distribution center), it is geared up with a set of make agents. According to the planning process model, production planning functions can be assigned to one or many make agents responsible for a part of the overall planning functions. This way, the planning functions can be distributed and assigned to make agents responsible for the planning of production cells which manufacture of a product family. The flexibility of this decomposition approach is necessary for the design of specific decision support tools that are specialized and adapted to the production process, as opposed to a generic one.

All agents are responsible for continuously monitoring their own environment and react to certain changes that may occur. Also, because agents interact with each other, their environment is also made of all messages received from other agents specifying a new or modified requirement plan, a new or modified replenishment plan, a contingency situation, or a high priority requirement to process. Although most agents in the reviewed literature are mainly reactive (i.e., they are designed to carry out specific tasks triggered when specific events occur), the agents envisioned in the proposed approach can also exhibit a pro-active behavior by not only reacting to changes in their environment, but also by initiating actions that could improve their performance. In this context, it appears that the notion of agent goals is fundamental as it guides the agents' decision-making behavior toward the achievement of their goals. The planning process model of an agent explicitly describes such objectives.

3.3 Collaborative environment

Beyond the description of the planning process, one of the main challenges of designing an agent-based supply chain planning system is to create a collaborative environment that allows each agent geared up with advanced planning tools to share and exchange in time partial or factual decisions, or constraints in order to make sure that the final planning decisions are made jointly with respect to the agents' constraints and their collective performance. As mentioned previously, Dudek and Stadtler (2005) propose such an approach and describe it from a planning process modeling perspective. From an implementation perspective, each agent must be configured so as to behave according to the collaborative planning process model, all the while being able to fulfill its local tasks and interact with local users.

In order to address both technological and planning process integration challenges, we rely on the concepts of conversation protocols that are commonly used in multi-agent systems. These protocols are used to specify how agents should collaborate through structured exchange of messages. Agents' collaborative behavior is then defined through internal workflows associated to each potential states of a conversation. At each of these states, each agent participating in the conversation must perform tasks to continue or stop the conversation. In our experimental planning platform, each agent has the ability to receive and send messages. Because planning the operations of a part of a supply chain in collaboration with other agents requires the ability to cope with various situations that cannot all be identified in advanced (i.e., at the configuration stage of the system), it may be necessary to use more advanced agents that can identify themselves which tasks to perform when facing a new situation. From an implementation perspective, conversation protocols are inspired from the FIPA-ACL3 standards for the interoperation of heterogeneous software agents.

4. APPLICATION TO THE LUMBER PRODUCTION PLANNING PROBLEM

The generic architecture presented in the previous section has been configured and specific operations planning agents have been designed in order to carry out the planning and coordination of the lumber production and distribution. The next sections introduce this particular application.

4.1 Introduction to the lumber supply chain

The lumber supply chain that is introduced here is a part of the entire forest product supply chain. In particular, it represents the lumber production and distribution part of it, also referred to as the first transformation. In brief, as it is usually the case in Québec, Canada, forest is harvested by small size entrepreneurs responsible for felling trees and for crosscutting them into appropriate length logs. Once in the mills, logs remain in the log yard until there are broken down into various sizes of rough pieces of lumber. A sawing process represents the use of a particular sawing pattern on a particular log class. The concept of sawing process allows the control the co-production predictability through the definition of more or less detailed classes of logs. Next, bundles of same dimension pieces of lumber (e.g., 2x3, 2x4, etc.) are then placed into kiln dryers according to the loading pattern corresponding to that dimension for various duration in order to decrease their moisture content to an appropriate level (standard or required by the client). Once dried, bundles are disassembled to be sorted and graded according to standard rules or customers specifications, then cut to length and planed. The overall finished products are made of various grades and various sizes of pieces of lumber that are assembled into homogeneous bundles. These bundles are then sent to their final customers, which include wholesalers, retailers, and industrial customers for a second transformation into prefabricated houses or house components.

In this typical example, a single company is responsible for orchestrating all operations from the forest to the customers. Such a company may also orchestrates the operations of many sawmills while presenting a single face to customers through a corporate office responsible for managing customer relationships and sales. Sawmill operations complexity arises due to its divergent production process, the current push-mode production strategy of the industry and the heterogeneous nature of the resource. These specific industrial characteristics make it difficult to anticipate with precision the production output mix and volume.

4.2 Application

In the context of this application, many decisions must be addressed and coordinated. Here, we illustrate only part of the global decision process and its relationships with the platform. The architecture described earlier and the specific application presented hereafter have been implemented with Microsoft .NET© and Ilog OPL Studio© (which includes Cplex© and Solver©) for the optimization part.

This application is dedicated to the planning of lumber production units. In those units, logs are transformed into lumber that is then dried and finished (i.e., planed). Next, finished lumber is either delivered to customers from an internal warehouse, or to an external warehouse. The design of the agent-based planning system for such a specific application is guided by the straightforward identification of homogeneous planning domains that can be found in the management of such supply chains: the planning of sawing, drying, finishing and transportation operations and the management of an external warehouse. Figure 3 illustrates the proposed design. In brief, sawmill planning unit is composed of generic agents: a planning unit manager agent (omitted in the figure), a deliver agent and a source agent. Then, specialized agents are introduced. A sawing agent, a drying agent and a finishing agent are respectively responsible for supporting the planning of sawing, drying and finishing operations. This specific internal structure is neither unique nor optimal. Although it is highly distributed, it is possible to design a more centralized structure with a single make agent responsible for the planning of all production operations. However, due to the complexity and the specificities of those problems, it seems difficult to take advantage of a single generic planning algorithm, as it is likely that only aggregated information could be handled with such an agent.

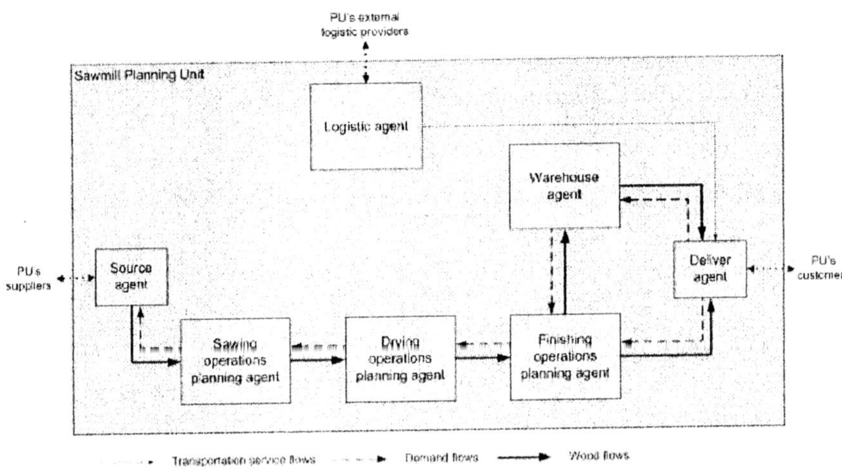

Figure 3 – Sawmill internal configuration

In brief, agents can calculate at any given time, a new requirement plan (i.e., a plan of its own needs to be sent to its supplier). The first requirement planning does not consider supply constraints and is referred to as the infinite sourcing capacity mode. The infinite sourcing capacity mode is meant to identify and communicate to the supplier the requirement plan that would best satisfy its local needs. In other word, it is seen as a target for the supplier that, in turn, trade-off the full satisfaction of its customer versus its own constraint and cost. Penalties for under or over satisfying the demand plan can be used in local planning models to allow such trade-offs. Then, finite sourcing capacity mode can be used either after the receipt of a new replenishment plan from its supplier in order to provide its customer with an accurate plan to fulfill its needs, or using the current replenishment plan in order to provide its customer with a quick answer to a request. In this context, agents' behavior model must be designed in order to explicitly describe the context of using these specific configurations of their local planning models. In other words, these behavior models represent the agents' knowledge of how and when to use specific configurations of their planning tools. Research is currently done in order to address the explicit modeling of this kind of knowledge. It is however outside the scope of this paper.

4.3 Validation

The validation of these developments was carried out with the collaboration of a forest product company. Real data was thus used to test the performance of the agent-based APS. More specifically, we developed a configuration of the distributed APS in order to address the planning of drying and finishing activities inside one plant. This configuration included different types of data, such as production processes, products, orders, on-hand inventory, selling prices, resource costs, forecasted supply, capacity and on-going work. This test was covering 100 products, distributed on two dryers and one finishing line, in a planning horizon of 6 weeks.

The first step of this validation was to model the drying and finishing processes with the partner's production manager. Loading patterns for dryers were known and available, but finishing processes were unknown. Work has thus been done to define in details these processes, which resulted in 20 finishing processes and 89 drying processes. The customer's order file and the on-hand inventory data were extracted from their ERP system. The sales team provided the data on final product prices and resource costs. Each week, the partner's production manager sent us the execution plan, including supply from the sawing line, daily capacity of the finishing line and on-going work. The needed information were then translated in XML format and introduced into the experimental platform. Approximately 80 exchange protocols, 100 tasks and 50 workflows were involved in the experimental planning platform.

We then generated production and logistics plans and presented these to the production manager, for comments. This interactive validation phase has permitted to review and adjust the planning parameters and algorithms. Moreover, by studying the real plan prepared by the manager, we were able to evaluate the performance of the platform in terms of number of late customer orders, production value, resource utilization, etc. These indicators, easily obtained by the platform, were precious to evaluate the performance of both plans and identify possible improvements.

This validation process took about one year and many corrections have been made on the platform. Now, plans generated by the platform offer considerable improvements, while compared to plans prepared by the partner's production manager. Also, the actual planning times for drying and finishing operations have been reduced dramatically.

5. CONCLUSION

This paper has presented a software architecture which aims at designing distributed APS tools for the forest products industry. The presented platform follows the double objective of providing the industry with advanced planning tools, but also with a means of studying the dynamic and performance of such tools working together. The first objective has been already addressed, and quantitative testing of these tools in a large scale context (i.e., real industrial test case) has been undertaken. Concerning the second objective, a research effort has been initiated. This effort focuses on the simulation part of the platform. In other words, this effort aims at designing the mechanisms and software components required to put such distributed advanced planning tools in a simulated environment in order to test them in a live (though simulated) environment.

Finally, another research direction that is currently investigated concerns the agents' architecture. Agents in such a context seem to require the ability to exploit planning tools in various configurations according to the situation. In this context, it seems necessary to provide agents with sufficient knowledge and competencies about the planning tools and their environment, in order to deal timely with different situations by using the different available tools in the most appropriate way.

6. ACKNOWLEDGMENTS

This work was funded by the FOR@C Research Consortium and the Natural Sciences and Engineering Research Council of Canada (NSERC).

7. REFERENCES

1. Caridi M, Cavalieri S. Multi-agent systems in production planning and control: an overview. Prod Plan Control 2004; 15(2):106-118.
2. Dudek G, Stadtler G. Negotiation-based collaborative planning between supply chain partners. Eur J Oper Res 2005; 163(3): 668-687.
3. Epstein R, Morales R, Seron J, Weintraub A. Use of OR Systems in the Chilean Forest Industries. Interfaces 1999; 29(1): 7-29.
4. Fleischmann B, Meyr H. Planning Hierarchy, Modeling and Advanced Planning Systems. In Handbooks in OR&MS, Elsevier, 2003.
5. Fox MS, Barbuceanu M, Teigen R. Agent-Oriented Supply-Chain Management. Int J Flex Manuf Sys 2000; 12:165-188.
6. Frayret JM., D'Amours S, Montreuil B. Coordination and Control in Distributed and Agent-Based Manufacturing Systems. Prod Plan Control 2004; 15(1): 42-54.
7. Frayret JM, Boston K, D'Amours S, Lebel L. The E-nabled Supply Chain - Opportunities and Challenges for Forest Business. Working paper of the CENTOR, Université Laval, 2005.
8. Parunak HVD. Industrial and Practical Applications of DAI. in Weiss, G. (Ed.), Multiagent Systems a Modern Approach to Distributed Artificial Intelligence, Cambridge (MS): MIT Press, 1999.
9. Rönnqvist M. Optimization in forestry. Mathematical Programming, Series B 2003; 97(1-2): 267-284.
10. Shen W, Norrie DH. Agent-Based Systems for Intelligent Manufacturing: A State-of-the-Art Survey. Knowl Inf Syst 1999; 1(2): 129-156.
11. Shen W, Norrie DH, Barthes JP. Multi-Agent Systems for Concurrent Intelligent Design and Manufacturing. London: Taylor & Francis, 2001.
12. Stadtler H, Kilger C. Supply Chain Management and Advanced Planning - Concepts, Models, Software and Case Studies. Berlin: Springer, 2000.
13. Stadtler H. Supply chain management and advanced planning–basics, overview and challenges. Eur J Oper Res 2005; 163(3): 575-588.
14. Tharumarajah A. Survey of resource allocation methods for distributed manufacturing systems. Prod Plan Control 2001; 12(1): 58-68.
15. Weiss G. Multiagent Systems A modern Approach to Distributed Artificial Intelligence. Cambridge (MS): MIT Press, 1999.
16. Stephens S. The Supply Chain Council and the Supply Chain Operations Reference (SCOR) model: Integrating processes, performance measurements, technology and best practice. Logistics Spectrum 2000; 34(3): 16-18.

PART A

MULTI-AGENT AND HOLONIC SYSTEMS IN MANUFACTURING

4 AN AGENT-OS FOR AUTONOMOUS COOPERATIVE SYSTEMS

Francisco Maturana[1], Raymond Staron[1],
Pavel Tichý[2], Pavel Vrba[2], Charles Rischar[1],
Vladimír Mařík[2] and Kenwood Hall[1]
Rockwell Automation
[1] 1 Allen-Bradley Drive, Mayfield Heights, OH 44124-6118, USA
[2] Pekarska 10a, 15500 Prague 5, Czech Republic
{fpmaturana, rjstaron, ptichy, pvrba, cmrischar, vmarik and khhall}@ra.rockwell.com

We are interested in providing an agent infrastructure for truly distributed control. Requirements include multiple language implementations so that this agent host environment can exist in real-time controllers. Our first infrastructure, the Autonomous Cooperative System (ACS) was built as a low priority control task within the operating system of the automation controllers. The environment added the functionality to host a set of agents. This ACS kernel evolved for a long time almost independently of the controller operating system. New releases of the controllers will cause complications in the integration of the software and adoption of new functionality. We need to evolve ACS to be seamless integrated with the controllers. These changes will affect the infrastructure. In this document, we discuss the implications and results of the transformation.

1. INTRODUCTION

The use of intelligent agent technology in industrial control has revealed very important discoveries for building a new breed of automation systems. It has been shown that intelligent agent systems can cope with complex requirements for managing distributed information because agents integrate information dynamically as a normal aspect of their operation. In particular, systems with physical and logical redundancy are excellent candidates to implement agents.

Knowledge from each automation component impacts the overall process locally and globally. System engineers need to oversee these impacts at design time to program the rules for controlling the process and for mitigating the negative impacts on the system when unforeseen scenarios occur. To execute successful production, the information infrastructure needs to be built with a lot of contingencies to prevent malfunctioning. Here is where we find the primary obstacle to flexible and balanced automation. In classical control terms, the integration of the system is subordinated to the maladies of islands-of-automation. However, there are cases where islands-of-automation are still a necessary structure.

Intelligent agent systems, however, will not solve all problems in automation and information systems, but they do provide a reference for distributing knowledge and information. There have been many contributions from the research community

Please use the following format when citing this chapter:

Maturana, F., Staron, R., Tichý, P., Vrba, P., Rischar, C., Mařík, V., Hall, K., 2006, in IFIP International Federation for Information Processing, Volume 220, Information Technology for Balanced Manufacturing Systems, ed. Shen, W., (Boston: Springer), pp. 29–36.

to characterize agents and their frameworks. According to those studies (Brennan, 2003), (Shen, 2001), (Wooldridge, 1995), agents can exist at opposite ends of the distribution spectrum. At one end, agents can be highly granular to act on behalf of a micro-process such as the opening and closing of a valve. At the other, they can be larger decision-making actors such as the type required in process planning and scheduling. Agents can be found anywhere within that spectrum. Each side of the spectrum imposes requirements on the communication and computing.

Some interesting aspects in building agent systems have been the design of agents and the encapsulation of the agents within adequate computing hardware. These considerations can be very critical in large heterogeneous systems in which the hardware and processes are dramatically different but interdependent.

The Foundation for Intelligent Physical Agents (FIPA) (FIPA, 1996) established the need for agent technology to link to the physical world. FIPA now has working groups involved in the manufacturing production and scheduling areas for solving the real-time control interface problem. The FIPA specification is a good start in the direction of a common framework for building agents. However, there is no available specification to make agents transit to the production environment without breaking legacy rules. Thus, what are the essential characteristics of the hosting devices to enable the operation of agents in different contexts?

To make a computing unit agent enabled, it is necessary to identify the middleware as a set of functions to be integrated with the device's operating system (OS). Moreover, the middleware need to be integrated with the hosting OS seamlessly to avoid interruptions during the execution of the control tasks. Agent computing is demanding on the CPU and the communications because of the agent-to-agent messaging and reasoning. Hence, to integrate agents with controllers, we need to consider the agent functionality as another part of the OS.

Our original agent environment, the Autonomous Cooperative System (ACS) (Maturana, 2002), (Staron, 2004), (Tichy, 2001), was implemented in C++ since it was intended for controllers that currently support this type of programming language. The agent environment was successfully tested on multiple industrial applications such as steel rolling mill factory, material handling, and shipboard automation. Most recently, the infrastructure has been used to model complex networks for power distribution, beverage distribution matrix, and water/waste water systems in simulation. Although ACS has been successfully deployed, the system presents overheads in computation time and memory consumption, two resources that have other critical demands on them in a real-time control system. Thus two goals for improving ACS are 1) reducing agent demand on these resources without sacrificing functionality or flexibility, and 2) minimizing the work effort needed to create new versions of ACS for future revisions of the controller firmware.

2. BACKGROUND WORK IN THE AREA OF AGENT CONTROL DEVICES

The concept of a real-time interface between the soft world of agents and the physical world of machinery is tied closely to some fundamental ideas in Holonic Control Devices (Brennan, 2003), (McFarlane, 2000), (Marik, 2001), (Vrba, 2005). This consortium defined some properties of such control devices, but not from the perspective of changes to existing real-time operating systems used in manufacturing. The agents occupying the lowest layer of an automation system require features necessary for enabling the agents to enter into negotiations about the performance of manufacturing tasks and to mutually coordinate the performance of those tasks. The ability to locate, join, leave and participate *in cooperation domains* is tied to the ability of the operating system to coordinate such tasks without interrupting time critical operation as well as providing an efficient message delivery, redirection and prioritization.

At the machine level, agents have temporal restrictions, i.e., hard and soft constraints, and as a result, agents must use their time to accomplish real-time activities. This condition restricts the time availability to perform other operations such as message parsing and queuing. For example, real-time agent activities can be divided into three main activities: (1) domain, (2) control, and (3) meta-level control. Domain activities include execution actions that achieve high-level tasks. The combined effect of execution actions produces a domain-level action (a process step). Control activities are classified as high-level goals and constraints on how to achieve them (e.g., scheduling) or those activities that facilitate cooperation with other agents. Meta-level control relates to learning patterns. Meta-level control tends to use patterns to optimize the agent search by reducing converge time.

Other definitions of agents consider them as purely responsive entities. In current research (Discenzo, 2002), (Maturana, 2002), (Shen, 2001), proactive capabilities have also been part of agents to learn and predict behaviors of the physical device. Real-time distributed control conditions affecting the agents is a new area which has only recently been investigated (Balasubramanian, 2000).

3. REQUIREMENTS FOR AN AGENT-OS

In recent years, we have been researching how to architect agent infrastructures for control. In this investigation, we have learned important characteristics of agent systems that are tied to the information processing requirements. Agents that are intended to process large amounts of information are better off operating in intermediate and upper levels of the enterprise, where larger blocks of memory are available and no time-critical operations need to be used. On the other hand, agents that are intended to interact with the physical device and whose responsibilities are to steer control and handle events (e.g., pump, breaker, conveyor, etc.) are better off operating in the controllers. There has been important work on agent systems for the upper levels of control (e.g., scheduling) (Leitão, 2002), (McFarlane, 2000). Little work has been done to apply these techniques to the control level. To integrate these areas, a simple and straightforward interface between the agents and the control functionality is needed. As a result, a multi-tier technique has been explored.

Agents have been integrated with controllers that are based on IEC 61131-3 programming language (IEC, 2001). Control systems based on the IEC-611499 specification are also very compatible with agents (Lewis, 2001). The role of any interface is to provide accessibility. The interface between the agent infrastructure and the controller OS allows agents to monitor the status of the controller and to perform reasoning about the process. This is the case in the original ACS (Maturana and Tichy, 2002). Figure 1 shows the agent functionality within the controller which was established as a layer on top of the controller's OS. The ACS layer interacts with the controller's firmware to access control functionality such as data table, programs, and communication. In this controller architecture, agent processes execute within the OS at the lowest priority.

To establish a path for a formal architecture, we will extract the lessons learned from ACS toward the formalization. ACS permits the downloading of agent classes into the controllers. The agent classes are instantiated to create specific agents. Each agent has a control program associated with it. The agents change the control behavior through an events and the data table of the controller. ACS is a low priority task initialized during the device's power up cycle. The agents are threads within the ACS task. These threads are assigned local priorities below the ACS task to avoid interferences with the device's priority. The agent threads are coordinated by a local scheduler. The agent processes are slow compared to the device's tasks since the ACS task executes when the controller is not executing higher priority activities.

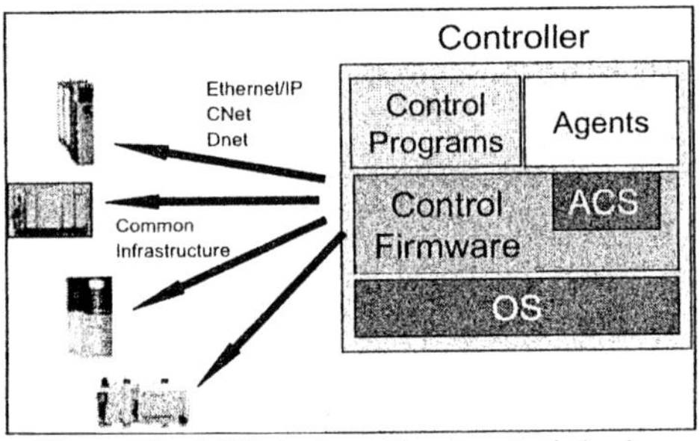

Figure 1: ACS Automation control device

The ACS infrastructure bears larger communication latency since its messages move through intermediate queues into the controller's queues during delivery and reception. The ACS infrastructure depends on ad-hoc anchors into the controller firmware, i.e., the code that connects ACS to the standard controller firmware has never been formalized. When the firmware is changed, the ACS infrastructure is forced to change to follow those changes. More often than not, updating the ACS system is not easy due to changes in the event handling and interrupt configuration.

ACS has obstacles in expandability and upgradeability. To considering it as a commercial product, we need to combine it with the controller's firmware using a leaner approach to avoid the need for continual reengineering.

4. AUGMENTING CONTROL FIRMWARE WITH AGENTS

There is a need for a specification relative to enabling agents natively in the controllers. The extensions to the controller to support ACS were a good starting point and it helped to develop the infrastructure to its full extent. Now, we need to learn how to incorporate the agent functionality into the controller's core.

The architecture of the agent-based control device is simplified by blending the agent functionality into the controller firmware. This change means a significant transformation and enhancements in relation to the style of information processing for the agent decision making process, handling of messages and events, agent composition and creation, and software maintenance and upgradeability.

In the new system, there is no subordinate OS to handle the agent functionality. Instead, the agent functionality follows controller firmware rules. There is no need for special anchors in the firmware to enable agents since each agent process is created as a task using normal priorities. The agent task shares its CPU time with other tasks and it is not segregated to the tail of the list. The controller firmware now schedules all tasks in the controller. The control firmware gains several abilities:

- The ability to *reason* about manufacturing tasks and their relationships to distributed control applications, and to acquire and share *knowledge* related to such reasoning: This ability relates to a planning engine capability. The planning engine needs message parsing and the ability to create concurrent contexts to coordinate the agents. Advanced hardware can bring more efficient capabilities (e.g., hyperthreading in *Itanium* or *XScale* processors).
- The ability to issue appropriate *management commands* to dynamically modify existing applications to perform new tasks or to recover from abnormal operations: This refers to the reactive and proactive behaviors that emerge from the control and/or agent level. The OS must permit the inclusion of auto-generative event-based communication. The ability to transmit events must be allowed between control and the agents.
- The ability to handle both intra- and inter-agent communication of events: Agents communicate with each other via an inter-task or inter-threading message delivery. Across devices, agents need agent location services provided by the system to find addresses and communication routes. Messages are encoded according to FIPA message representation specifications. Also, directory facilitator and agent management system agents must understand and use Semantic Language for agent registration, deregistration, and searching.

In the new architecture, there is no need to have preallocated memory to create the agents. This aspect was limiting in the previous architecture since the controller's image was preconfigured with a block of memory for ACS, with no possibility for adjustment during runtime. The memory block was partitioned to support executive actions, agent threads, and messaging queues. It was a limitation, because in embedded systems, memory is a precious resource. Without knowing the actual size to be used during operation, fixed allocations are inefficient. In the new architecture, a stack is allocated for each task from general memory; thus, memory is consumed only upon need as agents are created.

The subordinate threading model was replaced with tasks (a task is a thread but in the controller firmware context). Each agent is a task which can be created at different priorities level depending on the type of operation, e.g., to act swiftly on complex operations. Support functions such as message parsing and FIPA encapsulation of messages remain the same, but these are also created as tasks. Thus, the communication interface between the agents remains the same. Agents interact with control via events and the data table.

The agents are created in an Agent Development Environment (ADE) as binary objects (i.e., the agent class). The binary object is downloaded to the controllers using a backend loader, as shown in Figure 2. Each object has two entry points, one for the creation of the object instance and other for the initialization of the object's task. The task takes the instance as a parameter. After the task initialization, the object becomes an agent and immediately starts its activity. Multiple objects can be downloaded into one or many controllers. Multiple agents can be created from a single object type.

With the features above, there is no need to create a special firmware to run agents. It is only necessary to proceed with the downloading of the agent infrastructure (which has been decoupled from the controller's firmware) prior to downloading agents. Thus enabling a controller to run as an agent-base controller is a two step operation: (i) agent infrastructure downloading and (ii) agents downloading.

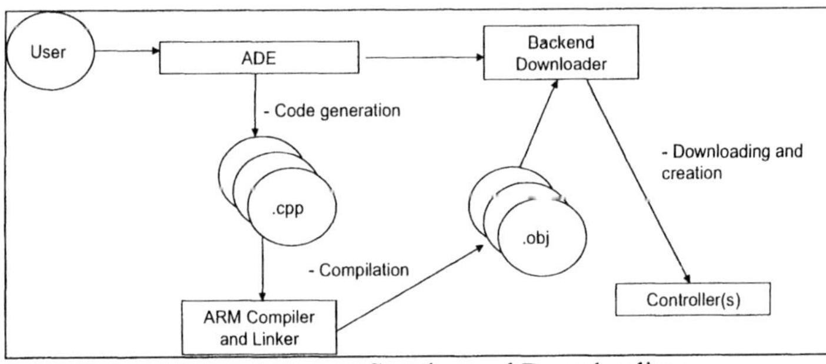

Figure 2: Agent Creation and Downloading

The agent infrastructure is stored on a computer as a library which is delivered to the target controllers by the ADE. The agent infrastructure is common to all controllers. The agents are application specific and these need to be compiled before downloading. There is a prototype system built and tested on two commercial controllers from Rockwell Automation (CompactLogix and ControlLogix L63/A).

5. COMMUNICATION ATTRIBUTES

The Agent-OS requires a unifying communication language, syntax and semantics, and communication transport stack. The message encapsulation is accomplished according to the FIPA specification. The transport encapsulation is based on the Common Industrial Protocol (CIP, 2001).

Figure 3 shows the evolution of a message. An agent emits a message to express some knowledge, desire, or intention. The FIPA layer transforms the message into a stream of data (a compact binary object or a string with XML or Lisp style of encoding). The CIP encapsulation fragments this stream into packets for transmission and attaches transport headers to it. The process of discovering a suitable destination to which to send the message is a linear association that matches the message request with agent capabilities (capability-to-agent). The receiving controller(s) applies packet redirection by looking into the CIP header destination field. The packets are reassembled and delivered to the FIPA layer for decompression and parsing, establishing the communication stack.

Agents can receive and send messages from/to agents or objects on either the same controller or another controller. Messages are delivered directly to the controller's communication queue via an interface for sending messages (*SendMsg()* and *Send()*, as shown in Figure 4). If the destination (Automation Controller ID,

ACID) is local, the message is posted with no fragmentation as a non-blocking action. If the destination is remote, the message goes through CIP encapsulation and it is staged into the communication task for remote delivery.

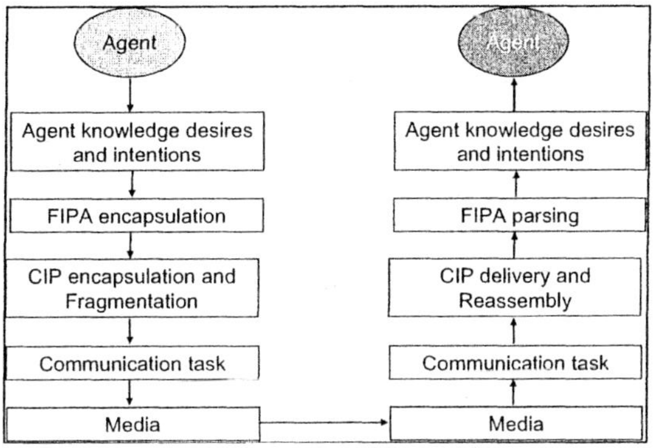

Figure 3: Message Transition

The CIP encapsulation learns the pattern of communication by establishing the frequency of communication between the originator and target controller. With this dynamic knowledge, the CIP encapsulation stimulates the communication task to create a dedicated channel for frequent communications. The less frequent communications retain their unconnected status. This is another important enhancement added to the controller firmware that helps to optimize bandwidth.

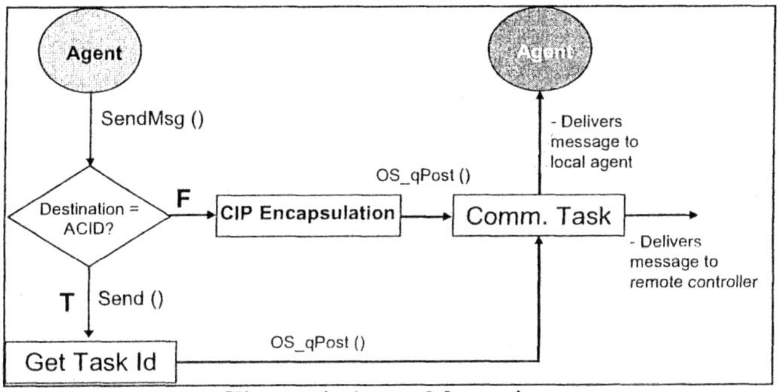

Figure 4: Agent Messaging

Each agent-based task blocks its activity until a new message is posted to its queue. The controller firmware provides a *OS_qPend()* function that is used by the agents to block for an infinite time until a message arrives. The agents access this OS function via an interface *ReceiveMsg()*. This implementation has proven to be efficient in absorbing messages as well as releasing the controller from cycling inside useless messaging loops. Also, an agent sometimes needs to send a message to itself but with delayed delivery. This functionality combines queues and delays.

The agents are retained until explicit deletion occurs by the user. Agents and control programs survive power cycles. The control firmware now has the capability to detect the departure of a controller using heartbeats. The controllers that depend on the missing capability carry out organizational reconfiguration to bypass the missing agents. This action triggers a communication spike until the controllers learn the new organizational knowledge.

6. REMARKS

Intelligent agent control is a good alternative to monolithic control and information because it is distributed. In this paper, we discussed the evolution of the ACS architecture into a formal approach for the creation of agent-based controllers. We identified the characteristics of the new system and how these will benefit and advance the state-of-the-art of controllers and distributed control. By absorbing the agent functionality into the controller's firmware, we open a path into a new realm in agent control systems.

7. REFERENCES

1. Balasubramanian, Sivaram, Brennan, Robert, Norrie, Douglas H. Requirements for Holonic Manufacturing Systems Control. DEXA Workshop 2000: 214-218.
2. Brennan R, Hall K, Marik V, Maturana F, Norrie D, "A real-time interface for holonic control devices", In: Marik V, McFarlane D, Valckenaers P: *Holonic and Multi-agent Systems for Manufacturing, Lecture Notes in Artificial Intelligence*, No.2744, Springer-Verlag, Berlin, 2003:25-34.
3. CIP: Common Industrial Protocol, 2001, http://www.ab.com/networks/cip_pop.html.
4. Discenzo F, Maturana F, Chang D. Managed Complexity in an Agent-Based Vent Fan Control system Based on Dynamic Reconfiguration, ICCS, Nashua, NH, June 9-14, 2002.
5. FIPA: The Foundation for Intelligent Physical Agents, 1996: http://www.fipa.org.
6. IEC (International Electrotechnical Commission), TC65/WG6, 61131-3, 2nd Ed., Programmable Controllers - Programming Languages, April 16, 2001.
7. Leitão P, Restivo F. A Holonic Control Approach for Distributed Manufacturing, In: Knowledge and Technology Integration in Production and Services: Balancing Knowledge and Technology in Product and Service Life Cycle, V. Marik, L.M. Camarinha-Matos and Hamideh Afsarmanesh (eds), Kluwer Academic Publishers, 2002: 263-270.
8. McFarlane DC, Bussman S. "Developments in Holonic Production Planning and Control", *International Journal of Production Planning and Control*, Vol. 11, No. 6, pp. 522-536, 2000.
9. Lewis RW. *Modelling Control Systems Using IEC 61499*, Institute of Electrical Engineers. 2001.
10. Marik V, Pechoucek M. Holons and agents: recent developments and mutual impacts, *12th International Workshop on Database and Expert Systems Applications*, IEEE Computer Society, 2001: 605-607.
11. Maturana F, Tichý P, Šlechta P, Staron R. "Using Dynamically Created Decision-Making Organizations to Plan, Commit, and Execute Control Tasks in a Chilled Water System". In Proceedings of the 13th International Workshop on Database and Expert Systems Applications DEXA 2002, HoloMAS 2002, Aix-en-Provence, France, 2002: 613-622.
12. Shen W, Norrie D, and Barthès JP. "Multi-Agent Systems for Concurrent Intelligent Design and Manufacturing". Taylor & Francis, London, 2001.
13. Staron R, Maturana P, Tichý P, Šlechta P, "Use of an Agent Type Library for the Design and Implementation of Highly Flexible Control Systems". 8th World Multiconference on Systemics, Cybernetics, and Informatics, SCI2004, Orlando, FL., July 18-21, 2004.
14. Tichý P, Šlechta P, Maturana F, Balasubramanian S. "Industrial MAS for Planning and Control". In (Mařík V., Štěpánková O., Krautwurmová H., Luck M., eds.) Proceedings of Multi-Agent Systems and Applications II: 9th ECCAI-ACAI/EASSS 2001, AEMAS 2001, HoloMAS 2001, LNAI 2322, Springer-Verlag, Berlin, 2002: 280-295.
15. Vrba P, Marik V. From Holonic Control to Virtual Enterprises: The Multi-Agent Approach. The Industrial Information Technology Handbook, 2005.
16. Wooldridge M, Jennings N. "Intelligent agents: theory and practice", Knowledge Engineering Review, Vol. 10, No. 2, pp. 115-152, 1995.

5 A HOLONIC APPROACH TO DYNAMIC MANUFACTURING SCHEDULING

Paulo Leitão[1], Francisco Restivo[2]
[1]*Polytechnic Institute of Bragança, Quinta Sta Apolónia, Apartado 1134, 5301-857 Bragança, Portugal*
pleitao@ipb.pt
[2]*Faculty of Engineering of University of Porto, Rua Dr. Roberto Frias, 4200-465 Porto, Portugal*
fjr@fe.up.pt

Manufacturing scheduling is a complex combinatorial problem, particularly in distributed and dynamic environments. This paper presents a holonic approach to manufacturing scheduling, which in opposite to traditional approaches, distributes the scheduling functions over several entities, combining their calculation power and local optimization. In this scheduling and control approach, the scheduling mechanism evolves dynamically to combine optimized scheduling, achieved by central entities, and distributed scheduling, improving its responsiveness and robustness.

1. INTRODUCTION

Manufacturing scheduling can be defined as the allocation, over the time, of jobs to machines, within a shorter temporal horizon and respecting a specific criterion, such as cost or tardiness. It is a complex combinatorial problem, more specifically a non-polynomial (NP) problem: the objective is to find the optimal sequence from the $j!^m$ possible scheduling sequences, where j is the number of jobs and m the number of machines. The manufacturing scheduling problem becomes even more complex when it takes place in an open, distributed and dynamic environment.

The scheduling problem has been widely studied, mainly due to its highly combinatorial aspects, its dynamic nature and its applicability in manufacturing systems [1]. Examples of such methods are heuristics, linear programming, constraint satisfaction techniques, Lagrangian relaxation, neighbourhood search techniques (e.g. simulation annealing or taboo search) and genetic algorithms.

Manufacturing scheduling is traditionally elaborated in a centralized manner using one of referred methods, often calculated off-line and considering that it is a static and deterministic problem. However, in an industrial manufacturing system the things rarely go as expected, mainly because: i) new tasks arrive continuously to the system, while scheduled tasks are cancelled, ii) certain resources become unavailable and additional resources are introduced, iii) unexpected events occur in the system, such as machine failures, operator absence, rush orders or unavailability of raw-materials, and iv) scheduled tasks may take more or less time than expected.

Please use the following format when citing this chapter:

Leitão, P., Restivo, F., 2006, in IFIP International Federation for Information Processing, Volume 220, Information Technology for Balanced Manufacturing Systems, ed. Shen, W., (Boston: Springer), pp. 37–46.

In such dynamic environments, the optimised schedule produced by the front office can quickly become unacceptable, requiring the dynamic re-scheduling, as fast as possible, and done in a short amount of time, to avoid the risk of degradation of the production productivity. Traditional methods don't fulfil the real dynamic re-scheduling needs mainly because they are inflexible and slow.

Agent-based and holonic manufacturing approaches suggest the implementation of distributed scheduling, since the scheduling algorithm maybe easily distributed over a number of entities which combine their calculation power and their local knowledge to optimize the global performance [2]. Unlike traditional manufacturing scheduling approaches, using centralised scheduler, in agent-based manufacturing scheduling systems, each agent can locally handle the schedule of its machine, operator, robot or station. The major advantages of the distributed scheduling are the improvement of reaction to disturbances and the parallel computation.

Some of the distributed scheduling approaches use traditional algorithms embedded in distributed entities, while others are based on emergent behaviour, like market-based and net protocol algorithms. Among others, Sousa and Ramos [3] proposes a dynamic scheduling system supported by a holonic approach, using forward and backward influence in the negotiation leading to the task allocation, to handle the temporal constraints and to solve conflicts, and Gou et al. [4] presents a holonic manufacturing scheduling approach using Lagrangian relaxation, where capacity constraints of a scheduling problem can be relaxed and replaced by a penalty cost. Markus et al. [5] proposes a market model to solve dynamic order processing and scheduling problems, and Sugimura et al. [6] models the manufacturing operations using an object-oriented approach and proposes a real time scheduling mechanism for assembly lines. Hino and Moriwaki [7] introduces a recursive propagation technique based on sending messages regarding schedule changes to agents responsible for subsequent tasks, and Logie et al. [8] extends this concept by limiting the focus of the agents to tasks within a specified time window.

This paper describes a holonic approach to the dynamic manufacturing scheduling, introduced by ADACOR (ADAptive holonic COntrol aRchitecture for distributed manufacturing systems) [9], which combines distributed scheduling, where holons negotiate the resource allocation using free market based techniques, thus achieving fast re-scheduling, with coordination entities, responsible for the introduction of global optimization. The holonic dynamic scheduling is supported by decision-making capabilities embedded in each distributed holon and cooperation mechanisms that support the evolution of the dynamic scheduling.

The rest of the paper is organized as follows: Section 2 presents the main principles of the proposed holonic scheduling architecture, focusing on the distributed components, dynamic adaptation model and distributed scheduling. Section 3 describes the cooperation mechanisms to support the dynamic re-scheduling and Section 4 describes the prototype implementation. Finally, Section 5 rounds up the paper with conclusions.

2. HOLONIC DYNAMIC SCHEDULING ARCHITECTURE

The idea beyond our scheduling approach is that a global optimized schedule should be generated whenever possible, and a fast re-scheduling should be used in case of disturbances, because, in this case, this is preferable to waiting a significant amount of time for an optimized schedule, which is likely to be not optimized again soon.

2.1 Holonic Manufacturing Components

The proposed approach is designed upon a community of distributed and autonomous control units, the holons, representing the manufacturing components. Three types of holons are identified to handle the scheduling and control at shop floor level [9], as illustrated in the Figure 1:

- *Task holons* that represent production orders launched to the shop floor to execute products, each one containing information about the production of the product, and about the progress of the production order execution.
- *Operational holons* that represent physical resources or operators available at shop floor, each one with a set of skills and knowledge.
- *Supervisor holons* that represent the logical coordination of a group of operational and/or supervisor holons, providing co-ordination and optimization services to the holons under their supervision, and thus introducing hierarchy in an otherwise decentralized system.

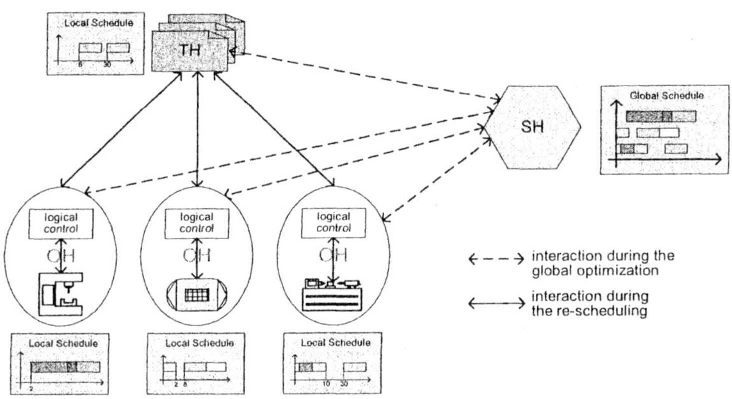

Figure 1 - Interaction between Distributed Manufacturing Components

Each holonic control unit has its own decision-making capability, performing, among others, control and scheduling functions. These embedded mechanisms are dependent of the holon's type, its behaviour and objectives.

The scheduling mechanism embedded in supervisor holons deals with the multiple machines and multiple jobs scheduling problem. Depending on the number of machines and jobs, the schedule produced by the top level supervisor holon may take a large computational effort and time, but it is an optimal plan since the supervisor has a global view of the system.

Each operational holon is responsible for its own schedule, built dynamically from its local knowledge, and using a scheduling mechanism addressing the problem of multiple jobs for a single machine. It achieves optimal local schedule plans, but due to the lack of global information, may not lead to an optimal global schedule.

The motivation of ADACOR holons to execute the manufacturing actions is regulated by a credits system. Each task holon when is launched receives a fund (π) to execute a production order and a penalty value for the delay. The task holons are responsible for managing the fund received to produce their production orders, without exceeding the initial fund or paying the delay penalty.

During the interaction to allocate the operations, task holons try to pay as less as possible and the operational holons try to receive as more as possible. After the negotiation, each task holon accepts to pay a price of ξ credits to the operational holon that will execute a certain operation and to receive a penalty of ξ credits from the operational holon if it does not fulfil the contracted due date.

Table 1 summarises the evolution of the credits of task and operational holons during their life cycles.

Table 1: Evolution of Credits during the Holon Life Cycle

Phase	Task Holon	Operational Holon
Resource allocation process.	Contracts the execution by ξ and the penalty by φ.	Contracts the execution by ξ and the penalty by φ.
Finish of an operation with success.	Pays the value ξ ($\pi \leftarrow \pi - \xi$).	Increases its credits by ξ ($\mu \leftarrow \mu + \xi$).
End of an operation with delay.	Pays the value ξ and receives the value φ ($\pi \leftarrow \pi - \xi + \varphi$).	Decreases its credits by φ and increases by ξ ($\mu \leftarrow \mu + \xi - \varphi$).
Operation cancelled (delay, failure, etc.)	Receives the value φ ($\pi \leftarrow \pi + \varphi$).	Decreases its credits by φ ($\mu \leftarrow \mu - \varphi$).

The global performance of operational holons in terms of credits is given by the sum of rewards received minus the penalties paid for the delays. These rewards and penalties reflect the *reputation* of the holon.

2.2 Dynamic Scheduling Model

The interaction process leading to the achievement of the manufacturing schedule has the following constraints:

- A part cannot be started until its preceding part(s) is finished.
- An operation cannot be started until its preceding operations are finished.
- Each machine can only process one operation at time t.
- A resource R_j possessing the set of skills S_j, has abilities to execute the operation o_{ik}, having a list of requirements $B_{ik} = \{B_{ikz} | z \in I\}$, if

$$B_{ik} \subseteq S_j \Leftrightarrow \forall x \in B_{ik} \Rightarrow x \in S_j$$

i.e., the resource has abilities to execute an operation if it fulfils all the requirements presented by the operation.

The dynamic scheduling model is the result of the dynamic interaction between task, operational and supervisor holons, combining the problem solving at the individual holon level and the coordination-negotiation schema at the system level, to produce a global manufacturing scheduling, which is simultaneously centralized in normal situations or locally produced in the presence of a disturbance.

The self-organization capability is the key concept to support this adaptive production control and scheduling mechanism. The adaptation is achieved by the self-organization of each ADACOR holon, contributing to the dynamic adaptation of the whole system. The self-organization is regulated by the *autonomy factor*, which fixes the level of autonomy of each holon, and evolves dynamically in order to adapt the holon behaviour to the changes in the environment where it is placed. The evolution is governed by a decision mechanism, and the overall efficiency of the self-organization is dependent on how the learning mechanisms are implemented, and on how new knowledge influences its parameters.

Figure 2 illustrates a small example about how the adaptive dynamic scheduling approach works. In normal operation, i.e. without the occurrence of unexpected disturbances, the holons are running in a hierarchical structure, with supervisor holons coordinating several operational and/or supervisor holons, and operational holons having low autonomy factor. Periodically, regulated by their internal clocks, supervisor holons generate manufacturing scheduling plans globally optimized.

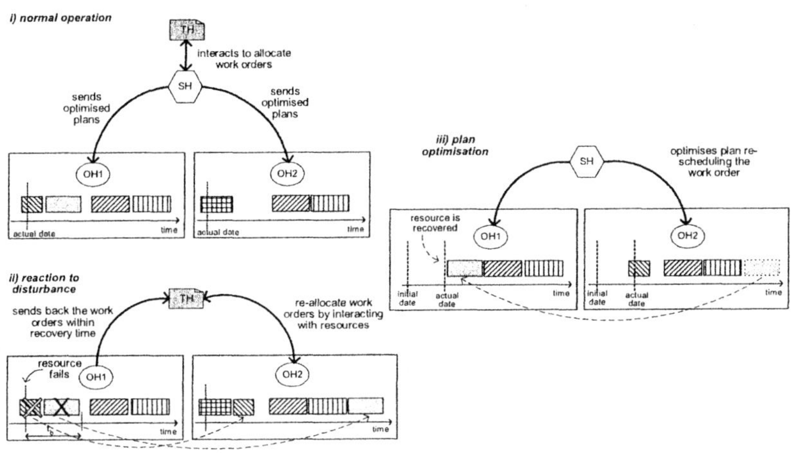

Figure 2 – Dynamic Holonic Scheduling

The optimised schedule plans are offered, as advices, to the holons under their coordination domains, which have the capability to accept or reject them. Normally, they follow the schedule advices proposed by the supervisor holons since they have low level of autonomy.

In turbulent scenarios or once an unexpected disturbance (e.g. a machine failure or deviation from plan) is detected, the system is forced to evolve to a heterarchical structure, characterized by totally decentralized decision-making mechanisms. In fact, the autonomy factor of each ADACOR holon is increased dynamically according to a function that takes in consideration its current value, the estimated time to recover from the disturbance, and the level of impact of the disturbance.

This transient state, which should be as short as possible, operates without the presence of coordination levels, the manufacturing scheduling being achieved in a distributed manner. The distributed scheduling results solely from the interaction between task and operational holons. The cooperation strategy built into each holon is therefore the key to the success of this approach.

After the transient phase, the system returns to a hierarchical structure.

2.3 Distributed Resource Allocation Schema

In the distributed resource allocation schema the computational complexity is related to find an optimal determination problem in combinatorial auctions. The distributed scheduling mechanism, introduced in this work, uses a resource allocation mechanism based on a multi-round Contract Net Protocol (CNP) [10], extending the original CNP schema with capability of apply the contract net schema several times, and capability to contract partial quantities.

In presence of operation announcements, each operational holon decides, based on its skills and capacity, its availability to execute the operation. In case of availability, the operational holon then calculates the price to be proposed to the task holon. The price may be calculated according to the following function,

$$p_{jik} = C_s + C_p \times d_{ik} + C_b \times \left(2 - e^{-\alpha \times \beta} \times (1-\gamma)\right)$$

which models the market laws, increasing or decreasing the final price in function of the actual load of the resource (reflected by the parameter β) and of the actual bid acceptance rate (reflected by the γ parameter, with $0 \leq \gamma \leq 1$). The holon uses the knowledge learned from the previous bids to adjust the final price: reducing the γ parameter if the acceptance rate is low or increasing it in the opposite case.

The task holon evaluates the proposals sent by operational holons to allocate the operation to the best bid. The decision procedure takes into account, among others, the proposed price, the location of the resource and the degree of confidence about the holon. The confidence degree reflects the trust that the task holon has in a operational holon, and is based on the knowledge learned from previous interactions. In case of an inconclusive evaluation, the task holon can start another iterative negotiation, re-formulating the bid parameters, for example the due date.

3. COOPERATION MECHANISMS FOR RE-SCHEDULING

The dynamic scheduling algorithm must respond dynamically and promptly to emergent and unexpected disturbances. An important design factor is the *size* of the disturbance that will activate the re-scheduling mechanisms described. Re-scheduling mechanisms can the be divided in: periodic re-scheduling, which considers all disturbances (usually many small disturbances) at once, generating new optimized schedules periodically, and event-based re-scheduling, which is more suited for bigger and single disturbances, like machine breakdowns or rush orders.

In the next sections, the several types of re-scheduling mechanisms (i.e. event-based and periodic) will be described.

3.1 Re-scheduling for Cancellation of Orders

The cancellation of a production order can be considered as a simple disturbance at shop floor level that only requires the local re-scheduling, in order to optimise the schedule. After generating a new schedule, the operational holon notifies the supervisor holon about its new schedule, allowing the synchronisation of both agendas and the optimization of the global schedule. It must be noticed that this type of disturbance may open free spaces in the agenda, allowing to execute earlier some operations that were eventually delayed.

3.2 Re-scheduling for Machine Breakdowns

In the case of occurrence of a machine failure, the operational holon determines the state of the machine and of the part after the failure, and estimates how long the downtime will be. The diagnostic can lead to different scenarios: the machine can become immediately available or stay out of service for a more delayed repair intervention, and the part may have been destroyed or not. If the part is not destroyed and the machine is ready to re-execute the operation, no action has to be performed; however, other scheduled operation(s) may become delayed, being then treated as a delay disturbance.

In the other cases, both operational and task holons have specific tasks to perform. The operational holon:

- In case of destruction of the part, removes the proper operations from its agenda, and notifies the task and supervisor holons about the occurrence.
- In case of machine breakdown, notifies the task holon about its impossibility to execute the operations in the scheduled dates.

The operational holon also executes a re-scheduling to optimise the plan.

The task holon can take two different actions:

- If the part is destroyed, re-allocates from the beginning all operations belonging to the production order.
- If the machine became unavailable, re-schedules the operations taking in consideration the information from previous resource allocation processes. The achieved allocation can lead to delays in the posterior operations, requiring an adjustment of the temporal window to execute each operation.

In both cases, the re-scheduling is performed using the distributed resource allocation schema.

3.3 Re-scheduling for Delays

An operation delay can occur after a disturbance, when the operational holon can not fulfil the scheduled due date of an operation. In this situation, the operational holon notifies the task holon about the delay, proposing a new date. The decision about the acceptance of the operation delay is dependent of the actual state of the operation. If the operation is already in execution, this notification is seeing as a warning of delay, being necessary to re-schedule all the posterior operations affected by the delay. If the operation is waiting for the execution, the task holon can try to find alternative resources to allocate the operation by asking other operational holons about their capacity to execute the operation.

Based in the proposals sent by the operational holons and in the estimated delay, the task holon decides if accepts the proposal for the estimated delay or changes the allocation to another operational holon. In this last case, the operational holon removes the operation from its local schedule, and triggers a local scheduling optimization. Additionally, the task holon re-schedules the posterior operations, adjusting their scheduled start and due dates.

3.4 Re-scheduling for New Rush Orders

A rush production order is an order, usually of high priority, that arrives to the system and must be processed immediately, since it has a near due date. As the schedule plans are elaborated periodically by the supervisor holons, the treatment of these kinds of orders causes a disturbance in the system.

In this situation, the rush task holon interacts directly with the operational holons to allocate it, using the distributed resource allocation mechanism. The problem appears if the rush production order has to be executed in a time window already occupied by other tasks, which requires a special negotiation to relax the other operations and to introduce the new ones.

Since each order has associated a priority, operational holons take this information in consideration. In the case that the rush order has maximal priority, i.e. that must be executed as soon as the current operation is completed, the operational holon tries to re-schedule, relaxing the operations that have minimal priority, i.e. those operations that can be delayed beyond the due date, to find capacity to execute the rush operation.

In case of impossibility to find capacity to allocate the rush production order, the task holon needs to negotiate with the task holons that have operations allocated to the resources occupying the requested time window to execute the rush order, trying to convince them to release some time window. This mechanism is based in the trade of credits units for rewards and penalties. The task holon can use the penalty value, that it will pay if does not fulfil the due date, to manage the problem of finding a time window to execute the rush operations.

In the negotiation process, the rush task holon interacts with other task holons, requesting the desired time window and offering a reward. Each one of the other task holons analyse the offer and in case that the reward covers the penalty that the task holon may to pay for the delay, it accepts; otherwise, it rejects the proposal. In case that all task holons reject the reward, the rush task holon must increase the reward value and make another offer. The task holon should repeat this procedure until one task holon accepts the offer or the offered reward value reaches the maximum value, which is equal to the penalty to be paid in case of delay.

In case that one task holon accepts the offer, it will notify the operational holons to decrease the priority to free the time window. In parallel, the rush task holon announces again the production order to the operational holons.

3.5 Re-scheduling for Optimization

After the execution of event-based re-scheduling, performed in a distributed manner, it is necessary to synchronise and optimise the elaborated schedule. The synchronization is required because supervisor holons don't know what kind of schedule was achieved during the distributed re-scheduling. For this purpose, lower-level operational holons notify the supervisor holon about its new schedule plan.

Since the current schedule was achieved in a fast but not optimised way, supervisor holons starts the optimization of the re-schedule plan achieved using the distributed scheduling schema. The elaboration of this optimized re-scheduling is performed in background and does not consider the operations included in a safe time window, as illustrated in the Figure 3.

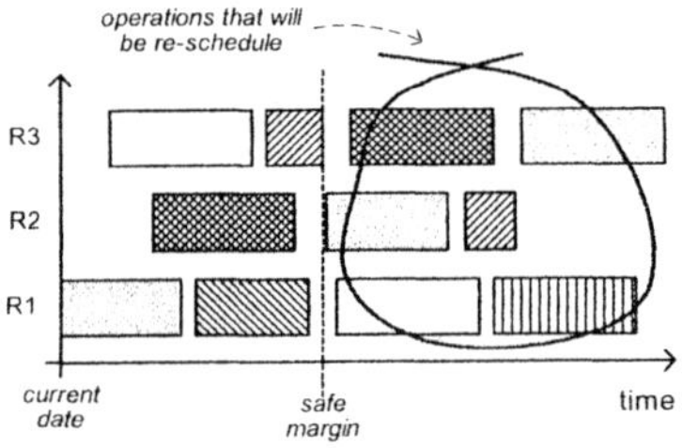

Figure 3 – Optimized Re-Scheduling after the Distributed Scheduling

This safe time window guarantees that the current schedule plan can be executed in the factory plant during the elaboration of the optimized re-schedule and is defined according the estimated time required to optimize the schedule plan.

4. PROTOTYPE IMPLEMENTATION

The proposed holonic manufacturing scheduling approach was implemented in a prototype, using agent technology, namely the JADE (Java Agent Development Framework) framework. All three types of designed holons were implemented being the communication between them performed encoding messages according the FIPA-ACL (Agent Communication Language) communication language.

The decision component embedded in an ADACOR holon uses a rule-based system, applying declarative knowledge expressed in terms of rules, to regulate the holon's behaviour. For this purpose, it is used the JESS (Java Expert System Shell) rule-oriented programming infrastructure. The decision component also uses procedural knowledge, embodied in procedures that are triggered as actions by some rules. The scheduling algorithm is an example of this type of knowledge.

In the prototype, the scheduling mechanism embedded in the supervisor and operational holon uses simple algorithms that guarantees rapid and reliable scheduling. As the ADACOR architecture is built upon functional blocks, similar to Lego® components, these scheduling algorithms can be easily modified in the future, by plugging more powerful scheduling algorithms. In a similar way, it was developed the required mechanisms to implement the distributed scheduling. For this purpose, it was implemented the mechanisms for the propagation of re-organization using ant-based techniques and the factor of autonomy.

Figure 4 illustrates the system prototype for the flexible manufacturing system from CIM Centre of Porto, described in [11]. It shows graphically the optimized schedule elaborated by a supervisor holon and the local schedule performed by an operational holon.

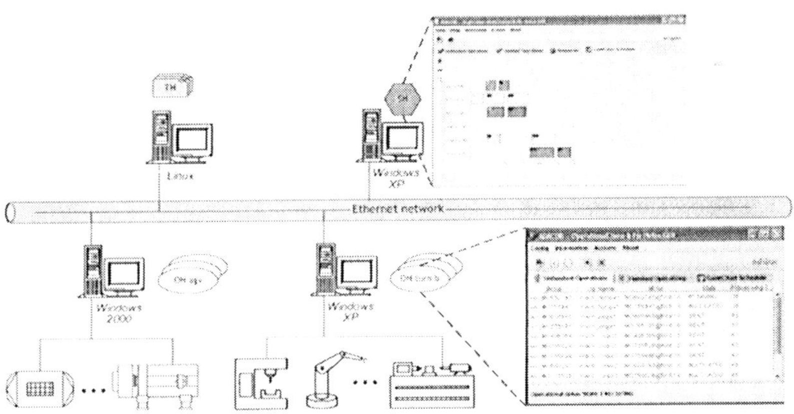

Figure 4 – The Central and Local Schedules in the Distributed ADACOR Entities

The prototype operation, showed in a first instance, the correctness and applicability of ADACOR control system, and particularity the holonic scheduling. It was also proved that the proposed holonic scheduling approach presents fast responsiveness and better flexibility, scalability and robustness, particularly for unexpected situations. A set of experimental results were presented in [11], where this approach was evaluated and compared with other two different control approaches, namely hierarchical-like and heterarchical-like ones. The results showed that the proposed approach has potential to improve the system performance, mainly combining agility and global production optimization. The adaptive and holonic control and scheduling approach plays a crucial role to support this performance.

5. CONCLUSIONS

Manufacturing scheduling is traditionally elaborated in a centralized manner and doesn't consider the dynamic re-scheduling. This paper presented a holonic approach to dynamic manufacturing scheduling addressing the improvement of the fast re-scheduling maintaining the global optimization. The architecture is based in the following main foundations:

- Distributed approach, with decision-making distributed by a community of autonomous entities, each one having partial knowledge about the problem.
- In normal operation the scheduling is achieved in a central manner, using coordination entities to achieve optimization, and in turbulent operation, the scheduling is elaborated in a distributed manner aiming fast re-scheduling.
- The dynamic adaptive mechanism allows the evolution of the overall system in order to combine centralised and distributed scheduling strategies.

At this stage, the objective is not to have complex scheduling algorithms but to achieve fast re-scheduling combined with global optimization, using simple local scheduling algorithms embedded in the holons. In further work, the embedded local scheduling mechanisms will be improved in order to achieve high quality scheduling in a timely fashion.

6. REFERENCES

1. Shen, W. and Norrie, D., An Agent-based Approach Dynamic Manufacturing Scheduling. Workshop Notes of the Agent-based Manufacturing Workshop at Autonomous Agents, 1998.
2. Bongaerts, L., Integration of Scheduling and Control in Holonic Manufacturing Systems. PhD thesis, Katholieke Universiteit Leuven, Belgium, 1998.
3. Sousa, P. and Ramos, C., A Distributed Architecture and Negotiation Protocol for Scheduling in Manufacturing Systems. Computers in Industry, 38(2), 1999, pp. 103–113.
4. Gou, L., Luh, P. and Kyoya, Y., Holonic Manufacturing Scheduling: Architecture, Cooperation Mechanism and Implementation. Computers in Industry, 37, 1998, pp. 213-231.
5. Markus, A., Vancza, T. K., and Monostori, L. A Market Approach to Holonic Manufacturing. Annals of CIRP, 45, 1996, pp. 433–436.
6. Sugimura N., Hiroi M., Moriwaki T. and Hozumi K., A Study on Holonic Scheduling for Manufacturing System of Composite Parts. In Japan/USA Symp. on Flexible Manufacturing, 1996.
7. Hino, R. and Moriwaki, T. Decentralized Scheduling in Agent Manufacturing System. Proceedings of the 2nd International Workshop on Intelligent Manufacturing Systems, Belgium, 1999, pp. 41-47.
8. Logie, S., Sabaz, D. and Gruver, W. A. Sliding Window Distributed Combinatorial Scheduling using JADE. Proceedings of the 2004 IEEE International Conference on Systems, Man and Cybernetics Netherlands, 2004, pp. 1984-1989.
9. Leitão, P. and Restivo, F., ADACOR: A Holonic Architecture for Agile and Adaptive Manufacturing Control. Computers in Industry, 57 (2), 2006, Elsevier, pp. 121-130.
10. Smith, R., Contract Net Protocol: High-Level Communication and Control in a Distributed Solver. IEEE Transactions on Computers, C-29(12), 1980, pp. 1104-1113.
11. Leitão, P. and Restivo, F., Experimental Validation of ADACOR Holonic Control System. In V. Marík, R. Brennan and M. Pechoucek (eds.), Holonic and Multi-Agent Systems for Manufacturing, LNAI 3593, Springer-Verlag, 2005, pp. 121-132.

6 TOWARD COLLABORATIVE SCHEDULING

Ana Almeida, Goreti Marreiros
Knowledge Engineering and Decision Support Research Group - GECAD,
Computer Science Department
Institute of Engineering – Polytechnic of Porto
Porto, Portugal
{ana, goreti}@dei.isep.ipp.pt

The scheduling process, usually involve the evaluation and selection of one alternative between a set of them. These decisions are not trivial, considering that they usually involve multiple, and sometimes conflicting, criteria. Particularly in scheduling which aim is to find the trade off between loading efficiency and delivery accuracy taking into account holding costs, tardiness penalties and expedition charges. Scheduling decisions should be taken in respect with the result of the integration of different criteria weighted according the several perspectives from manufacturing environment namely, production, commercial, and quality. So, scheduling is a multi-criteria decision problem; in practice different schedulers may agree as to the key objectives but differ greatly as to their relative importance in any given situation. The purpose of this paper is to address collaborative scheduling in complex dynamic manufacturing environment, presenting a collaborative scheduling approach which considers group decision support.

1. INTRODUCTION

Actual industrial environments deal with the globalization of the markets which caused a change in organizational perspective. Within the current scenario the organization tend to evolve to an organic system instead of a hierarchical one. The traditional hierarchical structure, functionality divided, is perceived as too rigid and slow, making it difficult to adapt quickly to changing environment requirements. There is a need to evolve to distributed structures with an horizontal functionality, where different functions collaborate in different processes.

Group decision-making represents an important role in actual organizations. The new economy demands that the decisions must be taken quickly however without damaging the quality of the decision-making process or its results. With the objective of making better decisions, more and more times decisions are taken by groups of individuals representing different organization perspectives, or even different organizations.

In the management field, most of the decisions we make involve the need to consult several persons discuss and argue for alternatives. For this reason Group

Please use the following format when citing this chapter:

Almeida, A., Marreiros, G., 2006, in IFIP International Federation for Information Processing, Volume 220, Information Technology for Balanced Manufacturing Systems, ed. Shen, W., (Boston: Springer), pp. 47–56.

Decision Support Systems emerged as a very important area in the domain of Management, providing Collaborative Frameworks for Decision Support. Scheduling is one of the areas where this kind of systems make more sense, namely due to the complexity and dynamic nature of the environment, and the different kind of actors involved on the process.

Scheduling decisions are often characterized by goals, roles, activities and resources that are dynamically changing, or uncertain. For improve competitiveness scheduling decisions should arise from the integration of the different production functions where each participating actor should collaborate in achieving a solution.

This paper addresses the interaction between the scheduling actors through the integration of the different kinds of knowledge in a global view of the system and the potential synergy in association with the collaborative activity of those actors taking in account multiple criteria which can improve the scheduling process. Considering this fact the option for a collaborative approach using the concept of Group Decision Support System (GDSS) plays an important role.

This paper is organized as follows. Section 2 provides a general approach to the scheduling process, focusing some scheduling techniques and Collaborative Scheduling. In Section 3 is discussed the relation between group decision and scheduling and group decision support systems are presented. The proposed approach to collaborative scheduling through group decision support and its main features are described in Section 4. Finally Section 5 presents some conclusions.

2. SCHEDULING

Scheduling problems are among the most important in operations management because they impact the ability of manufactures to meet costumers' demands and make profit. The problem of scheduling can be considered as the allocation of a set of tasks to a limited number of resources, with an objective of satisfying constraints and a set of criteria (Baker, 1974).

Over the last fifty years a great deal of research has been focused on solving the scheduling problem, resulting in a wide variety of approaches. So, in the literature it is possible to find many contributions trying to solve the scheduling problem. Those contributions can be classified according to the nature of the obtained results or the used resolution approach. In respect to the nature of the obtained solution we can distinguish between optimization or approximation methods. According to the resolution approach there are individual and collaborative processes. We will focus on collaborative scheduling.

2.1 Optimization Techniques

In optimization strategies there are the so called efficient methods, which solve a given problem optimally. The complexity of these methods increases polynomially with respect to the size of the input, because all possible solutions are considered. These methods build an optimum solution from the problem data by following a simple set of rules which determine exactly the processing order. French (1982) predicts that no efficient algorithm will ever be developed for the majority of scheduling problems.

Other kind of optimization strategies that find efficient solutions are the enumerative methods which generate schedules one by one using clever elimination procedures to verify if the non optimality of one schedule implies the non optimality of many others which were not yet generated, thereby preventing the need to search the complete space of feasible solutions. Here it can bc included mathematical programming techniques, like the Mixed Integer linear Programming (Manne, 1960), Lagrangian Relaxation (LR) approaches (Fisher, 1973) and decomposition methods (Ashour, 1967), or Branch and Bound techniques (Lawler, 1973) where a dynamically constructed tree representing the solution space of all feasible schedules is implicitly searched. The great formulation difficulty and the excessive computational effort required can be partially suppressed through problems decomposition and introduction of relaxations in order to simplify the initial problem. Although these strategies are of tremendous theoretical value, the majority of them is unable to achieve feasible solutions to many problems and therefore is of limited practical use (Jain, 1999).

2.2 Approximation Techniques

Approximation techniques do not guarantee achieving exact solutions, but they are able to attain satisfactory, near optimal, solutions with a reduced amount of computational effort, therefore they are more suitable for larger problems. Here we can distinguish between constructive methods which build a complete solution from scratch using the problem data, or iterative methods which can modify one complete solution to another by continually reordering the sequence of operations.

On the first group it can be included some techniques that had a great success in scheduling problems resolution, due to their ease of implementation and reduced computational requirement (French, 1982), like Priority Dispatching Rules (Conway et al., 1967); and Bottleneck based heuristics, namely the Shifting Bottleneck procedure (Adams, 1988).

On the second group are the Local Search or Neighbourhood techniques (Aarts, 1997) which philosophy is a spatial search for valid better solutions within a certain amount of time. Some examples are: Tabu Search (Glover, 1989); Simulated Annealing (Kirkpatrick et al., 1983); and Genetic Algorithms (Goldberg, 1988). Some other strategies have been used like Constraint Satisfaction techniques (Fox, 1987) which aim reducing the effective size of the search space by applying constraints that restrict the order in which operations are sequenced and start times are assigned to each operation.

2.3 Collaborative Scheduling

Through complexity and fashion how production scheduling problems were tackled in the past, we can actually conclude that a great disparity exists between the way that scheduling systems solve problems and the way human resolves them. While automatic-scheduling systems need complete specification of goals and scenario before beginning problem resolution, persons progressively learn with scenario and change their goals during planning and execution. Besides that, there are different evaluation criteria, which are many times contradictory, arising from the diverse manufacturing perspectives involved in the scheduling process. Furthermore, actual industrial environments are often geographically dispersed; in the scheduling environment unplanned events occur frequently requiring scheduling decisions to be taken constantly.

There is the need of a diverse range of technical capabilities, usually representing different manufacturing perspectives to work together, sharing their knowledge through a collaboration process to arise to a global scheduling solution.

The determination of feasible and mutually-acceptable schedules can be a major challenge (Kowalczyk et al., 2004). Moreover, automated scheduling methods, whatever their nature is, might not produce realistic schedules in environments where contextual information is inadequately represented, objectives are complex and unstated, and situations are dynamic and uncertain. Such issues can be addressed by domain experts (Wiers, 1997; Almeida et al, 2002). So, under these circumstances, human schedulers bring to the scheduling process their inductive and pattern recognition abilities. This problem evidenced the necessity to create collaborative scheduling systems, where a group of users and scheduling engines collaborate in plans generation, identifying candidate alternatives, and selection one of them, thus profiting the better of the two worlds. This form of collaboration provides a very powerful approach to multi-attribute, multi-criteria, decision support in complex manufacturing environments.

There are some works on distributed, cooperative or collaborative scheduling. An example is the work developed by Kawamura and his colleagues (Kawamura, 2000) which is a distributed cooperative scheduling system, where several scheduling agents negotiate among them to realize schedule adjustments among busy departments. Another different approach is presented by Murthy and his colleagues (Murthy, 1997) where autonomous agents work together to produce a set of candidate alternatives, and a human scheduler make the final decision interacting with the other agents.

3. GROUP DECISION

3.1 Scheduling and Group Decision

One approach to tackle multi-criteria decision problems involves assigning weights to different criteria, aiming to come to a unique decision depending upon the assigned weights. In a collaborative decision making process, which frequently involves many people, experts on different aspects of the problem, all the relationships arising from the different departments representing the diverse manufacturing perspectives must be considered, so a set of weighted criteria seems to be the most adequate. For instance, from the manager point of view the most important criterion should be the profit and from a quality control department the most important criterion is product quality. But as economic conditions change, the relative importance of different criteria may change (Murthy, 1997). This requires users to modify these weighting factors periodically, by changing the relative importance of each criterion.

Architectures that enable collaboration are useful when it is not efficient or possible to perform a task by a single agent or human. They provide mechanisms which allow several users to contribute with their knowledge to the system, participating on an equal basis in the selection of candidate alternatives. So a collaborative framework capable of integrate multi-criteria decisions, arising from the different actors involved in the manufacturing process, can be most adequate.

One way of enhancing collaboration between agents and humans is to produce not one but many candidate solutions, evaluated with respect to multiple criteria. This allows users to gain important insights into the tradeoffs between multiple competing objectives. They express their preferences by imposing weighting factors for different criteria.

The benefits of group work are several: Groups are better than individuals at understanding problems; People are more responsible for decisions in which they participate; Groups are better than an individual participant at detecting flaws in proposed ideas; A group has more knowledge than any one member individually; Synergy may develop so that the effectiveness of the group is greater than what could have been produced individually; Working in a group could stimulate the group members and consequently the process of decision making; Participants' differing knowledge and processing skills allow results that could not be achieved individually.

If the there are big advantages associated to group work, there are also several dysfunctions related to this theme: Time consuming; High costs; Improper use of group dynamics; Tendency to rely on some members the most of the work; Tendency to make incomplete tasks analysis and to choose compromise solutions of poor quality.

3.2 Group Decision Support Systems

The Group Decision Support Systems (GDSS) aims to reduce the loss associated to group work and to maintain or improve the gain (Holsaple, 2001).

The term GDSS emerged effectively in the beginning of the eighty-decade. According to Huber (1984) a GDSS consists of a set of software, hardware, languages components and procedures that support a group of people engaged in a decision related meeting. A more recent definition is from Nunamaker et al. (1997) and says that GDSSs are interactive computer-based environment which support concerted and coordinated team effort towards completion of joint tasks.

Some of the advantages of using GDSS are: Equal and anonymous opportunity to contribute with ideas and opinions; Allows parallel communication between group members; Helps the facilitator of the meting in the schedule management; Eliminates too big domination of some group members in the meeting; Provides automatically organizational memory; Makes it possible, to find out the common and dissenting preferences among the group members.

In the 80's most of the research in the GDSS area was focused in the synchronous/same-place dimension, several decision rooms were configured. In the last years, with the proliferation of Internet the research on GDSS has its focus on the different-time/different place dimension. Several web-based GDSS have been developed (Marreiros et al, 2004; Dennis et al, 1996), and others like for instance the GroupSystems, that initially were developed just to support configuration of decision room type, are now able to support remote decision making.

As it was referred on introduction, the scheduling process, involve the evaluation and selection of one alternative between a set of them. This are not trivial decisions, because they usually involve multiple and conflicting criteria. Actual organizations are dispersed around the world. GDSS seems adequate to support the scheduling process.

4. COLLABORATIVE SCHEDULING APPROACH

A scheduling system should provide a user support, to assist him in build, change and revising processes of the scheduling plans and not deciding for him. The user has intelligence and knowledge acquired along the years that are not to underestimate. Nevertheless, a scheduling system could have autonomous capacity, suggesting alternatives according to some claimed criteria.

The user provides intuition, a notion about goals and appropriate trade-off, and refined problem resolution strategies. The computer provides skill to manage details, to assign and schedule resources and operations, and to analyze quantitatively the suggested choices.

This approach considers multiple scheduling objectives in a global multi-criteria collaborative framework. It generates several scheduling alternatives by using autonomous agents which encapsulates different scheduling algorithms. Each scheduling alternative represents a solution regarding an objective such as, accomplishment of deadlines, minimizing throughput times, maximizing profitability, product quality, and minimizing manufacturing disruptions. It provides decision support considering the negotiation process of a group of users, each one of them with a different perception of the problem, effectively acting as a team to achieve a common and unique solution.

The proposed approach, represented in Figure 1, includes two modules; the Scheduling Module responsible for the creation of a set of scheduling solutions and the Group Decision Support Module which is in charge of the selection of a scheduling solution.

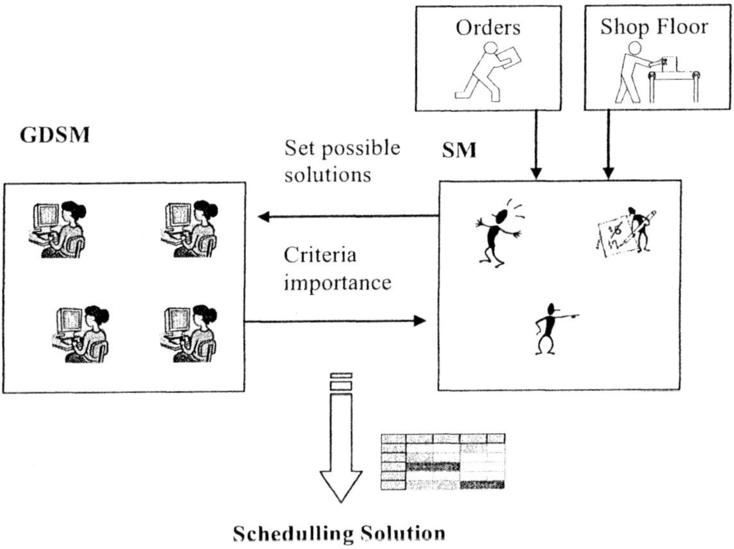

Figure 1 - Approach to collaborative scheduling through group decision support

4.1 Scheduling module

Agents and multi-agent systems are an important area of research and development in Artificial Intelligence. Agents are typically knowledge representation entities characterized by independence and autonomy. The software agents are entities that have the ability to plan, to establish their actions ahead of time, to develop appropriated problems solving strategies, to communicate, or to share resources.
In this approach agents have the possibility to follow events as they occur in the environment, interpreting and sharing knowledge or data.

The Scheduling Module (SM) consists of multiple problem-solving methods (called agents) working at the same time on a common problem, so it does not represent any single method or heuristic, but is rather an attempt to use multiple techniques by encapsulating individual algorithms as autonomous agents.

The scheduling agents are different autonomous agents each one of them embodying a particular scheduling algorithm, as it was referred previously. According the scheduling criteria or objective, there is a broad range and variety of scheduling methods. One can apply any of the techniques referred on section 2, or any other algorithm or heuristic.

In this approach the information agent, in accordance with the type of scheduling problem, sets a time window for the generation of the several scheduling alternatives. Also some criteria are settled; this way just the agents embodying algorithms respecting the established criteria will be triggered. Only the alternatives generated within the settled time window are considered for analyse and discussion by the GDSM module.

The setting agent settle on the criteria importance according with the global preferences of the GDSM members, in order to cover all the relationships arising from the different departments.

4.2 Group Decision Support module

The Group Decision Support Module (GDSM) will support the members of a scheduling meeting and the facilitator. This last one prepares the meeting and invites a group of people to participate, and to exchange different points of view, expertise and information, in order to choice the "best" solution from the set of scheduling solutions proposed by the SM. After the generation of a set of alternative solutions, by the SM, the group members will use this module to individually choose the preferred one.

The GDSM is composed by the following components: Setup, Management, Argumentation, Multi-criteria, Voting and Database (Marreiros et al, 2004).

The facilitator of the meeting is supported by the GDSM in the execution of several activities: General schedule meeting configuration (specifying dates and time for the start and the end, as well as the goal of the meeting); Selection of participants, invitation sending, confirmation of reception and participant replacement, if necessary; Definition of the member importance (number of votes); Definition of decision rules (voting rules: consensus, majority, qualified majority, maximum number of voting cycles, anonymity; rules for argumentation: minimum/maximum number of arguments for each participant; the arguments are or are not visible to all the participants)

The participants of a group decision scheduling module are also supported by the GDSM: indication of the set of criteria; establishing individual preferences (will help the individual scheduler agent to ranking the several scheduling alternatives); support the exchange of arguments between group members (each participant will therefore argue for the most interesting alternatives or against the worst alternatives, according to his/her preferences). By expressing their arguments, participants expect to influence the others' opinions and make them change their choices.

5. CONCLUSIONS

The attention to the individual needs of each customer is a driving force behind many changes taking place in every industry sectors, so a scheduling decision, must take into account the knowledge and experience of different points of view, allowing the consideration of broad issues of the company rather than focusing on scheduling tasks for a single process. In a collaborative architecture multiple solutions can be generated and shared with the scheduling responsible for different stages of production and with customer service representatives, serving as basis for negotiating a set of solutions to come to the best alternative for the enterprise.

The presented approach addresses many of the limitations of existing job-shop scheduling systems, by providing a decision support collaborative framework for scheduling. It allows the evaluation of scheduling solutions by experts with different points of view, facilitating the incorporation of their knowledge into the system.

Improving schedule quality, lead to delivery accuracy, improved quality and fast time to market which results on a increase of competitiveness and profit. However, just as important are the opportunities to change business processes that the presented approach affords by showing multiple good alternatives, some of which may violate some constraints. These solutions can suggest good opportunities to look for alternative means of production and can suggest when it would be profitable to negotiate to change customer requirements or business policies. For instance if the deadlines imposed by a particular costumer are always tiny and can never be accomplished, it may indicate that it would be better to negotiate longer deadlines. So, significant savings and improved customer satisfaction are expected.

6. ACKNOWLEDGEMENTS

The authors would like to acknowledge FCT, FEDER, POCTI, POSI, POCI and POSC for their support to R&D Projects and GECAD Unit.

7. REFERENCES

1. Aarts, E. and Lenstra, J. K. Local Search in Combinatorial Optimization, Wiley, Chichester, 1997
2. Adams, J., Egon B. and Zawack, D. The Shifting Bottleneck Procedure for Job Shop Scheduling. Management Science, Vol. 34, nº 3 Março, USA, 1988.
3. Almeida, A; C. Ramos and C. Sílvio. Toward Dynamic Scheduling of Manufacturing. The International Journal for Manufacturing Science & Production, Freund Publishing House, 2002.
4. Ashour, S. A Decomposition Approach for the Machine Scheduling Problem, International Journal of Production Research 6(2), 109-122, 1967.
5. Baker, K. R. Introduction to Sequencing and Scheduling, John Wiley & Sons, 1974.
6. Conway, R.et al. Theory of Scheduling, Addison-Wesley, Reading Massachusetts, 1967.
7. Dennis, P. et al. Using Internet to implement support for distributed decision Making. In Chapman & Hall, UK, pp. 139-159, 1996.
8. Fisher, M. L., Optimal solution of scheduling problems using Lagrange multipliers: part I. Operations Research 21, 1114-1127, 1973
9. Fox, M. S. Constraint - Directed Search: A Case Study of Job-Shop Scheduling, Research Notes in Artificial Intelligence, Pitman Publishing, London, 1987
10. French, S. Sequencing and Scheduling - An Introduction to the Mathematics of the Job-Shop, Ellis Horwood, John-Wiley & Sons, New York, 1982.
11. Glover, F .Tabu search – Part I", ORSA Journal on Computing, Vol. nº1, pp. 190-206, 1989.
12. Goldberg, D. Genetic Algorithms in Search Optimization and Machine Learning. Menlo Park, California, Addison-Wesley, 1988.
13. Holsapple, C.; Whinston, AB. Decision support systems: a knowledge-based approach. Thomson Learning, inc, 2001
14. Huber, G. Issues in the design of group decision support systems. Mis Quarteley, 3(8), 1984.
15. Jain, A. S. and Meeran, S. Deterministic job-shop scheduling: Past, present and future. European Journal of Operational Research 113 390-434, 1999
16. Kawamura, T et al. Development of a Distributed Cooperative Scheduling System Based on Negotiations between Scheduling Agents, In Systems and Computers in Japan, vol. 31, nº1, 2000.
17. Kirkpatrick, S.;Jr. C. Gellat and M. Vecchi. Optimisation by Simulated Annealing", Science, 220 (4598) pp. 671-680, 1983
18. Kowalczyk, R.et al..Towards supporting collaborative scheduling in adaptive supply networks with negotiation agents. Proc. Theory and Applications of e-Negotiations, Poznan, Poland, 2004.

19. Lawler, E. L. Optimal Sequencing of a Single Machine Subject to Precedence Constraints, Management Science, vol 19, 544-546, 1973.
20. Manne, A. S. On the Job-Shop Scheduling Problem, Operations Research, vol 8, 219-223,1960.
21. Marreiros, J. P. Sousa and C. Ramos, "WebMeeting - a group decision support system for multi-criteria decision problems". ICKEDS04, 2004.
22. Murthy, S. et al. Agent-based cooperative scheduling, In AAAI97 Workshop on Constraints and Agents., pages 112--117, 1997
23. Nunamaker, J.F. et al. Lessons from a dozen years of group support systems research: A discussion of lab and field findings, Journal of Management Information Systems, 13 (3), 1997.
24. Wiers, V. Human–Computer Interaction in Production Scheduling: Analysis and Design of Decision Support Systems for Production Scheduling Tasks, Ph.D. Thesis, Eindhoven University of Technology, Holland, 1997.

7

A FRAMEWORK OF AGENT-BASED SUPPLY CHAIN PERFORMANCE ANALYSIS SYSTEM

Guannan Qu, Zhiyi Fang
School of Computer Science and Technology, Jilin University
Changchun, 130012, P.R China
qgn_0316@163.com, zyfang@public.cc.jl.cn

Chao Zhang
College of Economy and Information, Jilin University
Changchun, 130012, P.R China
zhangchao_m47@sina.com

Supply chain management is a popular trend for intensive global competition, and the performance analysis plays an important role in improving supply chain. This paper introduces the basic of the supply chain management and performance analysis. We emphasize the importance of building the performance analysis system to evaluate the entire supply chain. Via reviewing the characteristics of supply chain and challenges in supply chain performance analysis, we design a framework of performance analysis system to meet the challenges in supply chain performance evaluation, and then give an application in this paper.

1. INTRODUCTION

Intensive global competition, faster product development, increasingly flexible manufacturing system, an unprecedented number and variety of products are the characteristics of today's global market (Simichi et al., 2000). Supply Chain Management, as a new mode of management, is becoming a more and more challenging task in such an uncertain world. Lambert and Cooper (2000) defined the Supply Chain Management (SCM) as the management of relationships across a supply chain to capture the synergy of intra- and inter-company business process which referred to as supply chain management.

Performance analysis for the supply chain is an important part of the SCM (Supply Chain Management), and it mainly aims at (Xu et al., 2004): (1) carrying on the appraisal to the performance of the entire supply chain, which provides the proof of making the strategy for core enterprise; (2) evaluating the business process, and bottleneck in the supply chain which will be found and conquered; (3) examining the

Please use the following format when citing this chapter:

Qu, G., Fang, Z., Zhang, C., 2006, in IFIP International Federation for Information Processing, Volume 220, Information Technology for Balanced Manufacturing Systems, ed. Shen, W., (Boston: Springer), pp. 57–66.

node in supply chain, and presenting the status of the node corporations, which can help to establish more reasonable symbiosis.

Currently, the design and development of supply chain performance analysis system exhibits two trends. First, some systems put the performance evaluation as a choice into the related part, such as sale performance, which is taken as a part of sale management module in supply chain. It does a good job on collecting information in time but against adapting diverse performance metrics, and it fails in building multi-criteria. The second trend is to build unattached performance system while ignoring its context--supply chain, which facilitates rebuilding and updating performance measures, yet makes it more difficult to gather dynamic data.

In this work, we combine these two trends and explore the performance analysis system, which correlates with supply chain management system. Although, it is autonomic, namely, the operation of managing criteria and calculating is independent, however, the information that is used to do analysis should be collected from the in time running of supply chain through communicating between agents, and the analysis result impacts the SCM structure and configuration directly so that adjustment and optimization of the SC could be achieved.

This rest of the paper is organized as follows: in Section 2, we analyze the characteristics of supply chain and challenges in supply chain performance analysis. The framework of the agent-based performance analysis system will be illustrated in Section 3. In Section 4, we use this framework in an application and demonstrate it by an example.

2. CHALLENGE OF CONSTRUCTING PERFORMANCE ANALYSIS SYSTEM

Historically, Enterprises have focused only on their resources, constrains, and policies to make decisions and reduce costs. With intense competition and reducing profit margins, this approach is no longer sufficient. They need to consider the interactions with their suppliers and customers and incorporate them into their decision-making process (Julka et al., 2002). The aim of performance analysis is analyzing and evaluating the process and the outcome of the entire supply chain using appropriate approaches.

In this situation, there are several challenges in constructing supply chain management system and its performance analysis system. The first challenge is from the characteristics of supply chain, since it is the information source of performance analysis. Secondly, how to evaluate the performance of entire supply chain comprehensively should be taken into consideration. Finally, the performance analysis system should adapt to different performance measures designing and continuously updating. These are now described in turn.

2.1 Characteristics of Supply Chain Elements

We identify the elements in a supply chain, their features, and the challenges associated with SCM, and these are the foundation of modeling the performance analysis system (Chai and Liu, 2001). The elements in supply chain can be classified as entities and flows. Entities include all manufacturers, logistics providers, electronic exchanges and all their internal departments that participate in the business process. These entities are essentially the operators in the supply chain. Flows are of three types – material, information and finance, and these are the operands in the supply chain. These entities have four common features:

1) Dynamic: The supply chains are more flexible now. In today's business environment, there are no obligations for companies to be part of supply chain for certain time period and they may join or leave based on their own interest. This changes the structure and flows in the supply chain. Information in the supply chain, such as prices, demands, technologies, etc, is also changing continuously.

2) Distributed: The elements are distributed across various geographical locations. The planning and operating systems used by an entity may also be geographically distributed. For example, there may be a dedicated inventory database residing at each warehouse of a manufacturer. The SCM related information might even reside as rules-of-thumb with the people responsible for performing the various tasks in the business process.

3) Disparate: The entities in a supply chain use different systems built on different platforms for planning and management of their business. Information pertaining to the various elements is also disparate in form.

4) Cooperative: Cooperating closely between the enterprises or enterprise interior is emphasize in supply chain, e.g. a certain affair possibly involves to design, manufacture, purchase, transportation and so on, and these need the relevant units' cooperating inevitably.

Unfortunately, information needed by performance analysis sourced from these dynamic, distributed and disparate entities and flows, passing through the actions and the reached results.

2.2 Comprehensive performance evaluation

Entities in supply chain will make their own decision, which are usually based on their own constrains. Most often, the decisions are taken independent of the interdependency factors contributing to the overall business process of the enterprise. So the performances are optimized locally within the departments but do not assure a global optimum for the enterprise.

Multi-criteria is an integrated evaluation system, and it is constructed by some criteria in terms of given authority. For example: Yahya and Kingsman (1999) use Satty's ananlytic hierarchy process (AHP) method to determine priority in selecting suppliers, as show as Figure. 1 (Liu et al., 2005).

The problem is to evaluate the comprehensive performance of suppliers. On the top level is the overall goal of selection suppliers. On the second level are eight criteria that contribute to the goal. On the third level are eight criteria that are decomposed into 13 subcriteria, and on the bottom level are to be evaluated in terms of the subcriteria of the third level.

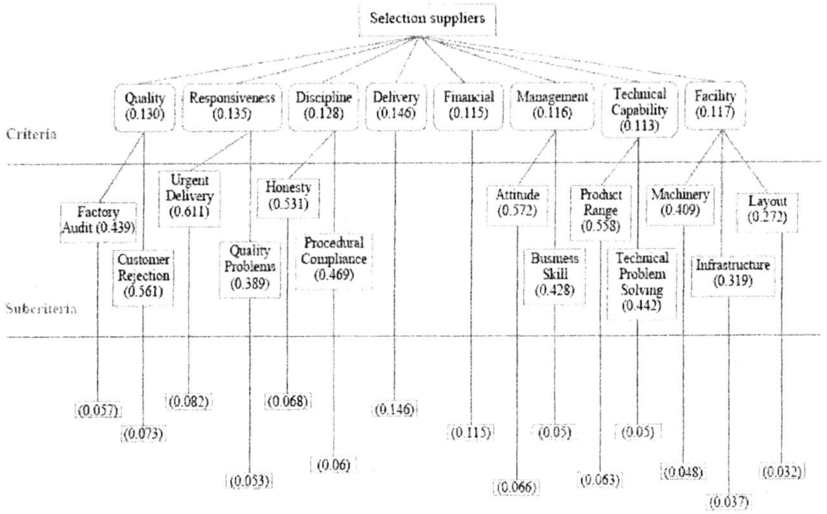

Figure 1. Hierarchy of selection suppliers

2.3 Performance Metrics

There are a large number of different types of available performance measures and a given supply chain could select appropriate performance analysis system for themselves (Beamon, 1999). Such as, for an electronic industry, which is interpreted as a fast-moving industry, how can its supply chain be measured? An appropriate performance analysis system should be selected.

On the other hand, with further study and practice on performance metrics, the issues of continuously updating the performance measures and the analysis systems also should be taken into mind.

Mike Bourne et al. (2000) summed up three phases of developing performance analysis systems:

1) The design of the performance measures;
2) The implementation of the performance measures;
3) The use of the performance measures

The design phase can be subdivided into identifying the key objectives to be measured and designing the measures themselves; implementation is defined as the phase in which data that enable the performance measures to be made regularly should be collect and process; the use of performance measures to test the analysis system itself on one side, on the other hand, it is used to analysis the performance of entire supply chain.

We have introduced the challenges of supply chain performance analysis; next section will address these critical needs by presenting a framework of agent-based supply chain performance analysis system, which is better founded on agent-based supply chain management system. This framework focuses on building a system which meets the demands of each phase of developing performance measures and making it easier to rebuild and update. It obtains information actively and timely from supply chain via agent, and simultaneously manages the criteria and multi-criteria templates independently.

3. FRAMEWORK OF THE AGENT-BASED PERFORMANCE ANALYSIS SYSTEM

In fact, performance analysis is part of supply chain management, for its information derives from the operation activities of supply chain and each entity, on the contrary, its analysis result takes effect on the supply chain.

We construct a framework of agent-based supply chain performance analysis system, Showed as Figure 2, to overcome the challenges mentioned in last section, and we will explain the main parts of this system in turn.

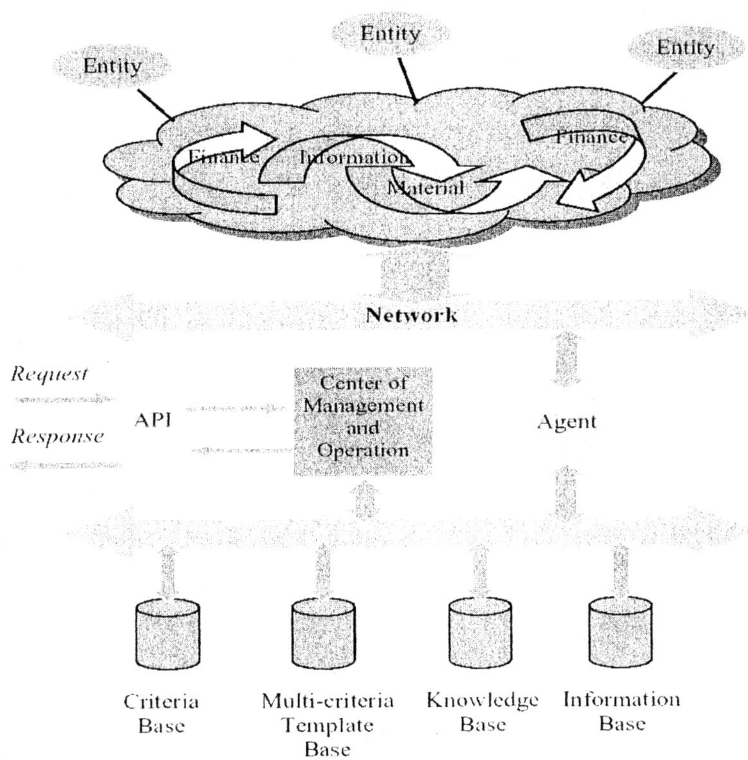

Figure 2. Framework of the agent-based performance analysis system

The whole performance analysis is base on a series of primary issues of the criteria, multi-criteria, analysis operations and criteria management, and information collection. The performance analysis of the supply chain could be reduced to several main parts as follow:

(1) Criteria

Criteria are the most elementary factors in supply chain analysis, and they are monomials. In another word, each type of criteria only appraises a given segment of the supply chain. For example, **Percentage of Repairing and Exchanging Purchase** is defined as the rate of the products that need to be returned and repaired in a given time, and it reflects quality of the products.

Managing these criteria seemed to be very simple, however, there are two issues to be settled. The first is the category of criteria. Base on the diverse evaluating methods and sustainable development of performance analysis, category of criteria may be added, deleted or modified, although it is relatively stable to a given supply chain.

The second issue is the operands of the criteria. For example, **Loyalty of Retailer** is defined as:

$$R_f = \frac{h}{m} \times 100\%$$

There are two variables in this expression – h and m. h is the number of the retailer who drops out the certain supply chain to seek new supplier in a given time; m is the sum of retailers initially. These operands are obtained from the elements of supply chain – entity and flow. Unfortunately, the entities of supply chain are dynamic, distributed and disparate. Given this reality, there is a clear need for an agent (Wooldridge and Jennings, 1994), which is capable of gathering the dynamic and disparate information required for performance analysis.

Figure 3 shows the class of criteria and its rules, and agent obtains data in terms of these rules, which are preserved in knowledge base.

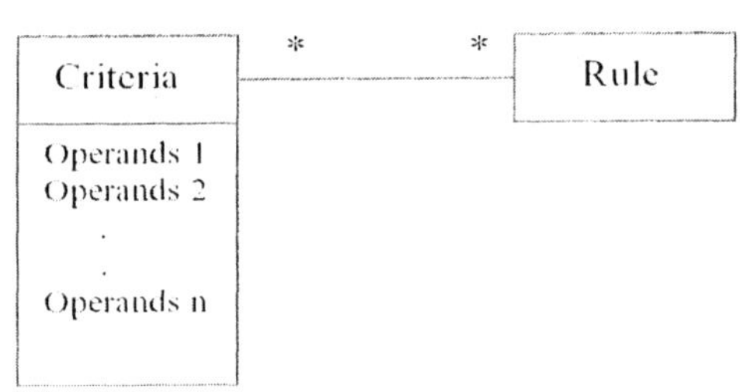

Figure 3. Class of criteria and its rules

(2) Multi-criteria

Multi-criteria is an integrated evaluation system, which is reconstructed by certain criteria in terms of given authority, according to some strategy. It is used to analyze the comprehensive result of the whole or integrant supply chain operation. The establishment of the multi-criteria needs the experts involving in this domain. Yet it is also needs to be added or cancelled flexibly.

However, we don't care how to develop the multi-criteria rather than identifying some templates of the multi-criteria as files. The experts have built these multi-criteria, and we just focus on managing and using these templates.

Figure 4 shows the relation between Multi-criteria template and criteria.

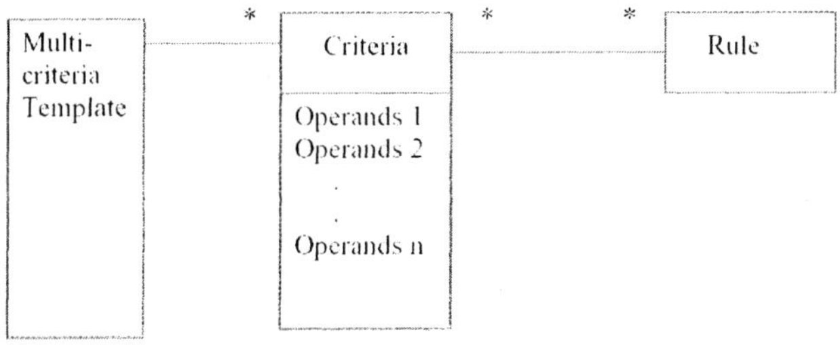

Figure 4. Relation between multi-criteria template and criteria

(3) Analysis Operations and Criteria Management

This part is the core of performance analysis. It answers for the request from the client or the application through API, and invoke other parts to accomplish analysis operation then return the result back to the relevant end. It also takes charge of managing and organizing criteria and Multi-criteria templates, make them related with knowledge, with which agent can communicate with other given agents to obtain information.

(4) Information Collection

The task of this part is simply gather data for performance analysis operation, yet it is so difficult to do. It should communicate with other business modules and gather information from the dynamic activities or results and preserved them to information base, prepared for the necessary of performance operation. Agent is a bridge of performance analysis system and other business modules.

4. APPLICATION

We use this framework of supply chain performance analysis system in the agent-based agile supply chain management system – a part of the project of APNM (Auto Parts Networked Manufacture System). JADE (Java Agent Development Framework) is used as a software framework to develop this agent-based ASCM and PAS.

4.1 JADE

JADE (Java Agent Development Framework) is a software framework to aid the development of agent applications in compliance with the FIPA 2000 specifications for interoperable intelligent multi-agent systems (Bellifemine et al., 2001), JADE is fully development in Java and it based of the principles of interoperability, uniformity and portability, easy to use and pay-as-you-go philosophy. It is an Open Source Project, and the complete system and the white paper can be downloaded from JADE Web site.

4.2 Performance Analysis System in ASCM

The agent-based supply chain management system in APNM comprises of Sale agent, Warehouse agent, Transport agent, Product agent, Purchase agent and Finance agent. Implementing an order can be divided into two phases, these are:

1) **Decision-making phase:** In this phase, decision center estimates the available resource by communicating related agents, or forecasts the performance through the Performance Analysis System, and then decide if the orders are able to be carry out or not.
2) **Implementing phase:** In this phase, the orders will be carried out, and the performance analysis system obtains information from the process by agent. Practice performance can be evaluated in this phase.

The model of agent-based ASCM with Performance Analysis System shows as Figure 5.

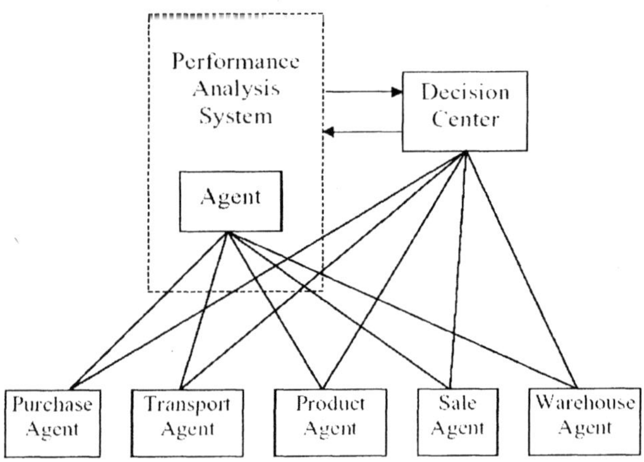

Figure 5. Performance analysis system in ASCM

4.3 Example

The process will be demonstrated by an instance – eligibility rate of material to be put in storage, which reflects the quality level of the goods provided by supplier.

Presuming Ri is the eligibility rate of the given material, items i represents certain supplier i, it needs parameters as: the gross quantity of this material to be put in storage (Yi) and quantity of disqualification among them (Di). The formula is:

$$\mathbf{Ri = (Yi - Di) / Yi * 100\%}$$

Criteria and its formulae are initialized in the Criteria Base and managed by Center of Management and Operation; meanwhile, the type of the parameters is sent to knowledge Base as a rule. Performance Agent is listening the process of the business, when the event – putting material to storage – happened, performance agent sends message to the warehouse agent according to the rules in knowledge base to ask for Yi and Di. The warehouse agent receives the request, and then sends back information with timestamp. Performance agent obtains information and reserves them to information base. It is convenient to pick up data in the given time.

While Center receives a request from client or application for eligibility rate of material to be put in storage in time segment T through API, it accepts and picks up the relevant data from Information Base in the light of its timestamp, and calculates using the formulae. Finally, the result will be returned.

5. CONCLUSION

Supply chain management has increasingly become the key strategic area which has direct impact over the success of any enterprise in today's highly competitive business environment, yet the performance analysis is the compass of the supply chain management, which evaluates the entire supply chain.

This paper has introduced supply chain performance analysis system, which is correlative with supply chain management but autonomic respectively. Criteria and multi-criteria are relatively flexible to different performance measures, so they demand autonomy. However, the data needed by performance analysis is sourced from the business of the supply chain. Information collection is vital, and agent is imported to settle this problem. We also have built the framework of the agent-based performance analysis system to meet the challenges in supply chain performance analysis, and use it in the project of Auto Parts Networked Manufacture System as an application.

6. ACKNOWLEDGEMENT

This work was supported by the project of Auto Parts Networked Manufacture System (National Development and Reform Commission People's Republic of China) and key laboratory symbolic computation and knowledge engineering of ministry of education.

7. REFERENCE

1. Beamon, B.M., "Measuring Supply Chain Performance", International Journal of Operations and Production Management, 19(3), 275-292, 1999.
2. Bellifemine, F., Poggi, A., Rimassa, G., Rimassa,G., "JADE: A FIPA2000 Compliant Agent Development Environment ", Proceedings of the fifth international conference on Autonomous agents, ACM, Montreal, Quebec, Canada, 2001.
3. Bourne, M., Mills, J., Wilcox, M., Neely, A., Platts, K., "Designing, implementing and updating performance measurement systems", International Journal of Operations & Production Management, Vol.20, No.7, 754-771, 2000.
4. Chai YT, Liu Y, Agile Supply Chain Management, Publishing House of Qinghua University, Beijing, China, 2001
5. Fox, M.S., Barbuceanu, M., Teigen, R., "Agent-oriented supply-chain management", International Journal of Flexible Manufacturing Systems, 12(2/3), 165-199, 2000.
6. Huo JZ, Evaluating Enterprise Innovation: Integrated Supply Chain Performance and Evaluation, Henan People's Publishing House, Henan, China, 2001.
7. Julka, N., Srinivasan, R., Karimi, I., "Agent-based supply chain management—1:framework", Computers and Chemical Engineering", 26, 1755-1769, 2002.
8. Lambert, D.M., Cooper, M.C., "Issues in supply chain management", Industrial Marketing Management, 29(1), 65, 2000.
9. Liu FH, Hai HL, "The voting analytic hierarchy process method for selecting supplier", International Journal of Production Economics, 97(3), 308-317, 2005.
10. Simichi, L.D., Kaminsky, P., Simichi, L.E., Designing and Managing the Supply Chain: Concepts, Strategies and Case Studies, Irwin/McGraw-Hill, Boston, 2000.
11. Suo HS, Jin YH, "The Application of Multi-Agent System in the Supply Chain Management", Computer Engineering and Application, 9-12, 2002.19, China.
12. Vrijhoef, R., Koskela, L., "The four roles of supply chain Management in construction", European Journal of Purchasing & Supply Management, 6, 169-178, 2000.
13. Wooldridge, M., Jennings, N., "Intelligent agents: theory and practice", Knowledge Engineering Review, 10(2), 155, 1994.
14. Xu Y, Chen JL, Li ZB, Wang ZX, "Study and Realization on Framework of Universal Performance Evaluation System for Supply Chain", Research for computer Application, 152-154, 2004.
15. Yahya, S., Kingsman, B., "Vendor rating for an entrepreneur development programme: A case study using the analytic hierarchy process method", Journal of Operational Research Society, 50, 916-930, 1999.
16. JADE Web Site, http://jade.tilab.com/

8

AN AGENT-BASED SIMULATION OF A JIT MATERIAL HANDLING SYSTEM

Qi Hao and Weiming Shen
Integrated Manufacturing Technologies Institute
National Research Council, Canada
800 Collip Circle, London, Ontario N6G 4X8, Canada
[qi.hao; weiming.shen]@nrc.gc.ca

Material handling is a loose loop in most assembly plants. Just-in-time (JIT) is a management philosophy that strives to eliminate sources of manufacturing waste by producing the right part in the right place at the right time. We propose to apply JIT principles to material handling in assembly plants. Material Kanbans are introduced as an effective means to control and balance the physical material/part flow in the plant floor. An agent-based simulation prototype is implemented using AnyLogic™. The flexibility of the agent-based approach facilitates the simulation of various "what-if" scenarios including different layout designs, objective parameters and dynamic situations in the plant floor.

1. INTRODUCTION

Material handling is a loose loop that is generally neglected in most production plants. From our observation, even in a well designed assembly line, in condition that the whole line is optimized in its layout, processes, buffering, scheduling, and operations, material handling is still laid outside of the scope of control. Managers spend their precious time hunting everywhere for missing parts and arranging their deliveries. They are unaware of material handling/delivery schedules and the related resource information (amount and utilization of resources, such as forklifts and drivers). As a result, material handling becomes the major barrier that results in low efficiency, production breakdowns, and low quality of a production system.

Just-in-time (JIT) is a management philosophy that could improve profits and return on investment by reducing inventory levels, reducing variability, improving product quality, reducing production and delivery lead times, and reducing other costs (such as those associated with machine setup and equipment breakdown). The pull mechanism, especially introduced by the Kanban control of JIT manufacturing, enables an optimized production process that benefits from the cutting down of production resources. For a plant that already operates under a pull mechanism, material handling should also employ a pull mechanism rather than a MRP-based push mechanism.

Please use the following format when citing this chapter:

Hao, Q., Shen, W., 2006, in IFIP International Federation for Information Processing, Volume 220, Information Technology for Balanced Manufacturing Systems, ed. Shen, W., (Boston: Springer), pp. 67–78.

This paper intends to propose a pull material handling system based on principles in JIT manufacturing. In such a system, materials transportation in the plant floor is considered as individual tasks. Material Kanban (M-Kanban) is introduced as a carrier of delivery tasks which is an effective means to control and balance the physical material handling flow in the plant floor. The main principle behind is that a task generated by a production station (cell) requires the occupation of an M-Kanban to be delivered.

Another technology used in this material handling system is agent. Agent technology is evolved from the research domain of Distributed Artificial Intelligence in 1990s. From its emergence, agent technology is widely recognized as a promising paradigm for the next generation of design and manufacturing systems (Shen et al., 2001). In the JIT material handling simulation system, multiple agents are implemented to facilitate a collaborative problem solving environment. For example, each transportation vehicle is encapsulated as an agent so that it is manageable on its own parameters and behaviors, such as velocity, local schedule, and the associated scheduling, routing and conflict resolving rules. The driver of a transportation vehicle can deactivate a vehicle from the system to take personal activities or when the vehicle malfunctions and needs a repair. Moreover, the allocation of transportation task is accomplished though the negotiation of a Kanban scheduling agent and a number of vehicle agents. With such capacities, the system is able to simulate very dynamic situations and get more accurate information of transportation resources in general.

The rest of this paper is organized as follows: Section 2 reviews the background knowledge and literature of this study; Section 3 identifies a sample JIT material handling problem and describes the corresponding requirements; Section 4 proposes an agent-based architecture of the JIT material handling system and discusses two major design aspects: production simulation and material handling simulation; finally, Section 6 draws our conclusions.

2. A TECHNOLOGY REVIEW

There are two classifications of production control systems, namely *push* and *pull*. Material requirement planning (MRP) systems and Kanban control systems are the two most popular implementations of the push and pull strategies respectively. In a push production, in order to buffer inaccurate forecasts, inaccurate lead times, inaccurate inventory records, variable production schedules or questionable bill of materials (BOMs), MRP generally incorporates safety lead times and safe stocks. However, in practice, MRP may result in a serious problem of excessive inventories (Shirk, 1998; Hopp and Spearman, 1996). Stock levels and lead times are amplified down throughout the supply chain, from the final distributor down to each hierarchy of suppliers.

In contrast, using a pull strategy, a JIT system uses underutilized capacity instead of buffer inventories to hedge against problems that may arise. Production is initiated in response to real customer orders and the removal of items from the final distributor buffers triggers production upstream to replenish exhausted inventories layer by layer. Krishnamurthy et al. (2004) quantitatively compares the performance of MRP and Kanban for a multi-stage, multi-product manufacturing system. They reached the conclusion that pull strategies are handicapped for manufacturing facilities that produce a number of different products with distinct demands and/or processing requirements, as well as for facilities that make highly engineered products in small batches (even one-of-a-kind) for their customers.

The ideal industries that JIT production applies include automobile because it is where the JIT concept originated. The automotive industry is characterized by low product variety, and high-volume production. In an automotive assembly line, although there are some sub-lines using push strategies (sometime it is called hybrid production), such as the body shop, paint shop, and engine line, however, once cars are lining up to be processed on the main assembly line, the production is under control of a pure pull mechanism. Buffers are set at offline sites of sub-lines to tickle uncertainties and better serve the optimized production rate of the main assembly line.

The concepts of JIT and Kanban are never new. JIT were firstly developed by Toyota in the 1950's and adopted in the United States in the 1980's. Lean manufacturing and lean enterprise, proliferating in western countries, are also evolved based on JIT principles. Many small and medium sized businesses have embraced these concepts along with some of the major corporations such as, Mercedes/Benz, Pratt & Whitney, Porsche and General Electric to name a few. Womack and Jones (1996) provide a thoughtful expansion upon their value-based business system based on the Toyota model. Along the way they update their action plan in light of new research and the increasing globalization of manufacturing, and they revisit some of their key case studies from the automotive, aerospace, and other manufacturing industries.

Many analytical, mathematical or experimental models are proposed to address the Kanban based operational planning and control issues (Berkley, 1992; Uzsoy and Martin-Vega, 1990). Simulation has been by far the methodology of choice in the majority of studies reported in the literature (Gupta and Al-Turki, 1998). Theoretically, the number of Kanbans and allocation of Kanbans in a system significantly affects the performance of a pull system. Instead of optimization of these Kanban arrangement which leads to a fixed number of Kanbans, Martins and Lewandrowski (1999) proposed a mathematical buffer stocks dimensioning approach using a dynamic kanban strategy. Gupta and Al-Turki (1998) compared the performance of a traditional kanban system (TKS) and a flexible kanban system (FKS). Through the simulation of two simple JIT models, they proved that FKS outperforms TKS under real-time manufacturing environments, such as sudden breakdown of a material handling system.

From application point of view, the researches of pull technologies could be classified in three categories: 1) production control; 2) inventory management; and 3) supply chain management (Kusiak, 2000). Material handling is a topic being previously researched in the literature (Gupta and Al-Turki, 1998; Askin, 1999; Venkataramanaiah et al., 2001), however, they all deal with the automatic material (specifically, the Work-In-Process, which belongs to the production line itself) transfer problem between production cells. The material we emphasize here does not refer to the WIP going through the assembly line, rather, it is the supply flow of material or parts subordinating to the main production line. None of known literature touched the topic of material handling from this aspect.

Moreover, in most plants, material inventories are only virtually under control of either a MRP II system or a Kanban system at the enterprise level, but not physically at the dynamic plant floor. Missing parts, wrong part delivered, parts not at right place at right time are common occurrence in almost all mainstream production plants, including GM, Ford, and Sterling Truck. Material handling is a frustrating problem faced by production managers. Production managers are feeling nervous everyday and are blamed for lack of ability to control the manufacturing process. As a result, analysis of material handling and dynamic simulation will be of great help to industries.

3. MATERIAL HANDLING SPECIFICATION OF A SIMPLIFIED ASSEMBLY LINE

Simulation of material handling in a pull production setting is the primary purpose of this research. A Kanban-based material handling will be investigated to make it in line with the pull production line. For the convenience of a common understanding, a sample scenario is chosen as the background problem, as shown in Figure 1.

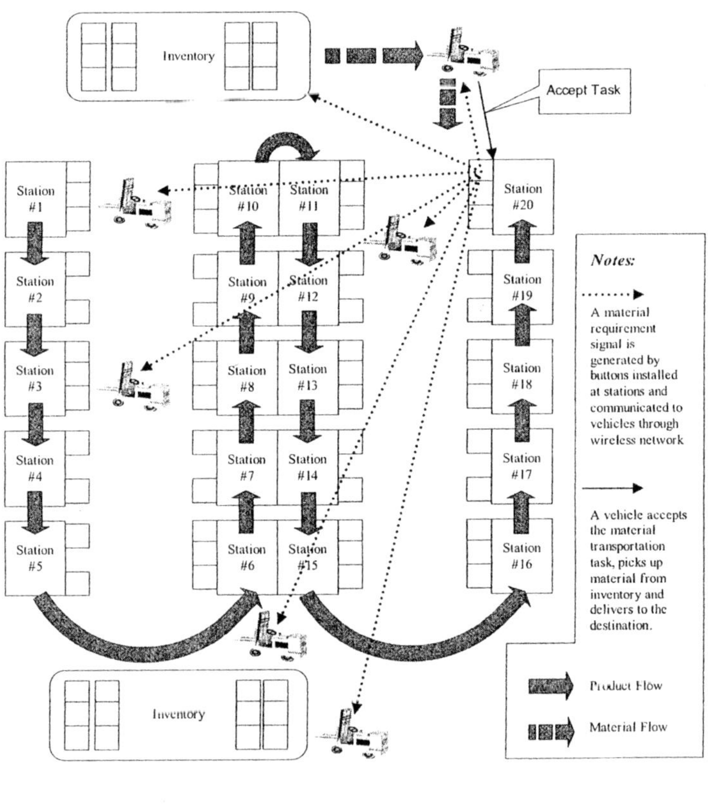

Figure 1 - A scenario of material handling simulation in an assembly line

The following assumptions are clarified:

- Production type: mass production of mixed product models.
- Production organization: U-shaped production/processing line, which integrates the manufacturing processes into a balanced and continuous material flow.
- Modular/station arrangement: the flow-of-products oriented production layout asks for separation of the whole process to manufacturing stations according to the optimized process rate and the granularity of material control.
- JIT production control: a pull control mechanism is applied for the control of production. Products are carefully sequenced before going on the line. There is no WIP buffer arrangement on the main assembly line. Production rate is a constant that is optimized by operation research practices.
- There are many transportation vehicles (illustrated by the icon of a forklift) moving around the whole plant mounted with wireless communication capacities and simple transaction systems for material handling.

The JIT-based material handling approach we proposed borrows similar principles from JIT-based production control and JIT-based inventory management in that: the right material is delivered from its inventory to the right production site, at the right time and in the right amount. Here, material transportation in the plant floor is considered as individual tasks. A task requires a material Kanban (M-Kanban) to be delivered. In figure 1, a material request signal is firstly generated by an assembly station running out of a part supply; after occupying a material Kanban, the request is broadcasted to a number of vehicles moving in the scope of a wireless network; then through negotiation, the task is confirmed by a vehicle and being delivered finally to the right station.

In our view, JIT material handling based on Kanban concept is not merely a pure event based system. In an event system, an event calls for a system response immediately; while in a Kanban based system, a generated event gets processed only after obtaining a physical object – M-Kanban. In other words, the processing (transportation) of a material requirement event holds until the system releases a M-Kanban and the event is qualified to occupy this free Kanban among all other events. Based on a Kanban control mechanism, we believe that the material handling system is able to reach a natural balancing of material requirements and transportation activities through delicate arrangement and management of Kanbans.

In addition to the standard function in a simulation environment, such as discrete event generation, simulation clock generation and an animation interface, this system should have special functional modules that try to model and simulate the dynamics in the plant floor. Three groups of functions make up the JIT material handling system: scenario generation, simulator, and graphical user interface. *Scenario generation* maintains a large variety of configuration information relating to: 1) static scenario, such as plant layout, moving tracks, and a central part inventory; 2) changeable simulation parameter, such as number of vehicles and number of Kanbans. *Simulator* is the core of the software in that it controls the simulation of both the production and material handling processes. Each transportation vehicle has a separate *vehicle simulation* module to make its own decisions, including task sequence, task schedule, moving control, loading and unloading operations, or even collision resolution decisions if necessary. *Station simulation* is a module to simulate a simplified production process taking into consideration only the consumption and replenishment activities of materials/parts at each station. *Kanban Simulation* manages buffering and circulation (life-cycle) of M-Kanbans. It makes two kinds of decisions 1) allocation of material requirement signals to empty M-Kanbans; 2) allocation of M-Kanbans to vehicles. In the simulator, basic simulation facilities such as timer and random number generator should be provided to simulate synchronize events or discrete events. *Graphical User interfaces* are supposed to timely update the graphical simulation, system event, exceptions, etc, and provide timely response upon users' requests for any kind of simulation and statistical data.

A distinctive feature of the designated JIT material handling simulation is "quasi-realism". The proposed simulation system possesses functions that surpass traditional simulations. The most distinguishing one is its ability to facilitate run-time reconfiguration. For example, the arrangement of assembly tasks to manufacturing stations could be adjusted during the execution of a simulation, so that the bottleneck (of the line) and system responses could be constantly changing. Another example is that each component is manageable not only in its configuration parameter, but also controllable in its behaviors individually (for example, each vehicle is able to make schedules, control status, and choose its own delivery route). In contrast, other simulation systems read a batch file before each simulation launch. It is difficult for people to analyze dynamic system behaviors by changing system configurations in separate simulation launches.

4. THE AGENT-BASED JIT MATERIAL HANDLING MODEL

We use agent technology to model major components in the JIT material handling simulation system. Agents are sophisticated computer programs that act autonomously on behalf of their users, collaborate across open and distributed environments, to solve a growing number of complex problems. There are four kinds of agents designed in the simulator.

- *Main Control Agent (MCA)*
  Main Control Agent (MCA) is responsible for simulation initialization, simulation termination, agent (thread) management, and thread synchronization. MCA also includes a timer and an event generator along with its main thread.
- *Station Agent (SA)*
  Station Agent (SA) is a running thread simulating material requirement activities at stations. It is dynamically generated and destroyed by the MCA. A simple production rate of the assembly line is set for all stations to consume their required parts in certain amounts. So, with the progress of one production step, the material balances at stations may reach the requirement levels or the urgent levels. In extreme occasion, materials may be exhausted which causes the whole assembly line to stop.
- *Kanban Schedule Agent (KSA)*
  Kanban Scheduling Agent (KSA) is a separate thread whose role is to take care of 1) the scheduling of material requirements to M-Kanbans, and 2) the assignment of M-Kanbans to vehicles. KSA is dynamically created and

destroyed by the MCA. It applies regular and emergent scheduling strategies. Regular scheduling is fulfilled by the negotiation carried out between KSA and participating VAs.

- *Vehicle Agents (VA)*
  Each Kanban assigned to a vehicle is confirmed by its VA and served by the VA through a series of actions. A vehicle agent is able to handle its local schedule, maintain its status, and controls its movement, repair, and resume actions. The threads for all vehicles are generated or destroyed by the MCA at the same time.

The four agents collaborate with each other and their interactive behaviors constitute the functionalities of the simulator. The MCA, located at the centre, functions as a core role for synchronizing. It generates two timely synchronizing events both for production and material handling process simulation. The *simulation event* is a regularly generated time event for pushing all simulation processes forward by one step. The frequency of simulation events is related to a specified speed set in the real-time mode for a simulation launch (or sometimes called an experiment). The *production event* is a regularly generated time event for pushing the production of products forward by one step. The time period for production event is set by the production rate of the assembly line. Two other Kanban events (*Free Kanban Event* and *Kanban Reschedule Event*) are used in material handling simulation and they will be further clarified in Section 4.2.

4.1. Production Simulation

By introducing a production event, a simple push mechanism with a constant production rate could be demonstrated. However, the realization of a pull mechanism needs much more effort. Figure 2 shows the principles of this pull control.

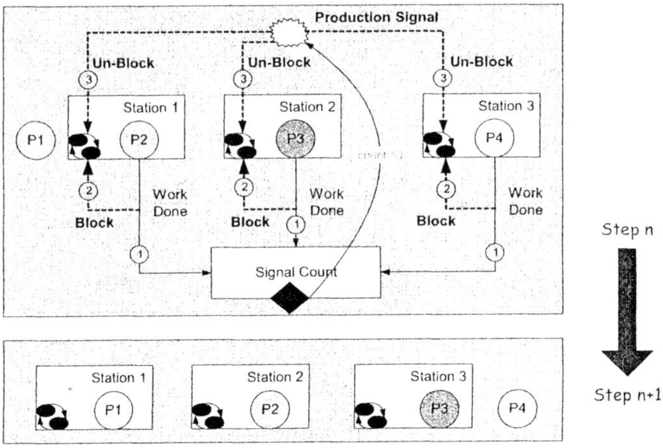

Figure 2 - Pull control of production simulation

Each station is doing some operations to the product on hand, say, P2, P3 and P4. Assume that the time required for the operations at a station is not a fixed number due to 1) complexity of operations on different products; 2) experience and tiredness of human operators; 3) availability of tools, etc. In this occasion, a flexible production rate is required so that when and only when all stations finish their duties, the production is "pulled" forward by one step. In the next step, station 1 will grab a new product P1, and P2, P3, and P4 will switch to succeeding stations, respectively.

In Figure 2, once a station finished its operations on the holding product, it sends a *Work Done* signal to a counter and then blocks its status. Until the count of signals reaches the total number of stations (for example, the count should be 3 in Figure 2), a *Production Signal* is generated which then unlocks all stations from blocking status. In this occasion, all stations "pull" a product from their previous stations when a production signal unblock their status from "blocking" to "normal".

This simple pull mechanism creates a dynamic production line that could predict system performance under various "what-if" scenarios, such as bottleneck shift, change of task/schedule, or even task re-arrangement. However, from a practical view, an assembly needs a constant production speed since physical equipment serving the line is not dynamically adjustable. The operations at stations need to be measured and balanced carefully using operation researches. Having this in mind, this is what simulation software targeted for - the purpose of simulation is to simulate the "what-if" situations, rather than a playback of real production settings.

Both push and pull control strategies will be simulated for a sample layout illustrated in Figure 1. A station agent is used to simulate the production process and generates material requests in line with the production progress. At each production step, a station will hold a product for a time period that is defined for the tasks assigned to it. The materials/parts used by these tasks are consumed by certain amounts set by these tasks.

4.2. Material Handling Simulation

Material transportation and delivery is the primary purpose of this simulation system. Material handling simulation could only be achieved based on the availability of production simulation, since the source of material transportation tasks is the material requirements accompanied with the production process. A material handling event is generated by the station agents. There are two levels of material handling events to represents two situations happened at stations: Material Requirement (MR) and Material Query (MQ). A *material requirement* is an event signaling a situation when the stock level of the referred material/part is low at a station and a full package of this material needs to be delivered to this station. A *Material query* is an event designed to signal a higher level of material shortage at a station. After a MR signal, the station continues to consume materials/parts at each production step. At a certain level, a MQ event could be generated and indicate that the material/part requested in the previous MR event has not been delivered to the station yet. The current balance reaches such an emergent level that the station staff and material management personnel need to know immediately the process status of this material. The system automatically tracks the queried material on behalf of the users and reports the status of material delivery.

In a pure event system, the generated MR or MQ are put into separate queues, and if possible, being processed immediately. In this system, we designed a material Kanban control mechanism on top of the event level. For the convenience of description, material Kanban is referred simply as Kanban in the rest of the paper.

Through correct choice of Kanban parameters and correct allocation of Kanbans in the layout of the production line, the material handling activities could reach natural balancing and optimization. A full material handling process can be divided into several steps (Figure 3):

1. Material requirement event, say MR1, is generated by a station and put in the MR event queue;
2. The MCA checks the Kanban Buffer regularly in each simulation circle and it identifies that an empty Kanban (designated by blue color) is available, say K1;
3. MCA generates a Free Kanban Event;
4. The KSA is triggered by the Free Kanban Event and picks MR1 from the event queue so that it could occupy K1. The status of K1 now changes to "ready-to-schedule" (designated by yellow color);
5. The KSA scans the Kanban Buffer and it finds out that K1 is a Kanban that has not been assigned to any vehicle;
6. KSA negotiates with all VAs and select one vehicle, say V1, to deliver K1. The status of K1 now changes to "assigned" (designated by red color);
7. V1 put K1 in its schedule;
8. V1 deliver the required material from the part inventory to destination station;
    a. V1 decided to withdrawn the task K1 from its schedule. This occasion is called "Kanban Re-schedule".
9. K1 is released by V1 and becomes an empty Kanban in the Kanban buffer;
10. V1 returns to the inventory site and picks the next Kanban from its schedule or becomes idle if there is no more task in its schedule.

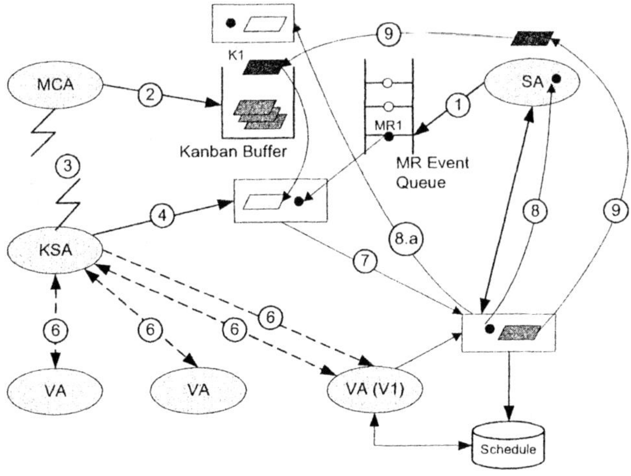

Figure 3 - Illustration diagram of Kanban handling

In above process, the Kanban buffer is a kind of data storage for all Kanbans currently circulating in the simulation environment. It has several "containers" (logical structure) to maintain Kanbans with different properties. The term container is used to analogize its physical appearance in Toyota plant in 1970's, when Kanbans are cardboard cards circulating in small areas and being collected using containers. The number of Kanbans in each container represents the Kanbans that are allowed to circulate in a certain area in the production line, generally an aisle. Of course, Kanbans and their circulation are implemented through electronic means. They are processed by a data processing system which dynamically changes the information and status they carried. A Kanban could be found in a number of states, including "free", "ready-to-schedule", "bidding", and "assigned".

Two major issues are involved in the material handling process: Kanban scheduling and vehicle control, which are realized by two kinds of agents respectively. A bidding process similar to the Contract Net (Smith, 1980) is adopted to do Kanban scheduling between the KSA and multiple VAs. So, bidding processes are designed both in KSA and in VA. The control of vehicle is a multi-facet mission including: 1) motion control- moving, loading/unloading, stop; 2) status control – idle, active, repair, pause, load/unload, delivery, empty load; 3) route control – route plan, route choice, return route, collision resolution; 4) bidding process control – giving a bid, responding to an awarded Kanban; 5) schedule control – local schedule, drop tasks, rearrange tasks; 6) Kanban status control – free, reschedule, delivery.

4.3. Implementation

A software prototype of the proposed JIT material handling simulation is implemented using a commercialized simulation tool - Anylogic™ of XJ Technologies™ (URL:http://www.xjtek.com/anylogic/). Among others, one important feature of AnyLogic™ is its agent-modeling capability. Even though this feature cannot fully meet our agent-based simulation needs, it helps quick deployments of complex simulation models in a professional way. Because of its open architecture and pure Java implementation, it will not be difficult for us to integrate AnyLogic™ with the multi-agent framework AADE (Hao et al., 2005) to create a more powerful agent simulation environment.

Figure 4 shows a snapshot of the prototype environment. A full working prototype will be available for demonstration at the conference.

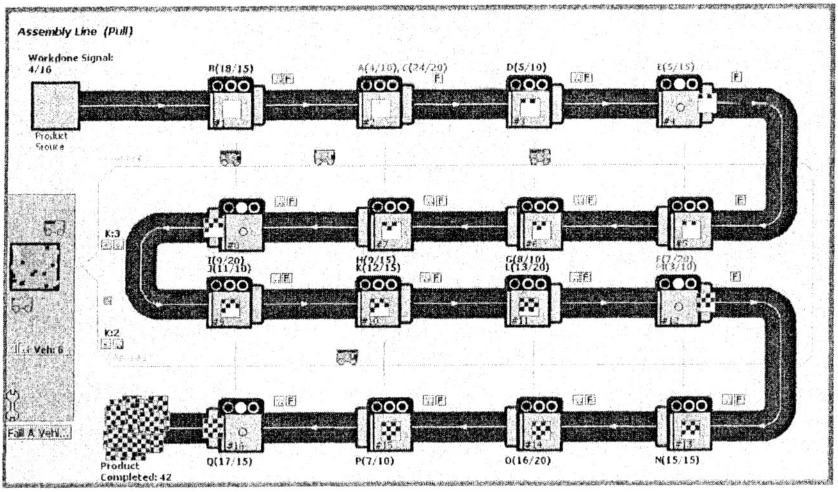

Figure 4 - A snapshot of the simulation prototype

5. CONCLUSION

Based on requirements of industrial partners, material handling has been recognized as a loose loop in most assembly plants. Just-in-time is a pervasive paradigm implemented in nowadays manufacturing plants that strive to survive in a global competition. The principles of JIT bring forward an optimized production environment and a mechanism for waste less inventory replenishment. However, material handling problem, especially the material/part supply at the plant floor level, is seldom addressed in research literature and in practices. In this paper, we propose a material handling simulation system that applies JIT principles. Material Kanban is an entity that carries a material request and represents a material transportation task. An agent-based simulation environment is designed and a prototype system is implemented using AnyLogic™. Many experiments will be performed based on the simulation model build for this purpose.

The JIT-Based material handling is expected to bring forward a number of advantages, such as optimization of stocks levels at production stations / cells, balancing of transportation load in the whole plant floor, obtaining manageability on material handling performance and accurate prediction and optimization of transportation resources. The major difference of this simulation from others is that the flexibility of the agent-based approach facilitates the simulation of various "what-if" scenarios including different layout designs, objective parameters and dynamic situations in the plant floor.

6. REFERENCES

1. Askin, RG. Manual and operator-based transfer batch sizes in JIT manufacturing. In Proceedings of Promat'99, Chicago, IL, pp.1-21, 1999. Reprinted in Progress in Material Handling Practice: 1999, Material Handling Institute, Charlotte, NC, pp. 167-87.
2. Berkley, BJ. A review of Kanban production control research literature, Production and Operations Management 1992; 1(4): 393-411.
3. Gupta, SM and Al-Turki, YAY. The effect of sudden material handling system breakdown on the performance of a JIT system. International Journal of Production Research 1998; 36(7):1935-60.
4. Hao, Q, Shen, W, Zhang, Z, An Autonomous Agent Development Environment for Engineering Applications, International Journal of Advanced Engineering Informatics, 2005, 19(2):123-134.
5. Hopp, W and Spearman, M. Factory Physics: Foundations of Factory Management. Irwin/McGraw Hill, Chicago, IL, 1996.
6. Krishnamurthy, A, Suri, R and Vernon, M. Re-examining the performance of MRP and Kanban material control strategies for multi-product flexible manufacturing systems. International Journal of Flexible Manufacturing Systems 2004; 16(2):123-50.
7. Kusiak Andrew. Computational Intelligence in Design and Manufacturing. John-Wiley & Sons, Inc., 2000.
8. Martins, K and Lewandrowski, U. Inventory safety stocks of kanban control systems. Production Planning and Control 1999; 10(6):520-29.
9. Shen, W, Norrie, DH and Barthes, JP. Multi-Agent Systems for Concurrent Intelligent Design and Manufacturing. Taylor and Francis, London, UK, 2001.
10. Shirk, WT. Material flow controls in a JIT environment. In Proceedings of the Technical Program: National Electronic Packing and Production Conference (NEPCON EAST'88), pp.325-332, 1988.
11. Smith, RG. The Contract Net Protocol: high-level communication and control in a distributed problem solver. IEEE Transactions on System, Man, and Cybernetics 1980; C-29(12): 1104-13.
12. Uzsoy, R, Martin-Vega, LA. Modelling Kanban-based demand-pull systems: a survey and critique. Manufacturing Review 1990; 3(3):155-60.
13. Venkataramanaiah, S, Krishnaiah, K and Pichandi, R. Management of WIP inventory in JIT environment under cellular layout – a case study. Journal of Scientific & Industrial Research 2001; 60(3):256-63.
14. Womack, JP and Jones, DT. Lean Thinking: Banish Waste and Create Wealth in Your Corporation. Simon and Schuster, New York, NY, 1996.

9 UNCERTAIN TEMPORAL SUPPORT IN MULTIAGENT MEDICAL SERVICE

Kamil Matoušek and Zdeněk Kouba
The Gerstner Laboratory, Department of Cybernetics
Faculty of Electrical Engineering, Czech Technical University in Prague
Technická 2, CZ-166 27 Prague 6, Czech Republic
{matousek,kouba}@labe.felk.cvut.cz

This paper describes an intelligent approach to represent information including uncertainly specified time periods by software agents in medical environment. Temporal reasoning capabilities are based on uncertainty ontology modeling and generalized Allen's interval relations. An example of temporal reasoning capabilities in data annotated by means of the temporal ontology is given.

1. INTRODUCTION

Current multi-agent medical systems range from hospital theatre scheduling systems to intelligent decision support systems. A survey of such existing systems can be found in (Foster, 2005). In order to perform their diagnostic and other useful tasks, the typical activities often include examination of patient's history for previous medical treatments, or gathering medical data together from different hospitals and other data sources.

Records of patients collected by medical specialists represent a significant source of data to be included in a digital knowledge base utilized by the agents. For knowledge sharing, ontology is a typical means. In practice, most current patient records are in form of text files containing some structured data such as body weight, blood pressure etc., but a significant part of the report consists of notes made by the physician.

In order to use the data for agent-based decision support among different medical laboratories and hospitals, it is necessary to convert as much information as possible into a structured form with added semantics. This can be achieved by annotation of medical records using ontologies and subsequent knowledge extraction. To capture as much detail from the patients' records as possible, several ontologies are necessary.

As an example, ontology describing the structure of medical records in the Institute of Biology and Medical Genetics at Charles University in Prague is being developed. OCML modeling language (Motta, 1999) is used for defining inference rules in the ontology modeling process. One of its parts, the *task ontology*, captures

Please use the following format when citing this chapter:

Matoušek, K., Kouba, Z., 2006, in IFIP International Federation for Information Processing, Volume 220, Information Technology for Balanced Manufacturing Systems, ed. Shen, W., (Boston: Springer), pp. 79–88.

the structure of records of examinations relevant for patients with neurofibromatosis. In addition, domain ontologies have to be used, e.g. *ontology of diseases*, *ontology of drugs* etc. in order to enrich the defined tasks with formalized concepts of the medical domain. In case of inherited diseases also an ontology describing *family relations* from genetic point of view was necessary.

It has been discovered that in addition to medical domain ontologies, an ontology capturing the imprecise and uncertain information contained in the medical records could significantly improve representational capabilities.

Existing medical standards introduce e.g. *temporal concept* in UMLS (Unified Medical Language system) as "a concept which pertains to time or duration", or HL7 standard includes the relation *temporally related to*. However, these descriptive elements are on a very high abstraction level and neglect the interesting detailed semantics of the specific terms.

2. CATEGORIES OF TEMPORAL STATEMENTS

Patients are located in space and time, while living in contexts and periods of their individual lives. Tracking changes of patients' health during time has always been a method that promised answering questions of professional physicians.

Concerning individual expressions of time, there is a wide range of precise, imprecise or vague, uncertain dating, which cause difficulties and inaccuracy in their use. Some reasons of inaccuracy of objects' dating may be e.g. the following:

- Data is not available (i.e. no written resources),
- it is a subjective information given by the patient,
- processes that lasted for a longer time are referred to as a single time point (e.g. feeling a pain).

In the domain of time, statements like "in childhood", "soon after the surgery" or "during the treatment in hospital XY" are corresponding examples. Our contribution is an attempt to deal with uncertainty in temporal assertions. The goal is to suggest a suitable and effective inference mechanism extension for ontology of medical records, which would yield sufficiently accurate localization in time.

Concerning people, the main temporal properties correspond to their lives and some events they encountered. Major time property is the duration of a time period (e.g. *convalescence period, pregnancy*), which could be expressed in terms of starting and ending time points. However, time durations may be relative as well (e.g. *for three months*) and thus having no exact starting or ending time. Time properties are not often expressed in a straightforward way as they may be inherent in the data. In this case, there are many expressions with different semantics (e.g. *tomorrow*, *at the beginning of the year*, *Monday*, or *June 5th*).

A categorization of temporal statements containing the most frequent expressions with respect to their accuracy with utility for the domain of our interest has been proposed in (Matoušek, 2004):

1. Precise statements. The whole data is available, maximum precision is reached (e.g. *January 12, 2006, 12:30:00*).
2. Statements with higher granularity. Data is available, but not so precise. It is necessary to distinguish instants and intervals (e.g. *January 12, 2006* can be seen either as an instant of higher granularity or as a 24 hour time interval).
3. Incomplete statements. Some information is missing for precise time identification. One may intentionally use this kind of statement for recurring temporal positions – regularly repeated instants (e.g. *January 12, 12:30:00*).
4. Uncertain statements with absolute specification of uncertainty (e.g. *between February 12 and February 13, 2006*).
5. Uncertain statements with relative specification of uncertainty (e.g. *around February 12, 2000, before 90s*).
6. Statements referencing other statements with temporal properties (e.g. *after the first surgery*, *during the last in-hospital treatment*).
7. Statements with unknown or missing information (e.g. *that time I was doing ...*).

Expressions related to the current time (e.g. yesterday, tomorrow) are supposed to implicitly belong to the category 6 above.

3. TEMPORAL SUPPORT

In order to support reasoning in the uncertain temporal domain some aspects including relations between time points and intervals, temporal granularity, temporal uncertainty and combining granularity values with other uncertainty factors have to be modeled. The following subsections describe them.

3.1 Time Relations

There are thirteen possible and mutually exclusive relations between two time intervals called *Allen's interval relations* introduced in (Allen, 1983). For graphical representation of the thirteen Allen relations see Figure 1 (moving from left to right corresponds to the passing of time).

For the intended inference capabilities, the supported time point (instant) relations are necessary. They are obvious but have to be stated. They include *before*, *equals* and *after* relations.

3.2 Time Granularity

Time granularity is a concept of the level of detail in which the time is considered (measured). Different time statements can refer to different time granularities. E.g. "May, 12, 2003" is a time statement with day granularity, while "In 2002" has the year granularity. With coarser time granularities their granularity values are defined, too. Time granularity defines its own unit scale for time positions.

Let us show an example of a relation between two time points with granularity (t^g). This is done as an extension of Allen's interval relationships for two time points with different granularity:

$$t_1^{g1} \textit{ precedes } t_2^{g2} \qquad \text{iff } \textit{End}(t_1^{g1}) \textit{ before Start}(t_2^{g2})$$

For time points with granularity common relations used to compare the finest time points were presented in (Matoušek, 2004).

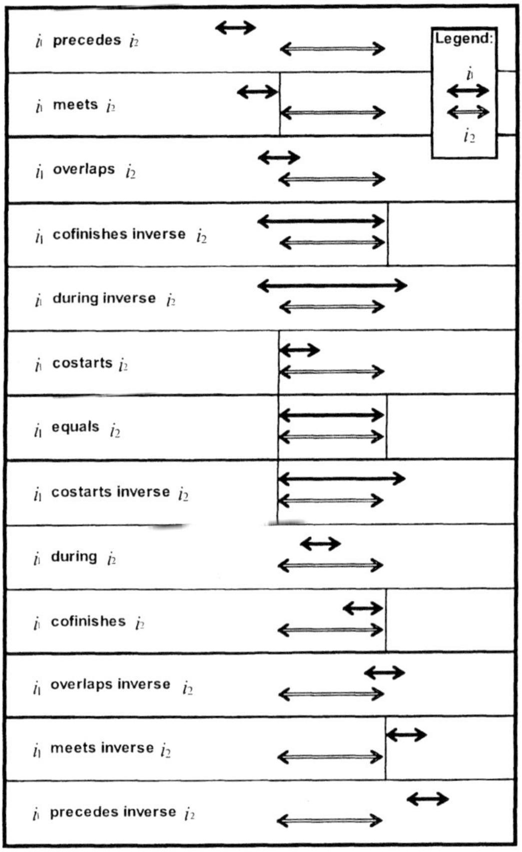

Figure 1 – Allen's temporal relations of time intervals

3.3 Temporal Position Uncertainty

As we have seen, there are various types of indeterminate expressions like: *now*, *(some time) before*, *(some time) after*, *at the break of a decade*, *during*, or even *unknown*.

We introduce *Time Uncertainty* as a concept, whose values represent individual uncertainty types falling into time intervals. Other approaches may consider e.g. also uncertainty in recurring temporal entities. The properties of time uncertainty are *FromTimePoint*, *ToTimePoint*, *BeforeRelTime* and *AfterRelTime*. The supported representations of time uncertainty are:

For *absolute* specification of a time range of uncertainty, given by specified time points, we introduce uncertainty as a property pair of two time points: *FromTimePoint* and *ToTimePoint*.

For *relative* time range of uncertainty, there is another pair of relative temporal position properties and their respective granularities: *BeforeRelTime*, *AfterRelTime*, and *BeforeGranularity*, *AfterGranularity*.

To clarify the usage of this time uncertainty representation, let us give here some examples:

Temporal statements like "*around the year 2004*", "*after 3rd of June 1991*", will use in its representation the *BeforeRelTime* and *AfterRelTime* together with uncertain location properties. On the other hand if an event happened in some time between "*The first examination and the last manifestation of a specific symptom*", then *FromTimePoint* and *ToTimePoint* will be used.

For uncertain time points we cannot easily define common relations used to compare the finest time points like before, equals, after. We can only express their sufficient conditions corresponding to certainty or necessary conditions corresponding to possibility. These possible and necessary bounds fulfill the rule that if one time point is *certainly before* another, then also the former is *possibly before* the latter.

Of course, an analogous property holds for the remaining pairs: <*certainly equal, possibly equal*> and <*certainly after, possibly after*>.

3.4 Combining Time Uncertainty and Granularity

As it is natural to combine uncertainty and granularity specifications, we support uncertain time points with granularity, which have specified both granularity and uncertainty.

An example of the possibility expression which is valid follows:

$$
\begin{aligned}
&{}^{u1}t_1^{g1} \textit{ possibly equals } {}^{u2}t_2^{g2} \\
&\text{iff } \textit{not}(\,\textit{Start}({}^{u2}t_2^{g2})\ \textit{after End}({}^{u1}t_1^{g1})\,) \\
&\quad \textit{and } (\textit{not Start}({}^{u1}t_1^{g1})\ \textit{after End}({}^{u2}t_2^{g2}))
\end{aligned}
$$

where ${}^{u}t^{g}$ denotes an uncertain time point with uncertainty u and granularity g. The uncertainty types under consideration are the *absolute* and *relative* time ranges of uncertainty (see section 3.3) and time granularity has been introduced in section 3.2.

For uncertain time points with granularity common relations used to compare the finest time points i.e. *before* and *after* expressed again in terms of their sufficient and necessary conditions can be derived e.g.:

$$
\begin{aligned}
&{}^{u1}t_1^{g1} \textit{ certainly before } {}^{u2}t_2^{g2} \\
&\text{iff } {}^{u1}t_1^{g1} \textit{ certainly precedes } {}^{u2}t_2^{g2}
\end{aligned}
$$

The possible and necessary bounds of common comparison relations fulfill the rule that if one time point is *certainly after* another, then also the former is *possibly after* the latter. The same property holds for the pairs <*certainly equal, possibly equal*> and <*certainly after, possibly after*>.

4 ONTOLOGICAL MODEL OF UNCERTAIN TIME

In this section we describe how the framework for temporal reasoning proposed in (Matoušek, 2004) can be practically used. We describe a calendar and time system that was used in our prototype of a temporal inference engine. We also provide a short description of a temporal reasoning system implemented in OCML language. It enables users to enter calendar data and to carry out temporal reasoning on a knowledge base containing temporal definitions. The temporal reasoning engine builds on the inference capabilities of OCML.

The basic class diagram of the time ontology in our temporal inference engine is depicted in the Figure 2.

The classes in the right frame correspond to the definitions in the framework for reasoning in the time domain. The main components of our temporal model are subclasses of *temporal-entity – time-point* and *time-interval*. The property *timeline-of* of *temporal-entity* is a sort of stereotype, which enables distinguishing different kinds of temporal entities. Any instance of temporal entity can have any number of timelines assigned to it. Query results can be constrained by including the name of a timeline that is the subject of interest in the query submitted to the temporal inference engine. Thus, timelines define a kind of namespaces. Beside classes, the model contains basic time point and time interval relations, rules, and functions that handle both granularity and uncertainty introduced as part of the framework.

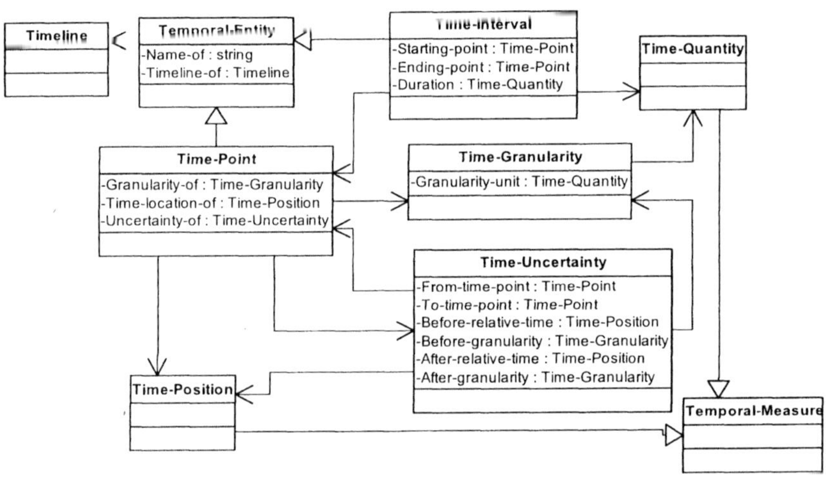

Figure 2 – Temporal ontology: core concepts

4.1 Calendar Dates and Time

In order to support calendar date and time specification, we introduce a subclass of *time-point*, denoted as *calendar-time-point* with the system of classes around it in the Figure 3.

The slots *century-of, year-of, month-of, date-of, week-day-of, hour-of, minute-of*, and *second-of* enable representation of individual parts of calendar date and time. The prototype inference engine provides internal routines, which are used to recalculate these slots into the internal representation of time locations and conversely back from the time location into these slots. When filling out only the slots with the respective information available or applicable, incomplete date specifications or those on a coarser granularity level can be used.

Some general constraints should always be satisfied, when working with temporal entities. One example is the property of transitivity of functions *before* and *equals*.

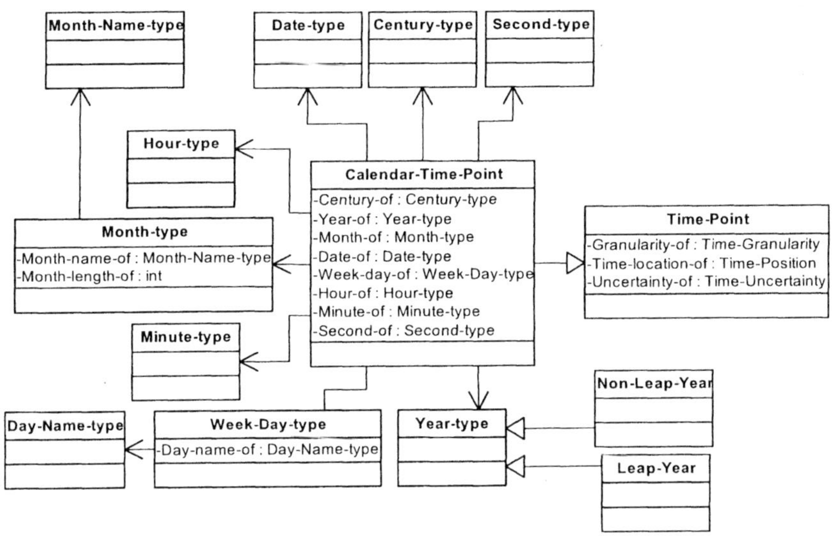

Figure 3 – Temporal ontology: calendar concepts

4.2 Domain Modeling and Inference Support

The ontology modeling style in OCML (Motta, 1999) enables to declaratively introduce classes and instances. More general relations can be also used as labeled n-ary relationships between OCML entities. Classes are considered as strictly unary relations (or predicates) and their slots (or attributes) as binary relations.

For example, the class *Time-point* is a subclass of *Temporal-entity*, which can have specified at most one time location, assigned at most one granularity type and one uncertainty type:

```
(def-class Time-Point (Temporal-Entity)
 ((Time-location-of :max-cardinality 1
             :type Time-Position)
   (Granularity-of :max-cardinality 1
             :type Time-Granularity)
   (Uncertainty-of :max-cardinality 1
             :type Time-Uncertainty)))
```

Moreover, procedural functions and backward chaining rules can be used to perform some necessary calculation or to introduce additional rules for inference. E.g. *get-location* function can be used in part of the *Before* relation definition:

```
(def-relation Precedes (?int1 ?int2)
 :constraint (and (Time-Interval ?int1)
                  (Time-Interval ?int2))
 :iff-def (and (Ending-Point ?int1 ?end)
           (Starting-Point ?int2 ?start)
           (Before ?end ?start)))
```

The proof system in OCML is capable to answer queries about the model using an algorithm for formula satisfaction with backward chaining like in Prolog (Derensart, 1996), i.e. depth-first search with chronological backtracking.

Practically, in our domain of interest, when comparing temporal entities, the relevant finest time points are being compared as necessary with respect to the given time point definitions including granularity and uncertainty. This enables the inference mechanism to properly reason about the individual temporal statements.

5. A SIMPLE PRACTICAL EXAMPLE

Let us show how the temporal inference engine can be used. We shall model the life and illness periods of a fictitious person, J. Hoffman. At the beginning the important time points have to be defined:

```
(def-instance John-Hoffman-birth Calendar-Time-point
        (       (date-of 14) (month-of 5) (year-of 1916)
                (granularity-of day-granularity)))

(def-instance John-Hoffman-illness-start Calendar-Time-point
        (       (date-of 26) (month-of 8) (year-of 1946)
                (granularity-of day-granularity)))

(def-instance John-Hoffman-illness-end Calendar-Time-point
        (       (date-of 29) (month-of 11) (year-of 1978)
                (granularity-of day-granularity)))
```

The following time intervals are related to the specified time points:

```
(def-instance Illness-John-Hoffman- Time-interval
        (       (starting-point John-Hoffman-illness-start)
                (ending-point John-Hoffman-illness-end)))

(def-instance Life-John-Hoffman Time-interval
        (       (starting-point John-Hoffman-illness-start)
                (ending-point John-Hoffman-death)))
```

If a statement is inaccurate, we may need to create an instance of time uncertainty. Here, we model the moment John Hoffman's knee got injured, which happened sometime around the year 1970. It is possible to include uncertainty parameterization as in the following case e.g. using *param-around-unc* instance of a newly defined class *time-parameter*:

```
(def-instance param-around-unc time-parameter((value-of 10)))

(def-instance Around-a-Year Time-Uncertainty
        (       (Before-relative-time param-around-unc)
                (Before-granularity year-granularity)
                (After-relative-time param-around-unc)
                (After-granularity year-granularity)))

(def-instance JH-Knee-Injury Calendar-Time-point
        (       (timeline-of John-Hoffman)
                (year-of 1970)
                (granularity-of year-granularity)
                (uncertainty-of around-a-year)))
```

Having defined the necessary facts, we may be interested in particular results using the inference engine. For example, having consulted data concerning all patient examinations and surgeries performed, the following query will retrieve the important time points between the first and the second surgery, which took place in the years 1965 and 1980 respectively:

```
(ocml-eval
        (findall ?a
                (and (timeline-of ?a John-Hoffman)
                        (and (before JH-first-surgery ?a)
                        (after JH-second-surgery ?a)))))
```

The returned result then finds (possibly among others) the John Hoffman's knee injury:

```
(JH-Knee-Injury)
```

6. CONCLUSIONS

Current results are giving useful means for agent's representation of even uncertain temporal aspects in medical patient's records including the uncertain facts. Using the proposed solution the medical ontology used by intelligent software agents was augmented and the first example results have been shown.

As a recommendation we suggest to extend the existing means of agent communication (e.g. KQML language performatives) to include the OCML language linkage in order to enable the exploitation of the presented solution. Alternatively, re-implementation in other reasoning frameworks is also possible.

In future the presented temporal ontology could be extended to cover a broader variety of patient-relevant data. Additional effort can be concerned with medical standard concepts mapping.

7. ACKNOWLEDGMENTS

The research is supported by the research grant of the Czech Ministry of Education, Youth and Sport: *Transdisciplinary Research in the Area of Biomedical Engineering II*, No. MSM6840770012.

8. REFERENCES

1. Allen JF. Maintaining Knowledge about Temporal Intervals. Communications of the ACM, 26 (11) : 832-843, 1983.
2. Deransart P, Ed-Dbali A, Cervoni L. Prolog: The Standard. Springer, 1996.
3. Foster D, McGregor C, El-Masri S. A Survey of Agent-Based Intelligent Decision Support Systems to Support Clinical Management and Research. In Proceedings of the 1st International Workshop, MAS*BIOMED'05, 2005, pp. 16-34.
4. Matoušek K, Uhlíř J. On Representing Uncertain Historical Time. Proceedings of DEXA 2004. Los Alamitos: IEEE Computer Society Press:105-109, 2004.
5. Motta E. Reusable Components for Knowledge Modeling. IOS Press, 1999.

10

COLLABORATIVE EVENT MANAGEMENT IN SUPPLY CHAINS: AN AGENT-BASED APPROACH

Pascal Forget[1, 2]
Sophie D'Amours[1]
Jean-Marc Frayret[1]
[1]*FOR@C Research Consortium*
Université Laval, Quebec, Canada
[2]*E-mail: pascal.forget@centor.ulaval.ca*

The development of integrated supply chains and the use of inter-organizational information systems have increased business interdependencies. Thus, the ability to deal quickly and seamlessly with everyday unplanned events is critical to maintain the overall performance of the supply chain. In order to develop tools to promote the collaborative management of such events, agent-based technology takes advantage of agents' ability to make autonomous decisions in a distributed context. Collaborative Event Management (CEM) is an approach designed to improve agility in a context where planning decisions are supported by a distributed advanced planning system (d-APS). This paper proposes an agent model geared with tools to collaboratively plan operations to deal with unplanned events.

1. INTRODUCTION

Events happen all the time. What is done is rarely done as planned. In manufacturing terms, we call these events, perturbations, disturbances, stochastic events, random events, uncertainties or contingencies. These events are in fact anything that moves production away from plan and requires planning adjustments. Machine breakdowns, sick leaves, supplier delays and planning errors are just a few examples of everyday events. The development of integrated supply chains and the use of inter-organizational information systems (IOIS) have increased business interdependencies. Managers know how to cope with events and solve the associated problems, but are they aware of the impacts of their decisions on partners? More precisely, are they taking into account the entire supply chain to make their decisions?

In fact, the majority of planning systems typically do not take events into account. The planning between different partners is done over a frozen horizon, with no disturbances, where variability is usually dealt with through inventory buffers and long lead times. If an event occurs, the plan is kept and actions are taken in the future to return as close as possible to the initial plan. As reaction time is generally

Please use the following format when citing this chapter:

Forget, P., D'Amours, S., Frayret, J.-M., 2006, in IFIP International Federation for Information Processing, Volume 220, Information Technology for Balanced Manufacturing Systems, ed. Shen, W., (Boston: Springer), pp. 89–98.

slow, this method generates misleading decisions between the event and the actual response. On the other hand, centralized systems usually deal with events by re-planning the whole plan or part of the whole plan, but in supply chains there are too many events and too many partners involved for re-planning at each change. Also, commercial supply chain event management systems usually only allows information about the occurrence of these events to be shared among supply chain partners, without any correction proposed. Distributed paradigms provide an interesting approach to permit local correction of the plan. In order to develop tools to promote the collaborative management of these events, agent-based technology provides a natural platform that takes advantage of the autonomy of agents and their ability to make decisions in a distributed context, using collaboration and goal-driven decisions following different events. An agent-based advanced planning and scheduling (APS) system could maintain a real-time plan by re-planning locally and allow for collaboration between agents to deal with events. Plan rectification in a short period of time leads the way to agile supply chains, fast reaction to events and increased global performance.

In this paper, we propose an agent-based model geared with tools to collaboratively plan production activities to deal with events in a distributed advanced planning system. We first provide a literature review on the use of agent-based technology in supply chains and existing agent-based collaborative schemes dealing with events. Next, we present the Collaborative Event Management (CEM) approach to improve the way events are handled in supply chains. Then, we introduce the agent-based experimental platform developed by the FOR@C Research Consortium, which is dedicated to supply chain planning for the forest industry. To profit from the advantages of CEM in a distributed environment such as the FOR@C platform, we detail a conceptual agent model showing suitable behaviors to deal with events in a distributed context. This takes advantage of the collaborative and goal-driven capabilities of agents. The North American lumber industry represents a perfect context for this technology. In fact, this industry is already highly distributed, where many business units interact in each production step. The main advantage of such technology in this industry is the large amount of stochastic events in many aspects of the supply chain, mainly due to the highly heterogeneous aspect of the resource, uncertain process output, production of co-products and by-products, price variation on spot market and demand variation on commodity markets.

2. LITERATURE REVIEW

2.1 Using agents in supply chain management

Gathering information in a centralized management system and redistributing plans has been the standard for decades in supply chain planning systems. Although they were first elaborated to deal with a single enterprise, these information systems have now expanded to the entire supply chain. Generally speaking, they offer good results and represent an advantage for synchronizing the planning of many production units when coupled with decision support systems, such as APS. However, when the supply chain involves many partners planning problems become more complex. Also, because of quantities of information only locally available and the time it takes to plan the supply chain, plans are sometimes not feasible and the supply chain shows low reactivity. In fact, traditional systems have not been developed to work in decentralized, dynamic and heterogeneous environments (Maturana *et al.* 1999).

In recent years there is a new trend of management systems emerging, distributing decisions all over the supply chain that drifts away from traditional centralized management systems. Seen as networks of suppliers, factories, distribution centers and retailers, supply chains are increasingly complex, making it more difficult for management systems to stay agile and react quickly. This new trend resulted in the development of planning systems with agent-based architectures. Agents are defined as intelligent software, with specific roles and goals, interacting with each other to make the most appropriate decision according to the situation, in order to carry out their part of the planning task. Distributed planning shows many advantages over central planning. For complex problems, sub-problems are easier to solve than centralized problems. Also, because decisions are distributed to different entities, reactivity to changes is increased and the feasibility of the plans is likely to be better. The challenge here is that plan performance is linked to agent collaboration capabilities to find good compromises. Agent-based technology has already been applied to different areas in supply chain management. For further details, the reader is referred to Parunak (1998) who presents industrial applications and case studies of agent-based systems, and Shen and Norrie (1999), who present descriptions of more than 30 research projects concerning on scheduling, planning and controlling. More recently, Caridi *et al.* (2004) present a survey and a classification of the different application domains of published multi-agent projects, denoting their degree of maturity.

2.2 Dealing with events

Companies in the same supply chain have business interdependencies since the behavior of one can influence another. In a highly dependent network of entities, when activities are tightly planned, events can have important repercussions throughout the supply chain. For example, a major mechanical breakdown in a strategic third tier supplier can halt supply for several days, which will have tremendous impacts on the whole supply chain translating in a delay for the final client. Another example is a quick change in demand pattern. When such change happens, every demand plan exchanged between each partner must be updated. If it is not done in a very short period of time, stocks will pile-up and money will be wasted. This has increased in recent years as supply chain integration systems are adopted, CPFR (*Collaborative Planning, Forecasting & Replenishment*) methodologies are followed and forecasts are prepared jointly. As stated in Frayret *et al.* (2004), interdependencies can vary greatly. The authors present a review of coordination structures, from centralized architecture to proper hierarchy, including heterarchy, to handle interdependencies. To permit a classification of events following their impact on the supply chain, Cloutier *et al.* (2001) present three levels of contingency management. These are local contingency management, local contingency management with need expression to partners and collaborative contingency management with eventual need expression to partners. Also, Davis (1993) presents a methodology developed with Hewlett-Packard for effectively addressing supply chain management, with interest on managing uncertainty in all supply chain processes.

2.3 Using agents to manage events

Agent-based planning systems have been proposed to manage supply chains and deal with events. Montreuil *et al.* (2001) present the NetMan architecture, an operation system for networked manufacturing organizations geared up to cope with events not previously authorized. Although they created an architecture able to manage unseen events, they do not present specific behaviors to solve problems following perturbations. Based on intelligent holons, Fletcher *et al.* (2001) present a conceptual architecture of a lumber processing system to improve flexibility and fault tolerance. The ExPlanTech architecture developed by the Czech Technical University (Pechoucek *et al.* 2005) is a multiagent approach for decision-making support and simulation for manufacturing process. This approach integrates coordination and negotiation to deal with distributed environment. These architectures outline the possibility of reaction to however they do not propose solutions.

Conceptual agent models have also been proposed to improve supply chain performance and react to events. The Agent Building Shell (ABS) presented by Fox *et al.* (2000) is a collection of reusable software components and interfaces needed for any agent involved in a supply chain management system, geared to handle perturbations caused by stochastic events in a supply chain. In this architecture, most of the efforts have focused on defining communication and collaborative aspects. The InteRRaP agent model provides an interesting approach using a layered architecture able to react and deliberate when confronting changes, using different capability levels (Muller, 1996). This architecture is made of three layers, which are used to build action plans, depending if an event requires a reactive response, local planning or collaboration for planning. The tri-base acquaintance model (3bA) presented by Marik *et al.* (2005) is a collaborative wrapper added to an agent, giving the possibility to deal with events in a global perspective instead of resolving problems only in a local view. This is done by using information about other agents without the need of central facilitator. The authors present an example of applications in supply chains and they define Social knowledge needed to increase the efficiency of agents. From this review, we intend to propose a general approach to deal with event in a supply chain context.

3. COLLABORATIVE EVENT MANAGEMENT

Authors use different definitions when they discuss events. Fleury *et al.* (1999) present random events as anything implying delays in production time. The authors enumerate classes of reasons why such events occur: human mistake, failure of resource, failure of equipment and foreseeable production mistakes. Davis (1993) identifies three distinct sources of uncertainty, which are suppliers, manufacturing and customers. With the objective of mapping all the events encountered in daily production, we interviewed planners, directors and researchers involved in lumber supply chain planning activities. Inspired by Davis's methodology, we divided the events into three different levels, such as demand variation, execution variation and supply variation, and we noted all impacts on the supply chain. This investigation made clear that events have important impacts on production plans and ultimately, on financial results. To cope with events, lumber companies use inventory buffers or long lead times. While reducing event impacts, they increase inventory cost and have poor delivery lead times. This shows the importance of understanding events more precisely and how to react to them in an efficient way.

To emphasize the importance of collaboration when dealing with events in manufacturing systems in supply chains, we introduce *Collaborative Event Management* (CEM). This approach represents our conception on how collaboration should be exploited to deal with events within any manufacturing system in a supply chain, not limited to the lumber industry. Due to interdependencies between business partners, there is a need to coordinate the planning processes to solve problems resulting from events in a timely and efficient way. In a CEM perspective, we represent manufacturing activities in four different phases (see Figure 1), which are the Planning phase, the Scheduling phase, the Shop floor / Simulation phase and the Adjustment phase, representing the interactions between two different planning units. The first two phases are handled by the support staff and the two last are done by the operational staff.

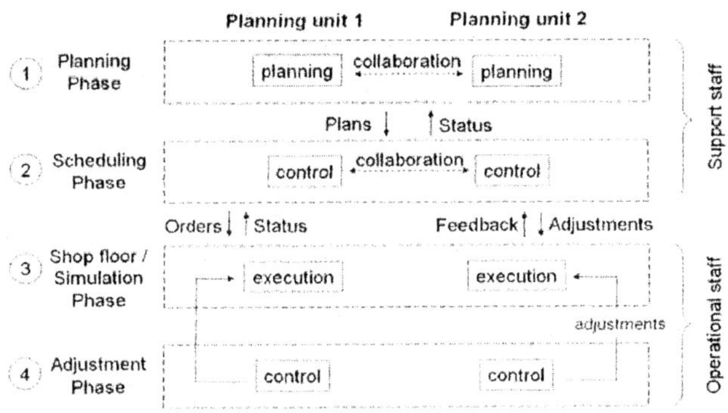

Figure 1 – Manufacturing phases in CEM

The Planning phase (1) includes the creation of the initial plan. From information exchanges between two different planners, they can collaborate and coordinate efforts for planning what product is needed and when. The Scheduling phase (2) dedicates resources to specific production tasks in a coordinated way. Status reports are redirected to the Planning phase, to make sure the initial plan is feasible with the resources available. The Shop floor / Simulation phase (3) is where the actual execution is done (or simulated). Orders from the Scheduling phase are transmitted and exact status is continually sent back to insure the plan is being followed. The Adjustment phase (4) works simultaneously with the third phase. It verifies resource status and monitors throughputs and reports to the Shop floor / Simulation phase for local adjustments. When adjustments are not possible because problems cannot be resolved at the Shop floor / Simulation phase without changing the plan, status reports are transmitted to higher levels. Because the last two phases are about local execution and control, there is no planning collaboration involved. Then, in the Scheduling or Planning phase, re-planning is executed, involving a single planning unit (locally) or many planning units (collaboratively).

CEM puts collaboration at the heart of the Planning and Scheduling phases. With extended collaboration protocols and anticipation of the impacts of their decisions, it is possible to propose problem solving techniques to face unforeseen contingencies.

Such an approach can smooth transitions in the supply chain, reducing safety stocks and lead times usually kept to cope with undesired impacts. Applied to an agent-based planning platform, CEM provides input to create agents with appropriate characteristics.

4. AGENT-BASED COLLABORATIVE EVENT MANAGEMENT

4.1 FOR@C experimental platform

For many years, the planning processes in the North American lumber industry have never been questioned. Due to the highly heterogeneous nature of the resource (i.e. trees) and the inherent complexity of forecasting production throughput the dominant thinking was to produce the maximum volume with the resource available (push production). Because of the commodity nature of the final product and the standards of sizes and grades, to take advantage of economies of scale, production is oriented towards large batches. This industry can be characterized by large inventories, low flexibility and low agility. The recent economical and international threats to the lumber industry have encouraged some researchers to rethink the planning processes in a way to quickly react to correct deviance from the plan, respond to demand, reduce inventory and exchange information promptly throughout the supply chain (Frayret *et al.* 2005). In order to compensate for the lack of control over the stochastic elements relevant to lumber production, it is necessary to increase the exchange of information between the different production centers and to quickly react in a coordinated manner to changes.

With the purpose of developing a new planning approach for the lumber supply chain, the FOR@C Research Consortium of the Université Laval (Quebec, Canada) has developed an experimental planning platform built on an agent-based architecture for APS, with interaction mechanisms inspired from FIPA (Foundation for Intelligent Physical Agents) standards. This architecture combines agent technology with Operational Research (OR), in order to take advantage of the ability of agent-based technology to integrate distributed decision problems, and the ability of OR to solve complex decision problems (Frayret *et al.* 2005). Because of the distributed context of the supply chain and the use of agents, this platform can be described more precisely as a distributed-APS (or d-APS), where the first issue is to plan and coordinate all supply chain operations. As shown in Figure 2, the agent-based architecture presented by FOR@C is based on the natural division of the planning domains, where the production activities are divided among specialized agents: a sawing agent, a drying agent and a finishing agent. Each of these agents is responsible for supporting the planning of its production center in terms of production output each day. Other agents are also part of the architecture, such as the deliver agent, source agent and warehouse agent. Currently, testing is in progress with the collaboration of an international forest products company. The test uses 15 agents, more than 600 products, and approximately 80 exchange protocols, 100 tasks and 50 task flows. This architecture is a major step toward an improved coordination process for planning requirements. The main objective is to develop specific planning methods for each planning agents, using collaboration to insure synchronization in the supply chain.

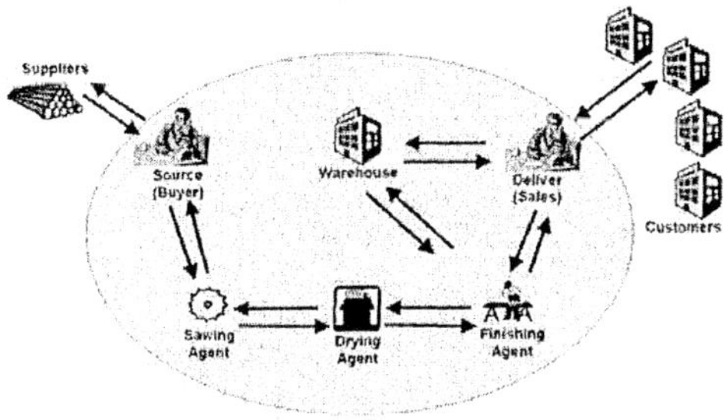

Figure 2 – Planning unit from the FOR@C platform

Each planning agent (such as sawing, drying and finishing agents) disposes of objects which are updated by local actions or actions from other agents. Actions are made possible by task flows, which are sequences of tasks, usually triggered by specific events. A standard task flow is the planning protocol (see Figure 3), triggered upon reception of a new demand plan from a client. This protocol is divided in two segments. The first is about modifying a requirement plan, creating a production plan with resource constraints and infinite supply, allocating demand to different suppliers and waiting for an answer. The second concerns receiving supply propositions, updating the production plan with a finite supply, allocating production to clients and modifying a replenishment plan. Optimization algorithms are deployed in the production planning (demand and supply propagation) and allocation tasks to suppliers and clients.

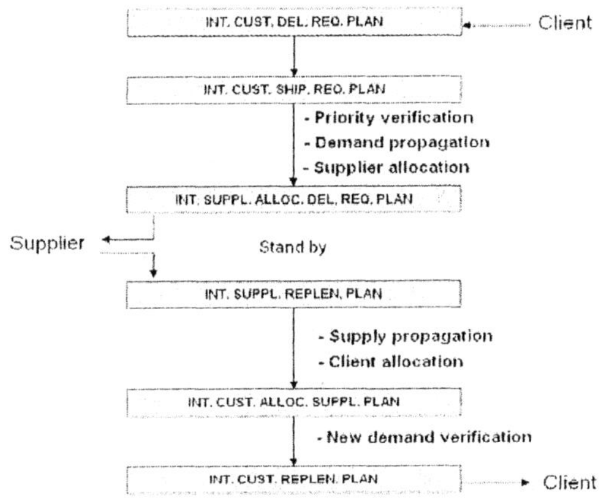

Figure 3 – Current planning protocol

4.2 Enhancement of current planning agents

Agent-based planning systems, such as the FOR@C experimental platform, represent a promising way to develop new planning systems and to apply the CEM approach. The next step is to develop an agent model to enhance the current planning agents of the experimental platform. Facing events, current agents use reactive task flows, triggered by specific messages. To deploy agents with behaviors adapted to different situations and environments, we must give the agent the possibility to make choices. Making clever choices requires sufficient knowledge and competencies to make good decisions. Also, the agent must understand the impacts of its decisions on itself, on other agents and on the whole supply chain.

4.3 A conceptual agent model for CEM

Our contribution in this paper is to describe a conceptual agent model that presents profitable behaviors to deal with events in a distributed collaborative context, using a CEM approach. Including similar concepts from previously presented models, we propose the *Sense, Think & Act* (ST&A) agent model (see Figure 4), integrating agent technology and OR tools. The model is composed of three distinct layers describing the different knowledge and competencies required to deal with events in supply chain planning. These layers can use different task flows, depending on their abilities to understand the impacts of their actions. In fact, layers represent what knowledge and competencies we wish to possess to face uncertain environment and be able to react efficiently to events. These layers do not depend on each other, since there is no strict passage from one to another for solving problems following events. In fact, it is an evolutive model, where the agent can have different intelligence levels in each layer, following its own evolution. Although presented as layers, the ST&A agent is basically an evolutive conceptual model which the architecture is not necessarily layered, such as the InteRRaP architecture. The ST&A conceptual model is similar, but its architecture can be completely different in its implementation.

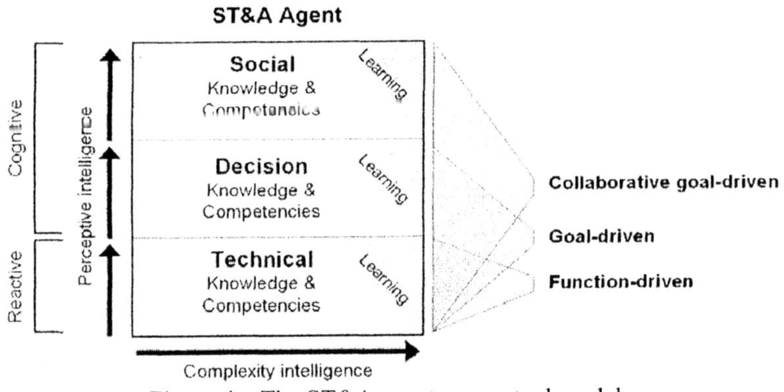

Figure 4 – The ST&A agent conceptual model

The Technical layer is basically a reactive layer, which includes all reactive tools, tasks and existing task flows, such as OR tools and algorithms, conversation and negotiation protocols, and queries. Goals in this layer are related to minimizing effort (computer processing) while maximizing results (optimization functions included in tools). Reactive actions to face events are known and used when needed. An agent strong in this layer but weak in the others would show a function-driven behavior. The current agents deployed in the FOR@C platform present such behavior. Facing events, they send a new demand plan to suppliers and then, a new supply plan to clients. This is where a superior reasoning behavior would become interesting, giving new possibilities for dealing with disturbances, other than just start a global re-planning protocol. The Decision layer evolves from the reactive behavior to a cognitive behavior. It includes the explicit knowledge of local goals and the progress toward these goals at any time. These goals are towards local optimality, where the agent is concerned primarily with its own performance. This layer also encompasses the knowledge of the impacts of its decisions on itself. Here, when an event occurs, the agent has the capability to choose which task, task flow, optimization algorithms or complete plan could fit better, according to its own goals. An agent strong in its decision layer and technical layer would present a goal-driven behavior. The Social layer also denotes cognitive behavior, but here local goals are replaced by global goals. The agent is now aware of the impacts of its decisions on other agents and on the whole supply chain. While choosing actions to correct deviation from plan, we want the agent to possess the ability to capture the entire potential of the network and be able to minimize impact on others. Again, the agent has the capability to choose which task, task flow or plan responds the best to its own goals. It anticipates its environment (like other agent goals) and network tactical decisions (specific product, client selection, supplier selection, etc.) It has the ability to use collaboration protocols with anticipation of other agent reactions. Agents covering the three previous layers are considered collaborative goal-driven agents. Learning is a specific competency imbedded in each layer, which gives the agent the potential to increase its knowledge in each layer. A specific behavior that showed good results in a situation could be learned and remembered for next time. However, this subject will not be detailed in this article, but will be studied in details in close future.

With the objective to observe the behavior of the ST&A agent in a CEM context, different event scenarios which confront planning agents have been developed, inspired from the lumber industry. This allows for comparing agent performances presenting different intelligence levels in each layer. Examples of scenarios are major dryer breakdown, out of stock and difference in harvest forecast. In a supplier/client relationship, we detailed the actions of these agents to solve the problem. The next step is to actually construct an agent architecture able to reproduce the behaviors denoted previously. Then, we will simulate different agent configurations by implementing them on the FOR@C agent-based platform.

4.4 Toward a Collaborative Intelligence

An agent showing collaborative behaviors, such as the collaboration-driven agent, uses *Collaborative intelligence*. Similar to the interpersonal intelligence in human intelligence (Gardner, 1983), we define *Collaborative intelligence* as the degree of knowledge and competencies toward using global goals to solve problems in planning activities. Indeed, this conceptual model represents what we consider to be required for a planning agent in a CEM context, as a complete collaborative goal-driven agent. Moreover, this model can represent a gradation tool to measure intelligence imbedded in any agent. In fact, each layer can be taken separately and scaled following degrees of achievement (as noted in Figure 4). We defined two axis of development; *Complexity intelligence* is the ability to deal with problem complexity and plurality of solutions, and *Perceptive intelligence* is defined as the degree of the environmental perception of the agent. This axis is scaled differently for each layer because achievements are different. Using this conceptual model, any planning agent could be positioned in the three layers and compared in its behavior confronting event management.

5. CONCLUSION

Even if events are considered normal, they can still cause much trouble and costs (following Murphy's Law). The close management of events is an important opportunity to reduce inventory and lead times, at the same time increasing client satisfaction by delivering the right product at the right time. We believe collaboration is the key for handling events in the best way possible in a supply chain context. Approaches like CEM can be an interesting avenue to develop new collaboration protocols. Collaborative agent architecture using the advantage of OR, such as the ST&A, can be a powerful tool to reach appreciated gains, when implemented in a distributed planning system. Following the conceptualization of the CEM approach and the ST&A agent intelligence level descriptions, future work is needed. Implementation and testing of different agent configurations into the FOR@C experimental platform will be executed, in real-world planning situations. In a different perspective, it will be of a great interest to increase research efforts on the learning competency, on its implications and impacts. Already, research is on going and results will be available shortly.

6. ACKNOWLEDGMENTS

This work was funded by the FOR@C Research Consortium and the Natural Sciences and Engineering Research Council of Canada (NSERC).

7. REFERENCES

1. Caridi M, Cavalieri S. Multi-agent systems in production planning and control; an overview. Production Planning and Control, 15(2):106-118, 2004.
2. Cloutier L, Frayret JM, D'Amours S, Espinasse B, Montreuil B. A commitment-oriented framework for networked manufacturing co-ordination. International Journal of Computer Integrated Manufacturing, 14(6): 522-534, 2001.
3. Davis T. Effective Supply Chain Management. Sloan Management Review Summer 1993: 35-46.
4. Fox MS, Barbuceanu M, Teigen R. Agent-Oriented Supply-Chain Management. International Journal of Flexible Manufacturing Systems, 12:165-188, 2000.
5. Fletcher M, Brennan RW, Norrie DH, Fleetwood M. Reconfiguring processes in a holonic sawmill, 2001 IEEE International Conference on Systems, Man, and Cybernetics, 1:158-163, 2001
6. Frayret JM, D'Amours S, Rousseau A, Harvey S. Agent-Based Supply Chain Planning In The Forest Products Industry. Working Paper Dt-2005-Jmf-1, Centor, Université Laval, 2005.
7. Frayret JM., D'Amours S, Montreuil B. Coordination and Control in Distributed and Agent-Based Manufacturing Systems. Production Planning and Control, 15(1): 42-54, 2004.
8. Gardner G. Frames of mind: the theory of multiple intelligences. New York: Basic Books, 1983.
9. Marik V, Pechoucek M, Stepankova O. Social Knowledge in Multi-agent Systems. Multi-Agent Systems and Applications, ACAI 2001, LNAI 2086, Springer, pp. 211-245, 2001
10. Maturana FP, Shen W, Norrie DH. Metamorph: An Adaptive Agent-Based Architecture for Intelligent Manufacturing. International Journal of Production Research, 37(10): 2159-2174, 1999.
11. Montreuil B, Frayret JM, D'amours S. A Strategic Framework for Networked Manufacturing. Computers in Industry, 42: 299-317, 2000.
12. Muller JP. The Design of Intelligent Agents: A Layered Approach. Lecture Notes in Artificial Intelligence, LNAI 1177, Springer, 1996.
13. Parunak, HVD. Practical and Industrial Applications of Agent-based Systems. Industrial Technology Institute, 1998.
14. Pechoucek M, Vokrinek J, Becvar P. ExPlanTech: Multiagent Support for Manufacturing Decision Making, IEEE Intelligent Systems, 20(1):67-74, 2005
15. Shen W, Norrie DH. Agent-Based Systems for Intelligent Manufacturing: A State-of-the-Art Survey. Knowledge and Information Systems, 1(2): 129-156, 1999.

11 DESIGN PATTERNS FOR DISTRIBUTED CONTROL SYSTEM BENCHMARKING

Karthik Soundararajan and Robert W. Brennan

Schulich School of Engineering, University of Calgary
2500 University Dr. N.W., Calgary, AB, Canada, T2N 1N4
Email: rbrennan@ucalgary.ca

In this paper, we describe the design and development of a simulation-agent interface for real-time distributed control system benchmarking. This work is motivated by the need to test the feasibility of extending agent-based systems to the physical device level in manufacturing and other industrial automation systems. Our work focuses on the development of hybrid physical/simulation environment that can be used to perform tests at both the physical device level, as well as the planning and scheduling level of control. As part of this work, we have extended the proxy design pattern for this application. This paper focuses on the resulting software design pattern for distributed control system benchmarking and provides examples of its use in our hybrid physical/simulation environment.

1. INTRODUCTION

Given the difficulty of practical manufacturing scheduling and control problems, recent research has moved away from traditional, analytical approaches that have been the domain of operations research for many years and towards new approaches that rely on artificial intelligence, holonic and multi-agent systems (Shen et al., 2001; McFarlane and Bussman, 2000). In order to make this research relevant however, it is important that realistic and industrially relevant test cases are available to address specifically the evaluation and stress of the performance of scheduling and control systems based on these new technological paradigms. As well, it is important that these test cases span the realm of research in this area from planning and scheduling systems to real-time control.

In order to address this need, a special interest group on benchmarks of multi-agent systems was established under the umbrella of the Network of Excellence on Intelligent Manufacturing Systems (IMS-NoE, 2004). This paper focuses on one aspect of the work being performed at the University of Calgary in this area.

In this paper, we describe a hybrid physical/simulated environment that is currently being developed for manufacturing systems control experimentation that incorporates both simulated and physical manufacturing devices. The objective of this work is to extend benchmarks of multi-agent systems to the full manufacturing

Please use the following format when citing this chapter:

Soundararajan, K., Brennan, R. W., 2006, in IFIP International Federation for Information Processing, Volume 220, Information Technology for Balanced Manufacturing Systems, ed. Shen, W., (Boston: Springer), pp. 99–108.

hierarchy: i.e., from device control to planning and scheduling. In order to accomplish this, an important aspect of the project involves developing an interface between the simulation software and the physical devices. In our work, we investigate the use a Tiny Internet Interface (TINI) board (Loomis, 2001) that runs Java programs (allowing us to develop local software), and has access various I/O (e.g., discrete/analog I/O, Ethernet). The link to the discrete-event simulation model is created using Arena® Real Time (Kelton et al., 1998) and Java socket-based communication. This paper focuses on the design and development of the Simulation-Agent interface for the hybrid distributed control system using design patterns.

We begin the paper with some background on the work on connecting simulation software with physical devices. Next, we introduce the notion of design patterns in Section 3 and follow this with details on the use of the proxy design pattern for the hybrid system in Section 4. In Section 5 we focus on the development of the Client-proxy and tests. Finally, we conclude the paper with comments on the future direction of this research.

2. THE HYBRID PHYSICAL/SIMULATION ENVIRONMENT

In the manufacturing domain, discrete-event simulation is a very powerful tool that can be used to evaluate alternative control policies. For example, discrete-event simulation has been used to evaluate agent-based scheduling approaches by interfacing agent-based or object-oriented software with a discrete-event simulation model of a plant to be controlled as is illustrated in Figure 1 (Brennan and O, 2004).

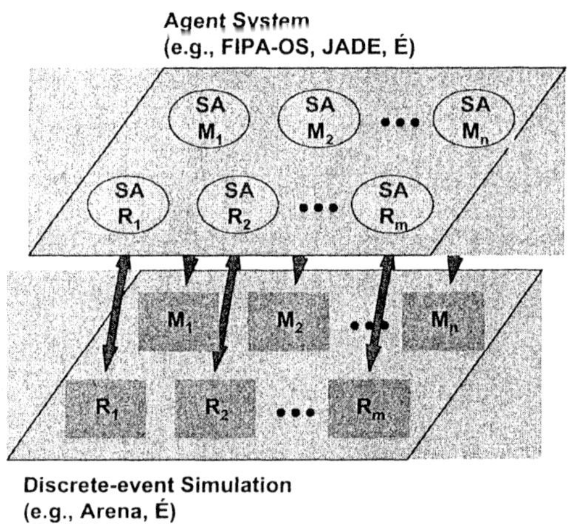

Figure 1 – Using discrete-event simulation to evaluate agent-based manufacturing systems

In this example, each entity in the discrete-event simulation model (such as a machine (e.g., M_1) or robot (e.g., R_1)) is represented by a corresponding software agent (*SA*) in the control module. For example, the software agents can be thought of as the "reasoning" part of the entity that is responsible for scheduling etc.

The reason for this direct correspondence between *SA* and entity is that (given recent advances in hardware and software) is it possible to have intelligent agent software running directly on a machine (e.g., computer numerical control (CNC), robot, etc.). This software can be thought of as the "brains" of the machine that can potentially allow it to act autonomously and/or cooperatively.

The next step is to have these *SA*'s run directly on the machines. This will allow us to test the real-time capabilities of the system (i.e., its ability to meet deadlines), and also allow us to incorporate extra functionality concerned with "execution". For example, *SA*'s (running directly on a machine) may be used to perform fault diagnosis and preliminary recovery services. *SA*'s may also be capable of reconfiguration (e.g., allowing new hardware to be added/removed/modified dynamically). This arrangement is shown in Figure 2.

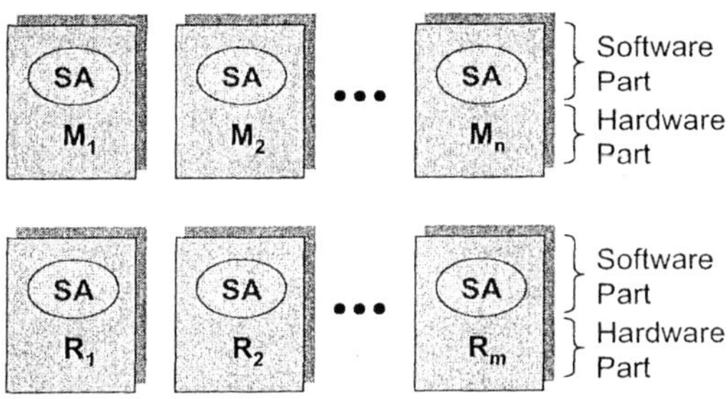

Figure 2 – Physical agents

From an experimental and research point of view, there are some problems with this second approach even though it represents the ultimate goal (i.e., industrial implementation) of the agent-based system. The main problem is that, given financial and space resources, most researchers using this approach are limited to experimenting with relatively small systems. As well, even if a large system is possible, it is debatable whether it would be a good time and cost investment. For example, we may only need a relatively small number of physical devices to test and validate the execution capabilities of our agent system. However, we would like to have a reasonable number of emulated machines to test the scheduling capabilities.

The former requirement (execution) is typically hard real-time (i.e., deadlines must always be met very quickly), while the latter requirement (planning, scheduling, and dispatching) is typically soft real-time (i.e., deadlines must be met on average and much more slowly). As a result, we need physical hardware to test the execution environment and could use a simulated environment to test the "higher reasoning" part of the system. Of course, physical hardware could also be used to test this latter part of the system.

A second problem (from a research perspective) with a pure physical system is that we lose many of the experimental benefits of discrete-event simulation software (e.g., statistical analysis, graphics, the ability to easily change the system configuration).

As a result, we suggest that a hybrid physical/simulated environment is developed for manufacturing systems control experimentation. In order to accomplish this, one aspect of the project involves developing an interface between the simulation software and the physical devices. One approach is to use a Tiny Internet Interface (TINI) board. This board runs Java programs (allowing us to develop local *SA*'s), has access to discrete/analogue I/O, and has Ethernet capabilities. The general idea is illustrated in Figure 3.

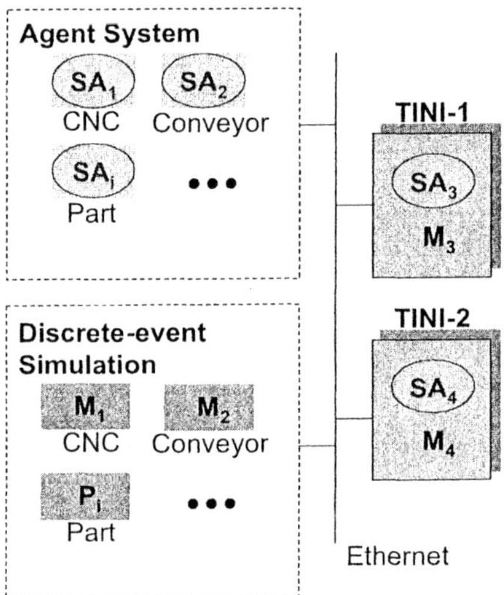

Figure 3 – A hybrid environment for experimentation

In this example, M_1, M_2, and P_i have software agents associated with them (SA_1, SA_2, and SA_i respectively). M_3 and M_4 are also part of the system, but they are real physical devices. For example, the TINI boards could act as controllers for robots or conveyors. Their agents (SA_3 and SA_4 respectively) communicate with the other parts of the system via a UNIX socket.

3. DESIGN PATTERNS

The hybrid prototype used for this research is comprised of Arena simulation software and a Java-based embedded controller (Soundararajan and Brennan, 2005). An important issue that was addressed during the application development was the transparency of the potential remoteness of the server (the Arena model) and the hiding and encapsulation of the means by which to contact such remote servers by the client. As a result, a design pattern that addresses this large-scale system design (the high-level architecture of the distributed system) concern was investigated.

The first step involved familiarization with the patterns literature related to distributed real-time systems. This was followed by the identification of the design pattern that addressed the distributed system design. Applying pattern matching for this architectural design issue led to the discovery of the "Proxy" design pattern for the hybrid system.

The proxy pattern allows clients to be decoupled from their server by providing a local proxy. The proxy is a local stand-in that allows clients to access the remote server. In this way, all clients of the server request information and services from the proxy. The proxy encapsulates the information necessary to obtain data and services from the actual server, and when a client requires information from the server, the proxy marshals this request to the server (Douglass, 1999). This design pattern is one of the distribution patterns. A common situation in which the proxy pattern is applicable is where "a remote proxy provides a local representative for an object in a different address space". The pattern allows us to vary how an object is accessed and its location (Gamma et al., 1995).

This design pattern was applied discriminately for the higher level architectural design and the structural elements of the hybrid system were organized to be consistent with the pattern. The following section describes the use of the proxy design pattern in the development of the FB interface.

4. THE PROXY PATTERN

The hybrid system is comprised of the Arena real-time discrete event simulation model and the function block application embedded in the TINI board. This real-time embedded system is distributed between the PC where the server is located and the embedded function block application on the embedded controller that forms the client. The design problem to be addressed here is to structure the distributed client-server communication.

The simulation model is devised using Arena and the real-time features are modeled in the system using specific function blocks (FB) for routing information to the external controller. The server uses socket abstraction over TCP/IP to talk to the client. In essence the entire simulation model is the server as it knows the whole of the system: i.e. the simulation model holds the information about the job entities, process etc, as well it has a database with its event calendar to track the workings of the model and records statistics. For a system comprising of one embedded controller the structuring required on the server side is the development of the simulation model and the verification and validation of the model.

Regarding the communication aspect on the server side, for the current setup of having one real-time controller linked to the simulation, the available Arena dynamically linked library file is well suited and need not be modified. When more real-time controllers are to be interfaced then the server proxy is required to have multi-threading features to deal with the multiple threads associated with the different real-time controllers. The server side of the hybrid model conforms to the server side structural elements of the proxy design pattern as described.

Therefore, for the hybrid system, it is the client side of the setup that warrants immediate use of the proxy design pattern. In this setup it is known that the server would be located in a remote address space. However, the location of the server though now at a static IP address could be relocated. As well the clients maybe deployed in an environment much different from that of the server's, such is the case in the hybrid model. The server is the Arena simulation model along with the server proxy and client applications are modeled as function block applications embedded in a Java-based embedded controller. The clients can be redeployed across different environments as well.

If the clients are intimately aware of the design details in such a setup then there would be problems when they are redeployed. By using the proxy design pattern we would be hiding and encapsulating the transparency of the remoteness of the server taking into account the client specific embedded controller environments. Along with this, using Java based embedded controllers is beneficial as Java based applications can be platform independent. Thus choosing Java based embedded control is beneficial when accommodating the redeployment of the clients. As long as the controller supports a java virtual machine with an Application Program Interface (API) then we could embed the developed client applications in them.

The proxy pattern adds a proxy between the Abstract Client and the Abstract Server. The pattern has two sides. In the first side, the Client-side Proxies subscribe to the Server-side Proxies, which publish the data under the command of the Concrete Servers. When the Concrete Servers call the send() operation, all the remote Client-side Proxies are notified of the new data (Douglass, 2003).

For the hybrid model a similar setup would serve well for the processing and data exchange involved. Only the deployment would be in a hybrid environment thus only part of the system is a real-time embedded controller. The data exchanged between the Arena model and the client application would be that information specific to jobs to be processed in the embedded controller, i.e., the emulator. The data exchanged is passed in the form of strings to the TINI function block application. In here the local proxy, namely the proxy function block will receive the information. This data/information would be extracted by local SIFBs (Service Interface Function Blocks), to process the information according to the nature of the job.

The hybrid system is designed and implemented conforming to the proxy pattern as illustrated in Figure 4. The pattern implementation in the hybrid system is unique since there is a simulation model and a real-time module. Therefore, the proxy pattern implementation has to be fine tuned to the specific environment. It is not possible to have the class relations in the implementation exactly as specified in the default template of the pattern but nevertheless the implementation does adhere to the proxy pattern specification (Soundararajan and Brennan, 2005).

For the hybrid setup, the pattern does a good job of isolating the subject from knowledge that the server may be remote. The client will be simplified by having the proxy, not having to deal differently with all the local FBs to subscribe data from the server. The proxy also encapsulates the knowledge of how to contact the server and what communication process it is using to talk to the server. Thus if the communication media changes or if the client environment changes the proxy can be changed to suit the needs appropriately. Also if the client environments do not support a Java virtual machine JNI's (Java Native Interfaces) can be built to serve the purpose.

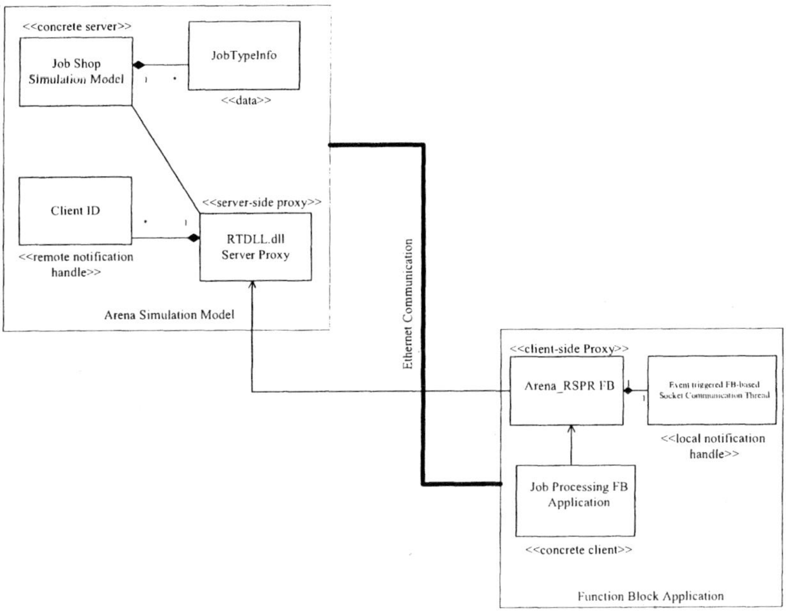

Figure 4 – Hybrid implementation of the proxy pattern

Since there is one client proxy instance than many client FB instances the traffic on the communication media is reduced. This benefit is not fully realized for the hybrid system as of now but when there would be multiple clients, this streamlining of the IPC would be very beneficial. Also, the subscription policy is event triggered due to the event triggered FB application. Although there is a continuous thread, confirm messages are used in the socket stream to request and receive messages from the server.

5. DEVELOPMENT OF THE CLIENT PROXY AND TESTS

The first step is to model the required SIFB, named "ARENA_RSPR", in the IEC 61499 (IEC, 2000) context by formulating the properties with appropriate event and data connections allowing for graphical representation and service representation. IEC 61499 specifies in general terms the way in which an interface to an SIFB is defined. A graphical representation of this is given in Figure 5.

The behavior of the SIFB is specified using Time-sequence diagrams. These diagrams help visualize the order in which the various messages or events occur. The Time-sequence diagram for the ARENA_RSPR depicts five transactions:

1. normal_establishment
2. unsuccessful_establishment
3. normal_data_transfer
4. server_initiated_termination

5. client_initiated_termination

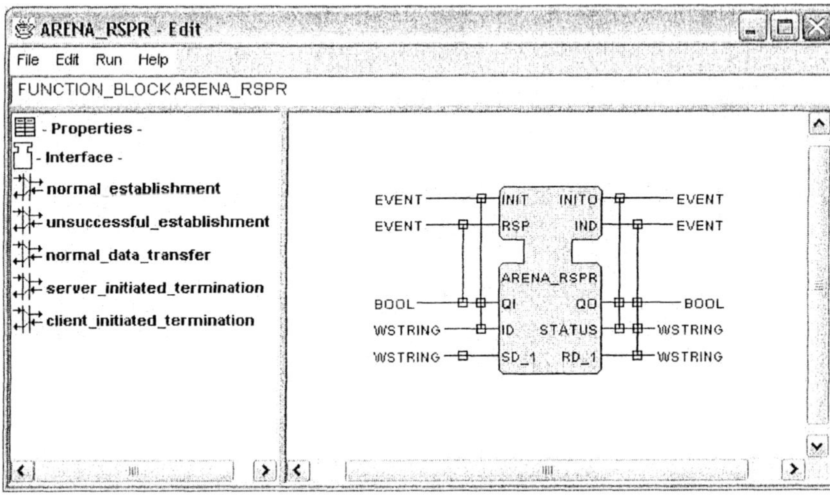

Figure 5 – Graphical representation of the client proxy

In this section, we will focus only on the "normal_data_transfer" sequence as shown in Figure 6, which specifies the data transfer between the server and the client. Arena sends the data to TINI which is acknowledged by the IND event output. When the TINI application has completed the machine execution the RSP event is triggered on the TINI and the confirmation message is received by the Arena model. The service sequence is as follows.

```
SEQUENCE normal_data_transfer
  ARENA.sendData(RD_1) -> TINI.IND+(RD_1);
  TINI.RSP+(SD_1) -> ARENA.receiveData(SD_1);
END_SEQUENCE
```

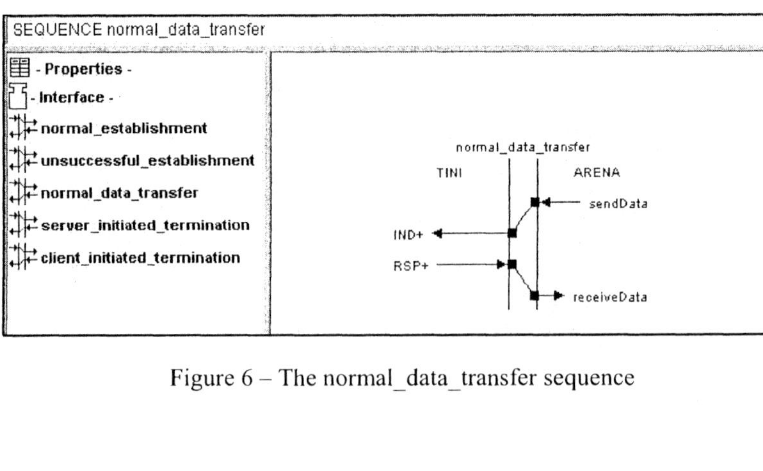

Figure 6 – The normal_data_transfer sequence

The ARENA_RSPR FB was developed as per the above specifications in Java using object-oriented principles. The process involved studying the abstract superclass for communication service interfaces in FBDK (Holobloc, 2006) and inheriting this class to develop the appropriate Arena class.

Figure 7 illustrates the program working with established client-server communication. In this case, the client proxy function block connects to the simulation model at IP Address 136.159.105.74 over port 4334.

The simulation model sends the process part instruction for job with TGID 2 to ARENA_RSPR FB and upon completion of the process instruction a confirmation message "0 2 0" is sent back to Arena. The next process instruction for job with TGID 3 is then sent to ARENA_RSPR and the confirmation message "0 3 0" is sent back to Arena model. The subsequent job to be processed, job with TGID 4 is then sent to the Arena model.

6. CONCLUSIONS

In this paper we have provided a summary of our work on the design and development of a simulation-agent interface for real-time distributed control system benchmarking with a specific focus on the design and development of the design pattern used for this interface.

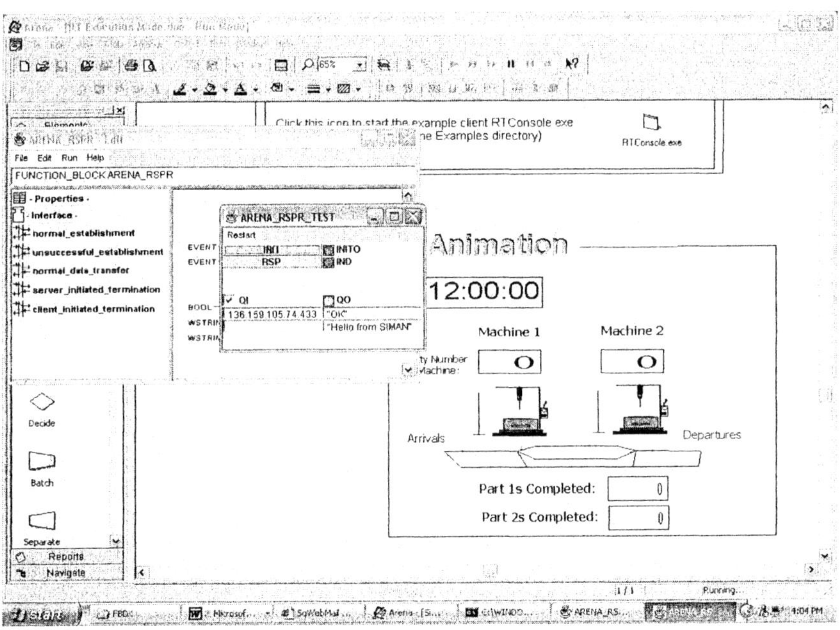

Figure 7 – Communication establishment

In this case, the proxy design pattern has provided a good framework for the large-scale design of the hybrid application. Organizing the structural elements of the hybrid system to the pattern has allowed us to model the client proxy's role and how it should allow for the client application to access the server for information.

7. REFERENCES

1. Brennan, R.W., and W. O, "Performance analysis of a multi-agent scheduling and control system for manufacturing", *Production Planning and Control*, 15(2), pp. 225-235, 2004.
2. Douglass, BP. Real-Time Design Patterns: Robust Scalable Architecture for Real-Time Systems. Addison-Wesley, 2003.
3. Gamma E, Helm, R, Johnson, R, Vlissides, J. Design Patterns: Elements of Reusable Object-Oriented Software. Addison-Wesley, 1995.
4. Holobloc, Website, http://www.holobloc.com, 2006.
5. IEC TC65/WG6 Voting Draft – Publicly Available Specification – Function Blocks for Industrial Process-Measurement and Control Systems, Part I, Part II and Part III, International Electrotechnical Commission, 2000.
6. Intelligent Manufacturing Systems, Network of Excellence in Intelligent Manufacturing, http://www.ims.org/projects/outline/noe.html, 2004.
7. Kelton, W, Sadowski, R, Sadowski, D. Simulation with Arena, McGraw-Hill, New York, 1998.
8. Loomis, D. The TINI Specification and Developer's Guide, Pearson, 2001.
9. McFarlane, DC, Bussmann, S. Developments in holonic production planning and control, Production Planning and Control, 2000; 11(6): 522-536.
10. Shen, W, Norrie, DH, Barthes, J. Multi-agent Systems for Concurrent Intelligent Design and Manufacturing, Taylor & Francis, 2001.
11. Soundararajan, K., Brennan, RW. "A proxy design pattern to support real-time distributed control system benchmarking", Proceedings of the 2nd International Conference on Applications of Holonic and Multi-Agent Systems, Copenhagen, Denmark, August 22-24, 2005.

12 INDUSTRIAL APPLICATION INTEGRATION USING AGENT-ENABLED SEMANTIC SOA: CAPNET CASE STUDY

Leonid Sheremetov and Miguel Contreras
Mexican Petroleum Institute, Eje Central Lázaro Cárdenas #152, 07730, México, D.F. Mexico
{sher, mcontrer}@imp.mx

The paper addresses the issues of industrial application integration in business processes using agent-enabled Service Oriented Architectures (SOA). In this paper, we show that agent-enabled SOA can play an important role for service integration. Our architecture combines Web services and intelligent agent technologies orchestrated by a business process management system. This architecture is grounded in a semantic service integration model and supported by the CAPNET agent platform tools. We describe the architecture and illustrate the approach by an industrial application scenario from petroleum wells' drilling.

1. INTRODUCTION

An important aspect of software in industry lies in its ever-increasing complexity. The often need to integrate heterogeneous applications composing business processes into corporate-wide computing systems, and even to extend that beyond company boundaries into the Internet, introduces new levels of complexity. In recent years Enterprise Application Integration (EAI) systems have evolved towards Service Oriented Architectures (SOA) (Haller et al., 2005). The SOA model is based on the principle that business functionality is separated and published as self-contained components, called services. Though Web Services (WS) as the most commonly used implementation of traditional SOA reduces the number of point-to-point adapters because every interface is based on WS Description Language (WSDL), it still leaves open the question of semantic interoperability of these interfaces. That is why recently a major effort has been invested in semantic extensions of SOA known as Semantic SOA allowing for scalable and controlled EAI (Sycara et al., 2003; Preist, 2004, Ackland et al. 2005).

Multiagent systems (MAS) are ideally suited for open and dynamic environments, where automatic group formation, multiagent adaptation, agent coordination, etc. could likely be fruitfully adapted for EAI (Jennings, 2001; Tesauro et al., 2004). Therefore there is a great opportunity to join the Web Services and Agent Services (AS) technologies under the SOA paradigm, exploring several

---

Please use the following format when citing this chapter:

Sheremetov, L., Contreras, M., 2006, in IFIP International Federation for Information Processing, Volume 220, Information Technology for Balanced Manufacturing Systems, ed. Shen, W., (Boston: Springer), pp. 109–118.

open topics such as business processes modeling, interaction of AS and WS, semantic description of services, and dynamic service negotiation and composition.

Motivated by the above considerations, we have developed an extension to the CAPNET agent platform (AP) permitting to make a step ahead towards the development of industrial applications for open environments (Contreras et al., 2004). The specific software engineering objectives of CAPNET are: to enable a distributed system to self-configure at runtime initialization; to develop agent templates that enable rapid application development through service composition within the system; to improve the efficiency of businesses through automation of business processes in accordance with the system's overall business objectives. We have developed several prototype systems as a concrete testbed to pursue these objectives. One of them, multiagent hybrid intelligent system (HIS) for Lost Circulation Problem (LCP) is described in the case study section of the paper.

The rest of the paper is organized as follows. First we describe briefly the application requirements that motivated the research and analyze the current trends of EAI architectures in order to formulate the basic requirements to agent-enabled SOA. Further we describe the extensions to the architecture of the CAPNET enabling the integration of services provided by agents in a SOA. Finally we illustrate the developed architecture by example of multiagent HIS called SMART-Drill followed by conclusions.

2. MOTIVATION FOR RESEARCH

An oil company is a multi-business organization, which produces, manufactures, markets and transports crude oil, natural gas and petroleum products. The interest in information technology (IT) for decision support of the operational activities lays in the fact that in oil companies like the Mexican *Petroleos Mexicanos* (PEMEX), senior personnel daily have to solve problems based on extensive data analysis and their experience gained through years of field work. Operation of the oilfield is composed of many complex industrial processes, drilling is one of them. LCP – also known as lost returns - stands for the absence or reduction of drilling mud pumped through the drillstring, which filtrates into the formation instead of flowing up to the surface. It is one of the most common problems: drilling fluid may flow freely into the shallow unconsolidated formations because of high permeability or just because of a broken tube. Drilling may continue, or the mud can be thickened and lost circulation material (LCM) added, in an attempt to cure the problem. Sometimes, the LCP is cured easily. But in most cases, the intervention of experienced petroleum engineers is required to find the most appropriate and efficient solution.

2.1 Application Requirements

Expert systems technology is a feasible option to support drilling operations (Garrouch and Lababidi, 2001). In (Sheremetov et al., 2005), a first desktop version of the HIS for LCP diagnostics was described. During the field-testing phase with the PEMEX Company, the following requirements were identified:

- The need for a real time system's feeding with data from distributed information sources: locating and typing data manually was "a torture" for the operators.

- The need to standardize input data description in order to exchange information with other drilling applications. Data should be described in XML-based Wellsite Information Transfer Standard Markup Language (WITSML) (WITSML, 2006) and be available in a server-based repository.
- The need to contact third parties in order to select the most appropriate solution.
- The need to enable a collaborative interaction between the well operator, petroleum and chemical engineers and project manager, since decision-making process is distributed.
- The need to operate over different oil & gas assets (North, South and Marine zones in the case of PEMEX) and over a common case-base.

Therefore, while developing a second version of the HIS, we considered a distributed architecture to overcome the limitations of the stand-alone approach. This new version was implemented over an agent platform and legacy WS, orchestrated by a business process management system. Even though there was no need for automatic discovery and dynamic service composition (but the need for dynamic instantiation), this solution implied several changes to the CAPNET AP allowing its agents to work within a heterogeneous IT environment in a service-oriented fashion.

2.2 Enterprise Application Integration Using Semantic SOA

Integration of IT is crucial for companies as only integrated information systems can deliver business values, such as efficient decision-making support, instant access to information, data integrity, along with decreased cost of software development and maintenance. Traditional EAI systems provided three types of integration levels (Ruh, 2000): process, transformation and transportation layers. Migration of EAI towards SOA changes these layers in order to provide: (i) a standardized way to expose and access the functionality of applications as services, (ii) an enterprise bus infrastructure for communication and management of services, including message interception, routing, transformation, etc, (iii) an integration architecture between the various services and existing and newly developed applications used in business processes and (iv) a specialized language (like the Business Process Execution Language for Web Services -BPEL4WS-) for composition of exposed functionalities of applications into business processes (Juric et al., 2006).

Semantics should be included as the fifth layer of the integration scheme since adding semantics to the service description, firstly, provides a formal description of the functionality of a service. This description allows the developer to base the manual integration on the knowledge about the meaning of the data. Secondly, the model permits decentralizing of semantics of different systems overcoming one of the principal drawbacks (centralized semantics) of traditional EAI. Finally, semantics bring closer the possibility of composing services dynamically by discovering them at runtime.

The centerpiece of semantic integration is an ontology that conceptualizes and codifies knowledge that can be mapped as a knowledge domain. In the context of this paper all the agents share an ontology of their domain of expertise - their *domain model* - that establishes the terminology for interacting with the agent and its services.

3. SEMANTIC SERVICE ORIENTED ARCHITECTURE OF THE CAPNET

The main objective of CAPNET is to bring an integrated infrastructure for MAS that covers the programming, deployment, administration and integration of agents with enterprise applications within the semantic SOA. In this section we present how this integration was achieved. Due to the space limitations, the integration details are covered at the architectural level only and not at the implementation level.

3.1 Making the Agent Platform Service Oriented

The requirement of integration of services provided by agents in a SOA where service composition can be achieved requires an AP to communicate effectively with agent service requestors and providers. However the current notion of services in the agent community is unstructured and does not provide the required elements for service composition which is crucial for expressing service relationships and defining service flows. Additionally agents do not expose standard interfaces that are consumable by the majority of non-agent software and as a result agent technology is not widely adopted in industrial applications, mainly because of its lack of support for integration with business processes. However even with these limitations, agents are ideally suited to representing problems that have multiple problem solving methods, multiple perspectives and/or multiple problem solving entities and incorporate semantics and knowledge management in a natural way.

An AP is a software architecture that controls and manages an agent community allowing the survival of an agent in a distributed and heterogeneous environment (FIPA, 2005). Based on the FIPA specifications that considers an AP as a set of four components - Agents, Directory Facilitator (DF), Agent Management System (AMS), and Message Transport System (MTS) - representing a set of logical capacities or services, it is possible to make a mapping between the concepts in APs and the features required for an architecture to be considered as Service Oriented. These concepts are: *services* provided by agents, *service provider* - the agents implementing services, *service consumer* (or requestor) - end-user application or another agent, and *service locator* - the DF that acts as a registry and allows for the lookup of service provider interfaces (agents) and service locations.

In order to implement this mapping and be able to establish relations, cross invocation and composition of agent and Web services, several extensions to a basic FIPA compliant AP should be addressed that include:

1. An agent service model compatible with the current standards for SOA.
2. A common model and related language for the semantic description of services.
3. A mechanism for publishing agent provided services in UDDI registries.
4. A mechanism for finding services published in UDDI registries.
5. A common transport protocol for communication.
6. A mechanism for exposing the functionality of agent provided services to Web service clients.
7. A mechanism that enables agents to consume Web services.

The architecture implementing these mechanisms within the CAPNET agent platform is addressed in the following section.

3.2 CAPNET Agent Service Model

The first requirement for the integration, to provide AS model compatibility with the current standards for WS, and its implementation implied the design of a new class of agent for CAPNET that was heavily based on the WS architecture for its specification. The architecture of the "component agent" (Figure 1) was designed as a .NET component that can be easily hosted either by the application that will use its services or by a CAPNET container that can provide it with mobility services.

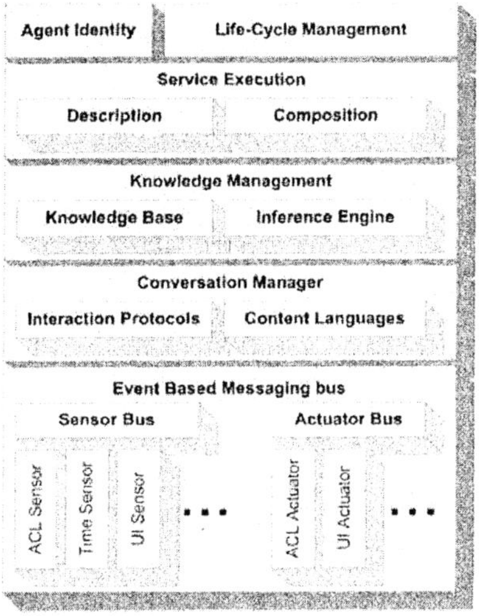

Figure 1 - CAPNET Agent's Architecture.

The "component agent" is composed of the following parts:

- An event based messaging bus that is connected to several components, which constitute the agent's sensors and actuators for the "outside world". Those sensors currently include (but are not limited to) one for receiving Agent Communication Language (ACL) encoded messages, a sensor for time awareness and a sensor for perceiving events generated by the user interface (UI) of the system where the agent resides. Currently, only two actuators are available but more can be implemented and plugged in: an ACL actuator that sends ACL encoded messages to the AP and an actuator that communicates changes in the agent and/or the system state to the UI of the system that contains the agent.
- A Conversation Manager that is capable of handling complex structured conversations based on FIPA Interaction protocols and the common patterns of service invocation (both synchronous and asynchronous) and using a content language for negotiation of services. The currently supported languages include: FIPA-SL0, FIPA-SL1 and propriety CAPNET-KRF.
- A Knowledge Management component that uses a Knowledge Base for representing the agent's domain (based on CAPNET-KRF) and an inference engine to handle this internal knowledge representation.

- A Service Description, Composition and Execution component that enables the specification of the services and their model (both syntactically and semantically) along with its particular implementation. This component allows for basic composition of services inside the agent by providing a specification language (a subset of BPEL) that handles sequence, parallel and loop service execution to form composed services. For the semantic description of services OWL-S is used to define the service profile, process and groundings.
- A component that handles the lifecycle of the agent in terms of the state transitions specified by FIPA and also in terms of its particular functionality. This component tracks the state of the agent and conducts its behavior based on the transitions of a finite state automaton fired by the inputs perceived by the sensors, the internal knowledge and the results of its services.
- A component that handles the agent identity and capabilities.

The second requirement deals with the semantic description of services and it is implemented by adopting the OWL-S ontology model and language allowing service description in terms of profiles, models, and groundings, where the service profile tells "what the service does", the service model tells "how the service works", and the service grounding specifies the details of how a client can access a service. In the case of the CAPNET AP it is a common practice to specify at least two service groundings for a particular service, where one maps to the SOAP interface of the platform and is meant to be used by external clients and another one intended for intra-platform use, maps directly to the agent identity and home address. This adoption of the OWL-S language required changes to be made to the agents' internal representation of services and to the DF service, which should be able to handle this extra information for storage and matching/retrieval operations.

Requirements 3 and 4 for the integration were also addressed by modifying the basic functionality of the DF agent that is now capable of storing service descriptions fully compatible/transformable with that of WS. If an agent requests to publish its service as a WS, the DF handles its registration with one or several UDDI directories. The DF is also capable of performing federated search for services on UDDI directories on behalf of the agents, providing them with service descriptions to enable direct service invocation by these agents using the platform's MTS.

The requirements 5, 6 and 7 are handled directly by the MTS, by the addition of a SOAP transport manager that implements the necessary mechanisms for SOAP based communication. It is implemented as a server extension for MS Internet Information Services (IIS) responsible for providing the access point for the services provided by agents, and to translate between AS and WS descriptions. For more details on the MTS description and the transport manager factory mechanism see (Contreras et al., 2004). The functionality of this transport manager includes ACL to/from SOAP encoding and transformation, service description and invocation conversion to/from WSDL and session handling for service invocation from the platform to external WS.

3.3 Domain Model

The primal intention of the CAPNET knowledge representation model is to provide a common and consistent symbolic representation for the agent domain. CAPNET Knowledge Representation Format (KRF) is based on the FIPA-RDF (FIPA, 2005).

This model solves the ambiguity in defining relationships among entities and introduces the possibility of expressing rules. KRF is used to specify the domain model; the service model is based on OWL-S descriptions.

The CAPNET KRF represents entities from the application domain by using structures known as Objects. Objects instantiate Resources. Properties set the value of some feature or attribute that belongs to the Object/Resource. A Property is composed of: a name, data type and value. If values for some property are restricted to a list of possibilities or valid ranges, then it is said that it has Constraints.

In CAPNET KRF the rules define relationships between known information (antecedents or premises) and information that can be concluded (consequents or conclusions). Antecedents and consequents are propositions that relate properties from resources to some value by one of the following operators: equal to, greater than, greater than or equal to, lower than, lower than or equal to and not equal.

CAPNET Knowledge Acquisition Tool (KAT) and CAPNET Expert System Shell (ESS) are complementary tools enabling knowledge acquisition and encoding in CAPNET KRF and fuzzy reasoning capabilities. CAPNET ESS can be used by the CAPNET agents as an internal (as in the "component agent") or external component. It implements conjunctive, disjunctive and additive algebras of strict monotonic operations on finite ordinal scales represented as multi-sets over the CAPNET KRF (Sheremetov et al., 2005). This representation gives the possibility to take into account the change of plausibility of premises in the rules and to differentiate and refine the uncertainties of conclusions.

4. CASE STUDY DESCRIPTION

The general solution scheme for the LCP consists of several stages. Once the problem is detected, the data stored in daily perforation reports (DPR), laboratory deposits and perforation program (PP) are analyzed in order to determine loss severity. In the case of a total loss severity, the action to perform is placing LCM; else, the solution consists of lowering pump rate and rate of penetration. If the problem persists, additional data from lab tests and the drill log are to be analyzed regarding the viability of lowering density as the next measure. In case of failure, two remaining measures could be applied: placing LCM or squeezing cement plugs. Finally, if none of these controlled the problem, critical measures must be applied.

The business process model developed for the orchestration of the Web and agent services follows the diagnostic LCP process model described above according to the business process scheme shown in Figure 2. The first step in the process is the specification of a user requests for the solution of a LCP problem. An iterative process takes place producing a possible diagnostics and a set of proposed measures at each iteration corresponding to one of the five phases of the problem solution. This diagnostic process involves the use of the IMP_SmartDrill_HIS service, which is actually an orchestration of several services provided by "normal programs" and agents, specified in BPEL4WS and implemented using MS Biztalk Server.

In the case study we modeled three main distributed information sources: DPR, PP and laboratory results. To get information from these sources we developed three agents that expose their functionality as Web services using the CAPNET AS-WS integration mechanism. Each asset has instances of these agents associated with it.

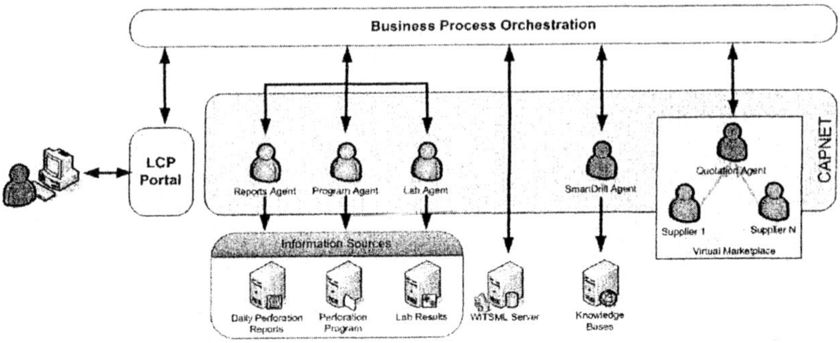

Figure 2 - Business process implementation scheme

The Reports Agent provides the DPR_WITSML_service responsible for converting a DPR format into WITSML objects. The Program Agent implements the PP_WITSML_service responsible for converting a PP format into WITSML objects and the Lab Agent implements the LR_WITSML_Service.

The WITSML Server is a legacy WS implementation of the API that is able to store, manage and retrieve WITSML objects from a central repository. This repository is used as the knowledge source for the inference process that would lead to a diagnostics.

Expert system module is implemented as a service provided by a single agent (SmartDrill Agent) that uses the CAPNET ESS as its engine, a set of knowledge bases as reasoning base and the knowledge from the WITSML Server as the source of facts for the inference. At each phase the agent may propose different measure for solving the problem: lowering pump rate and rate of penetration, lowering density, LCM, squeezing cement plugs or eventually taking extreme measures.

If the proposed measure involves the use of LCM or cement plugs, the Quotation Agent will attend a virtual marketplace for quotations of the required materials. The marketplace is implemented entirely by agents that represent different vendors of the materials that may be needed in an LCP problem. CAPNET agents that behave according to the FIPA-Contract-Net-Protocol and make proposals based on private catalogs currently are only simulating this marketplace. However no actual contracts are made, since the quotation agent only "collects" the proposals and reports the results to the LCP Portal.

When the system completes the iteration it would keep the state and ask the user if the problem has been solved. If this is not the case, it will proceed to the next iteration where the SmartDrill Agent will shift knowledge base and infer from a different set of rules, up to the fifth phase where the process ends. The application is developed as LCP Portal implemented as a set of MS Sharepoint WebParts indicating the current situation of the well, the relevant LCP data, and diagnostics and solution details (Figure 3).

From the server side the following tools were used:

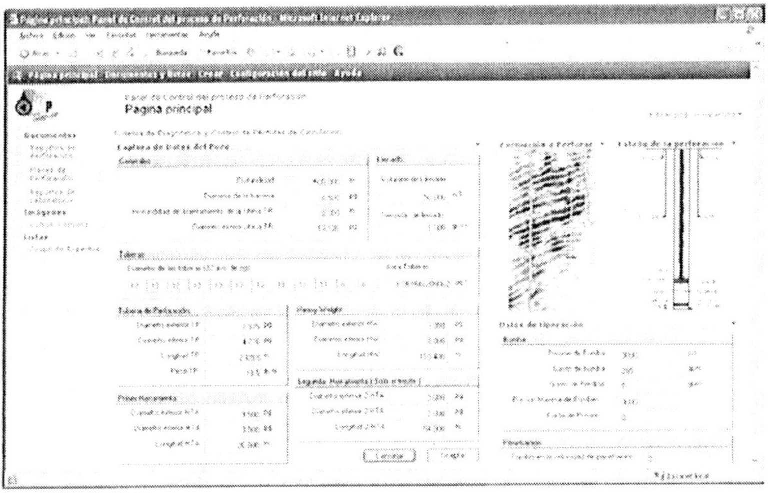

Figure 3 – LCP portal interface

- MS .Net Framework 1.1.
- MS SQL Server 2000.
- MS Biztalk Server 2004 for orchestration and choreography of BP.
- CAPNET Agent Platform.
- MS IIS Web server used as WS container for CAPNET and Sharepoint.
- MS Sharepoint Portal server.
- CAPNET Expert System Shell as the inference component for HIS agents.

5. CONCLUSIONS

In this paper, we descried a CAPNET service model featuring some innovative elements like seamless integration with the model of WS, the most widely adopted implementation of SOA. This integration allows taking advantage of the capabilities of WS-related standards such as service composition while maintaining the MASs' advantages for organization and dynamic coordination enabling easy integration of agent services in industrial applications. Compared to SOA kernel of the JADE (Bellifemine, 2001), CAPNET service model permits a transparent integration mechanism between AS and WS that does not require agents inside the platform to request the service from a local "gateway agent", instead a virtual agent reference is created by the DF when a search is performed and instantiated by the MTS for each invocation of an external WS.

The SMART-Drill application developed using the proposed approach has benefited from the agent technology for working with distributed knowledge sources, for fuzzy reasoning to get LCP diagnostics and for negotiation of services (using FIPA-CNP) by solution provider agents organized in marketplaces. Marketplaces also permitted to take advantage of role assignment techniques managed within agent organizations. Semantic capabilities of agents (domain model, KRF content language and ACL) were used mainly for the application description, while OWL-S extended Agent Service Description permitted agents to discover particular agent instances at the DF (data access agents for the assets). For the application at hand there was no need in automatic WS discovery mechanism and dynamic service composition.

Therefore a key contribution of this paper is twofold: (i) semantic SOA of an agent platform and (ii) a working demonstration of agent-enabled semantic SOA operating in real-life context of industrial applications. Several months ago, the second version of the SMART-Drill was also installed in PEMEX for field-testing.

6. ACKNOWLEDGMENTS

Partial support for this research work has been provided by the IMP within the projects D.00006 and D.00322. The last author would also like to acknowledge the support provided to him by the IMP Postgraduate Studies Program. Special thanks to J. Martinez for his invaluable contribution to the SMART-Drill programming.

7. REFERENCES

1. Ackland, R., Taylor, K., Lefort, L., Cameron, M. and Rahman, J. "Semantic Service Integration for Water Resource Management". In Proc. of the 4th International Semantic Web Conference, Gil, Y.; Motta, E.; Benjamins, V.R.; Musen, M. (Eds.), LNCS 3729, Springer, 2005.
2. Bellifemine F., Poggi A., Rimassa G., Developing multi-agent systems with a FIPA-compliant agent framework, Software - Practice And Experience, 2001 no. 31, pp. 103-128.
3. Contreras, M., Germán, E., Chi, M. and Sheremetov, L. Design and implementation of a FIPA compliant Agent Platform in .NET. J. of Object Technology, ETH Zurich, 3(3), March-April, 2004
4. Foundation for Intelligent Physical Agents – FIPA Specifications, (2005), http://www.fipa.org.
5. Garrouch A. and Lababidi H., "Development of an expert system for underbalanced drilling using fuzzy logic", J. of Petroleum Science and Engineering, vol. 31, pp. 23–39, 2001.
6. Jennings, N. An Agent-based Approach for Building Complex. Software Systems. Comm. ACM, 44, No. 4 (2001) 35-41.
7. Juric M., Sarang P. and Mathew B., Business Process Execution Language for Web Services Second Edition, Packt Publishing, pages 6-30, 2006.
8. Haller A., Gomez J. M., Bussler C. "Exposing Semantic Web Service principles in SOA to solve EAI scenarios". In Proc. of the WWW2005, May 10–14, 2005, Chiba, Japan.
9. Oberle D., Lamparter S., Eberhart A., Staab S., Grimm S., Hitzler P., Agarwal S., and Studer R. "Semantic Management of Web Services using the Core Ontology of Services". In Proc. of the W3C Workshop on Frameworks for Semantics in Web Services. 2005.
10. Preist C. "A Conceptual Architecture for Semantic Web Services". In Proc. of the International Semantic Web Conference (ISWC), 2004.
11. Ruh W. A., Maginnis F. X., and Brown W. J. Enterprise Application Integration: A Wiley Tech Brief. Wiley, 2000.
12. Sheremetov L., Batyrshin I., Martinez J., Rodriguez H., and Filatov D. "Fuzzy Expert System for Solving Lost Circulation Problem". In Proc. of the 5th IEEE Int. Conf. on Hybrid Intelligent Systems, Rio de Janeiro, Brasil, Nov. 6-9, pp. 92-97. IEEE, 2005.
13. Sycara, K. et al. Automated discovery, interaction and composition of semantic Web services. J. of Web Semantics. 1(1):27–46. 2003.
14. Tesauro G., Chess D., Walsh W., Das R., Whalley I., Kephart J. and White S., "A multiagent systems approach to autonomic computing", In Proc. AAMAS, 2004.
15. WITSML Web Site, http://www.witsml.org/, 2006.

13

A MULTIAGENT BASED CONTROL SYSTEM APPLIED TO AN EDUCATIONAL SHOP FLOOR

José Barata[1], Gonçalo Cândido[2], Filipe Feijão[3]
[1] *Universidade Nova de Lisboa / Uninova – jab@uninova.pt*
[2] *Uninova – gmc@uninova.pt*
[3] *Uninova – fnf@uninova.pt*

This paper addresses the design and implementation of a multiagent based control architecture to support modular reconfigurable production systems. The requirements for plugability of modules (manufacturing components) and product changes were considered and tested against an educational platform based on fischertechnik, which resembles a production system composed of several workstations connected by a crane and conveyors.

1. INTRODUCTION

Multiagent systems (Wooldridge 2002) are becoming more and more a paradigm recognised as important in the context of the manufacturing world, not only at the high level aspects of the Enterprise Information Technology (Camarinha-Matos and Afsarmanesh 2001; Lin et al. 1999) but especially at the lower level aspects that include shop floor control, which is the focus of the work being presented here.

It is now widely recognised that external conditions to production companies, such as the need to cope with short life cycles products, stricter quality requirements, low costs, and customized products, can only be solved if agile shop floors can be built (Barata 2005; Gunasekaran 2001; Vernadat 1999). Considering in particular the assembly domain this need is particularly emphasized in (Barata et al. 2005; Onori 2002a; Onori et al. 2003a; Onori et al. 2003b). What is required is not a solution which tries to accomplish all of the envisaged assembly needs within a closed unit (Flexible Assembly systems) but, rather, a solution which, being based on several reconfigurable, task-specific, process-oriented elements (system modules), allows for a continuous evolution of the assembly system: this is known as the Evolvable Assembly Paradigm (EAS) (Alsterman et al. 2004; Onori 2002b). The main points behind shop floor agility are: 1) the need for modular manufacturing components, 2) the need to cope with product changes, and 3) the need for better operational support.

The need for modular components is directly connected with the multiagent paradigm for two reasons. The first reason is related to modeling and abstraction. In

Please use the following format when citing this chapter:

Barata, J., Cândido, G., Feijão, F., 2006, in IFIP International Federation for Information Processing, Volume 220, Information Technology for Balanced Manufacturing Systems, ed. Shen, W., (Boston: Springer), pp. 119–128.

fact it is quite simple to make the connection between a module and a modular component. The second reason is related to the need for plugability. In fact if a modular system is being considered they must be plugged or unplugged at wish. Therefore a multiagent system seems to be ideal because by definition a multiagent system is adequate to plug and unplug agents (in this case component modules). Another aspect connected to a modular system is the fact that a particular assembly or production system is a composition of modules, which can then be also considered as a coalition of modules. This is another connection between production modules and the agent world. A very well known and established work is in fact the work on agent coalitions (Castelfranchi et al. 1992; Klusch and Gerber 2002; Pechoucek et al. 2002; Shehory and Kraus 1995), which can be applied also here (Barata 2005).

The need to cope with product changes is also directly connected to the multiagent paradigm by the considering that an agent is an entity which is autonomous, reactive, and with social ability. In this case the autonomy makes the agent responsible for its own acts which mean that each manufacturing component or module is individually responsible for the actions over the product. Adding or removing one of these modules does not have a big impact on the overall system structure since each module is a confined individual entity. It must be noted that being confined does not mean any interactions with other modules. In fact the social ability that characterise the multiagent world is fundamental to ensure that the modules interact among themselves.

The need for better operational support means all the tasks that are required to support the shop floor while it is operating. This include aspects such as online reconfiguration of production parameters, maintenance support functionalities, advanced diagnostic systems, advanced user interfaces, etc. All these requirements are also better solved if the multiagent paradigm is considered, because among other aspects it is important to model each manufacturing component or module as an entity that includes all relevant information (maintenance parameters, process related variables, etc) that can be used in supporting the aspects referred before. This shows how the connection between modular component and agents is done. Another important aspect at this level is the autonomy that each agent (module) must possess. In fact, each agent must support simultaneously different behaviours or actions. For instance, an agent representing a certain module participating in a coalition (system) must simultaneously do the following tasks: 1) answer requirements for executing some work (for instance, the request to transport one piece from one point to another received by a gantry robot), 2) keeping monitoring the internal sensors of the module, 3) answering requests from external users for internal parameters values, 4) interacting with other modules for advanced diagnosis functionalities, etc.

From the previous paragraphs the connection between multiagent systems and Evolvable (agile) Shop Floors were made and emphasised in order to create the motivation for the work described in this paper.

The objective of this paper is two fold. The first one is to describe a multiagent based control system that supports an agile shop floor. The second goal is emphasizing that educational platforms such as Fischertechnik (Fischertechnik 2006) if mimicking real production systems are perfectly adequate to test real production systems because it is possible to add or remove modules, without compromising the general goals, in a way that would be impossible with a real system.

The system used to test the multiagent based control is described in section 2, while the multiagent architecture is described in Section 3. The implementation aspects are presented in section 4 and finally the conclusions are described in Section 5.

2. SYSTEM DESCRIPTION

The production system used to test the multiagent based control system is composed of several machining centres, which are linked by conveyors and a gantry robot.

The MOFA France kit by Staudinger Gmbh (See Figure 1) simulates a flexible manufacturing system as a closed loop manufacturing circuit, achieving various possible situations based on generic manufacturing tasks.

1 - ToolMachine1 : Welder
2 - ToolMachine2 : Driller
3 - ToolMachine3 : Painter
4 - ToolMachine4 : Dryer
5 - MachineTable1 : Rotative Conveyor
6 - MachineTable2 : Slide Conveyor
7 - Conveyor Belt 1
8 - Conveyor Belt 2
9 - Conveyor Belt 3
10 - Conveyor Belt 4
11 - Robot - Crane
12 - Storage Area

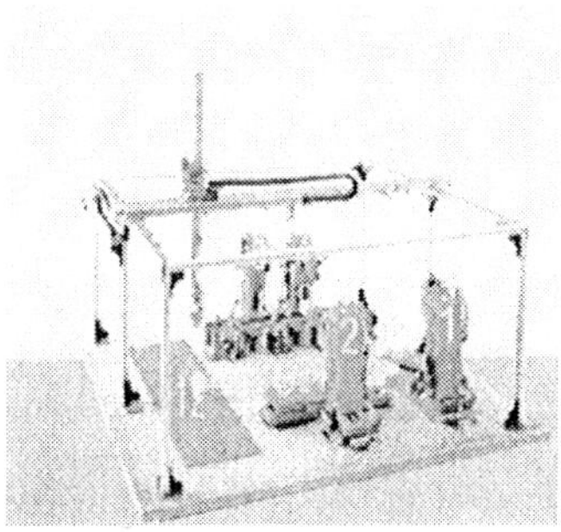

Figure 1 - MOFA France Kit

This kit has 4 machines (adaptable to any kind of manufacture tasks), a storage area, a cartesian robot (crane functions), local transporters and some material presence sensors.

The product which is stored in the storage area must be transported by the crane to the different machines. Some of the machines are also interconnected by conveyors. Some machines, such as machine 1, also include some critical resources such as a rotating table with two positions on the opposite sides of the table. This rotating table is a critical resource since it represents a kind of bottleneck for the machine 1. In the situation in which the machine is processing a piece but a position is vacant the crane can always bring a new piece. If, on the other hand, the table has already one finished piece and is processing one the crane should not bring a new piece before removing the finished one from the rotating table. This is more relevant if optimization of crane movements is to be considered.

Machine 2 is fed by a linear moving table with two places for pieces. This table can move between two positions. When the piece is being processed the other place is available for loading or unloading operations. This means that the crane can load or unload pieces from the machine 2 left or right hand side. One place of the table is always positioned in front of machine 2.

Machine 3 and 4 are connected by 3 bidireccional conveyors and the crane has several feeding places. It is possible therefore to connect pieces from machine 3 to 4 and vice versa using only the conveyors.

The pieces which represent pallets with parts that need to be transformed are very simple. In the case of this work each piece will be considered as an entity (agent) that includes the set of operations that need to be done.

A multiagent based control system was therefore suited in order to deal with critical resources, information exchange, parallel behaviours, sequential actions, as well as providing a clear and amicable graphical interface.

3. ARCHITECTURE

The main goal for the system being created is to provide a multiagent based architecture that could allow the addition and removal of new modules (manufacturing components) and control a system in which for each piece added with different process plans (this means different products) the system is able to realize the necessary operations without any new programming. The system must adapt to the new coming products represented by the pieces, which are also individually represented as agents.

3.1 System Architecture

The architecture needs to fulfill the previous requirements and provide a fully functional system, in addition to integrate several technologies.

The system architecture (see Figure 2) consists of several agents working together, trying to accomplish the production objectives. These agents have diverse abstraction levels according to their role on the environment, and they execute actions of distinct complexity levels. The roles of each production agent are known as skills since they identify what are the capabilities presented on each agent. These skills are important during system creation and also during system operation. The main agents involved are: Controller or Coalition Leader agents, Manufacturing Resource Agents (MRAs), and Agent Machine Interface (AMI).

The shop floor is composed of different Controller Agents, which represent different cell workstations that provide complex skills to other entities, such as pieces following its own process sequence. In the case of the considered scenario there is only one cell, and therefore only one coalition. Basically, these Controller Agents represent a group (or coalition) of manufacturing resource agents (also know as Basic Agents) whose high level requests are coming from the sequential order-build mechanism issued by the Controller Agent. This software mechanism tries to use all the available physical skills, setting up complex behaviours based on more simple ones, offered by lower level layer agents.

Manufacturing Resource Agents (MRA) represent a specific manufacturing component, such as robots, conveyors, machinery, etc. Using an agentified shop floor element, which encapsulates all provided skills, functions, interaction behaviours and internal status, it gives a clear view about a specific component, as well as giving a panoramic visualization of entire shop floor elements.

For some physical agents there is a need to use an AMI (Agent-Machine Interface), since physical agents developed are generic for each type of component and all the hardware connection is specific for each type of manufacturing equipment. So, if all the low level hardware communication is present on AMI and it can interact with its respective physical agent, there is no need of reprogramming. Instead, it is only necessary to configure agent parameters in the shop floor database. In a futuristic approach, these AMIs can be developed by module suppliers and delivered together with newly purchased equipment. In a simplistic form, the AMI connects the physical agent to the hardware, and then just executes orders from directly connected higher level agents. For legacy system, these AMI need to be created, in order to permit physical agent interactions.

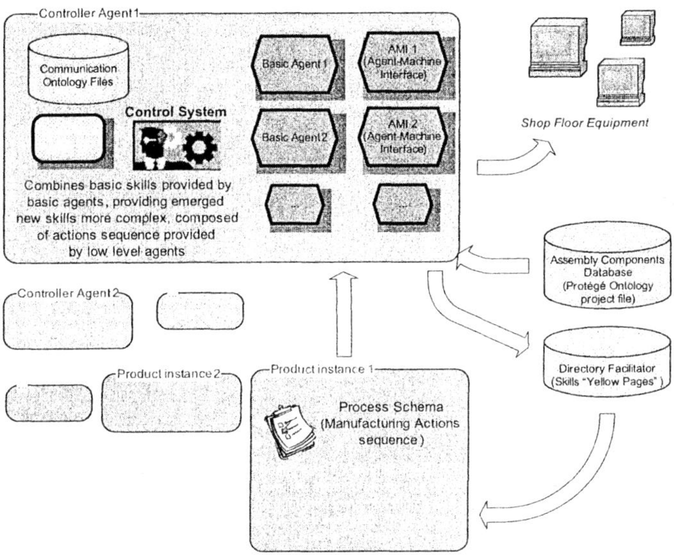

Figure 2 - System Architecture

The shop floor environment is commanded by each unique piece that wants to be processed. Each Piece Agent has its own custom production process pre-defined by user, and can be easily and rapidly changed thru a graphical interface. After the configuration phase, each piece knows its production objectives and behaves accordingly to it. Each piece agent goes thru its process plan order and asks the agent controller for the operations it needs until all of them are carried out.

All the interactions between agents use ACLMessages following FIPA (FIPA 2002) directives that permit necessary information exchange between them according to the immediate situation.

3.2 Manufacturing Resource Agent (MRA)

A MRA is nothing more than a shop floor component abstraction or module. It encapsulates the core of the physical component itself: skills, role and interaction behaviours.

The MRA reacts to external action requests, in order to execute its published skills, issuing all the necessary actions downwards in order to accomplish the requests received from the Controller agent that leads the coalition he is participating in. The MRA architecture is represented in Figure 3.

Again, all the interaction between agents follows previous methods, except the hardware equipment communication that is manufacturer dependent. This last type of communication needs original equipment interface libraries.

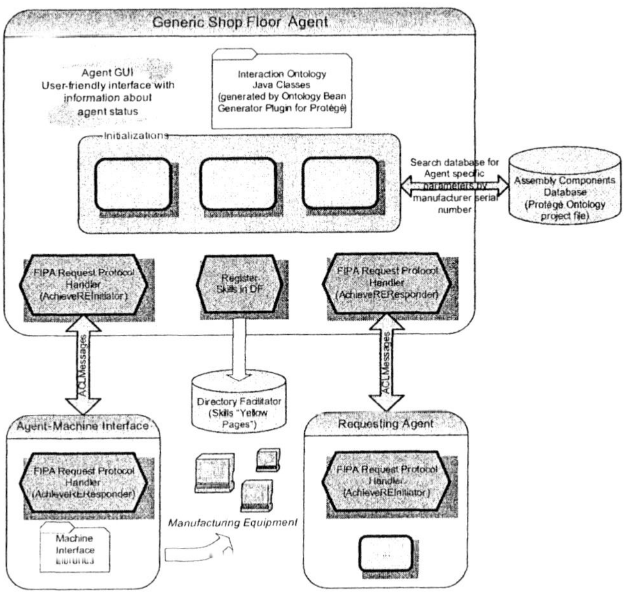

Figure 3 - Generic MRA architecture

3.3 Ontology

Taking into account the dimension and complexity of the system to implement, it was necessary to create a manufacturing components ontology, which takes into account its individual characteristics, as well as the relations established among them. The manufacturing components ontology defines the structure of the components in a production line as well as specific catalog information. At initialization phase, the MRAs search for configuration information in structures defined at this ontological level. MRAs get their configuration information from the database using their serial number.

Therefore, the ontology created, it is more correct in fact to call it Knowledge model because is a derivation from an ontology, is simple but generic. It supports the introduction of new components, adjustment of parameters and allows easy conformity. By using this process, it is easier to support fast alterations in the product, process and disposal of the assembly line, which is a crucial factor nowadays.

Skills are defined as basic or complex. A basic skill implements a service in its more elementary form. A complex skill is composed of basic skills or other lower level complex skills, supplying services more complex than only the sum of constitutive elements.

Skills have to be considered according to two different perspectives: one defines what the service is, the other defines how it will be executed. A skill can therefore be considered for configuration (what is the service) and for execution (how it will be executed). This information can be found on the skills database and allows other agents to know which type of interaction and necessary parameters are needed to get the action done.

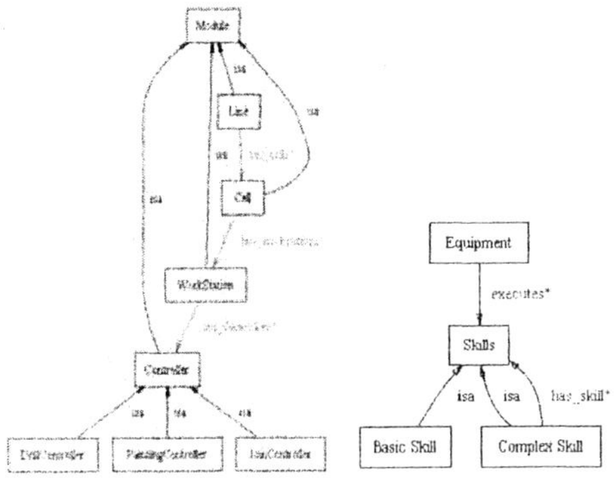

Figure 4 –Assembly Components Ontology relations examples

When an agent is launched, its first behaviour is reading the Assembly Database searching by component type and manufacturer serial name. Once the database entry is found, it contains all catalog information, AMI information, available skills and position on equipment tree. The agent collects that information, assumes a specific physical identity attached to real assembly component, registers skills and it is ready for interaction with other agents.

The Controller Agent database entry defines which agents it represents, which complex skills it affords and what are the interaction parameters. It contains all the relevant information about the coalition (group of MRAs) it is coordinating.

4. IMPLEMENTATION

4.1 Tools

The system was implemented using the JADE 3.3 (Java Agent DEvelopment Framework) (JADE 2001). JADE provides a distributed agent environment system that supports different functionalities such as FIPA specifications, compliant behaviours, and graphical tools for debugging and deployment phase.

For ontology and database development it was used the Protégé Ontology Editor and Knowledge Acquisition System (Protégé-2000 2000). This innate ontology development tool allows easy and prompt creations and updates because it is possible to integrate a Protégé project file in an ordinary java program. Instances of the MOFA France equipment were created in the Assembly Ontology and JADE behaviours were created in each MRA for initialization. In order to create the communication ontology used for agent interaction, it was used the Ontology Bean Generator *plugin* for automatic conversion from Protégé to Java files to be used by all agents.

All the created code was made on Java JDK 1.4.6, using Eclipse 3.1 Development environment added with JADE and Protégé libraries.

4.2 Hardware

Thc hardware connection to the multiagent system implementation is composed of several software layers giving consecutively higher level functionalities, which enables a final implementation isolated from low level system.

The connection to the MOFA cell was tricky because the equipment is also used to support classes using traditional ways of programming such as C++ or JAVA. Therefore all the outputs and inputs of the MOFA are connected to an Data Acquisition Card located in one PC. Hence it was required to guarantee concurrency at this lower level. It must be taken into account that the system would be simpler, at this level, if each module (component of the cell) included each own controller.

Using a previously developed Java class for MOFA France kit control in each AMI project, it is possible to have multiple connections to the hardware, enabling the concurrency the system was lacking. Whenever a request arrives, each AMI simply needs to call one of the functions belonging to the developed Java class.

4.3 Interaction

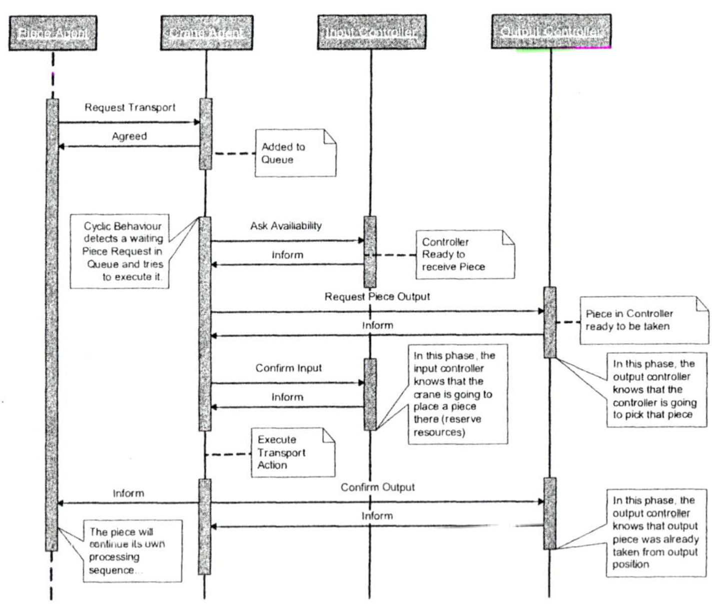

Figure 5: Crane Agent interactions

The cooperation between all the agents implies a robust interaction schema that supports information exchange and deals with error situations to avoid deadlocks and abnormal stopping situations. One of the most vital cases to deal with is the Crane Agent interactions as it consists of an extremely critical resource. In fact, no material can be transported between workstations if it fails (see Figure 5).

The above UML sequence diagram stands for a normal request for transportation made by a piece to the MRA crane agent from one origin (InputController) to a destination (OutputController). If some controller is occupied, the piece is remained on the queue and another request is chosen; then, its feasibility is checked and evaluated, according to its priority. With this approach of checking each request for transportation deadlocks are avoided.

All the messages exchanged follow FIPA Request Interaction Protocol, in SL1 language with communication ontology terms transporting information according to message performative and final objective. As an example, the Piece Agent that wants to be transported from Storing area to Join workstation interacts with Crane Agent using the next ACLMessage content example (see Figure 6).

```
((action
  (agent-identifier
    :name CraneMOFA1@PortatilGoncalo:1099/JADE
    :addresses (sequence http://PortatilGoncalo:7778/acc))
  (PieceAction
    :PieceStatus Incomplete
    :PieceRequest Crane_Transport
    :PieceArgs (sequence unknown Storing Join))))
```

Figure 6: ACLMessage content (PieceAgent to CraneAgent)

Due to space restrictions it is not possible to show all the interactions existing in the created system. However, they follow a similar pattern to this one, in which the major differences reside on the content of the SL1 language.

5. CONCLUSION

Due to space restrictions many important details related to the implementation were not included. The project behind this paper is working at the Electrical Engineering Department of the New University of Lisbon. The authors therefore believe the first goal enunciated at the beginning of the paper was achieved. In fact, a multiagent based control system is controlling the educational MOFA production system, which is composed of various heterogeneous manufacturing modules. Each of these modules was agentified and a multiagent based coalition was created that emulates the system.

The different tests made with different "products" with different process sequences proved the ability of the system to cope with these changes without any reprogramming. Moreover, it was also tested the addition of new models without implications in programming.

The second goal of proving that an educational platform can be useful for testing purposes were also achieved since the experience showed that a lot of effort can be saved if an educational platform is used instead of a real one. Of course, the authors recognise that the real system can always provide new challenges or “surprises” but to test the concept of product changes and process adaptability the use of such an educational platform is very recommendable.

6. ACKNOWLEDGMENTS

The authors would like to acknowledge the European Commission which has partially funding this work trough the IP EUPASS project and the students Luís Ribeiro, Nuno Veríssimo, Alexandre Silva, João Veríssimo, Rodrigo Guerreiro, José Santos e Rui Milagaia.

7. REFERENCES

1. Alsterman, H., Barata, J., and Onori, M. (2004). "Evolvable Assembly Systems Platforms: Opportunities and Requirements." Intelligent Manipulation and Grasping, R. Molfino, ed., IMG'2004, Genova, 18-23.
2. Barata, J. (2005). *Coalition Based Approach For ShopFloor Agility*, Edições Orion, Amadora - Lisboa.
3. Barata, J., Camarinha-Matos, L. M., and Onori, M. "A Multiagent Based Control Approach for Evolvable Assembly Systems." *INDIN 05 - 3rd International IEEE Conference on Industrial Informatics*, Perth - Australia.
4. Camarinha-Matos, L. M., and Afsarmanesh, H. (2001). "Virtual Enterprise Modeling and Support Infrastructures: Applying Multiagent Systems Approaches." Multi-Agent Systems and Applications, M. Luck, V. Marik, O. Stepankova, and R. Trappl, eds., Springer-Verlag, Berlin, 335-364.
5. Castelfranchi, C., Micelli, M., and Cesta, A. (1992). "Dependence Relations Among Autonomous Agents." Decentralized A.I. 3, E. Werner and Y. Demazeau, eds., Elsevier Science Publishers B. V, Amsterdam, NL, 215-227.
6. FIPA. (2002). "FIPA ACL Message Structure Specification." *XC00061F*, FIPA - Foundation For Intelligent Physical Agents, Geneve.
7. Fischertechnik. (2006). "http://www.fischertechnik.de."
8. Gunasekaran, A. (2001). *Agile manufacturing : the 21st century competitive strategy*, Elsevier, Oxford ; New York.
9. JADE. (2001). "http://sharon.cselt.it/projects/jade/."
10. Klusch, M., and Gerber, A. (2002). "Dynamic Coalition Formation Among Rational Agents." *IEEE Intelligent Systems*, 17(3), 42-47.
11. Lin, F. R., Tan, G. W., and Shaw, M. J. (1999). "Multiagent enterprise modeling." *Journal of Organizational Computing and Electronic Commerce*, 9(1), 7-32.
12. Onori, M. "Evolvable Assembly Systems - A New Paradigm?" *International Symposium on Robotics*, Stockholm, Sweden.
13. Onori, M. "Evolvable Assembly Systems - A New Paradigm?" *ISR2002 - 33rd International Symposium on Robotics*, Stockholm, 617-621.
14. Onori, M., Barata, J., Lastra, J., and Tichem, M. (2003a). *European Precision Assembly - Roadmap 2010*, Assembly-Net Project.
15. Onori, M., Camarinha-Matos, L. M., and Barata, J. (2003b). "European Assembly - Status Report." *Assembly Automation*, 23(1), 8-12.
16. Pechoucek, M., Marik, V., and Bárta, J. (2002). "A Knowledge-based Approach to Coalition Formation." *IEEE Intelligent Systems*, 17(3), 17-25.
17. Protégé-2000. (2000). "http://protege.stanford.edu."
18. Shehory, O., and Kraus, S. (1995). "Coalition Formation among Autonomous Agents: Strategies and Complexity." From Reaction to Cognition, C. Castelfranchi and J. P. Muller, eds., Springer-Verlag, Heidelberg, 57-72.
19. Vernadat, F. B. (1999). "Research Agenda for Agile Manufacturing." *International Journal of Agile Management Systems*, 1(1), 37-40.
20. Wooldridge, M. J. (2002). *An Introduction to Multiagent Systems*, J. Wiley, New York.

PART B

NETWORKED ENTERPRISES

14 DISTRIBUTED WORK ENVIRONMENTS FOR COLLABORATIVE ENGINEERING

H.-H. Erbe[1], D. Müller[2]
[1]*Center for Human Machine Systems, Technische Universität Berlin*
10623 Berlin, Germany, heinz.erbe@tu-berlin.de
[2]*artecLab, Universität Bremen*
28334 Bremen, Germany, mueller@artec.uni-bremen.de

Extended and networked enterprises distribute the design of products, planning of the production process, and manufacturing regionally if not globally. The usual face-to-face work is going to be replaced, at least partly, if not totally, by computer mediated collaboration. Awaited are reduced problems of resolution cycle time, increasing productivity and agility, and reduced travel to remote sites, enabling more timely and effective interactions, faster design iterations, and improved resource management. Mixed reality based work environments support distributed collaborative work between remote sites. An application for solving a control task collaboratively is described.

1. INTRODUCTION

Distributed design of products over remote sites is a strong demand of the industry for saving time and therefore costs. Centers for e-design got established at different universities, just to name Pittsburgh University and University of Massachusetts, Amherst (Nnaji, 2004), among others. The communication is mostly done with data-transfer using distributed CAE-systems like CATIA, i.e. a visual access to the products under the design process. Therefore with this trend to extend the designing and processing of products over different and remotely located factories the problem arises how to support an effective collaboration of the involved workforce. The usual face-to-face work is going to be replaced, at least partly, if not totally, by computer mediated collaboration. Collaboration demands a deep involvement and commitment in a common design, production process or service; i.e. to work jointly with others on a project, on parts or systems of parts. Enterprises investing in new information technology and communication infrastructures have also to consider the important issues in developing a culture and shared values that can facilitate the adoption of such technologies. Investment in advanced technologies may not necessarily result in improved communication by and between the employees. Often managers and the developers of IC - technology assume too much about the anticipated use of the technology by the employees. For most employees, interacting in a virtual mode via mediating technologies may be totally new and may cause anxieties. This loss of human contact could be balanced by maintaining continuous

Please use the following format when citing this chapter:

Erbe, H.-H., Müller, D., 2006, in IFIP International Federation for Information Processing, Volume 220, Information Technology for Balanced Manufacturing Systems, ed. Shen, W., (Boston: Springer), pp. 131–140.

communication as well as by holding occasional face-to-face meetings for information sharing and support. The workers or engineers have to negotiate a shared understanding. This is done partly through video conferences. But in the camera and monitor mediated world of videoconferencing the collaboration support is limited to the senses of sight and sound and eliminate the sense of touch. As a result, even in state of the art videoconference rooms using the highest quality equipment, the sense of co-presence enjoyed by individuals in the same room is never fully achieved.

CAVE-technologics (Computer Animated Virtual Environments) are capable of to supporting the feeling of an immersion into a common workspace. Available CAVE's at the market are very expensive. But cheaper versions with nearly the same performance are possible and therefore also affordable for small enterprises and training institutions. Integrating an analog-digital connection technology called Hyper-Bond into CAVE's to support tele-design and tele-maintenance can improve collaborative work.

The paper discusses first related developments, then mixed reality concepts helpful for distributed work, and a new development for collaborative task solving.

2. RELATED DEVELOPMENTS

The project "Future Workspaces" (2003), funded by the European Commission, developed the vision: Supported by CAVE's, engineers will be able to work seamlessly in their workspace environment with documents, scientific models and virtual prototypes, both alone and collaboratively with distant colleagues as if they were in the same room. Virtual and hybrid prototypes will be available as a means for engineers to design new products. They will access specialized services via intelligent and secure network infrastructures that can detect, predict and satisfy user demand at any time and any place through location- and device-independent applications, which are able to seamlessly migrate across network technologies. A step-by-step work timeline could be stored, allowing other users to understand the previous course of the work and thus be able to effectively carry on with the tasks in the work process. The diminishing cost of technological equipment will enable companies to implement technologically integrated spaces, housing large embedded displays, networked furniture, wireless devices for tracking people and remote access to supercomputers etc. The integration of technology with physical space will make the present computer systems and interfaces less visible or transparent in the future environment.

The Collaboration@Work-Report (2005) illuminates European future research on new working environments and practices. The report emphasizes the importance of technologies with a mediating role among a distributed workforce and as a glue to bring together diverse technologies (such as mixed reality) to support collaboration among people and by interaction with other artifacts like robots, actuators and sensors.

Schaffers et al (2005) considered mobile and networked workplaces (Figure 1). Different scenarios are described to challenge existing frames of mind in envisaging different types of workplaces than those already existing and to show plausible future directions for innovation. The underlying technology can be called ubiquitous or pervasive computing. The authors develop a critical perspective of ubiquitous computing or ambient intelligence because it could leave the users without control. Only in few cases is the focus of ambient computing on systems supporting humans in the understanding of what is going on at the level they choose, and supporting them in suggesting courses of action rather than automatic action.

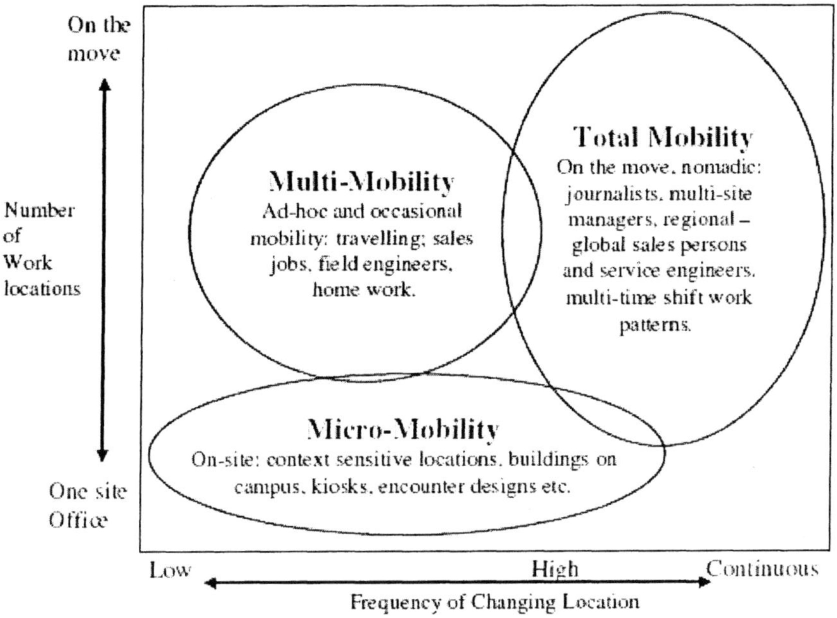

Figure 1 - Mobile workplaces categorization (Schaffers, 2005)

Participants of a workshop held in Aarhus, Denmark (2005) argued: there seems to be a need for a balanced view emphasizing how ambient systems need to be visible, how they can be deconstructed, how coherence can be achieved, how they can provide stability and understandability, and in particular how users can stay in control when dealing with a large number of autonomous components.

Bohn et al (2004) discuss economic effects. Despite the number of potential economic advantages, there are also substantial risks involved when relying on such technologies for large parts of an economy. The increasing automation of economically relevant aspects and the exclusion of humans as decision makers could certainly become a cause for concern. Under "normal" circumstances, automated control processes increase system stability – machines are certainly more dependable than humans for those who have to devote their whole attention to a particularly boring task. But situations that have not been anticipated in the software can easily have disastrous consequences if they are not directly controlled by humans. Other problems might arise from the intricate interplay of several automated processes, which might quickly escalate into an unanticipated feedback loop that gets out of control.

3. MIXED REALITY CONCEPTS

Mixed Reality environments as defined by Milgram (1999) are those in which real world and virtual world objects are presented together on a single display. Mixed Reality techniques have proven valuable in single user applications. Meanwhile, there has been done research on applications for collaborative users. Mixed Reality could be useful for collaborative distributed work because it addresses two major issues: seamlessness and enhancing reality.

When people talk face-to-face to one another, while collaborating on a real world task, there exist a dynamic and easy interchange of focus between the shared workspace and the human interpersonal space. The shared workspace is the common task area between collaborators, while the interpersonal space is the common communications space. In face-to-face conversation the shared workspace is often a subset of the interpersonal space, so there is a dynamic and easy change of focus between spaces using a variety of non-verbal cues (Billinghurst et al, 1999).

Ishii et al (1994) defines a seam as a spatial, temporal or functional constraint that forces the user to shift among a variety of spaces or modes of operation. Seams can be of two types:

Functional Seams: Discontinuities between different functional workspaces, forcing the user to change modes of operation. A functional seam exists when one has to shift the attention from one workspace to another.

Cognitive Seams: Discontinuities between existing and new work practices, forcing the user to learn new ways of working. A cognitive seam is that between computer-based and traditional desktop tools.

Functional and cognitive seams in collaborative work at distant workspaces changes the nature of collaboration and produces communication behaviors that are different from face-to-face conversation. For example, video-mediated conversation does not produce the same conversation style as face-to-face interaction. This occurs because video cannot adequately convey the non-verbal signals so vital in face-to-face communication, thus introducing a functional seam between the participants. Sharing the same physical space positively affects conversation in ways that is difficult to duplicate by remote means. Most current video conference equipment separates the user from the real world.

Mixed Reality concepts can also enable co-located users to view and interact with shared virtual information spaces while viewing or even manipulating the real world at the same time. This preserves the rich communications bandwidth that humans enjoy in face-to-face meetings, while adding virtual images normally impossible to see. Mixed Reality interfaces can overlay graphics, video, and audio onto the real world. This allows the creation of workspaces that combine the advantages of both virtual environments and seamless collaboration with the real environment. Information overlay may be used by remote collaborators to annotate the users view, or may enhance face-to-face conversation by producing shared interactive virtual models. In this way Mixed Reality techniques can enhance communication regardless of proximity and support seamless collaboration with the real world, reducing the functional and cognitive seams between participants. These attributes imply that Mixed Reality approaches would be ideal for multi-computer supported *cooperative* work (CSCW) applications. Roschelle & Teasley (1995) give a widely accepted definition of collaborative versus cooperative work: "We make a distinction between 'collaborative' versus 'cooperative' problem solving. Cooperative work is accomplished by the division of labor among participants, as an activity where each person is responsible for a portion of the problem solving. We focus on collaboration as the mutual engagement of participants in a coordinated effort to solve the problem together. We further distinguish between synchronous (i.e. working together at the same time) and asynchronous activity. Although we do not propose that collaboration cannot occur in asynchronous activity, we focus on face-to-face interactions, which can only occur as a synchronous activity."

When "cooperation" is exchanged for "collaboration," the requirements on Mixed Reality are stronger. Not only vision and sound, but also force and haptic rendering are demanded to providing real immersion in common as well as remote distributed workspaces. More importantly, users (workers, engineers) can see each other's facial expressions, gestures and body language, increasing the communication bandwidth. They can continue doing real world tasks while talking to collaborators in the conferencing space, and it is possible to move the conferencing space with the trackball so that collaborators do not block critical portions of the field of view. While Milgram & Colquhoun (1999) refer to mixed reality as "the merging of real and virtual worlds" such that "real world and virtual world objects are presented together within a single display," Benford et al (1998) broaden this to consider the joining together of whole environments. Milgram's approach, Benford argues, might be suited to a range of applications such as medical imaging, tele-surgery, machine maintenance, and the control of robots, for example. Benford's approach to mixed reality is more comprehensive. He developed the concept of creating transparent boundaries between physical and synthetic (virtual) spaces. Thus, instead of being superimposed, two spaces are placed adjacent to one another, and then stitched together by creating a "window" between them. This is close to a CAVE-like construction.

The approach of Bruns (2005) is close to Benford's. Bruns' concerns that most existing collaborative workspaces strictly separate reality and virtuality. For example, when controlling a remote process, one can sense and view specific system behavior, control the system by changing parameters, and observe the process by video cameras. The process, as a flow of energy - controlled by signals and information - is either real or completely modeled in virtuality and simulated. In mixed reality concepts distributed environments information flow can cross the border between reality and virtuality in an arbitrary bidirectional way. Reality may be the continuation of virtuality or vice versa. This bridging or mixing of reality and virtuality opens up some new perspectives not only for work environments but also for learning or training environments (Müller, 2005). The next section describes an application of this approach for collaborative distributed work.

4. COLLABORATIVE TASK SOLVING BETWEEN REMOTE SITES

Mejía et al (2004) reports on a collaborative e-engineering environment for product development in connection with the manufacturing process development of a Dry-Freight Van. Both parts were carried out at remote workplaces connected via the internet. AutoCAD and Mechanical Desktop were used for designing. The collaboration was supported by NetMeeting, and for the coordination MS project was used.

4.1 Connections with a Hyperbond

Bruns et al developed and tested a collaborative work scenario between remote sites (Bruns & Erbe, 2005, Müller, 2005). The task was to develop and test an e-pneumatic control circuit (Figure 2) for automatic welding operations: workpieces are fed and clamped; the welding operations are activated. When the welding is finished the clamps open and the workpieces are ejected.

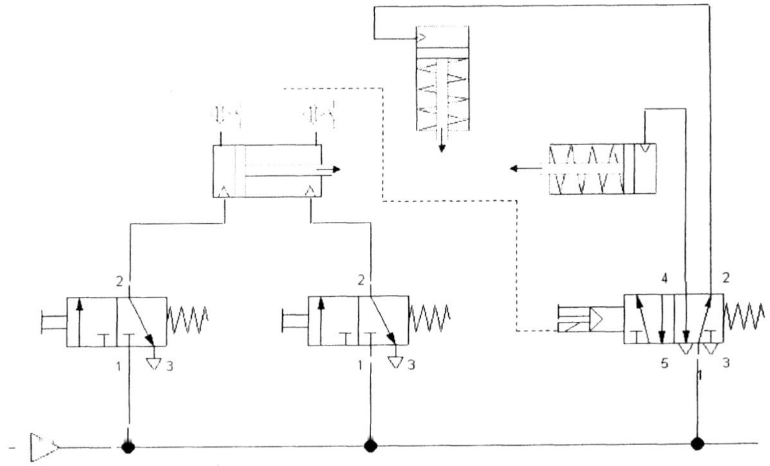

Figure 2. - E-pneumatic control circuit for automatic welding

Within this scenario three enterprises at different locations are involved to perform the following tasks: developing the control virtually, testing it at a real workbench, and manufacturing the device. To solve this task collaboratively work spaces are linked by the Internet. Figure 3 shows the real and the virtual workbench connected with analog-digital and digital-analog converters respectively. This connection is called a Hyperbond (Bruns, 2005). The virtual workbench can be accessed via the Internet so that engineers at remote locations are able to solve the task collaboratively. They have always video and sound feedback of the real workbench and their coworkers at the virtual workbenches. The engineers at the real workbench can also manipulate the components at the virtual workbench, and all can communicate visional and auditorial (Fig. 4 to 6).

When a solution of the control tasks is found at the virtual workbench the solution can be partly or completely exported via the Hyperbond interface to the real workbench. Also, the manufacturer of the welding device is connected to give his comments regarding the feasibility of a solution of the control task. The audio communication uses Internet telephony.

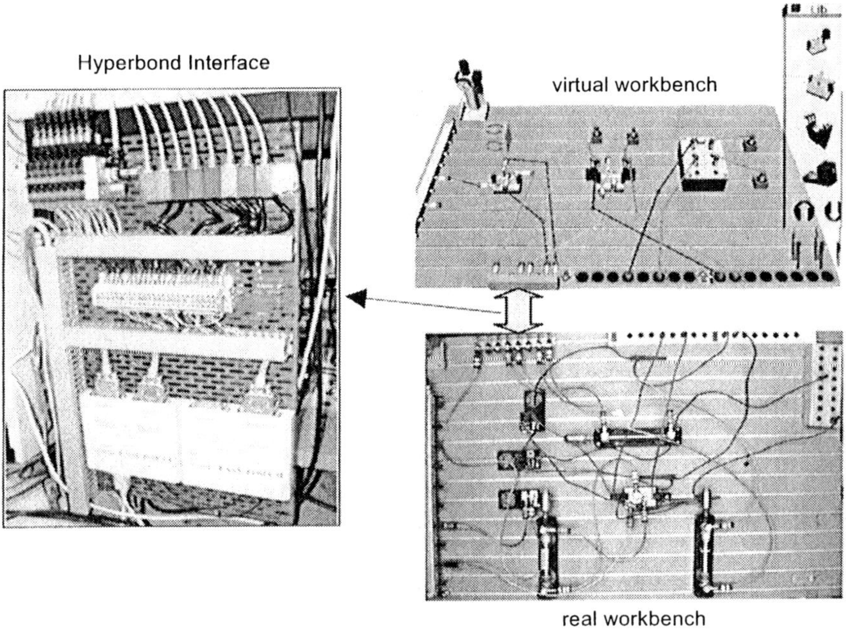

Figure 3 - Real and virtual electro-pneumatic workbench connected with a Hyperbond.

4.2 CAVE´s as workspaces

The connected workbenches are located in CAVE -like constructions. These consist of a scaffolding with canvases where the images of other workspaces with the persons working in it are beamed at. The architecture is shown in Fig. 4. The beamers are controlled by computers connected to the server.

The real and virtual workbenches were implemented as a Web Service to take advantage of the web technology (e.g. easy accessibility, platform independence). A central module is the Mixed Reality (MR) Server which realizes this Web Service. The Web Service itself processes HTTP requests and also manages the sessions of all remote users. Relevant data belonging to a certain work session is stored on the server, like virtual model data, support material and background information. The WWW front end consists of a HTML page including a Virtual Construction Kit (VCK) and a video stream window.

The VCK itself is a VRML based tool for assembling virtual worlds: by dragging and dropping objects from a library onto the virtual workbench new objects (e.g. cylinders, valves, and switches) can be added. Each of these objects has connectors which can be linked to other ones. Links can either be tubes (air pressure) or wires (electricity). Connections between real and virtual workbench elements were realized by the aforementioned Hyperbond technology.

Figure 4 – Sketch of CAVE's (scaffoldings with canvases, beamers and computers)

A Hyperbond consist of two parts: a real and virtual one. Thus any flow through this kind of object is automatically forwarded to reality or vice versa. Changes in reality are distributed by an updated simulation but also by a camera observing the real hardware. The virtual part of a running session can be stored on the server and reloaded later to continue the work task.

Figures 5 and 6 show the arrangements used for the test cases. The common virtual workbench and the real workbench (via video projection) are available via the Internet and visible at an enlarged screen or are beamed at canvases fixed at the scaffolding.

Figure 5 – Workspace with the real workbench and the virtual one in front, and the remote counterparts screened on the left and right canvases

The described mixed reality application for collaborative work is under further development for remote maintenance. Web based service for remote repair, diagnostics and maintenance (RRDM) is meanwhile widespread (Biehl et al, 2004). But it is almost limited to providing only audio and vision. Controls, drives, and sensorized machine parts can be diagnosed remotely through the manufacturers or service providers. For the maintenance of mechanical parts it might be of advantage to touch the parts. This requires force feedback and haptic rendering and is still under research (Yoo & Bruns, 2005). However, the above described application, a real workbench connected to virtual workbenches, can already be used for maintenance training between remote workplaces.

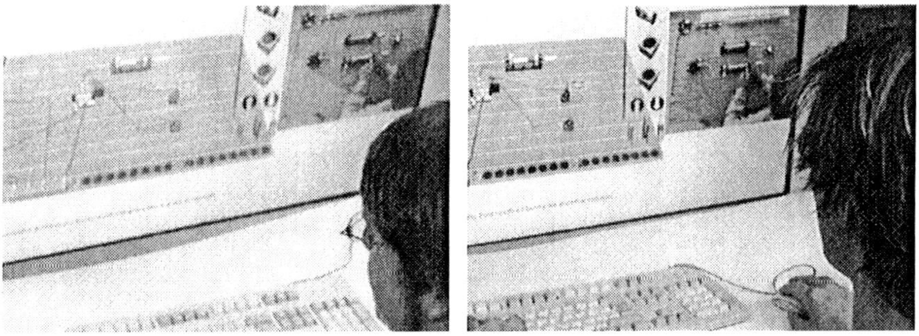

Figure 6 – Two remote workspaces connected with the remote workbench

5. CONCLUSION

Presently some research effort is being done to foster collaborative engineering between remote sites. One of the problems is to seamlessly connect the real to the virtual world. As an example the paper described the connection of a real workbench with the remote located virtual ones. Scaffolding with canvases represents a low cost CAVE, where the remote and local participants can immerse into a common workspace for solving a task. In comparison to other developments for supporting collaborative engineering, that mostly improves only the common work at CAD workplaces, the new approach allows that more than two participants at different remote sites can work together at the same time. They have always access to the real workbench. When some are sitting inside a CAVE and working at the real workbench connected with their colleagues at the remote CAVE´s and virtual workbenches, the test persons got nearly the feeling to working at the same site.

Workplaces providing not only vision and auditory perception but also haptic perception are still in its infancies. It is the goal for future research.

Although the presented development for collaborative work over distant sites has been explained with a simple example, the concept of Hyper-Bond was also developed for continuous events. When enhanced with force feedback it will be a useful tool for remote collaboration. The used CAVE technology has been developed in a project with students of applied informatics. It will be configured and tested to be usable for industrial sites.

6. ACKNOWLEDGEMENT

We would like to thank our colleagues at artecLab: Martin Faust (CAVE environment) and Yong-Ho Yoo (testing) for their contributions. Parts of our work were supported by the EU-Program Leonardo da Vinci, and ALFA/RexNet ALFA-II-465-A.

7. REFERENCES

1. Benford S., Greenhalgh C., Reynard G., Brown C., Koleva B., "Understanding and Constructing Shared Spaces with Mixed-Reality Boundaries", in: ACM Transactions on Computer-Human Interaction, Vol. 5, No. 3, September 1998.
2. Biehl M., Prater E., McIntyre J. R., "Remote Repair, Diagnostics, and Maintenance", in: Communications of the ACM, Vol. 47, No. 11, November 2004.
3. Billinghurst M., Kato H., "Collaborative Mixed Reality", in: Mixed Reality – Merging Physical world and Virtual World, Proceedings of the International Symposium on Mixed Reality (ISMR '99), March 19-21, 1999, Yokohama, Japan, 261-284.
4. Bohn J., Coroama V., Langheinrichs M., Mattern F., Rohs M., " Living in a World of smart everyday Objects - Social; Economic, and Ethical Implications", in: Journal of Human and Ecological Risk Assessment, Vol. 10, No. 5, pp. 763-786, October 2004.
5. Bruns, F. W., "Hyper-bonds – distributed collaboration in mixed reality", in: *AnnualReviews in Control*, Vol.29, No.1, pp. 117-123, March 2005.
6. Bruns F W., Erbe H-H., "Mixed Reality with hyper-bonds - a means for remote labs", to appear in: Control Engineering Practice, Elsevier Ltd., Oxford.
7. Ishii H., Kobayashi M., Arita K., "Iterative Design of Seamless Collaboration Media", in: *Communications of the ACM*, Vol 37, No. 8, August 1994, pp. 83-97.
8. Milgram P., Colquhoun H., "A taxonomy of real and virtual world display integration", in: Mixed Reality - Merging Real and Virtual Worlds, Y. O. a. H. Tamura, ed., Springer Verlag, Berlin,1999, pp. 1-16.
9. Mejía R., Aca J., Ahuett H., Molina A., "Collaborative e-Engineering Environments to support integrated Product Development", in: Emerging Solutions for future Manufacturing Systems, L.M. Camarinha-Matos, (ed.), Springer Verlag, NY, 2004, pp. 271-278.
10. Müller D., "Designing learning spaces for Mechatronics", in: MARVEL- Mechatronics Training in Real and Virtual Environments, Müller D. (ed.), Impuls Vol. 18, NA/Bundesinstitut für Berufsbildung, Bonn, 2005.
11. Nnaji, B., Wang, Y. & Kim, K.Y., "Cost effective Product Realization", in: Proceedings 7th IFAC Symp. Cost Oriented Automation. Elsevier Ltd., Oxford, 2005.
12. Roschelle J., Teasley S. D., "The construction of shared knowledge in collaborative problem solving ", in: O'Malley C.E. (ed.): Computer Supported Collaborative Learning, Springer Verlag, Heidelberg 1995, pp 69-97.
13. Schaffers H., Prinz W., Slagter R., "Mobile and location-aware Workplaces and global value Networks: a strategic roadmap", in: Collaborative networks and their breeding environments, Springer Verlag, NY, 2005, pp. 425-436.
14. http://www.aarhus2005.org/, Aarhus,Denmark
15. Project "Future Workspaces" (2002 – 2003), http://www.avprc.ac.uk/fws/
16. Collaboration@Work 2005 Report: http://europa.eu.int/information_society/activities/atwork/hot_news/publications/
17. Yoo, Y., Bruns, F. W., „Realtime collaborative Mixed Reality Environment with Force Feedback", in: Proc. 7th IFAC Symposium on Cost Oriented Automation, Elsevier Ltd., Oxford, 2005.

15 *MARKET OF RESOURCES* AS AN ENABLER OF (INTER-)ORGANIZATIONAL RECONFIGURATION DYNAMICS

Maria Manuela Cunha
Polytecnic Institute of Cávado and Ave, Higher School of Technology, Portugal
mcunha@ipca.pt

Goran D. Putnik
University of Minho, School of Engineering, Portugal
putnikgd@dps.uminho.pt

The Virtual Enterprise model is an emerging approach to answer to the new requirements of the business environment, relying on dynamically reconfigurable partnerships, with extremely high performances, strongly time-oriented while highly focused on cost and quality, in permanent alignment with the market, and strongly supported by information and communication technology, dictating a paradigm shift face to the traditional organizational models. Reconfiguration dynamics is a main characteristic of this model, which claims for supporting environments implementing a set of functionalities. Some existing technologies can partially support this organizational model, but the reconfiguration dynamics can only by assured by an environment able to managing, controlling and enabling networking and dynamics in virtual enterprise creation/ reconfiguration. The Market of Resources is an environment coping with these requirements. The paper presents and discusses its validity in terms of improving the potential inter-organizational reconfiguration dynamics.

1. INTRODUCTION

Enterprises are no longer confined to its four walls, and the networked value chain concept is extending to several organizational approaches, with extremely high performances, strongly time-oriented while highly focused on cost and quality, and permanently aligned with business opportunities, to answer to the new requirements of the business environment. Some of these approaches rely on dynamically reconfigurable partnerships in permanent alignment with the market, and strongly supported by information and communication technology, dictating a paradigm shift face to the traditional organizational models. The leading organizational model incorporating these characteristics is the Virtual Enterprise (VE) organizational model, characterized as a dynamic organizational model.

Please use the following format when citing this chapter:

Cunha, M. M., Putnik, G. D., 2006, in IFIP International Federation for Information Processing, Volume 220, Information Technology for Balanced Manufacturing Systems, ed. Shen, W., (Boston: Springer), pp. 141–148.

The implementation of the VE model should assure the required reconfiguration dynamics, which as we will see in the paper is dependent of (1) the reduction of reconfiguration costs and effort, i.e., requires a balancing between reconfiguration dynamics and reconfiguration time and costs and (2) the capability to preserve the firms' private knowledge on products or processes.

In the paper we discuss the VE reconfigurability requirement and the requirements of reconfiguration dynamics, introduce some environments to cope with these requirements and introduce the *Market of Resources* as a tool for managing, controlling and enabling networking and dynamics in VE integration.

This research makes two contributions: (1) to industry managers, it highlights the importance of dynamic organizational models, as the ultimate paradigm; and (2) to IS technologists: it alerts to the development of a new generation of electronic marketplaces, designated as market of resources, which supposedly should cope with the high reconfigurability of the VE model, at low transaction cost and preserving the firms' private knowledge.

2. VIRTUAL ENTERPRISE RECONFIGURATION DYNAMICS AND BUSINESS ALIGNMENT

For the last two decades, global competition has strengthened the significance of a company's ability to introduce new products, while responding to increasingly dynamic markets with customers rapidly changing needs, and thus claiming for shortening the time required to design, develop and manufacture, as well as for cost reduction and quality improvement. In the past a product could exist without great changes (adaptations, redesigns).

Reconfigurability, i.e., the ability of fast change face to the unpredictable changes in the environment (market), is a requirement of the VE to keep the partnership aligned with business requirements and is a consequence of product life cycle dynamics, i.e., business and market dynamics. This requirement implies the ability of (1) flexible and almost instantaneous access to the optimal *resources* to integrate in the enterprise; (2) design, negotiation, business management and manufacturing management functions independently from the physical barrier of space; and (3) minimisation of the reconfiguration or integration time.

Reconfiguration, meaning substitution of resources providers, generating a new instantiation of the network, can happen mainly from three reasons:

1. Reconfiguration during the network company life cycle is a consequence of the product redesign in the product life cycle, to keep the network aligned with the market requirements.
2. Reconfiguration as a consequence of the nature of the particular product life cycle phase (evolutionary phases).
3. Reconfiguration can happen also as a consequence of the evaluation of the resources performance during one instantiation of the network, or voluntarily by rescission of participating resources, wiling to disentail from the network.

VE dynamics considers a succession of network's states (physical configurations of the VE) along the time, i.e. the network reconfiguration dynamics.

Dynamics means precisely the intensity of change the VE is subject of. A VE is defined as a reconfigurable network to assure permanent business alignment, in transition between states or instantiations (configurations) along time, (Figure 1).

In (Cunha & Putnik, 2005a), the authors propose two parameters of *Reconfigurability Dynamics*: the number of requested reconfigurations per unit of time (*Reconfiguration Request Frequency*) and the time to reconfigure (*Reconfiguration Time*). *Reconfigurability dynamics* is directly proportional to the number of requests and inversely proportional to the time to make operational the reconfiguration (selection, negotiation and integration of resources in the VE). *Reconfigurability Dynamics* can be measured by a *ratio* between the frequency of reconfiguration requests and the reconfiguration time.

Ideally, reconfiguration time should tend to zero, and stable configuration durations should be dictated by business alignment needs, to keep VE performance at its maximum level. Due to reconfiguration cost and time, most of times dynamics is sacrificed by increasing the duration of stable and sometimes less performing configurations.

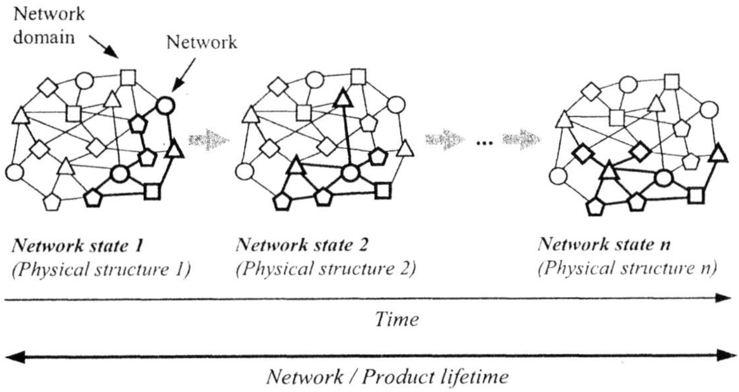

Figure 1 – Networking dynamics considers a succession of network's states along the time (Cunha & Putnik, 2005a)

3. NETWORKING AND DYNAMICS DISABLERS

The main critical aspects associated to the recent concept of dynamically reconfigurable global networked structures, i.e., the main factors against networking and reconfigurability dynamics are (1) the reconfiguration costs and (2) the leakage of private information.

In an ideal business environment, a firm makes an informed assessment of the relevant costs, benefits and risks of outsourcing *versus* internal procurement. If there exists a profitable opportunity to outsource a service or operation, the client and the suppliers enter into a contract with a full knowledge of the nature of the work, signing a complete and explicit written agreement covering all aspects of the outsourced service and payments. But in most contractual relationships things do not happen this way.

The costs of outsourcing are composed of both the explicit cost of carrying out the transaction as well as hidden costs due to coordination difficulties and contractual risks. The major costs associated with outsourcing include (1) the transaction costs and (2) the leakage of private information. In dynamic organisations, *transaction costs* are the firm reconfiguration costs, associated to partners search, selection, negotiation and integration as well as permanent monitoring and the evaluation of the partnership performance (Cunha & Putnik, 2003b).

A firm's private information is information that no one else knows, and gives a firm an advantage in the market. Most of the times, this private information is a core competitive advantage that distinguishes a firm from its competitors. Networking or partitioning tasks between resources providers increases the risk of losing control of such type of information, which only through complete contractual agreements could be safeguarded, and furthermore, through an environment assuring trust and accomplishment of the duty of seal. The implementation of the networked structures requires tools to enable the preservation of the firm's knowledge. When considering dynamically reconfigurable networks, the risk of leakage of private information increases.

The implementation of the dynamic organisations requires the existence of tools and environments that, by reducing reconfiguration costs and reconfiguration time, overcome these two disabling factors, allowing dynamics as high as required to assure business alignment.

4. RECONFIGURABILITY REQUIREMENTS

The implementation of VE as a dynamic model requires that a number of functionalities (such as the summarized bellow) be assured.

- The permanent alignment of the VE with the market requirements can justify a dynamic process of the VE performance evaluation and the analysis of reconfiguration opportunities.
- Reduction of reconfiguration costs, which is based on the existence of computer aided tools and flexible and almost instantaneous access to the optimal *resources* to integrate in the enterprise, negotiation process between them, selection of the optimal combination and its integration.
- All must speak the same language. Search of potential resources providers, negotiation, etc, requires a specific and unambiguous resources representation language; ebXML and other XML standards allow communication; processes specification, of which RosettaNet is a good example are also indispensable; Webservices are becoming a core technology. However, an environment assuring knowledge-based support is indispensable, as the user (or the VE owner) will not be able to manage VE reconfiguration.
- The ability of identifying reconfiguration opportunities, either due to unaccomplishment from partners, or because alternative resources providers offering better solutions or prices are identified, is important to keep alignment with the market and simultaneously to assure an increased performance and competitiveness of the VE.
- The reduction of the time-to-contract and risk minimisation in a contractual agreement, which is a prerequisite for the VE model implementation. To

reduce the contractualisation time, almost real-time contractualisation between the parties to integrate in the VE should be provided.
- The cost associated to the integration of an VE surpasses the sum of costs of making contacts, with the cost of overcoming distance, etc., it is also the opportunity cost because of taking a few more hours or days to locate resources or to reconfigure the VE. Speed is a fundamental characteristic, as one instantaneous physical structure (instance) of a VE may last only for a few days or even hours, so there it is necessary to act almost on real time.
- The ability to find the right potential partners to undertake further negotiation is essential. The domain for partners search should be large enough to provide good solutions.

5. ENVIRONMENTS TO COPE WITH THE INTER-ORGANIZATIONAL DYNAMICS REQUIREMENT

Value chains have been supported by a wide variety of technologies to communicate, but the pace of competition requires more intelligent and effective information and communication systems and technologies. Literature suggests that "traditional" Internet-based tools (such as WWW search engines, directories, e-mail, electronic marketplaces, etc.), can support some activities of VE integration, helping from procurement processes until the search of partners for a partnership, including electronic automated negotiation, electronic contracting, and market brokerage (Cunha & Putnik, 2003a; Dai & Kauffman, 2001; O'Sullivan, 1998; Wang, 2001).

Khalil and Wang (2002) have proposed ways for information technology to enable the VE model, by providing: (1) Web-based information systems, supporting B2B and B2C applications. (2) Sophisticated customer databases, supporting data mining, enhancing business intelligence and decision support. (3) Support for organizational learning (4) Groupware supported coordination and decision-making.

Several supporting infrastructures and applications must exist before we can take advantage of the VE organizational model, such as: electronic markets of resources providers, legal platforms, brokerage services, efficient and reliable global and intelligent information systems, electronic contractualisation and electronic negotiation systems, and decision support systems and tools.

To contribute to the reduction of search time in procurement and engineering, and to reduce transaction costs, manufacturers in several industries created electronic marketplaces, to pool their purchasing power and to develop technology platforms to exploit networked technologies. Electronic markets like *Covisint* (http://www.covisint.com) in the auto industry or *Elemica* in the chemicals (http://www.elemica.com) provide the environment to improve the VE dynamics.

In fact, several technologies and valuable applications have been developed that can support activities of the VE model, however, they do not cope with the requirements of the VE model, i.e., they do not implement the functionalities initially introduced and do not assure the high reconfigurability requirement.

6. THE MARKET OF RESOURCES

The *Market of Resources* (MR) is an institutionalised organisational framework and service assuring the accomplishment of the competitiveness requirements for VE dynamic integration and business alignment (Cunha & Putnik, 2005b; Cunha, Putnik, Gunasekaran, & Ávila, 2005). The operational aspect of the *Market of Resources* consists on an Internet-based intermediation service, mediating offer and demand of resources to dynamically integrate in a VE, assuring low transaction costs and reduced reconfiguration time (as demonstrated in (Cunha & Putnik, 2003b, 2003c)) and the partners' knowledge preservation.

The service provided by the MR is supported by (Cunha & Putnik, 2005b; Cunha et al., 2005):

- A knowledge base of resources and results of the integration of resources in previous VE,
- A normalized representation of information,
- Computer aided tools and algorithms,
- A brokerage service, and
- Regulation, i.e., management of negotiation and integration processes, as well as contract enforcement mechanisms.

The MR is able to offer (Cunha & Putnik, 2005b):

- Knowledge for VE selection of resources, negotiation, and its integration;
- Specific functions of VE operation management; and
- Contracts and formalizing procedures to assure the accomplishment of commitments, responsibility, trust, and deontological aspects, envisaging that the integrated VE accomplishes its objectives of answering to a market opportunity.

7. THE *MARKET OF RESOURCES* AS AN ENABLER OF RECONFIGURATION DYNAMICS

This section summarises some of the results of the validation of the *Market of Resources* using a cost and effort model specifically developed based on a demonstrator to simulate the MR performance, and its comparison with the utilisation of "traditional" Internet-based techniques (*WWW* search engines, *WWW* directories and e-mail) to support search, negotiation and selection of a set of resources providers to integrate or reconfigure a VE. These simulations were based on cost and effort models respectively for the MR and for the "traditional" Internet-based techniques. The cost and effort models, as well as results and discussion of this validation can be found in (Cunha & Putnik, 2003b, 2003c, 2006).

Consider K to be the number of required resources to integrate in a VE project (creation or reconfiguration). Figure 2 represents search and selection time (in minutes) corresponding to K = 2, K = 3 and K = 5, using the Market of Resources (represented by the line labelled "Market" in the legend of the figure) and using traditional internet-based technologies (the line labelled "Trad" in the legend of the figure), in function of the search domain dimension. A similar simulation was also undertaken to estimate cost instead of time.

For example, using the MR, it is possible to search, negotiate, contactualise a Virtual Enterprise creation/reconfiguration, for K=3 and an average search domain of 150 potential resources providers in around five hours. For the same situation, the value offered by the "traditional" Internet-based tools is ten times longer.

Based on analytical simulation results based on the already mentioned cost and effort models, the authors identified the domain of opportunities for the MR, in function of the number of required resources (K) and of the search domain dimension (Figure 3). The figure identifies the region where the *Market of Resources* presents increased efficiency when compared with the traditional Internet-based technologies, in function of the number of required resources (K) and the search domain dimension, considering time and cost parameters. Considering the time parameter, the MR reveals increased efficiency in almost all the situations (see the "Time line" in Figure 3).

The MR demonstrated increased efficiency in coping with complex products (especially K >= 3) and increasing with the solution space dimension.

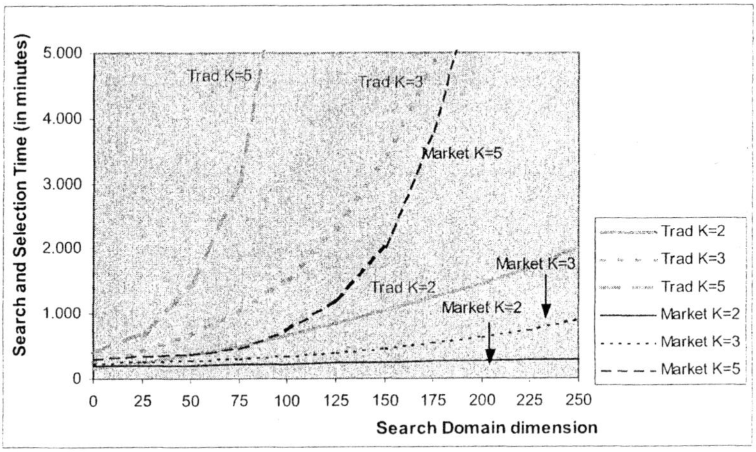

Figure 2 – Search and Selection Time (in minutes) in function of the Search Domain dimension, for different situations of the number of required resources (K)

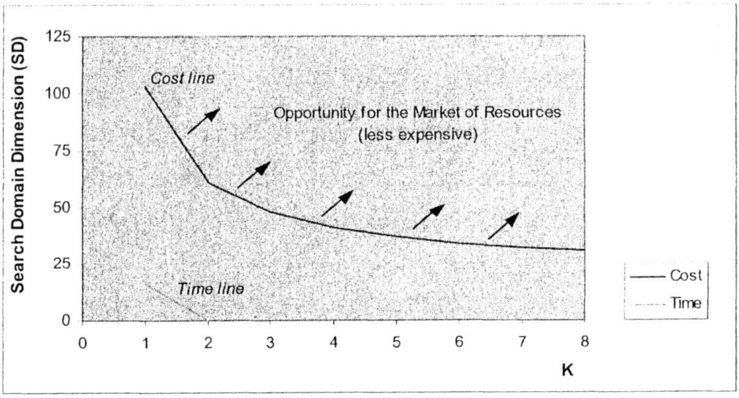

Figure 3 – Break-even points based on search and selection cost and time

8. CONCLUSIONS

The *Market of Resources* revealed the ability to support higher reconfiguration requirements than the traditional tools (due to the more reduced reconfiguration time and cost it allows) and its suitability increases with product complexity (here traduced by the number of required resources, K). Traditional tools only support simple products (one at each time) and do no support dynamics.

By reducing reconfiguration time and cost, the MR is an enabler of reconfiguration dynamics, an enabler of dynamic organizational models as the VE one. Obviously, dynamic organisational models are not general and "all-purpose" solutions. This model represents and adequate solution for highly customised products, with small series, in highly competitive and changing environments where permanent business alignment is crucial, i.e., situations where partnership stability is low (sometimes very low), dependency between partners is very weak and reconfiguration dynamics should be as high as possible, given the permanent monitoring of the structure to traduce the most competitive solution at every moment of the product life cycle.

9. REFERENCES

1. Cunha, M. M., & Putnik, G. D. (2003a). Agile/Virtual Enterprise Enablers: a comparative analysis. In D. N. Sormaz & G. A. Süer (Eds.), *Proceedings of Group Technology/Cellular Manufacturing - World Symposium 2003* (pp. 243-247). Columbus, Ohio, USA: Ohio University.
2. Cunha, M. M., & Putnik, G. D. (2003b). Market of Resources versus e-Based Traditional Virtual Enterprise Integration - Part I: A cost model definition. In G. D. Putnik & A. Gunasekaran (Eds.), *Proceedings of the First International Conference on Performance Measures, Benchmarking and Best Practices in New Economy* (pp. 664-669). Guimarães, Portugal (June, 10-13): University of Minho.
3. Cunha, M. M., & Putnik, G. D. (2003c). Market of Resources versus e-Based Traditional Virtual Enterprise Integration - Part II: A comparative cost analysis. In G. D. Putnik & A. Gunasekaran (Eds.), *Proceedings of the First International Conference on Performance Measures, Benchmarking and Best Practices in New Economy* (pp. 667-675). Guimarães, Portugal (June, 10-13): University of Minho.
4. Cunha, M. M., & Putnik, G. D. (2005a). Business Alignment Requirements and Dynamic Organizations. In G. D. Putnik & M. M. Cunha (Eds.), *Virtual Enterprise Integration: Technological and Organizational Perspectives* (pp. 78-101). London: Idea Group Publishing.
5. Cunha, M. M., & Putnik, G. D. (2005b). Market of Resources for Agile/Virtual Enterprise Integration. In M. Khosrow-Pour (Ed.), *Encyclopedia of Information Science and Technology* (pp. 1891-1898). Hershey, PA: Idea-Group Publisher.
6. Cunha, M. M., Putnik, G. D., Gunasekaran, A., & Ávila, P. (2005). Market of Resources as a Virtual Enterprise Integration Enabler. In G. D. Putnik & M. M. Cunha (Eds.), *Virtual Enterprise Integration: Technological and Organizational Perspectives* (pp. 145-165). London: Idea Group Publishing.
7. Cunha, M. M., & Putnik, G. D. (2006). Identification of the domain of opportunities for a market of resources for virtual enterprise integration, *International Journal of Production Research*, Vol. 44, No. 12, June 2006, pp. 2277-2298. Taylor & Francis
8. Dai, Q., & Kauffman, R. (2001). *Business Models for Internet-Based E-Procurement Systems and B2B Electronic Markets: An Exploratory Assessment.* Paper presented at the 34th Hawaii International Conference on Systems Science, Maui, HI.
9. Khalil, O., & Wang, S. (2002). Information Technology Enabled Meta-Management for Virtual Organizations. *International Journal of Production Economics*, 75(1), 127-134.
10. O'Sullivan, D. (1998). Communications Technologies for the Extended Enterprise. *Int. Journal of Production Planning and Control*, 9(8), 742-753.
11. Wang, C. X. (2001). Supply Chain Coordination in B2B Electronic Markets, *Proceedings of the 32nd Annual Meeting of the Decision Sciences Institute*. San Francisco, CA.

16

BUSINESS NETWORKS IN SMALL TEXTILE ENTERPRISES: THE CASE OF NOVA FRIBURGO-BRASIL

Américo Azevedo[1], Luísa Faria[2]

[1]*Faculdade de Engenharia da Universidade do Porto and Inesc Porto*
Rua Roberto Frias S/N, 4200-465 Porto, Portugal
[2]*Universidade de Aveiro - DEGEI*
Campus Universitário de Santiago,3810-193 Aveiro, Portugal
E-mail: ala@fe.up.pt, luisaffaria@gmail.com

Business Networking is an innovative business paradigm that can help companies to remain competitive in the market. Nevertheless, its practical implementation is very complicated because of the several dimensions that it involves. There are several cases of business network cooperation; however, each one has its particular characteristics that determine its success. This paper addresses the domain of collaborative networks, established by SME in textile industry, in the scope of an academic research project based on a case study research methodology and centered on a large industrial pole in Brasil..

1. INTRODUCTION

Manufacturing companies have to face constant change and permanent pressures. Competition is marked by volatile demand, shorter product life cycles, globalization, mass product customization and time to market. To meet today's challenges, companies are moving away from traditional functional structures, to process oriented approaches (Azevedo *et al.*, 2005). In that context, new ways of organizing production environments have been evolving, so that companies are able to address the increasing demand for flexibility, to improve the precision of on-time delivery and to reduce lead-times. Companies generally recognize that tight interaction and coordination among all the participants of their supply-chain is a key requirement for their continued survival. This new reality leads to networking with other companies as a natural approach, and gives rise to a new concept of cooperative value adding.

The main problem with the traditional business paradigm is that it assumes each company is an "island" i.e., an independent and self-operating entity. Such a view does not consider the fact that a company is instead part of a much broader matrix of business systems composed by customers, suppliers, products and global information. This leads to integration along different axes, namely geographical (physical) and functional (process). However, it can also be evidenced that in general terms small companies do not have enough conditions to compete in this

Please use the following format when citing this chapter:

Azevedo, A., Faria, L., 2006, in IFIP International Federation for Information Processing, Volume 220, Information Technology for Balanced Manufacturing Systems, ed. Shen, W., (Boston: Springer), pp. 149–156.

environment. One of the possible ways, discovered for these companies, was to start to cooperate between themselves, forming diverse types of alliances.

The aim of this paper is to address the domain of collaborative networks established by SME in the scope of an academic research project based on a case study research methodology. This paper underlines the necessity to consider several critical aspects that have to be fulfilled for network success, emphasizing the importance of trust competency and information technology infrastructure.

The reminder of the paper is organized as follows. After this introduction, the next section presents main problems and difficulties for business networks success usually considered by several researchers. The third section presents some successful experiences in the domain of collaborative networks of SME, showing that it is possible to get advantages from the cooperation and to conquer space in the market, including mainly the international market. The fourth section addresses the case study related to network cooperation in small textile enterprises of Nova Friburgo – Brasil. The last section contains the conclusions of the paper.

2. ENTERPRISE COOPERATION

The globalization process involved much more than the diffusion of technologies and productive processes, standardization of consumption and capital flows. Its bigger consequence was the reinforcement of the bonds of the economic interdependence between the companies and, for extension, between countries in the scope of the productive process - by strategic alliances and other networking forms.

Currently, however, the context of enterprise organization became completely different. The consumers have changed their behavior, demanding personalized products. When the consumer started to pull the production, the productive environment became more dynamic and uncertain, requiring high quality and conformity to the established standards.

In that context, co-operation motives can be cost and risk reduction, knowledge transfer or just the reduction of time to market (Eschenbacher *et al.*, 2001). Collaborative business environments can exploit the better cost position of the partner or use economies of scale. Even if this concept is now rather well known, few SME are actually involved in such alliances and their success rate after a couple of years seem low. Participating in a collaborative network means investing time and money, and both resources are scarce for an SME. The motivation to continue to actively participate clearly depends on the balance reached between efforts and benefits (Pouly *et al.*, 2005).

2.1 Cooperation Advantages

It is a fact that companies, very centered and vertically integrated, are not agile enough to adapt to the fast requirement of changes in the market. This search for more efficiency finishes in reaching contradictory objectives: the organizations need to be global and at the same time local; they must serve a widened market and also the more specific ones; they have to operate in a coordinate and synchronized form along the supply chain levels; and they simultaneously have to adapt to the unexpected market alterations, always guaranteeing a range of varied products with a reduced cost. It can be said that the main objective that makes the companies act on a linked strategy is to search, in the cooperative performance, for competitive advantages in the functions that add more value in the product, or in its value chain.

The successful network enterprise can bring the following main advantages: it allows the definition of joint strategies; it preserves the individuality and protects the information of the companies; it values brands and enables shared marketing strategies; it reduces production costs and investments risks; it intensifies the communication and the access to the information; it extends the productive scale and the market dimensions; it facilitates the credit access to the management qualification.

2.2 Requirements for the networking success

Despite the fact that enterprises networks are emerging as a valuable organizational instrument that can create considerable competitive advantage for small firms, unfortunately, its practical implementation is very complicated. Even if rational facts like the increase of market share, new business or cost reductions are key factors in evaluating the interest of a collaboration, "soft" factors like the exchange of experience and knowledge, human factor and sharing of common activities should not be underestimated (Pouly et al., 2005). Some studies estimate that up to 60% of the alliances fail to meet their initial objectives (Ellis, 1996), in a clear indication that there is still a lot of work to be done regarding the creation and management of these networks (Caldeira, 2004). This is due to the complexity of several dimensions that the concept of business networking implies and to some confusions and insufficiencies arising from the fact that the argument only started to be treated as a managerial and organizational topic a few years ago.

In general terms, the construction, running and the failure phases of the network depend on three basic aspects: trust, competency and ICT inter-organizational integration. (Casarotto Filho and Pires, 2001).

The trust perspective refers to the aspects linked to the cooperation between the companies, involving cultural aspects, of interest in such a way of the people as of the companies. The trust is fundamental in the business world, since the economic transactions involve risk, mainly the ones related with the unknown of the future events. If these risks will not be controlled, they can prevent businesses that would bring benefits for the companies of the network to become materialize. The building up of trust between organizations is one of the greatest challenges in the concept of collaborative networks. Moreover, transparency is a key success factor, opportunistic and selfish behavior will kill the collaboration very quickly.

The competency perspective deals with issues associated to the essential abilities of each partner. It involves material and physical aspects, as the installations and equipment and even though those incorporeal ones as the processes and know-how. Each member must be carefully selected during the construction phase to create value for all of them. In order to manage the creation of tighter cooperation within the network, a systematic identification of competencies' maps of the firms and an evaluation of their competencies levels should be followed.

The information technologies (IT) and namely the information flow, are vital in the cooperative environment between companies. However, traditional IT solutions have difficulties coping with enterprises networks, mainly because they were based on a centralized data base, located in one single site, and do not satisfy the distributed planning and control needs within a production network neither do they support the enterprise wide business process. Considering this, the development of an efficient information infrastructure that allows the companies to answer to this business dynamism can be considered an important competitive advantage. Among others things, the IT infrastructure, in particular, for collaborative production networks, should facilitate the negotiation of orders in real time, through a fast and more precise evaluation of the productive capacity, and should permit an optimized and synchronized activities planning of all partners and facilitate the control of the production orders flow giving more support to the decision making in order to solve problems referent to the occurrences that disturb this flow (early warnings).

Therefore, it can be inferred that the development of the information systems is directed towards the improvement of the inter-organizational processes management, giving particular emphasis to the customer operations and should be flexible enough to adjust itself to the structure of the enterprise network.

From an operative point of view, the main difficulty that has to be faced for networks efficiency is the design of network business processes that cover the needs of specific business purposes on one hand, and that reasonably fit with single business processes of network members on the other. Another crucial issue to point out as a possible barrier for network consistency is the definition of the most appropriate legal structure that formally identifies responsibilities and liabilities of network members with respect to stakeholders (Copani *et al.*, 2006).

3. SOME SUCCESSFUL EXPERIENCES IN THE WORLD

There are diverse known cases of network cooperation; however, each one has its particular characteristics that determine its success. That conclusion can be easily achieved after studying some industrial networks, namely, as example:

- The Virtual Factory – an organized network for regional co-operation in the manufacturing industry consisting about 30 enterprises out of Germany, Switzerland and Austria (Zheng and Possel-Dolken, 2002);
- Swiss Microtech – is a network of 8 independent SME active in the screw machining industry and USCO – is a network of 11 complementary SME active as suppliers to the machine industry (Pouly, 2005);

To give support to the case study research project considered, some known networks experiences have been studied, here briefly reported: Italy's experience as pioneer country in the formation of network cooperation between companies, and Denmark's experience mainly due to its peculiarity by the fact that the cooperation between companies has been "imposed", as a consequence of the great need to increase the companies' competitiveness in the country when faced with the threat of an increment in the external competition.

3.1 Italy's Experience

In Italy, the cooperation is characterized by the relations between micro and small companies, having been the result of the existence of a sufficiently favorable scene for its development: a great number of companies who had accepted the culture of cooperation very easily. For the Italian economy, the formation of the enterprise networks, also involving unions and the local government, keeps contributing to the increase of the competitiveness and the flexibility of the companies.

In the known region of the Emilia-Romagna there are hundreds of networks that are constituted by about 25 thousand small companies and offer a great deal of services, such as the creation and the development of services centers where market research of technology benefits all constituent companies of the networks. This is one of the main characteristics of the region, which refers to the creation of the *"consortia"* between companies, and the most common refers to the financial provision and marketing services (Best, 1990).

One of the generating factors of success for the Italian nets was the fact that the companies provide chances for the executives to meet and to gain mutual confidence, which expedited the collaboration and allowed the companies to take advantage of the chances of the market together.

3.2 Denmark's Experience

In the late eighties, Denmark faced a period of an increasing commercial deficit, high taxes of unemployment and low capitation of investments. It was pointed that the generator of the problem was the fact that the Danish companies were very inefficient and small, while the changes in the panorama business-oriented favored the wide scale operations of the multinationals companies.

For the fast development of the economy, the adopted solution was the prevalence of the small companies, but with flexible productive units. Therefore, these companies were joined in flexible nets, for which promoters were now needed.

The Commerce and Industry Ministry of Denmark established a general plan with the purpose of constructing a great number of companies' nets. To such, it started to promote the idea of cooperation by publishing information on the main concepts of cooperation and forming "brokers" for the enterprise network (private consulters who see in the cooperation networks a way of organizing the small companies in big groups), having in the mind the promotion of cooperative projects. This program was called "Strategy 92". Between 1989 and 1990, 3,000 of about the 7,300 Denmark's industrial companies had constituted network cooperation.

Therefore, the development of the Danish nets followed a different way than that of the Italian industrial districts. In Italy, the development of the flexible nets was the result of an evolution, with the support of the local governments. In Denmark's case, they developed a previous model just to create the cooperation.

4. CASE STUDY: NETWORK COOPERATION IN NOVA FRIBURGO, BRAZIL

The case study presented here is the result of a preliminary research study about the advantages and disadvantages of collaborative business networks for the increase of export levels in textile industry located in Nova Friburgo, state of Rio de Janeiro – Brasil. The capital focus will be the identification of main characteristics, organizational structure, main tendencies, and future strategies. Furthermore, in the scope of the research work, we plan to quantify the main advantages the cooperation brought for the companies, as well as for the region where it is located.

4.1 The Development of the "Network" in Nova Friburgo

With the increase of the international competition, the textile industry of Nova Friburgo entered a period of inevitable decline. The man power that was fired owned the know-how and so, it started to invest in machines and equipment opening its own business. For several identifiable reasons, the most viable alternative for these small companies was the confection of lingerie.

In 1998, the SEBRAE/RJ (Service for the Support to the Micron and Small Companies in Rio de Janeiro), together with the FIRJAN (Federation of the Industries of Rio de Janeiro), conducted a study that pointed the strong presence in the region of hundreds of small companies, very concentrated, and specialized in the sector of lingerie confection. The study recognized that most of companies were rather disorganized concerning operational processes and most of them do not even have any formal organizational structure. As a result, the efficiency and effectiveness levels of these companies were very low. Thus, a regional project, encompassing all local companies, was initiated aiming to create the adequate conditions to foster export trade levels through production networks and with of the mission to transform the region of Nova Friburgo into a recognized international pole specialized in lingerie domain. Therefore, the appearance of this specialized production network was not a natural development, having been induced, organized and fomented for diverse support institutions, with the prominence for the SEBRAE.

4.2 Productive environment

Currently, the industrial district of Nova Friburgo comprehends more than 4 thousands garment companies, and only about 600 are formally registered, corresponding to 25% of the Brazilian production of the segment, generating about 20 thousands ranks of work, with 8 thousands direct jobs and 12 thousands indirect ones. Most of the companies (68,5%) are very small (1 to 9 workers). In 2004, the turn-over of the pole was around 300 million USD (IntimaFriburgo, 2006).

Although the region hosts a significant concentration of very actively companies in same business domain, the existing pole still cannot be characterized as a cluster, mainly due to the low degree of cooperation between the companies, as much in the production as in the constitution of national and international commercialization channels or even in the technological cooperation. One of the causes of the relatively low level of organization and cooperation is the lack of qualification and the low level of education of the entrepreneurs. The majority of the entrepreneurs is outdated and not interested in the qualification programs promoted by the operating entities in the pole.

An important factor that makes the relevant participation of the pole's companies in the market more difficult, is related to the fact that, for textile sector, in general, there are a lack of barriers concerning the entrance of new competitors and mainly because the available brands are identified with low price and insufficient quality.

The confection sector of Nova Friburgo is intensive in man power, with lack of qualified staff, rendering the work directed toward the operation very difficult, creating barriers to the entrance of new technologies.

Despite the small representation, the external market is the main focus of the development projects of the pole. The objective is to increase the exportation for the countries which are already experts in the region's potential, and also to expand to other markets. According to Prochnik (Prochnik, 2002), the trusts are fundamental for the promotion of the exportations. The future strategy is that the consisting trusts launch a proper brand and a quality stamp to facilitate the products acceptance in the international market.

In accordance with the author, in 2002, there was little presence of business or services rendering between the companies. The ideal would be that, in the pole in question, there was the integration of the productive chain between producers and suppliers, inter-sector integration between the industrial and commercial areas, services and, still, solidarity and cooperation bonds between the companies. According to the study, in 2002, only 40% of the interviewed companies carried out some type of business.

The project implemented in the pole of Nova Friburgo is constituted by diverse programs that are part of the strategic development plan: the Strategic Program of Communication and Marketing; the Program for the Modernization of the Productive and Management Processes; the Program for Quality Increment and Design Improvement; incentives for better access to the markets; and simplified access to the credit.

4.3 Risks and Opportunities

The research study, developed so far, allows us to identify several main risks and difficulties for the Nova Friburgo's production network, namely: lack of information about the distribution canals and the final consumer, lack of politics of the product differentiation, technological difficulties of the small companies in improving its production and exporting, and the scarce participation of the entrepreneurs, workers and of the population in general in the project.

On the other hand, concerning the identified opportunities, we should underline the following ones: congruity and sharing of strategies and policy efforts for the institutions that support the network development, ample access to financial resources and great acceptance and future perspectives in the international market.

4.4 External Effects

It is important to point out that the proximity of lingerie's pole can represent important development opportunities for the other sectors of the region, as is the case of the metal-mechanic sector. However, Ferreira (Ferreira, 2002) evidenced that the interaction degree between the textile-confections sectors and the metal-mechanic one is still much reduced: only a very small number of the interviewed companies produces complementary accessories for the Lingerie's Pole. This lack of complementation between the local activities was pointed by the enterprise leaderships as one of the main problems for the development of the metal-mechanic sector. In special, the production of machines and components for the confection sector represents a great business chance for the local economy that must be explored. In accordance with the author, one of the attitudes that is being taken to increase the generation of jobs is to look for new forms to diversify the pole, creating new economic activities that can counterbalance the common cyclical variations of the region, as it is excessively specialized.

5. CONCLUSION

Although the research study is still running, the results achieved so far allow us to conclude that, despite the great expectations related to the development of the Nova Friburgo's Pole, it is essential for the project success that the sector entrepreneurs start to interact more between themselves and between the support institutions, so that the project can become self-sustainable.

An active participation of the institutions can be observed, mainly of the SEBRAE and the FIRJAN, in the implementation and coordination of the pole. These institutions acted as mediators and brokers of the local development, inducing the most direct joint between the entrepreneurs, in favor of the companies' modernization on the sector.

Finally, in spite of its initial development phase, this production network propitiates a development, not only for the companies of the confection sector, but also for the Nova Friburgo city as a whole. However, for this project to perpetuate and generate all its expected results, the companies must compulsorily change its individualistic culture and start to act in coordination with all members, or, with the predictable and planned retirement of the support institutions, it will not have any more "all" and it will be the end of the network.

6. REFERENCES

1. Azevedo A, Toscano C, Sousa JP. Cooperative Planning in Dynamic Supply Chains, International Journal of Computer Integrated Manufacturing, Vol. 18, No 5, 2005, p. 350-356
2. Best MH. Instituitions of Industrial reestructuring. U.K.: Polity Press, 1990.
3. Caldeira JC. Joint Ventures for Innovation & Competitiveness, in E-Manufacturing: Business Paradigms and Supporting Technologies, Ed. J.J Pinto Ferreira, Kluwer Academic Publishers, 2004
4. Casarotto Filho, N, Pires, LH. Redes de pequenas e médias empresas e desenvolvimento local: estratégias para a conquista da competitividade global com base na experiência italiana. São Paulo: Atlas, 2001.
5. Copani G, Bosani R, Tosatti LM, Azevedo A. A Structured methodology for business network design. Proceedings of ICE - 12th international conference on Concurrent Enterprising. Milan-Italy, 2006
6. Ellis C. Making Strategic Alliances Suceed, Harvard Business Review, 1996
7. Eschenbacher J, Kuck N, Weiser B. Business and legal issues in enterprise collaborations: a German perspective, Production Planning & Control, Vol 12,No.5, p. 488-503, 2001
8. Ferreira, M. S.. A Formação de Redes de Conhecimento nas Indústrias Metal-Mecânica e de Confecções de Nova Friburgo. Rio de Janeiro: UFRJ/IE, 2002.
9. IntimaFriburgo, site do Pólo de Moda Íntima de Nova Friburgo. Institucional. http://www.intimafriburgo.com.br, accessed on 20.01.2006, 2006.
10. Prochnik, V.. Perspectivas para a exportação de confecções: O caso do Pólo de Moda Íntima de Nova Friburgo. Rio de Janeiro: IE/UFRJ, 2002.
11. Pouly M, Monnier F, Bertschi D. Success and Failure Factors of Collaborative networks of SME, in Collaborative Networks and Their Breeding Environments, Springer, 2005.
12. Zheng L, Possel-Dolken F. Strategic Production Networks. Springer, 2002.

17

BUSINESS PROCESS BASED INTEGRATION OF DYNAMIC COLLABORATIVE ORGANIZATIONS

Guillermo Jimenez, Manuel Ocampo,
Nathalie Galeano, Arturo Molina
Instituto Tecnologico y de Estudios Superiores de Monterrey, Mexico,
{guillermo.jimenez, A00777498, ngaleano, armolina}@itesm.mx

A virtual breeding environment (VBE) is an association of enterprises willing to collaborate in a network to provide goods or services, described by business processes. Two major components of the VBE are an e-catalogue of competences and a repository of business process templates specifying the necessary roles that enterprises should play to participate in collaboration networks for providing a specific service. Process templates should be concretised for particular needs and enterprise capabilities. Which enterprises shall participate in and the committed resources have to be decided at moment of template concretisation: once identified the process template for a business opportunity different enterprises could collaborate. Specific responsibilities are assigned to enterprises using information from semantically defined capabilities stored in the VBE's e-catalogue. This paper describes an approach for template process definition and concretisation in a simple way; thus enterprises could be assigned to participate in a business process when their capabilities and availability of resources are found to be the most appropriate. The capabilities of a tool implemented to perform process -template concretization is described.

1. INTRODUCTION

A dynamic virtual organization is characterised by a limited lifetime, and the flexibility of individual participants to join or leave the group during its lifetime (Chadwick, 2004). Virtual Breeding Environment (VBE) could help the creation, operation and dissolution of organizations in dynamic environments (ECOLEAD D21.1, 2005). A VBE represents a pool of organizations and their supporting institutions that have both the potential and the will to cooperate with each other through the establishment of a long-term cooperation agreement. When a business opportunity is identified by one member (acting as a broker); a subset of these organizations can be selected, thus forming a dynamic Virtual Organization (VO). In this way, VOs comprise sets of independent organizations that share resources and skills to achieve their goals (Camarinha-Matos, 2004).

The VBE is a leading concept for the "European Collaborative networked Organisations LEADership initiative" (ECOLEAD), a project funded by the

Please use the following format when citing this chapter:

Jimenez, G., Ocampo, M., Galeano, N., Molina, A., 2006, in IFIP International Federation for Information Processing, Volume 220, Information Technology for Balanced Manufacturing Systems, ed. Shen, W., (Boston: Springer), pp. 157–166.

European Commission under Sixth Framework Programme (www.ecolead.org). ECOLEAD's vision is that "In ten years, in response to fast changing market conditions, most enterprises and specially the SMEs will be part of some sustainable collaborative network that will act as breeding environment for the formation of dynamic virtual organizations".

The VBE contains a Profiling and Competency Management System (PCMS) which is an e-catalogue of enterprises' profiles and competencies willing to collaborate in virtual organizations. When a business opportunity arises and a VO needs to be created, potential VO members are found using the PCMS (ECOLEAD D21.2a, 2005).

A VO is characterised by a particular set of processes that need to be performed to fulfil the VO's goals. Business processes are stored in the VBE as process templates specifying necessary competences that need to be concretised. These business processes define in an abstract way the business logic needed for collaboration among enterprises, but no participants are included in the process definition. Once concretised the business processes will be used to orchestrate the information exchange among enterprises' and their supporting applications. In order to distinguish between a template and an executable business processes the first one is called a Generic Business Process (GBP) and the second one is called a specific business process a (SBP).

This paper describes an approach and associated tools for the definition of generic business processes and their concretisation into executable business processes, necessary to construct VOs specific to collaboration requirements as demanded by a business opportunity. The roles defined in GBPs are assigned to specific enterprises and their applications linked using Web Services (Hass, 2004) for interoperation among VO participants.

2. VIRTUAL BREEDING ENVIRONMENTS AND VIRTUAL ORGANIZATIONS

It is far less costly and much more effective to quickly build a VO in a breeding environment context than through a generalized partners' search. In other words, VBEs substantially contribute to increase the level of preparedness of their members for participation in potential collaborative processes. Clearly, the organizations need invest in getting prepared for collaboration within the VBE; therefore in order to make this investment worthwhile, the participation of organizations in the VBE must be beneficial. With the provided support through the VBE, it is foreseen that the VOs of the future can go far beyond the buy-sell activities, what is the state of the art for the current VOs (ECOLEAD 21.1, 2005). In ECOLEAD's concept, Virtual Organizations are networks of independent, dynamic and geographically dispersed traditional organizations with common goals.

A critical VBE component for VO creation and operation is the Profiling and Competency Management System (PCMS); it is a repository of organizations' profiles and competencies willing to collaborate in virtual organizations. When a business opportunity arises and a VO needs to be created, potential VO members are found using the PCMS. The mission of PCMS is to simplify integrating SMEs in a VBE and foster cooperation among SMEs members of the VBE by sharing their

individual profiles of competencies, in order to achieve collaboratively new business opportunities (ECOLEAD D21.2a, 2005).
The PCMS could be considered as the knowledge base of companies belonging to the VBE. It contains enough information to find enterprises capable of participating in a business opportunity. Such information will be used for VO members to integrate their applications to business processes. Next section describes the architecture used to make application integration in a VBE context happen.

3. PROCESS BASED INTEGRATION FRAMEWORK

3.1 Runtime Architecture

Realizing process based integration requires the definition of an architecture which specifies components for process definition, a collaboration gateway, and enterprise's application linking. Figure 1 shows the architecture developed in order to enable seamless integration of application systems as services to business processes.

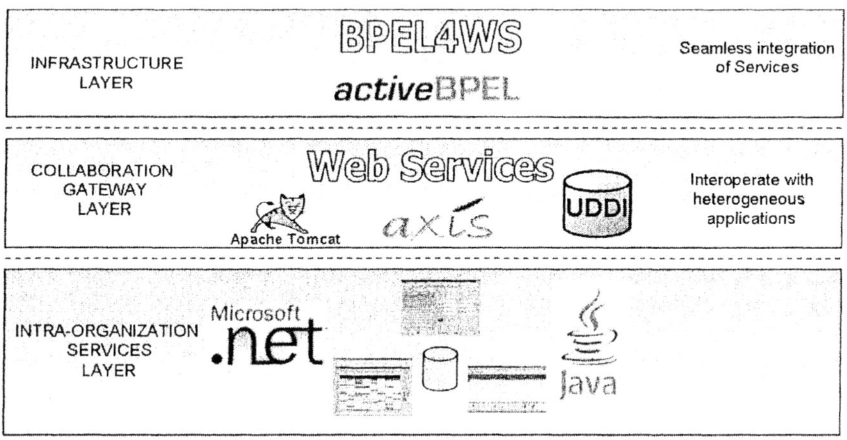

Figure 1. Process-based Architecture (adapted from ECOLEAD D61.2, 2005)

The architecture consists in three layers:

- Infrastructure layer. Manages business process execution; this layer enables users create, execute and keep track of business processes. The business process engine "activeBPEL" (www.activebpel.org) could be used for these tasks.
- Collaboration gateway layer. In this layer resides the Web services that allow information interchange between enterprise applications and the business processes, and it also contains the Web services registry for Universal Description, Discovery and Integration (UDDI) (www.uddi.org). A Web Service container for this layer could be "Axis" (www.apache.org/axis).
- Intra-organization services layer. The enterprises' applications are in this layer. Enterprises enable data access by creating Web services that behave

as intermediaries to the applications, databases or end users and obtain the information needed.

The business process engine (activeBPEL) and the Web Service container (Axis) are supported by the Apache TomCat servlet container which resides in the Collaboration gateway layer. The following subsections show how integration is achieved using this architecture.

3.1 Generic Business Processes

A Generic Business Process (GBP) is a process template specifying a set of steps that define the business logic of operations among SMEs forming a VO. A GBP specification should:

- allows definition of business processes: order of execution of the constituent activities, the roles required to be played by partners, the messages exchange among these partners, and the fault and exception handling mechanisms;
- enables future integration with applications from different enterprises;
- provides an integration driver for communication between applications and processes using Web Services.

A modelling language that fits the requirements of a GBP specification is BPEL4WS (Business Process Execution Language for Web Services). It combines the features of a block structured process language (XLANG - Microsoft) with those of a graph-based process language (WSFL - IBM's).

A BPEL4WS process specification describes a flowchart. Each element in the process is an activity. Activities are connected through links to form directed acyclic graphs (Wohed, 2003). Three different types of files are necessary to describe a business process using BPEL4WS:

- a WSDL file for any Web service that will be used by the process;
- a Business Process (BPEL) file;
- a WSDL file for the process (.i.e, the world will see the process as a web service).

The last point deserves special attention because it makes possible to integrate a process to other processes. Web Services are becoming a feature necessary to every programming language. A standard to orchestrate WSs, such as BPEL4WS facilitates the integration of applications developed with different programming languages and platforms. BPEL4WS is becoming a popular process definition language and as a consequence there are several open and proprietary process engines that implement the standard.

GBPs should be created to satisfy specific needs that VOs may have. GBP modelling is the responsibility of GBP providers. Figure 2 shows a very simple request for quotation process modelled using activeWebflow Professional (active-endpoints.com); the process flows as follow:

1. The process is instantiated when a request for quotation arrives, or in other words when the Web service that describes the process is invoked, it receives data about a product.

2. The assign activity "Data_from_customer_to_providers" translates the product data to a format that enterprises A and B can understand.
3. Activities "Quote_enterprise_A" and "Quote_enterprise_B" invoke the corresponding Web services to get a quotation for the product.
4. The assign activity "Quotation_info_for_assesment" transforms Quotations coming from Enterprise A and B to a format understandable by the assessment activity.
5. Activity "Make_Quotation", invokes the assessor Web service and obtains the quotation ready to be sent to the customer.
6. "Reply_Quotation" delivers the quotation to the customer.

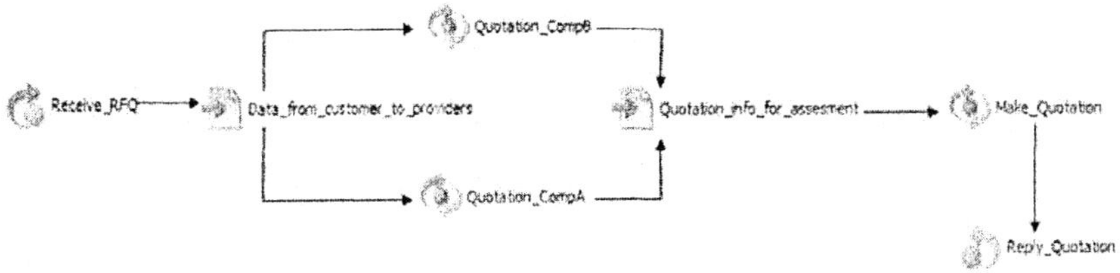

Figure 2. Request for quotation process

GBP providers have to test their processes within diverse scenarios, which imply GBP concretization, deployment and test as a SPB for every test scenario. SBPs are deployed on an ActiveBPEL server; this is performed by creating a BPR package which will contain: the BPEL process; the process deployment descriptor (PDD) which contains Web services endpoints and their relation to activities in the process; and a WSDL file for each Web Service used in the process (strongly recommended not necessary) including the WSDL for the process itself and the WSDL catalogue.

3.2 A GBP Adaptation Tool

A GBP is just a template whose parameters will be the real Web services necessary to interchange information with the applications systems. Once all parameters are concretized, a Specific Business Process (SBP) is produced. A SBP specifies concrete Web services therefore it is executable in a business process engine. Concretising SBPs from GBPs by hand could rapidly become a cumbersome task when multiple sources should be specified. This is roughly a mechanical task which uses information from the GBP definition, and from UDDI.

A software generator tool (GSPA – Generic to Specific Process Adaptor) was implemented assisting users to perform the necessary adaptations of GBPs to produce concrete SBPs, by guiding them in the specification necessary to concretize template processes. Figure 3 shows how GSPA works: first it analyzes GBPs looking for invoke activities that do not have a Web service associated. For every empty slot, it uses the UDDI to find out which Web services are available; then analyses the description files to resolve if they are compatible with the ones needed by the invoke activities.

Finally, GSPA integrates existing Web services to the target invoke activities by adding the endpoint URL and in case it is necessary, adds assign activities to transform data format from/to process/Web services. If Web services do not exist, the GSPA could create them along with Web forms (e.g., browser windows) to show the services input data and allow capturing the data that will be replied. The persons responsible of a VO can always decide in configuring the process manually, so they can choose which Web service will satisfy an invoke activity in a process and also set those particular configuration parameters needed by a process.

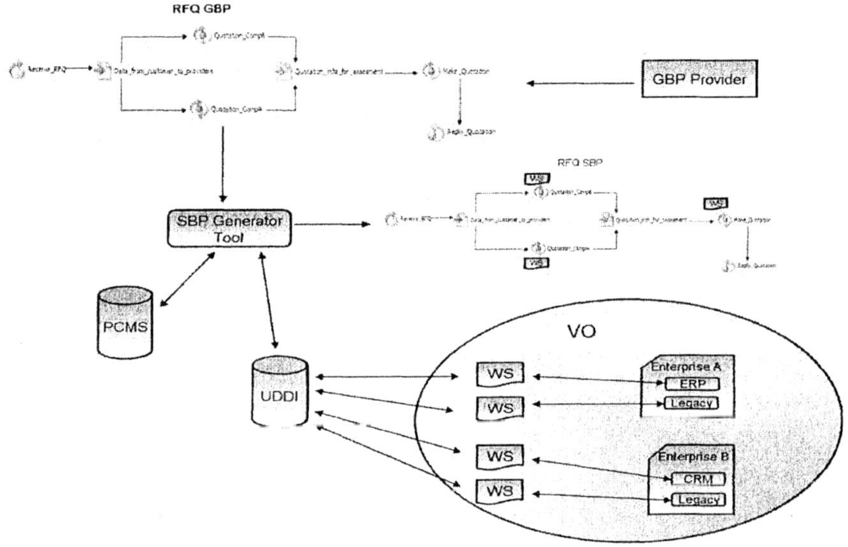

Figure 3. Specific Business Process generation

Using as example the Request for quotation process shown in Figure 2, and the diagram in Figure 3, the previous steps to create a SBP from a GBP could be described in detail as follows. The first user task is to provide GSPA with paths to the BPEL and PDD files. GSPA then will parse the PDD and find three partner links to be replaced. Then it goes to parse the BPEL to look for the invoke activities that use the partner links just found on the PDD, and can show the user which messages the process expect to send/receive to/from the Web service. The user will see a list of three invoke activities that require setting up Web services for each. Some details in the invoke activities, like the input and output messages, are gathered and shown to the user.

The second task a user should perform is to provide the companies names that will participate in the process and their Web Services information. GSPA will search for Web Services belonging to the companies and match them with the ones needed by the process. The user will finally decide the matching between invoke activities and the companies' Web services; but GSPA will make easier these tasks by performing a business process context search in the UDDI and parsing those companies' Web services that are associated. Then GSPA compares the invoke activities messages against the Web services messages to find out whether they are compatible, and finally suggests the instantiation.

The third user task is to detail the SBP. GSPA shows the matching suggestions and also the services that didn't match but are available. The user can decide whether the selected services satisfy the particular requirement. The user can also make changes. For instance, do not use the suggestions and select other Web services from the ones available, or in the worst case, provide the address location of the Web service she prefers to use. After all invoke activities have an associated Web service, the SBP will be ready for deployment.

The final user task is to generate the BPR package for deployment. To do so, the user just has to let GSPA know the activeBpel server location where the package will be deployed. GSPA will generate a WSDL file for each Web service used in the process, make the necessary changes to BPEL and PDD files to work with the selected Web services, create the WSDL catalogue, create the BPR package and deploy it to the server specified. A customer can start the deployed process by invoking it as a Web service.

In the described way GSPA makes SBP concretization a straightforward four step process (for the user) and also makes it look almost like a plug and play process. Figure 4 shows the SBP generation process in detail, with the roles that every participants play.

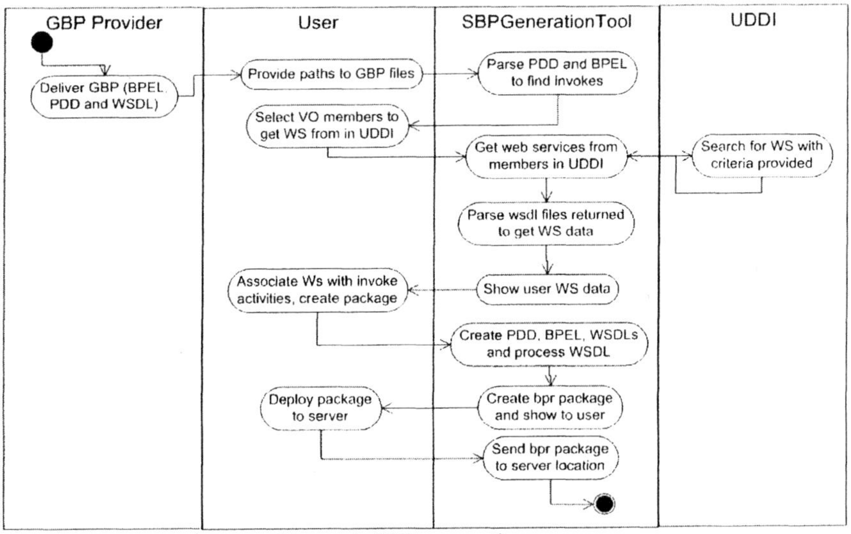

Figure 4. SBP generation process

As a corollary to the approach described, several things can be noticed. First of all, each VO partner can focus on their business problems without concerning other partners. A GBP provider worries on how to provide good-quality business processes to their users, with the best practices and as generic as possible, so those processes satisfy user needs when participating in VOs. The GBP includes in its specification a way to communicate with it so an enterprise can generate compatible Web services.

On the other hand, enterprises forming part of a VO enable access to their applications by creating Web services compatible with Business Process of interest. The SBP generator tool makes easier to particularize a GBP to a certain VO. As long as the GBP provider gives a correct process and the information from the VO is available, the GSPA is able to concretize a SBP from a GBP.
GSPA was designed with simplicity in mind, thus the SBP Generation Process is a simple one. One goal was make the user interface intuitive and following the process shown in Figure 4 and described above. Figure 5 shows the first screen of GSPA.

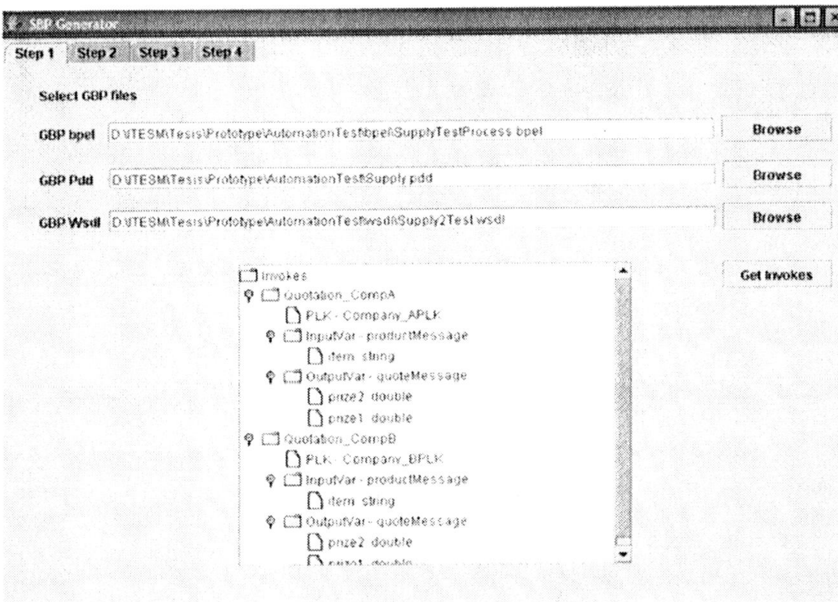

Figure 5 SBP Generator Tool – Step1

Using GSPA the user will be able to do the following tasks:

- Step1 tab. Load GBP bpel, pdd and wsdl, and analyse the GBP in order to find which invoke activities need a Web Service associated.
- Step2 tab. Search the UDDI for Web Services belonging to VO members.
- Step3 tab. Associate the invoke activities specified in Step 1 tab with the Web Services found in Step 2; and generate the SBP deployment package. See Figure 6.
- Step4 tab. Finally on this tab the user can deploy the SBP to the server.

The GBP that is being instantiated in Figure 5 and 6 is the Request for quotation process which was shown in Figure 2 and previously described.

Another interesting feature of GSPA is that if the user needs to generate several SBPs from one GBP, she just needs to go to the Step2 tab, change the invoke activities – Web Services associations and deploy the process in the Step4 tab.

GSPA is already functional but still needs some improvements. For instance, very important is a simpler approach to provide users with a more consistent invoke activity – web service suggest association.

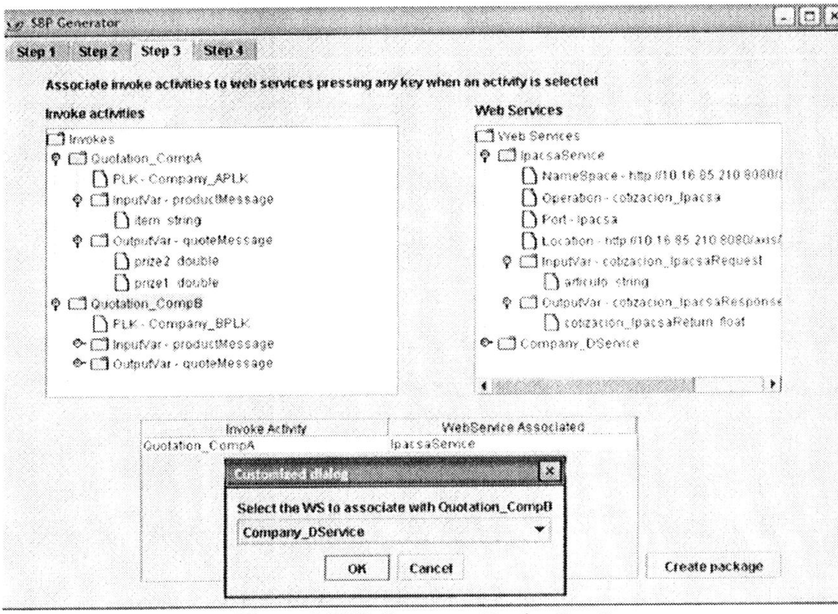

Figure 5 SBP Generator Tool – Step3

GSPA has been tested in processes more complex than the one described here, one of them is a Production Sales Forecast Process. This process involves several SBPs from two GBPs: the Production Forecast Process and the Sales Forecast Process. In this last experiment we enjoyed the versatility of BPEL. It was straightforward to compose processes from existing ones, which gave us a big spectrum for integration possibilities.

There is more to a VBE than business processes concretization. Many issues need to be solved for the VBE to be effective as a means to promote the dynamic creation of VOs. Different legal, sociological and technical aspects need to be set out. However, technical propositions are necessary to understand the advantages that a VBE should provide in order to be more effective. The research presented in this paper is an example of how information in a VBE could be used to automate VO creation. Still more research needs to be conducted to solve other problems faced at VBE and VO management.

4. CONCLUSIONS AND FURTHER WORK

Virtual Breeding Environments constitute an ecosystem necessary for creating and increasing trust among enterprises thus they are able to collaborate on networks of partners offering goods or services. However, VBEs could be more useful if extended with tools to simplify dynamic VO creation.

This paper showed how the information stored in a VBE could be augmented including additional details about applications that member enterprises export as services and templates of processes describing the operation of a VO. Using such information, an integration tool was developed to produce concrete processes necessary to manage a VO's operation.

The research reported cover just part of the whole problem for dynamic VO creation and management. There are several other necessary functions to manage the operation of a VO and decide what information has to be recorded at the time when a VO is dissolved. These requirements will be part of future research we need to conduct.

5. ACKNOWLEDGMENTS

This work was supported by the ECOLEAD (European Collaborative Networked Organisations LEADership initiative) project and the Instituto Tecnologico y de Estudios Superiores de Monterrey (ITESM), Campus Monterrey. The authors thank the contributions of ECOLEAD partners. Also, the authors thank Active Endpoints and the Microsoft Research Lab at ITESM, who provided us academic licences of ActiveBpel Designer and Microsoft UDDI servers, respectively.

6. REFERENCES

1. Camarinha-Matos, L. VOSTER Report on Infrastructures and Interoperability for VOs. VOSTER. Deliverable D44. 2003
2. Camarinha-Matos, L. Afarmanesh, H. Collaborative Networked Organizations. A research agenda for emerging business models, pages 5-8. USA: Springer. 2004
3. Chadwick , D., Dimitrakos, T., Kleese-Van Dam, K., Damian, R., Matthews, B., Otenko, A. Multilayer Privilege Management for Dynamic Collaborative Scientific Communities, UK Workshop on Grid Security Experiences, July 2004.
4. ECOLEAD internal document deliverable D21.1, Edited by Afarmanesh, H. Galeano, N. Camarinha-Matos, L. D21.1 Characterization of Key Components, Features, and Operating Principles of the Virtual Breeding Environment. 2005
5. ECOLEAD internal document deliverable D21.2a, Edited by Ermilova, E. Galeano, N. Afsarmanesh, H. D21.2a Specification of the VBE competency/profile management . 2005
6. ECOLEAD internal document deliverable D61.2, Edited by Arana, C., Ma del Mar, R. Global approach for a horizontal infrastructure architecture. 2005
7. Hass, H. Brown, A. Web Services glossary. Retrieved November 17, 2005, from http://www.w3.org/TR/ws-gloss/#defs. 2004
8. Nayak, N., Bhaskaran, K., & Das, zR. Virtual enterprises-building blocks for dynamic e-business. Information Technology for Virtual Enterprises, 2001; 29-30, 80 – 87.
9. Wohed , P., Van Der Aalst, W.M.P., Dumas, M., Hofstede, A.H.M. Analysis of Web Services Composition Languages: The Case of BPEL4WS. 22nd International Conference on Conceptual Modeling (ER 2003), volume 2813 of Lecture Notes in Computer Science, pages 200-215. Berlin: Springer-Verlag. 2003.

18

MANAGING COMPLEXITY IN INDUSTRIAL COLLABORATIONS WITHIN TOOL & DIE INDUSTRY

Günther Schuh, Alexander Sauer, Sebastian Döring
Laboratory for Machine Tools and Production Engineering (WZL)
RWTH Aachen University, Germany
G.Schuh@wzl.rwth-aachen.de
A.Sauer@wzl.rwth-aachen.de
S.Doering@wzl.rwth-aachen.de

Joining collaborations and maintaining relationships within these has become a major concern for managers in industrial companies. This change to a large extent arises due to the globalisation of markets and the ongoing specialisation of companies, fostered by the possibilities of information technology and data-communication. However, such a structural change requires adaptations by companies to fit the characteristics of industrial networks. In particular, the increasing complexity of collaborations in highly dynamic environments oftentimes is underestimated. This paper shows an approach to cope with this increasing complexity by the application of principles of complex systems from various sciences to collaborative enterprise networks regarded as socio-technical systems, considering the tool and die industry as example.

1. INTRODUCTION

Today, industrial companies are challenged by a highly dynamic environment, which requires to develop and manufacture products at a high level of flexibility and quality for low costs. Companies are forced to specialise in order to minimise product complexity and reduce production costs (Schuh, 2006). As the complexity of a single task decreases the more partners are working on it, participating in collaborations and maintaining relationships within these has become a major concern for managers in industrial companies. Although collaborations are assisted by modern information technology and data-communication, both the complexity of coordination and, as a result, the resources needed for the control load of coordination, increase. It is proceeded on the assumption that there is an optimal point of lowest total complexity with a corresponding number and organisation of the partners, depending on the type of problem to be solved collaboratively.

However, studies reveal that the optimal collaborative system setup is difficult to find: currently, the failure rate of collaborative projects in manufacturing industry is near 50%. Although there are multiple reasons for this, the underestimation of the system's complexity might chiefly be responsible for this. Regarding complexity,

Please use the following format when citing this chapter:

Schuh, G., Sauer, A., Döring, S., 2006, in IFIP International Federation for Information Processing, Volume 220, Information Technology for Balanced Manufacturing Systems, ed. Shen, W., (Boston: Springer), pp. 167–174.

managers often ignore to match the type of the networks system setup with the type of task to be solved. In the case that they do take a matching into consideration, there is a lack of knowledge about network complexity that may lead to insufficient results. Reliable models and instruments for managers to handle both the internal complexity of networks and the complexity of the environment are not available, yet. Moreover, there exist multiple unforeseen, emergent network effects in elasticity and controllability as well as in overall network and production system behaviour, which increase the system's complexity. A practical example for the mentioned developments and problems is given by the present situation of the tool and die industry.

2. THE TOOL AND DIE INDUSTR

2.1 Today's Situation

Tool and die making has become an important but critical function within a demanding field of tension (Eversheim, 2002). Established between product development on the one hand and manufacturing and assembly on the other hand, tool and die making contributes both sides up- and downstream the value chain. Concerning the product development, it provides know-how for the specification of parts and the development of efficient production processes. For manufacturing and assembly, the tool and die making process provides productivity and operational availability. Therefore, the tool and die making process is a key process for realising shortened time-to-market goals and competitive cost structures (see Figure 1).

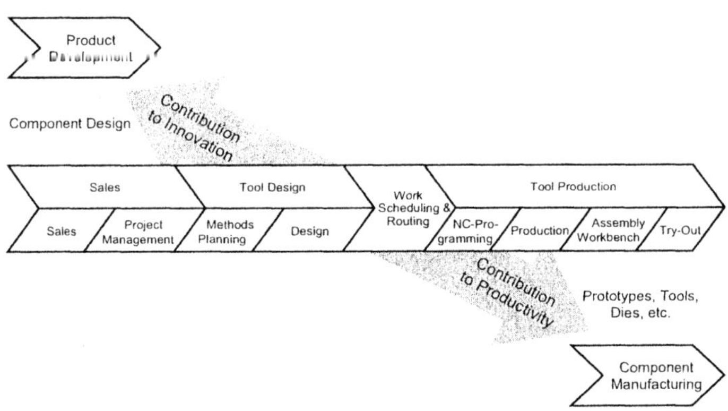

Figure 1 – Contributions of a tool and die shop to adjacent value chain processes

Due to its key position within the value chain, the tool and die shops make various contributions to the value and economic success. The first type of earnings is made by tool construction and manufacturing. In addition, productivity is provided by maintenance and repair of tools and dies. Thus, from individual repair and maintenance orders of flat-rates, the second type of earnings can be made. While maintenance usually requires a medium- or long-term planning and consequently is considered to be quite foreseeable, repairs mostly fall due unexpectedly. To decrease wear and improve stability, tools and dies could be produced at a higher level of quality. This would minimise the likeliness of a breakdown on the one hand, but, at the same time, increase the expenses for manufacturing, decrease the flexibility and extend the time-to-market on the other hand. As a consequence of the higher manufacturing expenses, the competitive pressure that is mainly caused by new entrants from Eastern European Countries and the Far East, increases. If a tool and die shop decides to produce at a lower level of quality, more spare capacity has to be kept in the shop in order to guarantee productivity by quickly reacting to tool or die breakdowns. This inevitably leads to underutilisation and, as a result, to additional personnel costs. However, companies have to spend an availability premium caused by unexpected production breakdowns or underutilisation of their production means in the tool shop (Fricker, 2005). Collaboration seems to be a strategic option to minimise additional costs and availability premiums.

2.2 Collaborations As Strategic Option

A promising approach for tool and die shops to cope with the mentioned trade-offs is to collaborate in regional networks. Sharing capacities enables the collaborating companies to highly reduce their spare capacity and consequently decrease the rate of underutilisation. By joining individual competencies an optimal tool supply can be achieved at a higher level of quality. Moreover, the Intellectual Capital of all participating partners in industrial collaborations can be significantly increased by application of the optimal partners and a balance of power within the collaboration (Sauer, 2005). In addition, the flexibility to react to changing customer needs can be increased and the time-to-market can be shortened.

There exist various approaches and models supporting such collaborations – oftentimes called virtual organizations (e.g. Sydow, 1992; Schuh, 1998; Goranson, 1999). However, collaborations with other tool and die makers – even with competitors – require adaptations by companies in order to fit the characteristics of industrial networks. Although the conditions for collaborations have been improved during the last years – especially in terms of information technology and data-communication – the management of tool and die shops needs to tackle the increasing complexity of networked structures (Colotla, 2003). This can be pointed out with an example:

The Plastics-Cluster Upper Austria is an industrial network initiated by the Austrian State with approximately 300 partners from the industrial sectors tool and die making, plastics machine engineering and plastics processing. Most of the participating enterprises are small and medium sized and operate as suppliers for the automotive industry. The network targets the pooling of competencies to improve the innovation capacity and to strengthen the competitive position of the participants. The network is provided by a systematic cluster management for a continuous change process support in the fields of information and communication, qualification, collaborations and projects as well as marketing and public relations. Even though the cluster is well-developed in its normative and strategic alignment, there is a lack in the development of tactical and operative management. In particular, the project planning and realisation is allocated to the project partners without any systematic support by the management. However, this support is essential to cope with the complexity emerging from collaborative order processing.

3. MANAGING COMPLEXITY

Reducing and managing complexity mostly aims at structuring organisations and implementing organisational changes. An example for such an approach can be found in modular product configurations (e.g. Dekkers and Sopers, 2001; Schuh et al., 1998; Schuh, 2005) or Release Engineering (Schuh, 2004). Regardless of how companies build on existing capabilities that are present in available resources and current structures, alternatives for coping with external changes remain limited. Adapting tool and die shops and their networks to the dynamics of the environment requires more than one-time interventions that are seeking for stability. Companies have to increase their complexity handling capability, which means the ability to cope with the changes in their environment and the associated complexity pouring in (Boswijk, 1992). Further on, they have to build on existing capabilities for new situations or incorporate new knowledge in order to create new capabilities for a better survival in the global competition. However, the applied strategy must be chosen carefully; the world of management has been overfed with theories that might have been adequate to at least some enterprises dealing with contemporary challenges of industry, but not to others (Micklethwait, 1996) – for example Business Process Reengineering or Core Competencies. All these theories have in common that their foundations stem from a variety of presuppositions pertaining to different factors that might directly influence the rate of success of an organisation at one place and time. Direct transferences of these approaches to networked enterprises regularly fail as they lack problem-oriented interdisciplinary inferences. A new perspective to advance research in industrial collaborations can be achieved by incorporating findings from different fields of sciences dealing with complex systems.

3.1 Collaborations As Socio-Technical Systems

Several approaches of General Systems Theory exist that aim at a specification of generic organisational concepts. However, none of these systems theories have been adequately implemented in the domain of networks, yet. Most methodologies apply systems theories in order to model organisations from a cybernetic point of view and combine these theories with a socio-technical approach for the design of new organisational structures. The systems theories might require some further elaboration by means of the adoption of theories for complex systems, networks and biological models. Concerning this level, the validity of the design approach should be scrutinised. The design approach has the characteristics of static, one-time interventions, which industrial companies have to avoid due to their severe effects on organisations. The review of other theories, such as complex systems theories, networks theories and biological models, can facilitate the identification of the companies' structures and their arrangement in networks, which is required for adapting to environmental changes and continuous change.

Human-influenced complex networks have common properties, which are hardly in line with existing cybernetic approaches. The lack of network-orientation within such systems theories becomes obvious, considering the fact that most companies nowadays act in such networks - here one must draw the conclusion that existing approaches remain hypothetically and are not capable of representing the reality of networking companies. According to Milgram, the so-called small-world property states that the average path length in the network is relatively small compared to the system size (Milgram, 1967). In particular, this is true for the scattered tool and die branch that is characterised by individual human relations between regional enterprises in order to hedge capacities.

Another property of complex networks is clustering, i.e. the increased probability of node pairs being connected with neighbour nodes that are also connected. Therefore, increased efforts were made to identify other measures of complex (enterprise) networks (Fricker, 1996). The clustering tendency particularly can be found in the tool and die industry. There are for example three established clusters in Europe, located in Italy, Portugal and Germany, and one upcoming cluster in Czechia. Beyond these clusters, most tool and die shops belong to companies of the production industry that they mainly support internally. There exist only a few independent tool and die shops scattered outside the clusters.

The most important property is the distribution of degrees, i.e. the distribution of the number of links between the nodes. It has been pointed out that several real world networks have scale-free distributions, often in the form of a power law. In these networks, a huge number of nodes have only one or two neighbours, whereas a couple of them are multiple-connected. The three specific properties of complex networks mentioned hardly appear in the original systems theories such as the Applied Systems Theory.

While a number of models have been proposed to generate networks with different combinations of the three properties, each of these models describes a process that ends up in a network having the desired properties. Less effort has been devoted to the design of a dynamic system that would not only generate but maintain such a network. While there exist only few model approaches (Friedli, 2000; Schwaninger, 2000), most of them are based on the assumption that the system size or the number of links increases. However, in some industries, such as the tool and die industry, the system size and number of links also decrease as a consequence of a consolidation of the sector. Therefore, advances in network theories should focus on the dynamics of socio-technical systems accounting for the typical properties of complex networks.

3.2 Collaborations As Complex Systems

In the early 1980's, the paradigm of self-organisation emerged and opened a new branch for the description and control of complexity (Jost, 2004). With the increasing number of elements in artificial systems their control became increasingly complex (Bar-Yam et al., 2003). As a result, the deterministic top down approach to systems control became inefficient, or even impossible, in particular for highly dynamic environments. However, it is assumed that in the field of complexity simple and comprehensible laws exist. The field of study for complex systems holds the assumption that the dynamics of complex systems is founded on universal

principles that may be used to describe disparate problems ranging from particle physics to the economics of societies (Kauffman, 1993). The development of complexity science means a shift in scientific approach having the potential to profoundly affect business, organisations and government. The objective is to comprehend complex systems whilst considering the questions: Which principles govern their behaviour? How do they adapt to change? How can they learn efficiently? How can they optimise their own behaviour?
The term complexity can represent two meanings that are relevant to this research:

- As an expression of structure, mostly internally oriented; either being part of networks or of an individual system;
- As an expression of emergence, associated with new behaviour and complexity emerging from environment.

To cope with emergence, different entities might develop different types of complexity handling capability; under these conditions, balance will hardly be achieved. Within the scope of this research only paradigms that address the dynamics of industrial networks and the environment will be chosen for an elaboration.

Agent-Based Modelling is a new and special branch of computer simulation that emerged as a methodology for studying complex systems (Buchanan, 2004). Agent-Based Models consist of agents, which have states and behavioural rules as well as an environment. In the environment, which is either spatial (e.g. a rectangular grid) or non-spatial (e.g. an abstract trading community), interactions among agents take place. The interactions can either be direct where the action immediately changes the state of the partner, or indirect when the action changes the environment which, in turn, causes the partner's state to change. Traditional social sciences, especially classical economics, have very strong assumptions concerning the rationality of agents. Most Agent-Based Modelling use bounded rational agents that have only local, limited information, and limited ability and time to process that information. This is comparable to the real-life situation in the tool and die industry: small enterprises with highly restricted ressources and rationality that do not know much about the world market.

Both complexity sciences and network sciences are two sides of the same coin for future research in different disciplines. Only if a profound and interdisciplinary understanding of complex adaptive systems is gained, quantum leap improvements in handling and purposefully using these systems will be attained. In a close interaction of both approaches the fields can mutually benefit from each other's experience, knowledge and approved solutions. Thus, the potential progress in both disciplines, complexity science and network sciences, may not only be additive but multiplicative. Regarding the tool and die industry that means to focus on the essential aspects including network, collaboration and relationship management.

3.3 Evolutionary Approaches

The progress in the science of complexity has also affected models in evolutionary biology. Especially, the models of developmental pathways and co-evolution deserve closer attention with respect to industrial networks. Environmental changes (concerning market domains and technology) can be considered by means of evolutionary biological models. The most accredited models that describe the interaction between organism and environment are:

- The NK-model based on Fitness Landscapes (Kaufmann, 1993);
- The Evolutionary Stable Strategies, application of game theories to the domain of biology (Meszéna, 2001).

A preliminary study of evolutionary mechanisms and their meaning for organisational development reveals the importance of the criteria of sustained fitness, optimisation and mutation in order to reach a local optimum and evolvability. This means the capability to penetrate the new product-market combinations and disperse in combination with bifurcation processes (Dekkers, 2004). Regarding several current scientific elaborations on strategic management, the importance of the fitness concept becomes obvious (Fricker, 2005; Frick, 2005; Sauer, 2005). Fitness not only levers synergy potentials as shown in production industry, but also increases intellectual capital within collaborative processes. However, the strategic position and the business model have to support the offered products and services as backbone.

Several approaches exist in literature to describe the evolution of collaboration. During the past decade, advances have been made in game theories, the descriptions of co-evolution, altruism, etc. within the domain of evolutionary biology. These advances can be transferred to the domain of organisations and networks (Dekkers, 2004), yielding more appropriate models to describe collaborations; in turn this might lead to a higher effectiveness of collaborations and a more purposeful development of cooperation. Additionally, a more effective collaboration will result in adaptations by agents in networks to the dynamic environment.

4. CONCLUSION

What kind of correlation exists between type and complexity of the collaborative problem and the most suitable underlying network structure for solving it?

With the field of complexity research still being a patchwork of scattered insights, a pragmatic and interdisciplinary approach holds the potential of yielding valuable insights into complexity modeling in today's networked production industry. The most common approaches focus on the complexity of structures having a static character; this links to the most common system theories. In our opinion, the dynamic dimension of complexity, found in recent progress of various sciences, will fit the characteristics of industrial networks. The scientific objective should be to provide a properly designed framework-of-thought for the implementation of complexity management infrastructures, which rely either on state-of-the-art information technologies or on new insights into architecture and characteristics of complex systems.

The implementation of this framework enables companies to react more flexible when market opportunities arise, thus increasing their competitive position, and to manage the networks they participate in more adequately. The adaptation to changing environmental conditions and the drive for innovation and fast product development will benefit from the research results. New paradigms for industrial networks will stretch beyond the traditional issues of trust, power, and supply chain management. Although the development of tools will be the next but one step, the results should guide companies' management of network dynamics, their higher degree of specialisation, the development and implementation of technologies, and their development of appropriate long or short term relationships. This will reflect on both the optimisation of the supply chains and the speed of innovation and product development. As shown for the tool and die industry, there is a strong demand for such new guidelines to meet growing world-wide chances, risks and challenges. Particularly, industry sectors such as the tool and die industry with small, broadly scattered and highly specialised companies, can benefit from new guidelines for the management of complex industrial networks.

5. REFERENCES

1. Bar-Yam, Yaneer; Tucker Brian; Furness, Colin; Olsen, Jesse; McGuirl, John; Oztas, Nail; Millhiser, Will. Complex Social Systems: Rising Complexity in Business Environments. Discussion paper, Cambridge: New England Complex Systems Institute, 2003.
2. Boswijk, Hendrikus Krijn. Complexiteit in evolutionair en organisatorisch perspectief; het zoekennaar balans tussen vermogens en uitdagingen. Rotterdam: Erasmus Universiteits Drukkerij, 1992.
3. Buchanan, Mark. It's the Economy, Stupid. New Scientist 2004; 182: 34-37.
4. Colotla, Ian.; Shi, Yong-Jiang; Mike J. Gregory. Operation and performance of international manufacturing networks. International Journal of Operations and Production Management 2003; 10: 1184-1206.
5. Dekkers, Rob; Sopers, F.P.M. "Logistic Control and the Order Entry Point." In: Hanus, D.; Talácko, J., ed. Prague: [CD-Rom] Proceedings 16th International Conference on Production Research, 2001
6. Dekkers, Rob. Adapting Industrial Organisations to the Dynamics of the Environment. Rotterdam: 2004.
7. Eversheim, Walter. et al. Was die Spreu vom Weizen trennt. Werkzeug- und Formenbau 2002; 1: 8-11.
8. Frick, Lutz. Erfolgreiche Geschäftsmodelle im Werkzeugbau. Dissertation: RWTH Aachen University, 2005.
9. Fricker, Achim. Eine Methodik zur Modellierung, Analyse und Gestaltung komplexer Produktionsstrukturen, Dissertation: RWTH Aachen, 1996.
10. Fricker, Ingo Christian. Strategische Stringenz im Werkzeug- und Formenbau. Dissertation: RWTH Aachen, 2005.
11. Friedli, Thomas. Die Architektur von Kooperationen. Dissertation: University of St. Gallen, 2000.
12. Goranson, H.T. The Agile Virtual Enterprise. Westport: Quorum: 1999.
13. Jost, J. External and internal complexity of complex adaptive systems. Theory in Biosciences; 123: 34-37.
14. Kauffman, Stuart Alan. The Origin of Order. Oxford: Oxford University Press, 1993.
15. Meszéna, Geza.; Kisdi, Éva.; Dieckmann, Ulf; Geritz, Stefan A.H.; Metz, Johan A.J. Evolutionary Optimisation of Models and Matrix Games in the Unified Perspective of Adaptive Dynamics. Selection. 2001; 2: 193-210.
16. Micklethwait, John; Woolridge, Adrian. The witch doctors. London: Heinemann, 1996.
17. Milgram, Stanley. The Small World Problem. Psychology Today 1967; 2: 60-67.
18. Sauer, Alexander. Machtbasierte Entwicklung von Intellectual Capital in Kollaborationen der produzierenden Industrie. Dissertation: RWTH Aachen, 2005.
19. Schuh, Günther; Millarg, Kai, Göransson, Asa. Virtuelle Fabrik – Neue Marktchancen durch dynamische Netzwerke. München: Hanser, 1999.
20. Schuh, Günther. Release Engineering – An Approach to Control Rising System-Complexity. Annals of CIRP 43/1, 2004, 167-171.
21. Schuh, Günther. Produktkomplexität managen. München: Hanser, 2005.
22. Schuh, Günther. Changen Management – Prozesse strategiekonform gestalten. Berlin: Springer, 2006.
23. Schuh, Günther; Schwenk, Urs; Speth, C. Komplexitätsmanagement im St. Galler Management-Konzept. Io Management 1998; 3: 78-85.
24. Schwaninger, Markus. Das Modell lebensfähiger Systeme. Discussion Article No. 35, Universität St. Gallen, 2000.
25. Sydow, Jörg. Strategische Netzwerke – Evolution und Organisation. Wiesbaden: Gabler, 1992.

19

A MULTIAGENT-BASED COMPLEX SYSTEMS APPROACH FOR EFFICIENT PARTNERING IN VIRTUAL ENTERPRISE

Toshiya Kaihara
Kobe University, Department of Computer and Systems Engineering
kaihara@cs.kobe-u.ac.jp

Nowadays a sophisticated match-making mechanism is necessarily for appropriate collations in virtual enterprise (VE). Virtual market based match-making operation enables effective partner search in terms of products allocation by distributing the scheduled resources according to the market prices, which define common scale of value across the various products. We formulate the VE match-making model as discrete resource allocation problem, and propose a complex market-oriented programming framework based on the economics of complex systems. Three types of heterogeneous agents are defined in the complex virtual market. It is described that their interactions with micro behaviour emerge a macro order of the virtual market, and the clearing price dynamism can be analysed in economic terms. The applicability of the framework into resource allocation problem for VE is also discussed.

1. INTRODUCTION

In recent years consumers' demands have turned to be diversified, and enterprises must produce appropriate quantity of goods with reasonable price considering consumers' needs, such as fashions, quality, lead time, and so on. The introduction of information & telecommunications technologies and more recently, distributed object computing technology has enabled the creation of functioning Virtual Enterprises (VE), that do not have the geographic and structural restrictions that have traditionally constrained conventional enterprises. These technologies have enabled people to interact and collaborate effectively over distance as part of VE [1].

Traditionally, marketing, distribution, planning, manufacturing, and the purchasing organizations operated independently. These organizations have their own objectives and these are often conflicting. The result of these factors is that there is not a single, integrated plan for the organization - there were as many plans as businesses. Clearly, there is a need for a mechanism through which these different functions can be integrated together. Although cooperation is the fundamental characteristic of VE concept, due to its distributed environment and the autonomous and heterogeneous nature of the VE members, cooperation can only be succeed if a proper management of dependencies between activities is in place just like Supply Chain Management [2][3].

Please use the following format when citing this chapter:

Kaihara, T., 2006, in IFIP International Federation for Information Processing, Volume 220, Information Technology for Balanced Manufacturing Systems, ed. Shen, W., (Boston: Springer), pp. 175–182.

On the other hand, market price systems constitute a well-understood class of mechanisms that provide effective decentralisation of decision making with minimal communication overhead. In a market-oriented programming approach to distributed problem solving, the optimal resource allocation for a set of computational agents is derived by computing general equilibrium of an artificial economy [4] [5]. Market mechanism can provide several advantages in the partnering process in VE.

So as to facilitate the optimised VE management with e-Commerce infrastructure, a sophisticated business matching mechanism is required to manage such a large-scaled environment. Since the matchmaking place is a kind of pure market in terms of its structure, the idea of VE combined with virtual market (VM) must be promising [3] [6].

In e-commerce et al., the diversification of consumers' needs makes it difficult for traders to make appropriate decision makings. So, in this study, we focus on subjects about how prices of the goods are decided and how the goods are dealt in among the enterprises that compose VE. Then we try to construct a VM based on the model of `Economics of Complex Systems` [7]. Economics of Complex Systems has several features: increasing returns to scale, bounded rationality, self-organization, one-to-one trading, and so on.

2. ECONOMICS OF COMPLEX SYSTEMS

Economics of complex systems is new approach in economics to combine economics and complex systems [7]. Within this framework it is possible to construct an economic system starting, bottom-up, by its most elementary ingredients. An economic system is visualised as a large number of interacting agents whose individual actions as well as the interactions among them are explicit enough to be put into algorithmic terms. Although this approach bears the advantage of imposing weaker restrictions than a purely mathematical one, it is still necessary to greatly simplify the real situations. The challenge is that the essential features that are responsible for the emergent behaviours of the system, do not get lost. A successful model, in spite of being a heavy abstraction of real economic systems, allows discussing basic, stylised facts and working as true laboratories in which extreme conditions can easily be simulated and studied.

The main motivation of the economics of complex systems is to study the self organizing driving forces that act within an economic system. The hope is that learning about them can also provide information about the mechanisms that drive economic systems towards or away a stationary situation. The search of a global stable configuration and the process of self organisation are the two main emergent properties to clear about. A particularly relevant ingredient to model the relaxation of an economic system off equilibrium is both, the expectations and the adaptive capacity of its economic agents. Most of the robustness of the system as a whole can certainly be attributed to the memory that agents have of their previous experiences in deciding their future attitudes in their economic transactions. Adaptation on the other hand provides the necessary plasticity and change of individual behaviours to absorb changes and shocks. Learning and adaptation may therefore be considered as the basic element to model the self-organizing features of an economic system, as well as its robustness - or equivalently - its bounded homeostasis.

We formulate VE model as discrete resource allocation problem based on negotiated transaction, and propose a complex VM framework based on the economics of complex systems in this paper.

3. AGENT FORMULATION

3.1 VM structure with complex systems

VM consists of three types of agents in this study, such as producer agent, customer agent, and intermediate agent, because the VM provides auction environment for enterprises in VE.

- Producer agent: players who produce and supply exchanging resources in VM
- Customer agent: players who buy and consume the resources in VM
- Intermediate agent: players who provide auction field, and intermediate the trades between producer and customer agents. As a consequence, individual match makings are established between a set of producer and customer agents. As a basic study, the Intermediate agents are assumed never to try to gain profit during the matching process in this study.

Negotiations are occurred just between producer agent and intermediate agent, or customer agent and intermediate agent in this study. Thus all the tradings in the VM are based on negotiated transaction, which is completely different from the intensive transaction in stock exchange market proposed in micro economics.

An example of the proposed VM structure is shown in Fig.1. Market environment is divided into finite number of small grids where only one agent is located inside. Initially all the agents are located randomly, and they are assumed not to move their locations as a basic study. Both the producer agent and the customer agent behave individually without any contact to other agents except the intermediate agent. Only the intermediate agent has a transactional scope, and the intermediate agent is able to make communications or negotiations with other agents inside the scope. The negotiation is carried out in one to one relationship between producer agent and intermediate agent, or customer agent and intermediate agent as negotiated transactions. The scope corresponds to information transmission space, i.e. information distance, in practical situations.

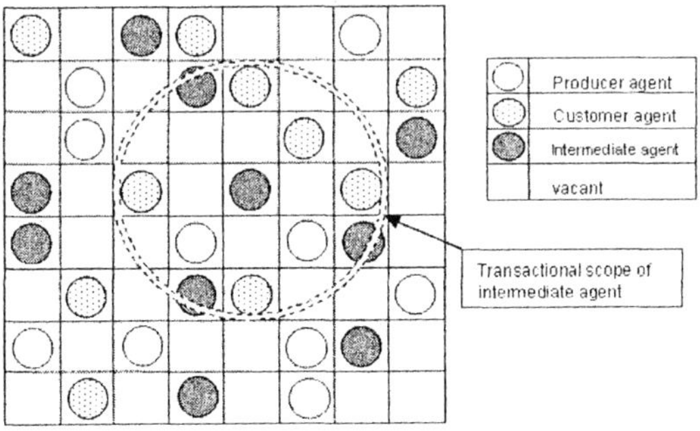

Figure 1 – An example of complex VM

3.2 Producer agent

Producer agent tries to pursue its profit by producing and selling goods x from its own capital. We formulate the producer agent (s) as follows:

(a) Objectives: maximise its capital (M_s^t) at present (t)

$$M_s^t \rightarrow max \tag{1}$$

(b) Behaviour

Step 1: Produce goods x by the amount O_{sx}^t, that calculated previous Step 5, and increase its capital (G_{sx}^t).

$$G_{sx}^t = G_{sx}^{t-1} + O_{sx}^t \tag{2}$$

Step 2: If the estimated profit ($EPro_{sx}^t$) is not negative, decide to sell all the stock and go to Step3 (Case 1). Otherwise give up to sell and go to Step 4 (Case 2).

$$EPro_{sx}^t = P_{sx}^t - Co_{sx} \tag{3}$$

where P_{sx}^t : recommended price, Co_{sx} : cost price

Step 3: Carry out transactions with the intermediate agent, and modify its capital (M_s^t) according to the transactions. The details are described in 3.4.

$$SL_{sx}^t = \sum_d \{(MP_{sxd}^t - Co_{sx}) \times n_{sxd}^t\} \tag{4}$$

$$M_s^{t+1} = M_s^t + \sum_x SL_{sx}^t \tag{5}$$

$$G_s^{t+1} = G_s^t - \sum_d n_{sxd}^t \tag{6}$$

where SL_{sx}^t : total sales, MP_{sxd}^t : clearing price, n_{sxd}^t : amount of trade

Step 4: Calculate the recommended price for the next step.

$$P_{sx}^{t+1} = \begin{cases} P_{sx}^t \times \alpha_{sx} & : Case1 \\ P_{sx}^t + \pi & : Case2 \end{cases} \tag{7}$$

where α_{sx} : price modify parameter, π : price increase parameter

■ α_{sx} : producer agent modifies this parameter according to the amount of trade. If the total amount of trade is equivalent to the prepared stock, the agent increases the price for expecting higher profit at the next step. On the contrary, if it has no offer from intermediate agent, it decreases the price.

Step 5: Calculate the production amount so as to maximise its profit by solving NLP problem as follows:

$$maximise: \Pi = P_{sx}^{t+1} \times O_{sx}^{t+1} - (Co_{sx} \times R_{sx}^{t+1} + C_s) \tag{8}$$

$$s.t.\ O_{sx}^{t+1} \le f(R_{sx}^{t+1}) \tag{9}$$

where R_{sx}^{t+1}: material amount, C_s : fixed cost

The parameter Π means profit, and f is the production function. Producer agent is formulated to maximise its profit locally by solving the above NLP.

(c) Withdrawal conditions

If the agent exhausts its capital, then it withdraws from the market.

3.3 Consumer agent

Consumer agent is assumed to try to purchase goods x and sell it to the market underneath in supply chain. The underneath market isn't modelled precisely in this study and it is assumed as fully stable as a basic research. We formulate the consumer agent (*d*) as follows:

(a) Objectives: maximise its capital (M_d^t) and stock (G_d^t) at present (t)

$$M_d^t \rightarrow max \wedge G_d^t \rightarrow max \tag{10}$$

(b) Behaviour

Step 1: Increase capital M_d^t with sales income(C_{Md}).

$$M_d^t = M_d^{t-1} + C_{Md} \tag{11}$$

Step 2: Carry out transactions with the intermediate agent, and modify its capital (M_d^t). The details are described in 3.4.

$$PY_{dx}^t = \sum_s (MP_{dxs}^t \times n_{dxs}^t) \tag{12}$$

$$M_d^{t+1} = M_d^t - \sum_x PY_{dx}^t \tag{13}$$

$$G_{dx}^{t+1} = G_{dx}^t + \sum_s n_{dxs}^t \tag{14}$$

where $\sum_x PY_{dx}^t$: total cost, $\sum_s n_{dxs}^t$: amount of purchased goods

Step 3: Calculate the desired purchase price for the next step.

$$P_{dx}^{t+1} = P_{dx}^t \times \alpha_{dx} \tag{15}$$

where α_{dx} : price modify parameter

■ α_{dx} : consumer agent modifies this parameter according to the amount of trade. If the total amount of trade is equivalent to its requirements, the agent decreases the price for lower expenditure. On the contrary, if it has no offer from intermediate agent, it increases the buying price.

Step 4: Calculate the requiring amount in the next step by solving the following simultaneous equations.

$$M_d^{t+1} = \sum_x P_{dx}^{t+1} \times H_{dx}^{t+1} \tag{16}$$

$$P_{dx1}^{t+1} \times H_{dx1}^{t+1} / P_{dx2}^{t+1} \times H_{dx2}^{t+1} = MRS_{x1,x2}^{t+1} \tag{17}$$

where MRS : marginal rate of substitute

(c) Withdrawal conditions

If the agent exhausts its capital or one of any goods, then it withdraw from the market.

3.4 Intermediate agent

Intermediate agent corresponds to auctioneer in e-market place, and provides partnering environment for the attendees. The intermediate agent has transactional scope, and the scope V_i^t is dynamically spread followed by logistic function.

$$V_i^t = (A/2) \times \{\exp(C_{V_i}t)/1 + \exp(C_{V_i}t)\} \tag{18}$$

where A: diagonal length of the market, C_{V_i} :constant value

All the agents inside the scope are considered as applicants for match-making algorithm. We applied 4-heap algorithm[8] for the match-making process.

The behaviour of intermediate agent is as follows:

Step 1: Call for the market to all the agents inside the scope.

Step 2: Collect the information from all the agents who attend the market by their own decision.

Step 3: Proceed the match-making based on 4-heap algorithm. Clearing price between producer agent s* and consumer agent d* is calculated as Eq. (19).

$$MP^t_{s^*xd^*}(=MP^t_{d^*xs^*})=(P^t_{s^*x}+P^t_{d^*x})/2 \tag{19}$$

Step 4: Report the transactional results to all the applicants.

Step 5: Negotiated transactions are finally carried out in VM, based on the match-making process.

4. EXPERIMENTAL RESULTS

4.1 Experimental parameters

We constructed small but principle complex VM as a basic study to analyse distinctive characteristics. The parameters are shown in Table 1. U(min, max) and N(μ, σ^2) mean uniform distribution and normal distribution, respectively in Table 1. Simulation trial is 100 times at each result described in this chapter.

Table 1 Experimental parameters

	Producer agent		Consumer agent
M_s	10000	M_d	20000
P_s	U(400, 500)	P_d	U(450, 550)
O_{s1}	10	G_{dx}	N(50. 25)
O_{s2}	15	μ_{d1}	5
Co_{s1}	5	μ_{d2}	6
Co_{s2}	4	C_{md}	5000
π	10		
C_s	100		

4.2 Price dynamism and scope effects

Dynamic transactional price transition of producer agent in goods 1 is shown in Fig. 2. That figure shows that the converged transactional prices are emerged in the VM, although it has no positive convergence mechanism which Walras market has. The converged prices are acquired through adaptive micro-macro interactions between agents and market, and that means the complex VM successfully attains practical market mechanism in reality. There happens some autonomous selection amongst agents, and only the appropriate number of agents is able to survive in this market according to the power balance between supply and demand.

The VM isn't stable system in nature, and these prices aren't always converged. A number of simulation trials showed us that the transactional scope of intermediate agent has great effects on the price stability.

The Fig. 3 shows the relationship between the scope parameter C_{V_i} and coefficient of variation of the converged transactional price (goods 1).

It is obviously shown that the price is stabilised as c_{V_i} increases, i.e. the scope enlarges rapidly. Especially c_{V_i} is around 0.1, the coefficient of variation decreases sharply, and the price stability is improved rapidly. The fact implies that the relaxation of the constraints on agent visibility draws the VM with complex systems closer to classical Walrasian market, which has positive convergence mechanism inside. It has been confirmed that the price convergence of the proposed VM is manageable by the scope factor.

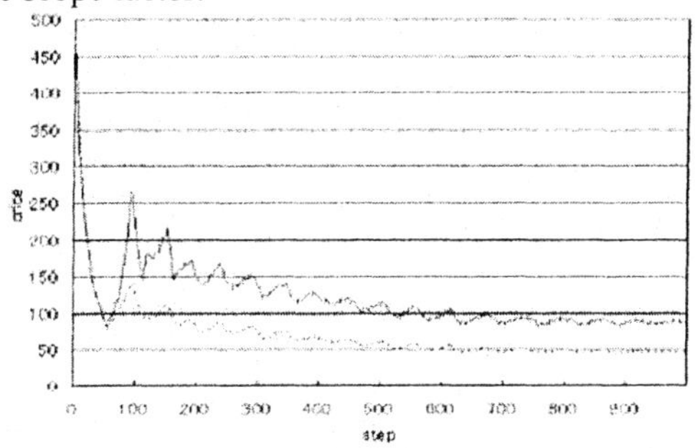

Fig.2 Price transition of producer agent (goods 1)

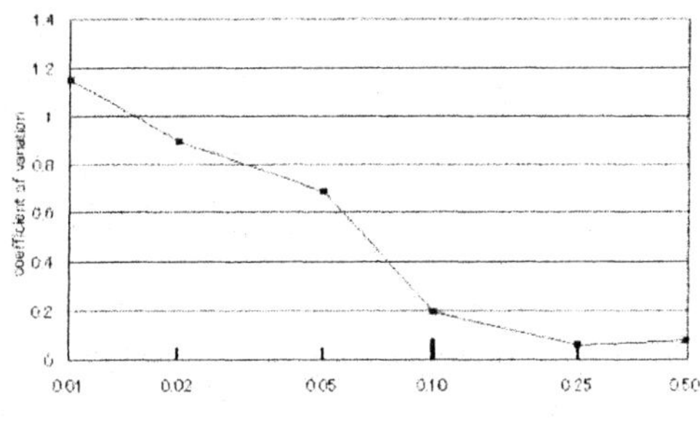

Fig.3 Scope and price convergence (goods 1)

4.3 Sensitive dependence on initial conditions

Finally we analysed the proposed market in terms of the dependency on initial conditions, which complex systems have in general.

Fig.4 illustrates the final budgets (at t=1000) of customer agents in the market. (x, y) means the existing position of each consumer agent in this figure. It has obviously observed that the budget distribution is very large, and the budget difference is deeply connected to the initial conditions of the market, such as positions, budget and price modification parameter. An accidental small difference at first emerges great diversity as the consequence of agent interactions finally. We have confirmed that our VM obtains Sensitive Dependence on Initial Conditions (SEDIC).

5. CONCLUSIONS

In this study, we focused on the subject about how prices of the goods are decided and how the goods are dealt in among the enterprises that compose the supply chain.

Then we have constructed an "Artificial Market" in computer, and the macro dynamism of the market have been analyzed from the micro point of view. The artificial market has been constructed based on the model of Economics of Complex Systems, which has several unique features compared with conventional Walras market.

At First, the basic market model, which abstracts and simplifies an actual market, has been defined and constructed, and we have simulated this model. Some distinctive characteristics of the price dynamisms have been observed in this simulation, and the decision mechanism on the market price has been analyzed by the behaviour of each agent. It has been clarified that the price is fully converged in some cases, although the market doesn't have specific mechanism for the convergence that Walras market holds. Additionally, we have found out that the spread speed of intermediate agent's view has affected market's activity and stability. And SEDIC has been observed in this model. Moreover, it has been observed that the balance of rising / lowering price parameters between supply agents and demand agents is important to stabilize market.

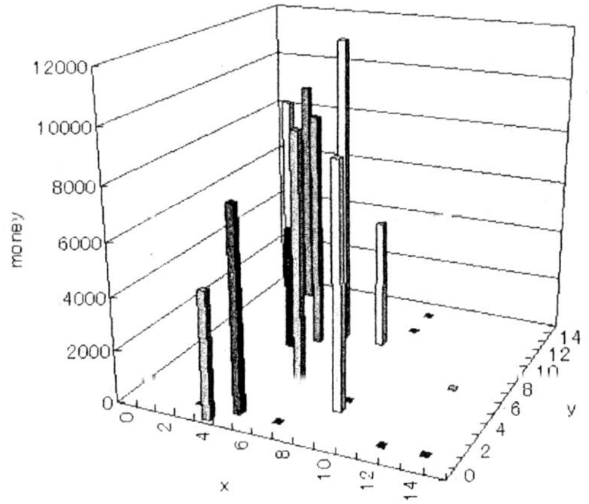

Fig.4 Final landscape of consumer agents

6. REFERENCES

1. Camarinha-Matos, L. M. et al., The virtual enterprise concept, Infrastructures for virtual enterprises, Kluwer academic publishers, Boston, pp.3-14, 1999.
2. Fisher M. L., Making supply meet demand in uncertain world, Harvard Business Review, May/Jun, 1994.
3. Kaihara T., Supply Chain Management based on Market Mechanism in Virtual Enterprise, Infrastructures for Virtual Enterprises, Kluwer Academic Publishers, Boston, 399-408,1999.
4. Shoven J. B. and J. Whalley, Applying General Equilibrium, Cambridge University Press, 1992.
5. Kreps, D. M., A Course in Microeconomic Theory. Harvester Wheatsheaf, New York, 1990.
6. Kaihara T., Supply Chain Management with Market Economics, International Journal of Production Economics, Elsevier Science, Vol. 73, Issue 1, 5-14, 2001.
7. Shiozawa, Y., An Introduction to Economics of Complex Systems (in Japanese), Seisansei Shuppan, Tokyo, 1997.
8. Wurman P. R., Walsh W. E., Wellman M. P., Flexible Double Auctions for Electronic Commerce: Theory and Implementation, Univ. of Michigan Artificial Intelligence Laboratory, 1997.

20 DESIGN FOR PRODUCT LIFECYCLE MANAGEMENT

Wenlei Zhang[1,2], Yushun Fan[3]
[1] *Shenyang Institute of Automation, Chinese Academy of Sciences, Shenyang, P.R. China,*
[2]*Gradute School of the Chinese Academy of Sciences, Beijing, P.R. China, zwl@sia.cn*
[3]*Dept. of Automation, Tsinghua University, Beijing, P.R. China, fanyus@tsinghua.edu.cn*

Product lifecycle management (PLM) is a concept that aims at integrating the various processes and stages involved during a typical product lifecycle for the extended enterprise. PLM technologies endeavor to offer a powerful collaborative platform to support distributed product development. In order to maintain the integrity of product definition data throughout the life of the product, and to manage business processes used to create, manage, disseminate, share and use the information, this paper first explores the connotations of PLM; then presents a conceptual modeling framework including a four-tier-architecture; furthermore models product lifecycle in the stages of requirement analysis, conceptual design, engineering design, manufacturing and services; and finally proposes an integration framework to support interoperability of distributed product data sources.

1. INTRODUCTION

The Product Lifecycle Management (PLM) concept holds the promise of seamlessly integrating all the information produced throughout all stages of a product's lifecycle to everyone in an organization at every managerial and technical level. Most of global leading providers and consultant organizations of manufacturing automation solutions, including UGS, IBM, PTC and CIMdata etc, consider PLM is the key to collaboration and cooperation. "*The strategy of PLM is to build up extended enterprise based on web and support all the suppliers, partners and trusted customers to capture, manage, evaluate and utilize all the related information.*" [1]

As the backbone of PLM approach, product information model enable the description, dissemination and sharing of product data in the distributed manufacturing environment. So, the product modeling technologies have won the extensive recognition in both academia and industry. In the last two decades, product modeling technologies for specific stages have been enormously explored. Bidarra and Brownsvoort give the product engineering design models based on semantic featuring modeling technology [2]. Jiao offers a generic bill-of-material and operation (GBOMO) data model to manage variant product structures of the product family [3].Simon presents a product lifecycle data acquiring technology (LCDA), which employs sensors and micro-processors to collect product operation process

Please use the following format when citing this chapter:

Zhang, W., Fan, Y., 2006, in IFIP International Federation for Information Processing, Volume 220, Information Technology for Balanced Manufacturing Systems, ed. Shen, W., (Boston: Springer), pp. 183–192.

data dynamically to build product operation data model [4].Shehab etc. presents a knowledge-based intelligent product cost evaluation system, which helps designers evaluate product manufacturing cost [5].

With the increasing competition in the global market, the manufacturing industry need make agile responds to changes of market. Modes of enterprises federation, web manufacturing and collaborative manufacturing demand enterprises adopt information technologies to maintain and management distributed product data uniformly [6]. And product lifecycle model offers a conceptual mapping mechanism for them to link data of product design, process, quality, cost, sales and operation. And it allows the related distributed data to be acquired and processed. The Axiomatic Design (AD) method proposed by Suh is one such approach that provides a systematic lifecycle guideline for evaluating the acceptability of designs [7]. In AD, the design process is considered a series of domains and their mappings, and the domains mainly include: {CSs}, {FRs}, {DPs} and {PVs}. Jiao and Tseng further establish the FBS-views product model [8], and the mappings among the views push the product variant design along the product lifecycle. And bill-of-materials can represent the product information by a common and simple data format [9, 10]. Single Source of Product Data (SSPD) is presented to explore the logic relations of product data, aiming to build a logic uniform data source of physically distributed product data, so the access and operation of product data can be derived from the single source [11]. And web-based product lifecycle modeling technologies have been broadly researched [12-13]

But PLM solutions stress that product data should be created, exchanged and shared across all the stages along the product lifecycle. And the definition and management of a global product information model is the key to implement PLM. However, the above modeling technologies have been rare researched. So this paper devotes to exploring the approaches of building an integrated, reconfigurable and consistent product lifecycle information model, which is addressed by the framework of global PLM model, stage models and their evolution modes.

2. THE PLM DEFINITION AND SYSTEM ARCHITECTURE

CIMdata gives the classic definition of PLM as: (1) A strategic business approach that applies a consistent set of business solutions that support the collaborative creation, management, dissemination, and use of product definition information; (2) Supporting the extended enterprise (customers, design and supply partners, etc.); (3) Spanning from concept to end of life of a product or plant; (4) Integrating people, processes, business systems, and information. [1]

We present our understanding of PLM values as Figure 1 shows, and this figure shows that

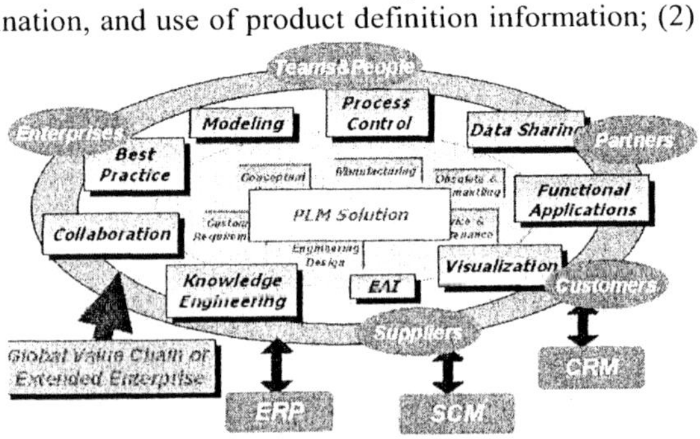

Figure 1 – Collaborative platform of PLM

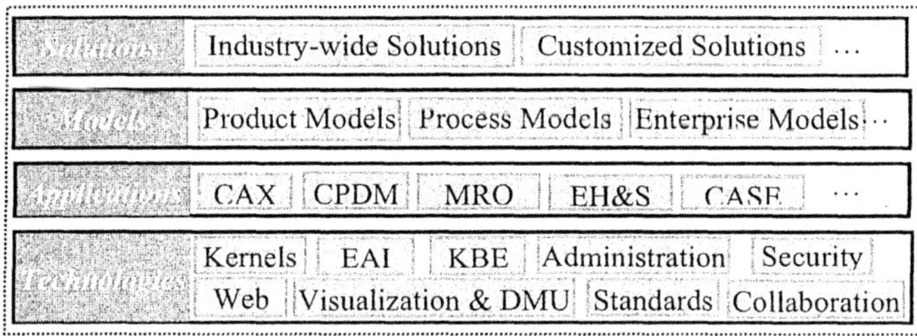

Figure 2- System Architecture of PLM

PLM solutions endeavor to encapsulate foreign data, processes and heterogeneous information systems. So a collaborative platform is built up for the global value chain to create and share knowledge assets.

Based on elaborate researches on the related cases, solutions and literatures, we build up system architecture of PLM solution as Figure 2 shows.

So the following sections of this paper will discuss a conceptual modeling framework dedicating to resolve these core problems. This framework demonstrates the mechanism to create and transform data across each stage along the product lifecycle, while keeping data integrity and smooth processes.

3. GLOBAL PRODUCT LIFECYCLE MODEL

We divide the product lifecycle into 5 main stages, as Figure 3 shows: (1) requirements analysis, to capture and analyze customer requirements (CRs); (2) conceptual design, to develop a conceptual product plan; (3) Detailed design, to design parts, assemblies and product structure etc.; (4) manufacturing, to produce the product; (5) services, to offer the technical support, maintenance, overhaul and repair for products.

These 5 stages are not segregated from each other, but overlapped or intertwined partly. For example, in detailed design stage, the engineers may need to collaborate

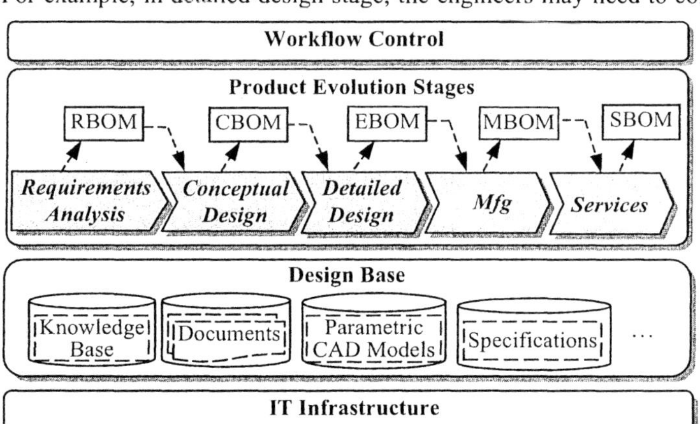

Figure 3- Global product lifecycle model

with customers to acquire clear requirement.

And the integral flow of data and businesses are consistently created and evolved along this life line. Each stage contributes to this flow. BOMs can be easily read and stored by database systems, so most of formats of data and files are organized and indexed by them. Most of their data can be partly or totally derived form the design base, which affords single source of product data for knowledge reusing. And the product evolution flow needs to be controlled by workflow system, so the businesses can be triggered and performed automatically. IT infrastructure includes hardware, software, and Internet technologies, underlying representation and computing languages, and distributed objects and components.

3.1 Requirements Analysis

To capture and understand customer needs effectively and subsequently transfer them to design specifications is one of essential premises for successful product design [14]. As the initial stage of the product lifecycle, Requirements analysis aims to transform the product from the customers' viewpoint to the designers' one.

Figure 4 shows the model of requirements analysis in three processes:

(1) ***Capturing***, the requirements can be addressed in terms of "Information", "Features" and "Sketches". ① "Information" is designed to capture general CRs like customer features, main functions, and general needs etc; ② "Feature" means engineering CRs, which needs customers to collaborative design with enterprises to describe the sizes of parts, assembling conditions or process demands etc; ③ "Sketch" describes technical CRs by means of a serial of sketches, addressing the functional or technical view of the product in the customer's mind.

(2) ***Translation***, the original requirements data need to be quantified, evaluated and transformed to formal CRs specifications, which are organized by RBOM-Requirements BOM. QFD-quality functions deployment and AHP- analytical hierocratic process are the routine methods to analyze requirements quantitatively.

(3) ***Management***, the CRs should be managed along the product lifecycle to ensure that the product is the exact one customer needs, and make agile responses to the CRs changes.

Web-based configuration, real-time conference and 3D collaborative design are the

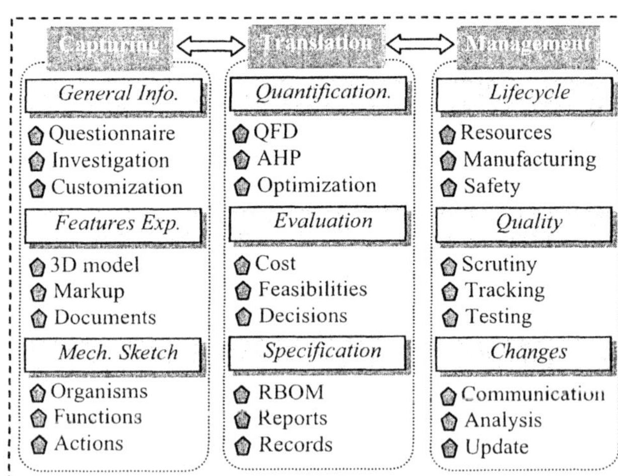

Figure 4 - Requirements analysis model

necessary tools to support CR analysis. And this stage can derive questionnaire templates, CR topology structure, functional and structural data of product and customer models from the design base.

3.2 Conceptual Design

Conceptual design is responsible for the project design to transform CRs specifications to a systematic product developing specification. This specification decomposes the design tasks, schedules and resources according to the CRs. And conceptual design is the key to innovation and production efficiency and effectiveness [15].

As Figure 5 shows, a conceptual product structure need to be constructed in the functional, technical and physical view. The decomposition tree of product functions is closely related to RBOM and generated by mapping CRs to functional modules of product family. Then based on the knowledge and matching rules in technical domain, the design parameters (DPs) coupling to each functional module is deduced, and finally a technical view is derived. By seeking assemblies, components or parts to satisfy these DPs, the physical product structure can be formed.

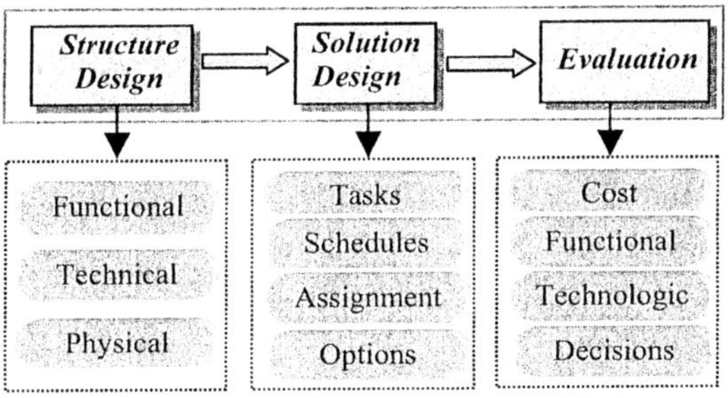

Figure 5 - Conceptual design model

And generic product structure, components or parts base, configuration rule base function as the design base. Further, the technical tasks of design, manufacturing and services need to be specified according to the conceptual product structure. There are perhaps a set of conceptual product solutions, and each of them is designed for specific consideration. These solutions describe the product developing scheme in terms of tasks, targets, engineering demands, schedules and arrangements etc. and the cost, feasibilities and difficulties of these optional solutions should be evaluated for decision. And a conceptual product BOM (CBOM) affords as a design specification for the downstream stages.

3.3 Detailed Design

Most of the new products are designed by resorting to existing technologies or similar products. Modular and structured design methodologies are a deliberate attempt to reduce product development cycles. In fact, if properly used, methodologies provide much more than incremental improvements in products, described as discontinuous jumps in a product's evolution s-curve. Now modularity, product platform and product family architecture are extensively accepted as the basis for unburdening the commonality and the knowledge from the product design and process [16][17].

So we present a product family architecture model that can best satisfy the above demands. As Figure 6 shows, platform, modules, structure and domains are the essentials to address the PF and the product design based on it.

(1) ***Modules,*** within the module, a collection of components or parts are closely integrated and form a physical building block, which bears the standard interface and implements specific functions.

(2) ***Platform***, a platform is constructed by key modules and bears the key technologies, product line policies and market strategy of a product family. A platform acts as the common architecture of the related product family, so by plugging the different functional modules into it, diversified products can be easily derived even though they possess the same core-platform.

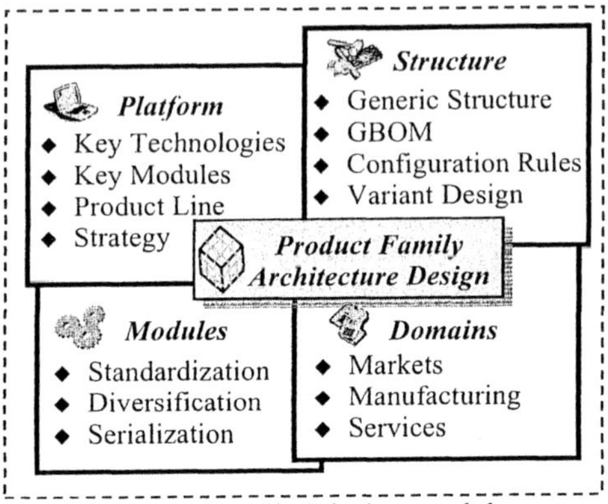

Figure 6 - Detailed design model

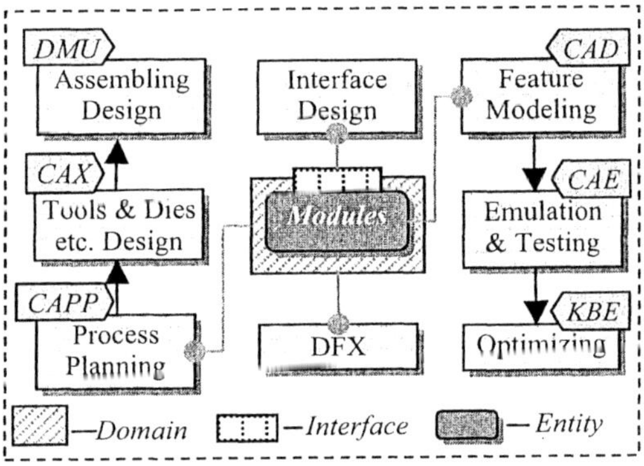

Figure 7 - Module design

(3) ***Structure***, a generic product structure is responsible for organization of all the related data and files of a product family. And all modules are classified according to their interrelationships and assigned as nodes in the generic product structure tree. GBOM- generic BOM expresses the data relationships of this generic structure in different views such as functional GBOM and engineering GBOM etc.

(4) ***Domains***, Domains are special data views covering each stage of product lifecycle. So the knowledge base for each stage is unburdened to form the foundation and mechanism, and it can be reused and reconfigured to produce the new products along the lifecycle.

Based on this model, product design takes different forms: ATO – assemble to order, ETO – engineering to order, MTO – manufacturing to order, and RTO – research to order. To ATO/ETO/MTO, it would be easier to design products by customization or partly modification based on PF architecture.

To RTO, a totally new product needs to be developed from the beginning. And this new product can grow to be a new product family. Following the conceptual design, the detailed design tasks are mostly focused on the modules and their last assemblies. The Figure 7 illustrates the modules design mode.

And an EBOM-engineer BOM will be produced to transfer the engineering data to the next stage.

3.4 Manufacturing

In manufacturing stage, a physical product is gradually produced and assembled based on the detailed design specifications. Three intertwined essential flows string all the activities and functions of this stage, including information flow, material flow and energy flow. A P3R (Product - Process - Plant - Resources) integrated model is presented as Figure 8 shows, which attempts to support concurrent development, manufacturing and production processes.

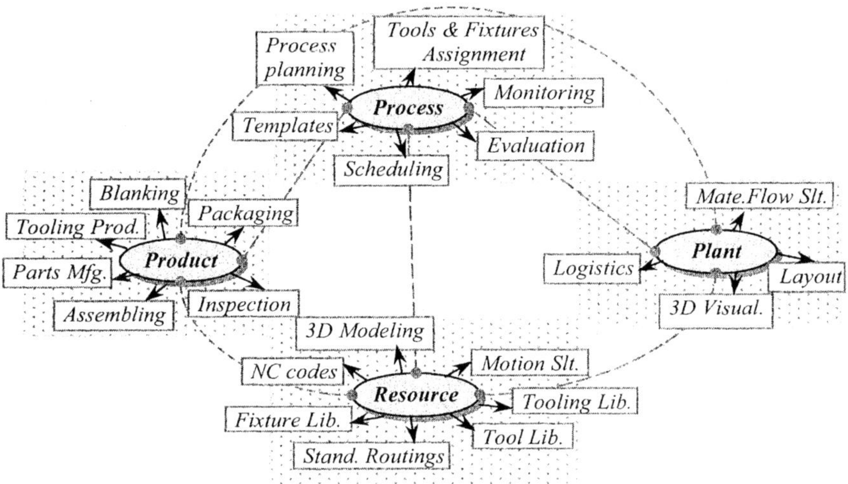

Figure 8 - Manufacturing model

In addition to managing the data lifecycle of the product components and assemblies, P3R model can help manage the processes of manufacturing product assemblies, machining individual components, laying out the facility in which the process will be executed and deploying the resources used by the processes.

MBOM - manufacturing BOM is produced to manage the manufacturing data.

3.5 Services

As the final stage of product lifecycle, especially after delivering products, the services of technical support, maintenance, repair and overhaul need to be offered for customers. And they can be performed or managed based on a common base, which is a part of the design base.

This common base can be modeled as Figure 9 shows, which mainly includes five sections: platform, knowledge, resources, for-services data and services modules.

(1) ***Services platform***, the strategies, policies, common solutions and core data of services are drawn out and designed here. And the services for product variants can be derived from it [18].

(2) ***Services modules***, Parts and components are marked with the services properties, which are accessories, alternate, reclaimable and standard. Services modules also include modular services contents such as the training courses and manuals, so enterprises can reuse them across the product families and along the product line.

(3) ***For-services data***, the services data can be addressed by Services BOM, which may be a series of BOMs due to the complexity of the services. For instance, manufacturers need a dismantling BOM for the dismantling and reclaiming process after products retired.

(4) ***Resources***, the services resources store and manage e-files of manuals, the records of the customers and their purchasing reports and services log etc. and even the repair tools and equipments can be classified as resources. Customers and companies can get supply from this base for the services.

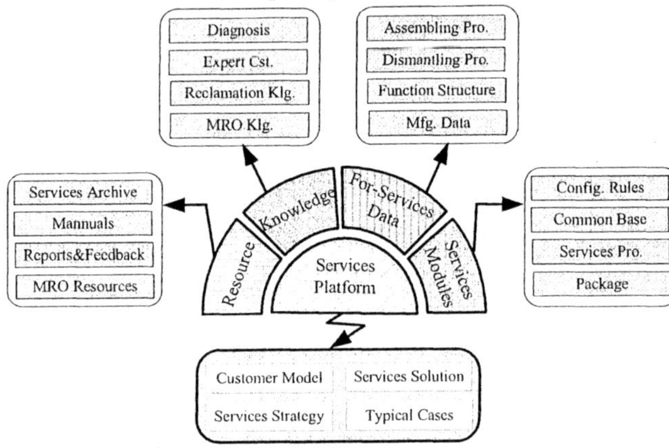

Figure 9 - Services model

(5) ***Knowledge***, the common knowledge may be offered for customers to deal with the difficulties or problems from the products practicing, such as FAQs or online diagnosis system. For the special products, enterprises may need expert systems to analyze the troubles. And to MRO and green manufacturing, knowledge plays a very important role.

4. A J2EE BASED PLM PROTOTYPE SYSTEM

Based on the above product lifecycle modeling framework, we developed a J2EE based PLM system, i.e. Tornado PLM, which is now successfully adopted by several Chinese manufacturing enterprises.

Tornado PLM employs an integration framework based on J2EE, as Figure 10 shows. It intends to unlock essential

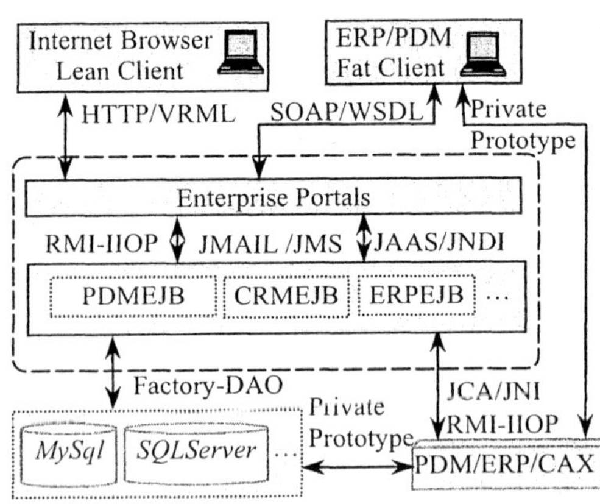

Figure 10 - A J2EE based integration framework

information within core business systems (PDM, ERP, CRM, CSM) across the entire value chain providing the product centric data integration of disparate systems along with PLM centric functionality & behavior.

By Tornado, designers can accomplish web based product lifecycle developing from customer requirements capturing to product services support. And Figure 11 gives an interface of this system for illustration.

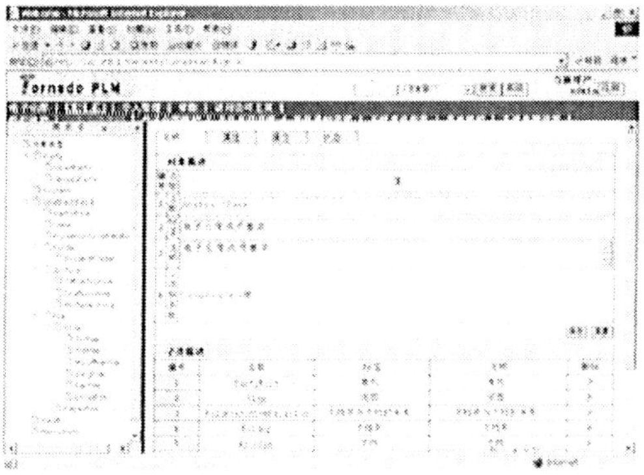

Figure 11 - Tornado PLM system

5. CONCLUSION

The PLM concept promises to provide support for the product's entire lifecycle, from the product requirements to the disposal of its last instance. The volume, diversity, and complexity of information describing the product will increase correspondingly.

This paper makes a proposal for a conceptual framework for product lifecycle modeling that can manage, generate, serve, and reuse all the product information throughout the entire lifecycle. We divide the product lifecycle into five main stages: requirement analysis, conceptual design, detailed design, manufacturing and services. All these stages are discussed in order to demonstrate the consistent business and data flow evolved across the product lifecycle. Methodologies for producing and management of the product information and businesses are presented. And especially we intend to develop a common base that can support single, steady, uniform and consistent product lifecycle developing. And a J2EE based integration framework is proposed for collaborative product development.

6. ACKNOWLEDGEMENT

This paper is funded by the Major State Basic Research Development Program of China (973 Program), No. 2002CB312200.

7. REFERENCES

1. CIMdata, "Product Lifecycle Management, Empowering the Future of Business", www.cimdata.com, 2002.
2. Bidarra R, Brosvoort WF. Semantic feature modeling. Computer-Aided Design, 2000, 32: 201-225.
3. Jiao JX, Tseng MM, Ma Q, et al. Generic Bill of Materials and Operations for High Variety Production Management. Concurrent Engineering: Research and Applications, 2000, 8(4): 297-321.

4. Simon M, Bee G, Moore P, et al. Modeling of the life cycle of products with data acquisition features. Computers in Industry, 2001, 45: 111-122.
5. Shehab EM, Abdalla HS. Manufacturing cost modeling for concurrent product development[J]. Robotics and Computer Integrated Manufacturing, 2001, 17: 341-353.
6. Rezayat M. The enterprise-web portal for life-cycle support. Computer-Aided Design, 2000, 32:85-96
7. Suh NP. The Principles of Design. New York: Oxford University Press, 1990.
8. Jiao JX, Tseng MM, etc. Product family modeling for mass customization. Computers ind. Engn, 1998, 35: 495-498.
9. Trappey AJC, Peng TK, Lin HD. Object-oriented bill of materials system for dynamic product management. J Intelligent Manufacturing, 1996, 7(5): 365-371.
10. Du J, Gu P, Jiao Y. Generic BOM processing system for Assembly-to-order production. Journal of Systems Science and Systems Engineering, 2001, 10(1): 1-6.
11. Glowen R. DCAC/MRM paves way to simpler production system. Boeing Frontiers Online, December 2003. http://www.boeing.com/news/frontiers/archive/2003/december/i_ca2.html
12. Wang CE, Chu CB, Yin CW. Implementation of remote robot manufacturing over Internet. Computers in Industry, 2001, 45(3): 215-229.
13. Rezayat M. Knowledge-based product development using XML and KCs, Computer-aided design, 2000, 32: 299-309.
14. Tseng MM and Jiao JX. Computer-Aided Requirement Management for Product Definition: A Methodology and Implementation. Concurrent Engineering: Research and Application, 1998, 6(3), 145-160.
15. Tay FEH and Gu JX. "Product Modeling for Conceptual Design Support", Computers in Industry, 2002, 48(2), 143-155.
16. Jiao JX, Tseng MM, Dufty VG and Lin F. "Product Family Modeling For Mass Customization", Computers ind. Engn, 1998, 35(3-4), 495-498.
17. Meyer MH and Lehnerd AP. The Power of Product Platforms: Building Value and Cost Leadership, The Free Press, New York, USA: 1997.
18. Zhang WL and Fan YS. "Extending Product Family Approach to Support Services", Proceedings of the Second IEEE International Conference on Services Systems and Services Management, Beijing, P.R. China, June. 4-7, 2005, pp. 735-740.

21

TOWARDS A REFERENCE MODEL FOR COLLABORATIVE NETWORKED ORGANIZATIONS

Luis M. Camarinha-Matos [1], Hamideh Afsarmanesh [2]
[1] *New University of Lisbon, PORTUGAL, cam@uninova.pt*
[2] *University of Amsterdam, NETHERLANDS, hamideh@science.uva.nl*

A growing number of collaborative networked organizations can be found in industry, services, and research. However, the lack of a reference model that could synthesize and formalize the base concepts, principles, and recommended practices, is an obstacle for an easier and more consistent development of the area. Therefore a reference modeling approach is proposed considering multiple modeling perspectives. Examples are given and steps for further research are identified. Establishing a reference model in this area is a long term endeavor; this contribution is aimed as just a step in this process.

1. INTRODUCTION

Collaborative networks are emerging in a large variety of forms, including virtual organizations, virtual enterprises, dynamic supply chains, professional virtual communities, virtual organization breeding environments, collaborative virtual laboratories, etc. [1], [2], [3]. Although not all, most forms of collaborative networks imply some kind of *organization* over the activities of their constituents, identifying roles for the participants, and some governance rules. Therefore, these can be called manifestations of collaborative networked organizations (CNOs). A large body of empiric knowledge related to collaborative networked organizations is already available, but there is an urgent need to consolidate this knowledge and build the foundations for a more sustainable development of this area.

Lack of reference models for collaborative networked organizations or even to some of their manifestations (such as virtual enterprises) is a common concern found in the literature, being also pointed out as an obstacle for a more consistent development of the area [4]. The difficulties are found namely in the terminology and associated meanings, what leads to frequent misunderstandings among members of this research community with a different original background.

Establishing a reference model for a new entity is not an easy task when only limited background inputs are available. In this context the reference model shall play a guiding / visionary role. Once established, the reference model defines a common basis for understanding and explaining (at least at a high level of abstraction) the different manifestations of the paradigm. It shall facilitate the

Please use the following format when citing this chapter:

Camarinha-Matos, L. M., Afsarmanesh, H., 2006, in IFIP International Federation for Information Processing, Volume 220, Information Technology for Balanced Manufacturing Systems, ed. Shen, W., (Boston: Springer), pp. 193–202.

development of particular models for specific CNOs (Figure 1). These particular models will drive the implementations and serve also to simulate / evaluate concrete networks.

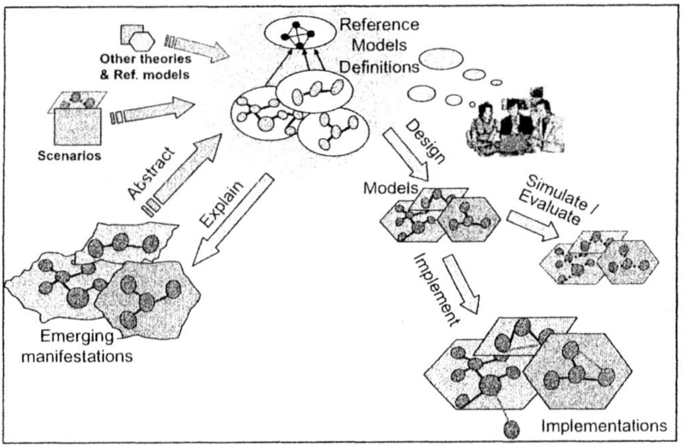

Figure 1 – Reference models in a context

2. EARLY CONTRIBUTIONS

When attempting to establish a reference model it is fundamental to consider the potential inputs and partial contributions from previous works. In fact some previous and ongoing projects have tried to contribute to reference models of some manifestations of collaborative networks, namely for Virtual enterprises / virtual organizations (VE/VO).

Figure 2 illustrates the diversity of sources which can potentially be used as inputs to this activity. As shown, there are two main streams:

- Enterprise-centric stream, which starts from the extensive past modeling activities at enterprise level and try to incrementally extend / adapt such models to the context of networks of enterprises.
- Network-centric stream, which puts the emphasis primarily on the networks and their properties, rather than on the characteristics of the individual elements.

These streams are not totally disjunctive and several initiatives show in fact partial elements of the two perspectives.

The approaches to modeling very much depend on the dominant background of people involved in each initiative. Three main groups or "schools" encompass most of the past VE/VO related developments:

i) Enterprise modeling, based on the underlying "culture" represented by the Zachman framework, GRAI-GIM, PERA, CIM-OSA, GERAM, and related developments [10].
ii) Organizational / management school, which departs from traditional organizational structures such as supply chains and the corresponding SCOR model, and tries to reason about emerging organizational patterns in new collaborative forms.

iii) VE/VO ICT-based projects, which put a strong emphasis on the ICT tools and infrastructures to support collaboration. A large number of projects have been carried out in this area that, although showing a "fragmented" and mostly ad-hoc approach, contribute with partial elements to better understand CNOs, their modeling needs and possible approaches.

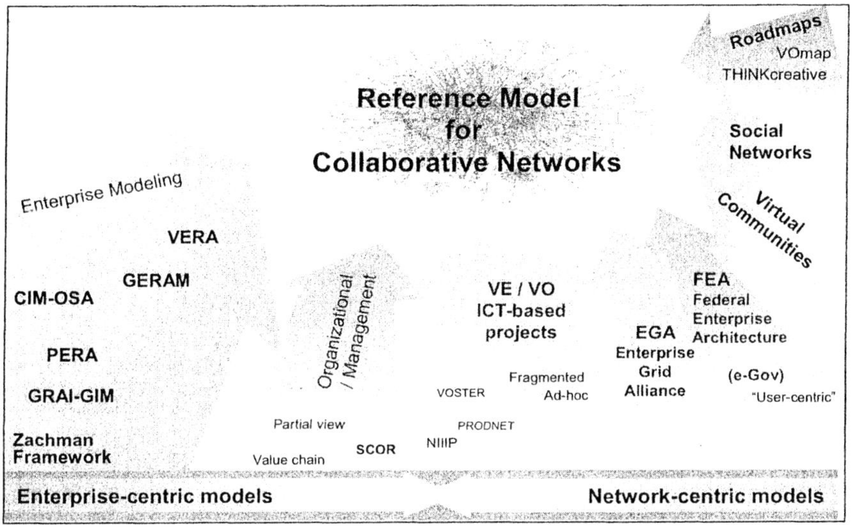

Figure 2 – Main inputs to the design of a CNO reference model

Some survey works analyzing early contributions namely in the areas i) and iii) above can be found in the literature, such as [11], [12]. The PRODNET project [4] or the VITE model [6] are examples of ICT-driven initiatives. An example survey under perspective ii) was conducted in the VOSTER project [9], which also included some analysis of ICT developments and common practices on VE/VO implementation [4]. Other areas of interest include:

iv) Grid community, which has been moving towards virtual organizations and is now trying to consider a business perspective, as in the case of the Enterprise Grid Architecture initiative [7].

v) E-Government, which represents a wide area but that has some common elements when it addresses the cooperation among different governmental organizations, as illustrated by the Federal Enterprise Architecture [8].

vi) Social networks and virtual communities are areas that although not yet offering much in terms of reference models, have developed considerable background in terms of the basic properties of networks with a strong basis on graph theory.

vii) Collaborative networks roadmapping initiatives such as THINKcreative, VOmap and others which have contributed to the identification of the research challenges in the area [2].

Figure 3 tries to put into a simplified historic perspective some of the key initiatives and events that represent a substantial input to a better understanding of collaborative networks and therefore offer base material for the elaboration of reference models for CNOs.

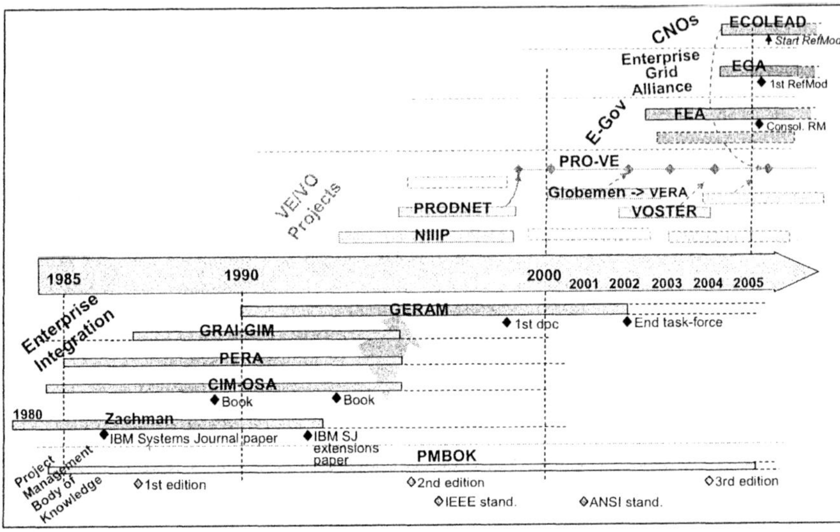

Figure 3 – Towards a CNO reference model - A simplified historic perspective

The lower half of the diagram in Figure 3 includes major representatives of the enterprise integration and modeling area that were particularly active in the 80s and 90s. A parallel initiative, from a different area but that can also give some hints for some cases of virtual organizations, is the Project Management Body of Knowledge.

The upper half of the diagram shows initiatives that are more directly related to collaborative networks. Of particular relevance here is the heritage of a large number of VE/VO projects. VOSTER represented an attempt to synthesize part of this heritage. The PRO-VE series of conferences and the corresponding proceedings are also playing a major role in the consolidation of knowledge in the area and contributing to establish some (progressive) consensus, important elements towards the definition of reference models.

3. BASE TERMINOLOGY

The establishment of reference models for CNOs is frequently pointed out as a major need for the consolidation and sustainable development of the area. However it seems that there is not much consensus on what this term exactly means. In fact it seems that it represents quite different things for different people and consequently it raises quite different expectations regarding its utility. It is therefore necessary to revisit the concept of reference model and its purpose.

The clarification of the base concepts is however not that easy as the literature in this area is full of confusing terminology. To refer only a few, it is common to find terms such as reference architecture, reference framework, architectural framework, system architecture, etc. often used with similar or largely overlapping meanings. Without the aim of giving a "final" definition, the following working definitions are suggested:

- **Model**: A model is an abstract representation of an environment, system, or entity in the physical, social, or logical world.

Typically a model refers only to some aspects of the phenomenon being modeled, and two models of the same phenomenon may be essentially different. This may be due to: different requirements, differences in conceptual approaches, esthetic preferences, and also different past experiences. Therefore, users of a model need to understand the model's purpose and the assumptions or limits of its validity. There can be models at various levels of abstraction, from very abstract theoretical constructs, to (detailed) representations very close to the modeled entity or implementation.

- **Framework**: In general a framework is a structure for supporting or enclosing something else. In the modeling area, a framework can be seen as an "envelope" that might include a number of (partial) models, collections of templates, procedures and methods, rules, and even tools (e.g. modeling languages).
- **Reference model**: A reference model is a generic abstract representation for understanding the entities and the significant relationships among those entities of some area, and for the derivation of other specific models for particular cases in that area. Preferably a reference model is based on a small number of unifying concepts and may be used for education, explaining, and systems' development.

  A **CNO reference model** is thus a generic conceptual model that synthesizes and formalizes the base concepts, principles and recommended practices for collaborative networked organizations. It is intended as an authoritative basis (guide) to streamline or facilitate the creation of focused models for the various manifestations of CNOs as well as architectures and implementation models for particular systems development. A reference model is generic and not directly applicable to concrete cases but rather provides the basis for the development (derivation) of other models closer to those cases.
- **Architecture**: An architecture is an abstract description of a specific system, i.e. a particular model that even at a logical level tends to indicate the system structure, functions of its components, their interactions, and constraints, and can be used to develop the system. Architecture is focused on "building a system" and must be complete at its level of abstraction; therefore not all models are architectures. Although there is a difference between engineering and architecture (compare with roles of civil engineer and building architect), to some extent the architecture depends on engineering principles and available technology. An architecture can be formulated in a descriptive or in a prescriptive style. Descriptive style defines an enumeration of design elements and formal "arrangements" between them. Prescriptive style establishes constraints, namely by limiting the possible design elements and their "arrangements".
- **Reference architecture**: A reference architecture aims at structuring the design of architectures for a given domain by defining a unified terminology, describing the functionality and roles of components, providing template components, giving example architectures, and defining a development methodology. It corresponds to architecture as a style or method in the sense that may represent a coherent set of design principles to be used in a specific area. The reference architecture is the basis for designing the specific architectures for particular instances of systems in the class of systems covered by the reference architecture. In the CNO domain, a reference architecture for VO management systems would represent the "structure" and principles to be followed by particular architectures of concrete

VO management systems. The concept of reference architecture also induces the creation of generic re-usable "building blocks".

Based on the definition given above, two main "anchors" can be associated to a reference model: Authority and re-use (Figure 4).

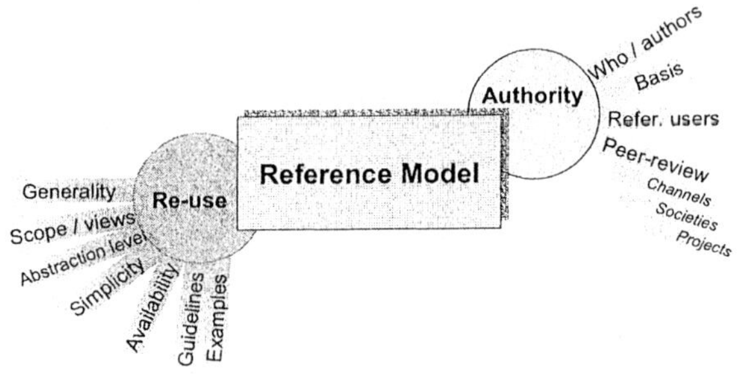

Figure 4 – Key anchors in a reference model

Establishing a model as an authoritative reference depends on a number of factors, including the authorship, i.e. the reputation / prestige of the involved contributors, the adopted bases and referenced sources, the list of early adopters or reference users, the quality of the peer reviewing process, and also the dissemination channels, professional societies and projects involved in its dissemination. Re-usability of the elements of a reference model, with the objective of streamlining the design and development of particular models, also depends on a number of factors, including: the generality of the model, its scope and covered views, the abstraction level and simplicity, the forms of availability / easiness of access to supporting information, the existence of guidelines for use and examples of application to typical cases.

It is also important to distinguish between reference models and standards. Both share some common aspects (Figure 5), namely aiming at simplifying the creation of new systems and providing some stable conceptual background or building blocks. Regarding the process, both start with building consensus but then they evolve into different directions.

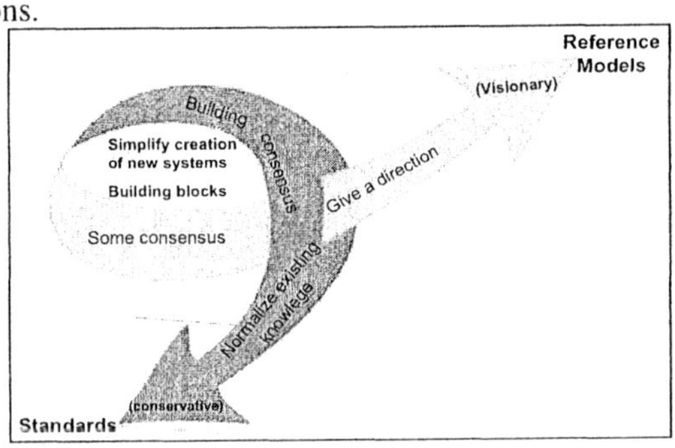

Figure 5 – Reference models and standards

4. REFERENCE MODELING APPROACH

A complex entity such as a CNO can be observed and analyzed through different lenses or perspectives. Each lens can provide complementary elements that help in achieving a better understanding of the paradigm. It is however important to note that lenses might also cause distortions. Particularly if one tries to explain all aspects of CNOs through the perspective of a single lens, not only it leads to dangerous over-simplifications, but even introduces some misconceptions. Therefore a holistic perspective is needed.

Most of the previous publications towards a reference model for a CNO (or some of its manifestations) are either technology-biased (e.g. [11]), or business-biased (e.g. [9]). A holistic approach, combining both perspectives would guarantee a better alignment of business and technology. On the other hand, we shall not ignore other aspects such as culture, values, norms and principles, trust, etc. that can represent another dimension – the "style" of the CNO (a term borrowed from the area of architecture / civil construction). These aspects are less addressed in previous modeling works but shall be considered in a holistic reference model for CNOs.

When modeling a CNO, it is important to consider both its internal and external aspects (Fig. 6).

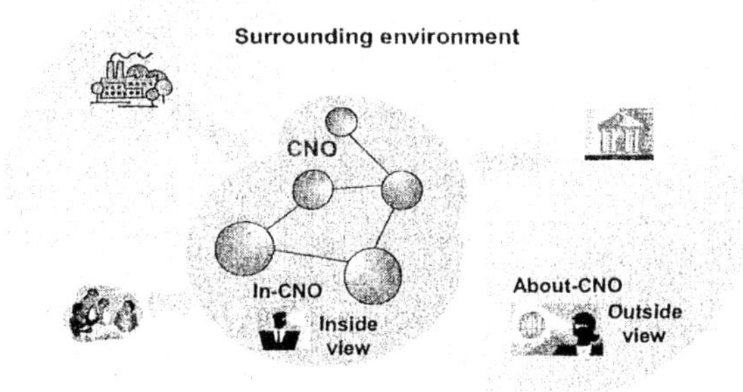

Figure 6– Modeling perspectives

Therefore in ARCON (A Reference Model for COllaborative Networks), being developed within the ECOLEAD project [5], [1], these two subspaces comprehensively cover the internal (In-CNO) aspects of CNOs as well as the external (About-CNO) aspects that are related to the logical surrounding of CNOs.

In-CNO perspective. This perspective aims at providing of an abstract representation of the CNO from inside, namely the identification of a set of characteristic properties that can together capture the elements constituting CNOs. Building In-CNO abstract representation is challenging due to the large number of distinct and varied entities, roles, concepts, functionality, rules and regulations, etc. inside the CNOs. To better characterize these aspects, four dimensions are proposed and defined to cover the In-CON perspective, as follows:

- **Structural dimension.** This perspective addresses the structure or composition of the CNO in terms of its constituting elements (participants and their relationships) as well as the roles performed by those elements and other characteristics of the network nodes such as the location, time, etc. This perspective is used in many disciplines (e.g. systems engineering, software engineering, economy, politics, cognitive sciences, manufacturing), although with different "wording" and tools.

- **Componential dimension.** This dimension focuses on the individual tangible/intangible elements in the CNO's network, e.g. the resource composition such as human elements, software and hardware resources, information and knowledge. Not all these elements are "physical" in a strict sense but rather represent the "things" of which the network is built of. Furthermore, elemental dimension also consists of ontology and the description of the information/knowledge.
- **Functional dimension.** This perspective addresses the "base operations" available at the network and the execution of time-sequenced flows of operations (processes and procedures) related to the "operational phase" of the CNO life cycle.
- **Behavioral dimension.** This dimension addresses the principles, policies, and governance rules that drive or constrain the behavior of the CNO and its members over time. Included here are elements such as principles of collaboration and rules of conduct, contracts, conflict resolution policies, etc.

About-CNO perspective. This perspective aims at reaching an abstract representation of the CNO as seen from the outside, i.e. which characteristic properties the CNO reveals in its interaction with its "logical" surrounding environment. A CNO as a whole might interact with, influence, and be influenced by a number of "interlocutors", e.g. customers, competitors, external institutions, potential new partners. The interactions between the CNO and these external entities are quite different, the same as the way each of these entity groups looks at the CNO. In order to better characterize these differences, the following modeling dimensions are proposed for the external or About-CNO perspective:

- **Market dimension**. This dimension covers both the issues related to the interactions with "customers" (or potential beneficiaries) and "competitors". The customers' facet involves elements such as the transactions and established commitments (contracts), marketing and branding, etc. On the competitors' side issues such as market positioning, market strategy, policies, etc. can be considered. The purpose / mission of the CNO, its value proposition, joint identity, etc. are also part of this dimension.
- **Support dimension**. Under this dimension the issues related to support services provided by third party institutions are to be considered. Examples include certification services, insurance services, training, external coaching, etc.
- **Societal dimension**. This dimension captures the issues related to the interactions between the CNO and the society in general. Although this perspective can have a very broad scope, the idea is to model the impacts the CNO has or potentially can have on the society (e.g. impact on employment, economic sustainability of a given region, potential for attraction of new investments) as well as the constraints and facilitating elements (e.g. legal issues, public body decisions, education level) the society provides to the CNO development.
- **Constituency dimension**. This perspective focuses on the interaction with the

universe of potential new members of the CNO, i.e. the interactions with those organizations that are not part of the CNO but that the CNO might be interested in attracting. Therefore, general issues like sustainability of the network, attraction factors, what builds / provides a sense of community, or specific aspects such as rules of adhesion and specific "marketing" policies for members, are considered here.

In addition to these perspectives, a CNO model can be defined at multiple levels of abstraction. Currently three levels are considered in ARCON:

- **General concepts level** – that includes the most general concepts and related relationships, common to all CNOs independently of the application domain.
- **Specific modeling level** – an intermediate level that includes more detailed models focused on different classes of CNOs.
- **Implementation modeling level** – that represents models of concrete CNOs.

Fig. 7 combines the addressed perspectives into a single diagram.

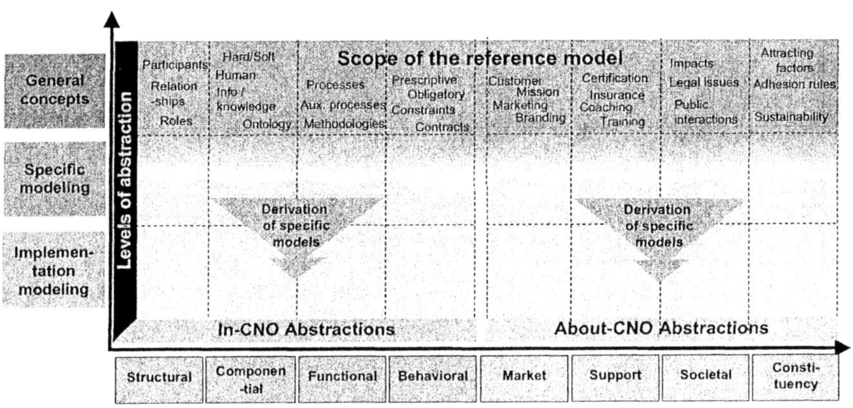

Figure 7 – A modelling framework for CNOs

Given the definition of reference model presented above, the scope of a CNO reference model covers mainly the "General Concepts" level and it might represent some elements of the "Specific Modeling" level. In other words, at the current stage of development of the CNO area, the first priority for a reference model is to consolidate the most general concepts that are common to all types of CNOs. With the progress in the area it will make sense to progressively consolidate more specific models for each major class of CNOs.

In terms of representation, a generic tool like UML, or another standard representation formalism (perhaps an ontology representation languages) that is proper for human understanding can be adequate for representation of the General Concepts level. For the other levels it will be necessary to consider other modeling tools and theories (e.g. set theory, graph theory, Petri nets, deontic logic, complexity theories, multi-agent systems, federated systems, etc.).

Another fundamental perspective to consider in the framework of CNO reference modeling is its life cycle. The following main stages can be considered: Creation, Operation, Evolution, and Metamorphosis or Dissolution. This is the subject of our ongoing work on development of ARCON, and will be addressed in future papers.

5. CONCLUSIONS AND FUTURE WORK

The elaboration of a comprehensive reference model for CNOs is a very important element in the consolidation of existing knowledge in this area, and a basis for its consistent further progress. A number of early attempts have been made in various research projects, but frequently biased by a single perspective.

ECOLEAD is attempting to contribute to a more holistic reference model for CNOs considering multiple perspectives. Nevertheless it is clear that the establishment of a reference model capturing the variety of CNOs and their complexity is a long term endeavor that needs to start with a careful analysis of the current baseline and definition of related reference modeling frameworks. This paper presented our preliminary results in this direction as developed in the framework of an ongoing effort to establish A Reference Model for COllaborative Networks (ARCON).

6. ACKNOWLEDGEMENTS

This work was funded in part by the European Commission through the ECOLEAD project. The authors thank the contributions of their partners in this project.

7. REFERENCES

1. Afsarmanesh, H.; Camarinha-Matos, L.M. – A Framework for Management of Victual Organization Breeding Environments, in Proceedings of PRO-VE'05 – Collaborative Networks and their Breeding Environments, Springer, Valencia, Spain, 26-28 Sept 2005.
2. Camarinha-Matos, L.M.; Afsarmanesh, H. (Editors) – Collaborative Networked Organizations – A research agenda for emerging business models, Springer, 2004.
3. Camarinha-Matos, L.M.; Afsarmanesh, H. – Collaborative networks: A new scientific discipline, J. Intelligent Manufacturing, vol. 16, N° 4-5, pp439-452, 2005.
4. Camarinha-Matos, L.M.; Afsarmanesh, H.; Ollus, M. (Editors) – Virtual Organizations – Systems and Practices, Springer, 2005.
5. Camarinha-Matos, L.M.; Afsarmanesh, H.; Ollus, M. - ECOLEAD: A holistic approach to creation and management of dynamic virtual organizations, in Proceedings of PRO-VE'05 – Collaborative Networks and their Breeding Environments, Springer, pp. 3-16, Valencia, Spain, 26-28 Sept 2005.
6. Chalmeta, R. – Virtual transport enterprise integration, *Transactions of the SDPS*, Vol.4, N.4, Dec 2000. http://www.sdpsnet.org/journals/vol4-4/Ricardo-4.pdf
7. EGA - Enterprise Grid Alliance Reference Model, 13 Apr 2005. http://www.gridalliance.org/en/workgroups/ReferenceModel.asp
8. FEA – FEA Consolidated Reference Model, May 2005, http://www.whitehouse.gov/omb/egov/documents/CRM.PDF
9. Katzy, B.; Zhang, C.; Loeh, H. – Reference models for virtual organisations, in Virtual organizations: Systems and Practices, L. M. Camarinha-Matos, H. Afsarmanesh, M. Ollus (Editors), Springer, 2005.
10. Noran, O. – A mapping of individual architecture frameworks (GRAI, PERA, C4ISR, CIMOSA, ZACHMAN, ARIS) onto GERAM, in Handbook on enterprise architecture, P. Bernus, L. Nemes, G. Schmidt (Ed.s), Springer, 2003.
11. Tolle, M.; Bernus, P. - Reference models supporting enterprise networks and virtual enterprises, International Journal of Networking and Virtual Organisations 2003 - Vol. 2, No.1 pp. 2 – 15. http://www.inderscience.com/storage/f161185104291237.pdf
12. Tolle, M.; Bernus, P.; Vesterager, J. - Reference models for virtual enterprises, in Collaborative business ecosystems and virtual enterprises, L. M. Camarinha-Matos, Editor, Kluwer Academic Publishers, 2002.

22 AGENT-BASED INTELLIGENT MEDIA DISTRIBUTION IN ADVERTISEMENT

Raafat Aburukba[1], Hamada Ghenniwa[1], Weiming Shen[1, 2]
[1] *Department of Electrical and Computer Engineering*
University of Western Ontario, London, Ontario, Canada
[2] *Integrated Manufacturing Technologies Institute*
National Research Council Canada, London, Ontario, Canada
roaburuk@engga.uwo.ca; hghenniwa@eng.uwo.ca; weiming.shen@nrc.gc.ca

Networked electronic displays provide unique and effective capabilities for businesses and industries to communicate with their consumers. Today there are many solutions for static distribution of media-contents. However, delivering dynamic media-contents remains a major challenge. The overall goal of this work is to provide an intelligent media-content distribution, by which business intelligence is utilized to strengthen the business-customer relationships and increase profitability. This paper proposes a multi-agent approach to model the dynamic scheduling problem in media-content distribution. The proposed approach is validated through a prototype implementation.

1. INTRODUCTION

Improved consumer addressability, especially in electronic-based markets, allows companies to send focused promotional messages to specific customers, facilitating a targeted advertising approach. Networked electronic displays provide unique and effective capabilities for businesses and industries to communicate with the public.

Today there are many solutions for static media-content distribution in content-based network domains such as advertisement where a human user schedules media-contents to specific media displays at a specific time. The current approach requires a tremendous amount of work specially if there exist a large number of media-content and displays distributed across the country or the world. Also, if we consider the dynamic behavior of the content-based domain such as, media displays breakdown, the arrival of new media-contents, introduction of new media displays, and media-contents removal, then previous schedules might become infeasible due to the unforeseen changing situations. Therefore, finding coherent schedules in the content-based network environment where disturbances may occur at anytime is a major challenge.

In this work, we look at the content-based network problem as a scheduling problem, where scheduling can be described as the allocation of tasks to capable resources. Most scheduling models can fit in a 4-element structure, which consists of resources, tasks, constraints, and objectives:

Please use the following format when citing this chapter:

Aburukba, R., Ghenniwa, H., Shen, W., 2006, in IFIP International Federation for Information Processing, Volume 220, Information Technology for Balanced Manufacturing Systems, ed. Shen, W., (Boston: Springer), pp. 203–212.

- **Resources**: physical/logical devices capable of processing tasks;
- **Tasks:** a set of operations which are physical/logical processes, to be executed by a resource;
- **Constraints:** a set of conditions which must be satisfied. Constraints may be operation-based, task-based, resource-based, or a combination of both;
- **Objectives:** decision criteria which can be modeled as to maximize or/and minimize some function.

To illuminate confusion we will use tasks and resources for media-contents and media displays respectively in the rest of this paper.

In this work, we focus on an instance problem in content-based networks where specific locations of resources are used to display third party contents. By third party, we mean that a business rents a specific time slot from a content-based network provider to display their contents. This paper models the scheduling in the distributed content-based network environment and deal with unforeseen events such as resource addition and removal, and task additions and removal. We present an architecture that provides intelligent scheduling and handles changes in the media-content distribution environment. Previous work on dynamic scheduling has been investigated, and we strongly believe that agent-orientation is an appropriate design paradigm for distributed scheduling in the content-based network environment. The application of the agent for dynamic scheduling integrates the following:

- A negotiation protocol for scheduling to allow agents to cooperate and coordinate their actions in order to define a feasible schedule that satisfies both resource and task objectives and to handle the presence of real-time events.
- Flexibility and scalability where it is possible to dynamically introduce new resources and tasks, or remove existing resources and tasks without interrupting currently scheduled tasks in the environment.

The remainder of this paper is organized as follows: Section 2 describes related work; Section 3 introduces the scheduling problem model; Section 4 provides our dynamic scheduling model and approach; Section 5 introduces our implementation, and Section 6 provides a brief conclusion.

2. RELATED WORK

Previous work is done to transport multimedia contents under heterogeneous clients and network connections. The work in 7 is based on an approach that provides a descriptor schemes. Those descriptor schemes are used for matching multimedia contents with respect to the user profile for personalized video delivery on the web. The matching determines which parts of the video to include and what hotspots and hyperlinks to provide to the user. Their application supports personalization of individual sequences within an interactive video presentation, as well as per video-clip payment of interactive video. Another work in 12 combines video indexing techniques to parse TV News recordings into stories, and information filtering techniques to select stories to construct automatically personalized TV news programs. They formalize the selection process as an optimization problem. Experiments showed that a simple heuristic can provide high quality selection with little computation.

Generally, scheduling problems which are concerned with finding optimal schedules are subject to a limited number of constraints, and have a combinatorial explosion of possible solutions and are generally NP-hard 56. Early scheduling approaches involved finding the optimal schedule, but they were typically not efficient when the problem size grew. The majority of the research work in the scheduling area has been devoted to either simplifying the scheduling problem to the point where some algorithm can find optimal solutions, or to devising efficient heuristics for finding approximate solutions. The solution methods form two distinct classes: exact methods and heuristic methods. Exact methods guarantees finding an optimal solution. Such methods include: mathematical programming, dynamic programming, and branch and bound. Heuristic solutions in the other hand do not guarantee the optimal solution, but typically assure some degree of optimality when compared to the optimal. Some of the meta-heuristic techniques include: Simulated Annealing, tabu search, and genetic algorithms. Meta-heuristics are high-level heuristics that guide local search heuristics to escape from local optima.

Dynamic scheduling refers to the ability to adapt the schedule to the new situation by using appropriate actions to handle each event 17. The work in 11 used simulation to investigate the performance of various schedule repair heuristics (such as: no rerouting, queue rerouting, arrival rerouting, all rerouting) for unexpected resource failures in dynamic job shops. The experimental results showed that the proper selection of a good schedule repair heuristic is based not only on the system characteristics (utilization, resource down times, and frequency of resource failures) but also on the material handling system in terms of speed and the number of material handling system devices. Another study in 8 presents a simulation based analysis of different proposed dispatching rules for scheduling in job-shops with resource breakdowns. The results of the simulation study revealed that the relative performance of scheduling rules can be affected by changing the breakdown parameters. A work in 13 presents several rescheduling strategies are proposed for process time variations, resource breakdowns, and new task arrivals in a dynamic environment. Monitoring the environment is performed periodically and either rerouting to alternative resources or order splitting policies are activated in response to unexpected disruptions.

Agent technology has recently been applied to solve distributed scheduling problems including distributed manufacturing scheduling, transportation scheduling and computing load balancing **2019**18. Many agent-based scheduling in distributed systems adapt economic-based approaches such as contract-net **19**3 which is based on the tender economic model. More complex approaches include combinatorial auctions 16 for many-to-many negotiation problems.

In a content-based network environment such as advertisement, it is highly desirable to employ dynamic scheduling. The dynamic nature of scheduling can be viewed as:

- the process of updating an existing schedule in response to a disruption 21
- the process of generating a new feasible schedule upon occurrence of a disruption 1

3. MODELING THE SCHEDULING PROBLEM

3.1 Scheduling Problem in Content-Based Networks

In a content-based networks domain, a system with m resources can be denoted by $\mathbf{R} = \{R_1, R_2, ..., R_m\}$. A resource is a physical/logical device that is capable of executing some operations. In this context, a resource can be formally defined as a set of operations $R_i = \{o_i^1, o_i^2, ..., o_i^{n_i}\}$, where n_i is the number of operations that R_i can process. An operation is a physical process to be executed by a resource in the content-based domain such as media-contents or tasks. The overall operations that can be performed by the system $\mathbf{R}$ is equal to $Y_i^m R_i$. Partially Overlapping (PO) systems consist of resources, each of which is capable of performing a specific set of operations that may overlap with those of other resources. For these systems, the following two conditions must hold:

1) There exist at least two resources $R_i, R_j \in \mathbf{R}$, $1 \le i, j \le m, i \ne j$, such that $R_i \, I \, R_j \ne \phi$;
2) There exist at least two resources $R_k, R_l \in \mathbf{R}$, $1 \le k, l \le m, k \ne l$, such that $R_k \ne R_l$.

Given a set of n tasks in the content-based networks, denoted by $\mathbf{T} = \{T_1, T_2, ..., T_n\}$, where task $T_j \in \mathbf{T}$ (*j=1...n*) is formulated as a set of operations $T_j = \{o_j^1, o_j^2, ..., o_j^{n_j}\}$, where n_j is the number of operations belonging to T_j. The tasks need to be performed by $\mathbf{R}$. The scheduling process involves the allocation of tasks among resources at specific times. Each task T_j may have its operations processed by a single resource R_i from a group of resources, in which each resource can process all operations of T_j. In partially overlapping systems structure 10, the instance problem in content-based network domain focused on this paper falls under the OR-structure where a task is to be processed by any resource in $\mathbf{R}$. The content-based network scheduling problem in PO systems has the following structure:

- Resource capabilities are sufficient. All operations of the tasks can be processed in the system, that is $Y_{j=1}^n T_j \subseteq Y_{i=1}^m R_i$
- Task T_j is to be processed in a resource R_i that has the capabilities to process all the task's operations (PO system OR-structure)
- Tasks have objectives to be achieved in the domain
- Resources have objectives to be achieved in the domain
- No pre-emption. Each task, once started, must be completed before another task may be started on a resource
- No resource may process more than one task at a time
- Each task has a hard start time, and end time
- tasks can have sequence constraints
- tasks can have location constraints

3.2 Modeling Content-Based Scheduling Problem

In this section, we model a content-based instance of scheduling tasks to resources owned by a content-based network provider to meet the tasks' and resources' objectives. In this instance, we focus on two objectives, maximizing the profit and minimizing the cost. These objectives are useful for the content-based network business owners to maximize their profit, and for the tasks' business owners to find the minimum cost to display their tasks.

3.2.1 Maximizing the Profit

The basis of modeling the profit is based on the formula: *Profit=Revenue – Cost.* The cost in this formula is based on the price it costs the business to provide the media-content distribution scheduling service. This cost is created by the media-content network provider based on an analysis to their expenses. Here, we will focus on increasing the revenue since the cost in the profit function is presented as a constant value. To increase the revenue, we maximize the resource utilization, since by having the resource idle, the content-based network business provider will not have any revenue, and therefore, the profit will decrease. In this instance, each time period to promote a task differ from other times which is set by the content-based network providers. For example: the cost to display a task during rush hour in busy streets is more expensive than displaying a task after mid-night. The mathematical formulation to maximize the resource utilization is:

$$\max\left\{\sum_{j=1}^{n} x_{ij} p_{ij}\right\} \quad (1)$$

s.t

$$\sum_{j=1}^{n} x_{ij} p_{ij} \le T_i \quad (1.2)$$

$$\sum_{i=1}^{m} x_{ij} = 1 \quad (1.3)$$

$$x_{ij} \in \{0,1\} \quad (1.4)$$

In the above formulation, n presents the number of tasks, and m presents the number of resources. T_i is the completion time of tasks scheduled on R_i, and $\sum_{j=1}^{n} p_{ij}$ is the total processing time of n number of tasks in resource R_i. x_{ij} is a variable that is set by constraint 1.3 and it can be either 1 or 0. If $x_{ij} = 1$, task j is to be processed by resource i.

3.2.2 Minimize the Cost

The second objective we are looking at in this work is to minimize the cost. The mathematical formulation to minimize the cost is:

$$\min\left\{\sum_{i=1}^{m} c_{ij} x_{ij}\right\} \quad (2)$$

s.t.

$$\sum_{i=1}^{m} x_{ij} = 1 \quad (2.1)$$

$$x_{ij} \in \{0,1\} \quad (2.2)$$

In this objective function, *m* represents the number of resources, c_{ij} is the cost price to execute task *j* in resource *i*. x_{ij} is a variable that is set by constraint 2.1 and it can be either 1 or 0. If $x_{ij} = 1$, task *j* is to be processed by resource *i*.

4. CONTENT-BASED SCHEDULING APPROACH

4.1. Business Rules

The Business Rules Group 2 describes a business rule as a statement that defines or constrains some aspect of the business; a business rule is intended to assert business structure or to control or influence the business's behavior. A rule engine evaluates and executes rules, which are expressed as if-then statements. The power of business rules lies in their ability both to separate knowledge from its implementation logic and to be changed without changing source code 14. A rule is composed of two parts, a condition and an action: When the condition is met, the action is executed 14.

In this work, we use the concept of business rules to capture precise business logic in resources, tasks, and time to govern the behavior of the scheduler. We propose three types of business rules:

- **Resource Business Rules** are related to the resource location such as marketing zone. In our proposed solution, each resource and task has a business rules associated with it. Assigning business rules to both tasks and resources will govern the scheduler to find the right location for the task.
- **Tasks Business Rules** are related to tasks such as baked products. In our proposed solution, each task is associated with tasks business rules. This association governs the scheduler to produce tasks sequence or order.
- **Time Business Rules** are related to time such as breakfast time. In our proposed solution, time intervals, and tasks are associated with time business rules. This association governs the scheduler to assign the task at a specific time interval.

4.2. Proposed Solution

In a dynamic environment, scheduling usually involves complex and non-deterministic interactions between different participating tasks and resources. We strongly believe that agent-orientation is an appropriate design paradigm for scheduling in a dynamic environment. Indeed, such a paradigm is essential to model:

- The dynamic structure of the environment where it is possible to dynamically integrate new tasks, and resources, or remove tasks and resources without disrupting schedules previously established and re-initialising the environment.
- Multiple and distributed objectives and constraints: each resource can be geographically distributed, where each resource can have its objective and constraints, and solve the scheduling problem locally. Hence, distributed and autonomous scheduling systems are more appropriate than centralized and non-autonomous scheduling approaches.

A key aspect of agent-orientation is the ability to design artefacts that are able to perceive, reason, interact and act in a coordinated fashion. Therefore, unlike the traditional way of having a centralized scheduler, an agent based scheduling system supports distributed scheduling where each resource agent can find a schedule that satisfies its own objective. However, agents collectively find a schedule that satisfies both task and resource objectives. In our proposed approach, we model scheduling as distributed, autonomous and collaborative problem.

- **Distributed:** By distributed we mean that each resource is geographically distributed and responsible for finding a solution of the scheduling problem based on its objective of maximizing the profit.
- **Autonomous:** Each resource is independent from other resources. By independent we mean that when a resource finds a schedule that satisfies its objective to maximize the profit, the schedule does not depend on other resources.
- **Collaborative:** Multi-objectives between tasks and resources

Our proposed solution models the scheduling problem as a multi-agent system. We introduce three types of agents:

- **Interface Agent:** this agent gathers the information of the tasks and sends a request to the manager agent. The information gathered includes: tasks features, associated business rules, and the objective to be achieved.
- **Manager Agents**: represent the resources within its domain. Manager agents are capable of generating a local solution to the schedule based on its objective.
- **Broker Agent:** provides the scheduling functionality to produce a global solution that satisfies the tasks' objective function.

Given the nature of the content-based environment of being distributed and independent, we divide finding a solution to the multiple objective scheduling problem into local and global solutions.

- **Local Solution:** this includes the manager agent view of the solution based on the resource based objective to maximize the profit.
- **Global Solution:** this includes the integral solutions generated by the broker agent in the proposed architecture within the context of the task based objective to minimize the cost.

To govern the scheduling search, we associated resources, tasks, and time to business rules. The resources are associated with resource related business rules. Time intervals are associated with time related business rules. Tasks are associated to resource, time, and task business rules. Having these business rules allow tasks to request for service at specific desired locations, at a specific time, and at a specific order or sequence.

4.3. Proposed System Architecture

The proposed architecture for distributed dynamic scheduling in content-based networks shown in Figure 1 involves the following agents: Interface Agents, Manager Agents, and the Broker Agent. The proposed multi-agent architecture allows manager agents to achieve their objective in scheduling locally. In addition, the manager agent and the broker agent cooperate using the contract-net negotiation approach 3 in order to find a global schedule. Focusing on the objective of minimizing the cost, and maximizing the profit, we present a solution to find a schedule that satisfies both task and resource objectives.

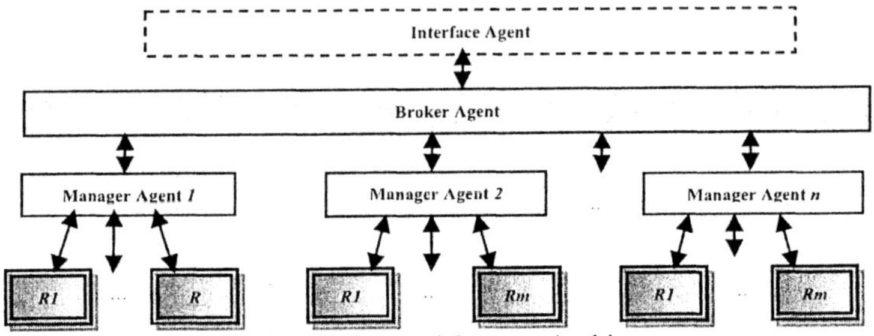

Figure 1 - High-Level System Architecture

Upon an existence of a resource within a domain, the manager agent is notified. The manager agent stores the resource existence and the resource capabilities in its knowledge base. The manager agent informs the broker agent with any updates. The information sent to the broker agent includes: the resource capabilities, and business rules of all the resources attached with the manager agent. The broker agent stores the information for the manager agents in its knowledge.

In case of a resource removal, the manager agent has knowledge of all scheduled tasks in the resources, and therefore,

- The manager agent removes the resource from its knowledge.
- The manager agent reschedules unprocessed tasks in the removed resource within its domain based on tasks' business rules.
- In the case of the manager agent can not find a schedule within its domain, it returns unscheduled tasks back to the broker agent to reschedule the specific tasks. The process of rescheduling these tasks is the same as scheduling new tasks mentioned below.

Upon receiving the tasks by the broker agent, the broker agent generates a match based on the task, time and resource business rules, as well as on the manager agents' capabilities. The match results include the resources capable to schedule the task, and the time interval to schedule the task. The broker agent interacts with the capable manager agents and the cost to execute the tasks in the resources at the specific time slots. Manager agents have the knowledge of the resources in its domain, as well as the scheduled tasks. The manager agents find the optimal schedule based on its objective to maximize the profit. The interactions between the broker agent and manager agents are as follows:

- The broker agent sends a request to all potential manager agents to schedule each task.
- Each potential manager agent finds a local schedule that maximizes the profit.
- Each potential manager agent submits a price to the broker agent.
- The broker agent collects the prices from manager agents and finds the global schedule based on minimization the cost objective.

In the case to remove a task, the system reacts as follows:

- The interface agent sends a request to remove the task to the broker agent.
- The broker agent sends a request to the manager agent that is scheduled to process the task.

- The manager agent cancels the task from the resource.
- The manager agent sends a confirmation to the broker agent.
- The manager agent reschedules unprocessed tasks based on its objective to maximize the profit.

5. IMPLEMENTATION

A prototype system has been implemented to provide scheduling in a distributed content-based environment. We used the JADE, a FIPA-complaint platform 9 in our agent implementation. Each individual agent (interface, manager, and the broker) in our proposed architecture is modeled using coordinated intelligent, rational agent (CIR-agent) 4. The internal structure of each agent is composed of knowledge and capabilities.

In general, the knowledge of the agent is stored in its local database. Each agent has access to its local relational databases via the Java Database Connectivity (JDBC). This allows the agents to query and access information from its local relational databases through a common platform-independent interface. The interface allows Java programs to interact with any SQL-compliant database. However, the capabilities of all agents in the environment include communication, and reasoning. Generally with all agents the communication capability allows the agents to exchange messages with other elements of the environment, including users, agents and objects. The implementation of this component takes advantage of the existing Agent Message System in JADE. The communication component of the agent's is equipped with an incoming message inbox, and message polling can be both blocking and non-blocking, with an optional timeout. Messages between agents are based on using FIPA ACL, in which different ACL performative are supported; the communication component is implemented as a set of object-oriented classes that inherit the *jade.Core.Agent* and *jade.lang.acl.ACLMessage* existing class of the JADE platform. The agent's reasoning capabilities include the problem solver, and interaction devices such as the assignment device. The problem solver in both the broker and the manager agents include the algorithm of the revised simplex optimization technique that finds the exact optimal value of the objective under the linear inequality constraints 15. As mentioned in Section 2, exact solutions are not efficient to solve large problems. However, the focus of this work is not to implement an efficient optimization algorithm. We have implemented the revised simplex optimization algorithm as a proof of concept in both the broker agent to find the global solution and the manager agent to find the local solution.

6. CONCLUTION

This paper presents an agent-based approach for intelligent media-content distribution. It considers the dynamic behavior of the system such as resource additions or breakdown, and task additions or removal. An important advantage of using the broker agent is that it allows the system to operate robustly when confronted with a resource or task agent's appearance or disappearance. Also, in a dynamic environment, the proposed broker agent can help tasks to locate distributed resources in an open environment. The proposed system can monitor any possible changes within the environment and performs scheduling based on these changes.

7. REFERENCES

1. Abumaizar, RJ, Svestka, JA. "Rescheduling job shops under random disruptions". International Journal of Production Research, 1997; 35 (7): 2065-2082.
2. Business Rules Group. URL: http://www.businessrulesgroup.org/
3. Davis, R, Smith, R. "Negotiation as a Metaphor for Distributed Problem Solving". Artificial Intelligence, 1983; 20: 63-109.
4. Ghenniwa, H. and Kamel, M. "Interaction Devices for Coordinating Cooperative Distributed Systems", Journal of Intelligent Automation and Soft Computing, 2000.
5. Gordon V, Proth J-M, Chu C. A survey of the state-of-the-art of common due date assignment and scheduling research. European Journal of Operational Research, 2002;139:1-25.
6. Graham, R.L., Lawler, E.L., Lenstra, J.K., and Rinnooy Kan, A.H.G., Optimization and approximation in deterministic sequencing and scheduling: A survey, Annals of Discrete Mathematics 5, 1979; pp. 287-326.
7. Hjelsvold Rune,Vdaygiri Subu,Léauté Yves, "Web-based Personalization and Management of Interactive Video", Proceedings of the 10th International Conference on World Wide Web, Hong Kong, 2001, pp. 129-139.
8. Holthaus, O. Scheduling in job shops with machine breakdowns: an experimental study. Computers & Industrial Engineering,1999; 36 (1), 137-162.
9. Java Agent DEvelopment (JADE) Framework. URL: http://jade.tilab.com/
10. Kamel, M, Ghenniwa, H. "Partially-Overlapped Systems: The Scheduling Problem". In Design and Implementation of Intelligent Manufacturing Systems, H. Parsaei and M. Jamshidi, Eds., Prentice-Hall, 1995, pp. 241-274.
11. Kutanoglu, E. and Sabuncuoglu, I.. Routing-based reactive scheduling policies for machine failures in dynamic job shops. International Journal of Production Research, 2001;39 (14), 3141-3158.
12. Merialdo, B, Lee, K. T, Luparello D, and Roudaire J. "Automatic Construction of Personalized TV News Programs," ACM Multimedia, 1999.
13. Nof, S.Y., Grant, F.H., Adaptive/predictive scheduling: review and a general framework. Production Planning and Control 2 (4), 298-312. Society,1991; 41 (6), 539-552.
14. Qusay H. Mahmoud. Getting Started with the Java Rule Engine API (JSR 94): Toward Rule-Based Applications, 2005. http://java.sun.com/developer/technicalArticles/J2SE/JavaRule.html
15. Ronald L. Rardin. "Optimization in Operations Research", pp. 236
16. Sandholm T. "Algorithm for optimal winner determination in combinatorial auctions", Artificial Intelligence, 2002; No. 135, pp. 1-54.
17. Saucer, J. "Knowledge-Based Systems Techniques and Applications in Scheduling". In Knowledge-Based Systems Techniques and Applications, TL Leondes, ed., San Diego, Academic Press, 1999.
18. Shen W. "Distributed Manufacturing Scheduling Using Intelligent Agents", IEEE Intelligent Systems, 2002, Vol.17, No. 1, pp. 88-94.
19. Sousa, P. and Ramos, C. A distributed architecture and negotiation protocol for scheduling in manufacturing systems. Computers in Industry, 1999;38 (2), 103-113.
20. Van Brussel, H., Wyns, J., Valckenaers, P., Bongaerts, L. and Peeters, P. Reference architecture for holonic manufacturing system: PROSA. Computers in Industry, 1998; 37 (3), 255-274.
21. Vieira, G. E., Hermann, J. W. and Lin, E. Rescheduling manufacturing systems: a framework of strategies, policies and methods. Journal of Scheduling, 2003, 6 (1), 36-92.

23 A SERVICE COMPOSABILITY MODEL TO SUPPORT DYNAMIC COOPERATION OF CROSS-ENTERPRISE SERVICES

Jing-fan Tang
College of Computer Science & Technology, Hangzhou Dianzi University,
tangjf@hziee.edu.cn

With the development of web services related technologies, more and more enterprises adopt web services to encapsulate their business systems to be published on Internet. Due to the different semantic of the web services, it brings much difficulty to implement the dynamic cooperation of the cross-enterprise services efficiently. This paper introduces a service composability model to support the dynamic cooperation of cross-enterprise services by utilizing the semantic rules of the internal composability. The WSDL descriptions of the web services are extended with semantic capability, which enable the cross-enterprise services to be composed and cooperated automatically according to the synthetic comparison of the semantic features. Furthermore, a policy driven negotiation process is also proposed to enable the cooperation of cross-enterprise services to achieve win-win.

1. INTRODUCTION

With the rapid development of technologies on web services and service-oriented architecture, more and more enterprises utilize web services to encapsulate the internal business process to provide web applications for all kinds of clients. The cooperation process of cross-enterprise services can be seen as services composition process. A web service will find a composable service to be composed and achieve cooperation. The big challenge for composing the cross-enterprise services is to enable the description of each service to be understood by each other, such as the functionalities, interfaces, etc.

The semantics of web services is crucial to enabling automatic service composition. It is important to insure that selected services for composition offer the "right" features. Such features may be syntactic (e.g., number of parameters included in a message sent or received by a service). They may also be semantic (e.g., the business functionality offered by a service operation or the domain of interest of the service). To help capture Web services' semantic features, the concept of ontology is used. Ontology is a shared conceptualization based on the semantic proximity of terms in a specific domain of interest (McIlraith, 2001). Ontologies are increasingly seen as key to enabling semantics-driven data access and processing (Bussler, 2002). They are expected to play a central role in the Semantic Web,

Please use the following format when citing this chapter:

Tang, J., 2006, in IFIP International Federation for Information Processing, Volume 220, Information Technology for Balanced Manufacturing Systems, ed. Shen, W., (Boston: Springer), pp. 213–222.

extending syntactic service interoperability to semantic interoperability (Horrocks, 2002).

In this paper, we introduce a service composability model to support the dynamic cooperation of cross-enterprise service by utilizing the semantic rules of the internal composability. Through the composability rules registered in an enhanced UDDI center, an automatic matchmaking process will be executed to find out the candidate services for the cooperation. In order to achieve the win-win cooperation among the cross-enterprise services, a policy driven negotiation process is adopted to choose the optimum service to establish the final cooperation.

The rest of the paper is organized as following. Section 2 introduces the related work. Section 3 describes the service composability model. Section 4 discusses the automatic matchmaking of the cross-enterprise services through the composability rules. Section 5 introduces the negotiation process, which is policy-driven. Section 6 concludes.

2. RELATED WORK

There has been some research on the service composition to support cross-enterprise business integration and cooperation (Vaggelis, 1999; Yang, 2001; Alan, 2002). And, the automatic composition of web services is a recent trend (McIlraith, 2001; Berners-Lee, 2001). This would include the automatic selection and interoperation of web services. Automatic composition is slated to play a major role in enabling the envisioned Semantic Web (Weikum, 2002).

A novel approach has been presented in (Zhou, 2005)to support the dynamic establishment of virtual enterprise based on the dynamic web service composition. A service execution plan will be dynamically built to aim at the target of the virtual enterprise through the composition of web services. The development and deployment of web services is separated to support the dynamically deploying and binding of the web services at run time. The members in the virtual enterprise will be located through the dynamic web service discovery based on DAML+OIL logic reasoning, and selected through the dynamic web services negotiation with multi-steps protocol to organize the virtual enterprise at run time.

Based on the research in (Zhou, 2005; Brahim, 2003; Tang, 2004a; Tang, 2004b; Tang, 2005], we will introduce a service composability model to support dynamic cooperation of cross-enterprise services in this paper.

3. SERVICE COMPOSIBILITY MODEL

3.1 Semantic description of web services

In order to support dynamic cooperation of cross-enterprise services, the description of each service should be understood by other services. An emerging language for describing operational features of Web services is WSDL (Web Service Description Language). WSDL is being standardized within the W3C consortium. WSDL has gained considerable momentum as the language for web service description. However, WSDL provides little or no support for semantic description of Web services. It mainly includes constructs that describe web services from a syntactic point of view. To cater to semantic web-enabled web services, we should extend WSDL with semantic capabilities, which would lay the groundwork for the automatic selection and cooperation of cross enterprise web services.

Before introducing the semantic description of web services, we give some definitions as following (Brahim, 2003):

Definition 1 – *Message.* A *message M* is defined as a tuple (P, T, U, R) where:

- *P* is a set of parameter names.
- *T*: *P* → *DataTypes* is a function that assigns a data type to each parameter. *DataTypes* is a set of XML data types.
- *U*: *P* → *Units* is a function that gives the unit of measurement used for each parameter. *Units* is a taxonomy for measurement units.
- *R*: *P* → *Roles* is a function that assigns a business role to each parameter. *Roles* is a taxonomy for business roles.

Definition 2 – *Purpose.* The *purpose* of an operation op_{ik} is defined by a tuple *(Function, Synonyms, Specialization)* where *Function* is op_{ik}'s business functionality defined within a given taxonomy, *Synonyms* is a set of alternative function names, and *Specialization* is a set of characteristics of op_{ik}'s function.

Definition 3 – *Category.* The *category* of an operation op_{ik} is defined by a tuple *(Domain, Synonyms, Specialization)* where *Domain* is op_{ik}'s area of interest defined within a given taxonomy, *Synonyms* is a set of alternative domains, and *Specialization* is a set of characteristics of op_{ik}'s domain.

Definition 4 – *Quality.* The *quality* of an operation op_{ik} is defined by a tuple $(Fees_{ik}, Security_{ik}, Privacy_{ik})$. $Fees_{ik}$ is the dollar amount needed to execute op_{ik}. $Security_{ik}$ is a boolean that specifies whether op_{ik}'s messages are securely exchanged. $Privacy_{ik}$ is the set of input and output parameters that are not divulged to external entities.

Definition 5 – *Operation.* An operation op_{ik} is defined by a tuple $(Description_{ik}, Mode_{ik}, In_{ik}, Out_{ik}, Purpose_{ik}, Category_{ik}, Quality_{ik})$ where:

- $Description_{ik}$ is a text summary about the operation features.
- $Mode_{ik} \in$ {"one-way", "notification", "solicit-response", "request-response"}.
- In_{ik} and Out_{ik} are the input and output messages, respectively. $In_{ik} = (\Phi, \Gamma_{ik})$ and $Out_{ik} = (\Phi, \Gamma_{ik})$ for notification and one-way operations, respectively.
- $Purpose_{ik}$ describes the business function offered by the operation (cf. Definition 2).
- $Category_{ik}$ describes the operation's domain of interest (cf. Definition 3).
- $Quality_{ik}$ gives the operation's qualitative properties (cf. Definition 4).

After giving above definitions, we will introduce the semantic description of web services as following:

Definition 6 – *Web service.* A *Web service* WS_i is defined by a tuple $(Description_i, OP_i, Bindings_i, Purpose_i, Category_i)$ where:

- *Description$_i$* is a text summary about the service features.
- *OP$_i$* is a set of operations provided by WS$_i$.
- *Bindings$_i$* is the set of binding protocols supported WS$_i$.
- Purpose$_i$ = {Purpose$_{ik}$(op$_{ik}$) | op$_{ik}$ $\in$ OP$_i$} is a set of WS operations' purpose.
- Category$_i$ = {Category$_{ik}$ (op$_{ik}$) | op$_{ik}$ $\in$ OP$_i$} $\cup$ {Category$_i$ (WS$_i$)} is a set of WS$_i$ operations' categories.

3.2 Composability model for web services

A major issue in the cooperation of cross-enterprise services is whether they can understand each other and can be composable. For example, it would be difficult for an enterprise service to invoke an operation of other enterprise service if there were no mapping between the parameters (e.g., data types, number of parameters). So, we want to identify two sets of composability rules to compare syntactic and semantic properties of web services (Brahim, 2003).

Syntactic rules include:

1) *mode composability*, which compares operation modes, and
2) *binding composability*, which compares the binding protocols of interacting services.

Semantic rules include:

1) *message composability*, which compares the number of message parameters, their data types, business roles, and units;
2) *operation semantics composability*, which compares the semantics of service operations;
3) *qualitative composability*, which compares qualitative properties of web services.

The detail definitions for the composability rules are introduced as following:

Definition 7 – *Mode composability*. Two operations op$_{ik}$ = (D_{ik}, M_{ik}, In_{ik}, Out_{ik}, P_{ik}, C_{ik}, Q_{ik}) and op$_{jl}$ = (D_{jl}, M_{jl}, In_{jl}, Out_{jl}, P_{jl}, C_{jl}, Q_{jl}) are *mode composable* if (i) M_{ik} = "notification" and M_{jl} = "one-way"; or (ii) M_{ik} = "one-way" and M_{jl} = "notification"; or (iii) M_{ik} = "solicit-response" and M_{jl} = "request-response"; or (iv) M_{ik} = "request-response" and M_{jl} = "solicit-response".

Definition 8 – *Binding composability*. Two services WS$_i$ = (D_i, O_i, B_i, P_i, C_i) and WS$_j$ = (D_j, O_j, B_j, P_j, C_j) are *binding composable* if $B_i \cap B_j \neq \Phi$.

Definition 9 – *Message composability*. Two operations op$_{ik}$ = (D_{ik}, M_{ik}, In_{ik}, Out_{ik}, P_{ik}, C_{ik} Q_{ik}) and op$_{jl}$ = (D_{jl}, M_{jl}, In_{jl}, Out_{jl}, P_{jl}, C_{jl}, Q_{jl}) are *message composable* if:
(i) $\forall p \in In_{ik}$, $\exists p' \in In_{jl}$ | p is data type compatible with p' and $U(p) = U(p')$, and $R(p) = R(p')$.
(ii) $\forall p \in In_{jl}$, $\exists p' \in In_{ik}$ | p is data type compatible with p', and $U(p) = U(p')$, and $R(p) = R(p')$.

Definition 10 – *Operation semantics composability.* We say that $op_{ik} = (D_{ik}, M_{ik}, In_{ik}, Out_{ik}, P_{ik}, C_{ik}, Q_{ik})$ is *operation semantics composable* with $op_{jl} = (D_{jl}, M_{jl}, In_{jl}, Out_{jl}, P_{jl}, C_{jl}, Q_{jl})$ if (i) P_{ik} is *compatible with* P_{jl} and (ii) C_{ik} is *compatible with* C_{jl}.

Definition 11 – *Qualitative composability.* We say that $op_{ik} = (D_{ik}, M_{ik}, In_{ik}, Out_{ik}, P_{ik}, C_{ik}, Q_{ik})$ is *qualitatively composable with* $op_{jl} = (D_{jl}, M_{jl}, In_{jl}, Out_{jl}, P_{jl}, C_{jl}, Q_{jl})$ if:
(i) Q_{ik}.Fees $\geq Q_{jl}$.Fees; and
(ii) (Q_{ik}.Security = true) => (Q_{jl}.Security = true); and
(iii) Q_{ik}.Privacy $\subseteq Q_{jl}$.Privacy.

Based on the composability model, we propose an efficient model to support the automatic cooperation of cross-enterprise services.
Figure 1 shows the architecture of our model:

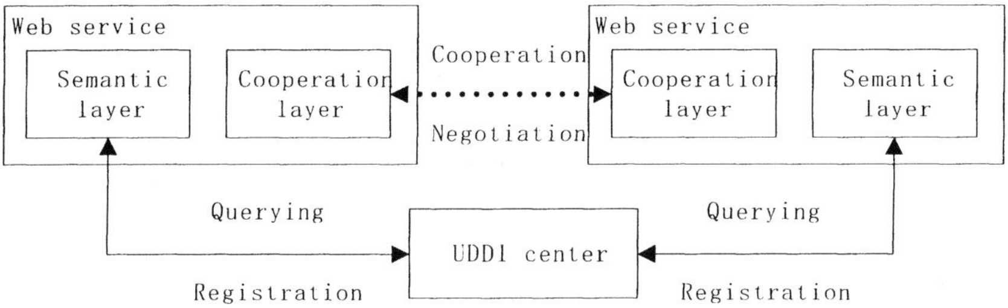

Figure 1. Cooperation model

In this model, there are two layers defined in the web service: one is the semantic layer and the other is cooperation layer. The semantic layer is responsible for defining the semantic rules for the web service. It will register the semantic description of the web service to the UDDI center and query the information of the composable web service for cooperation through the semantic description. Composability rules will be used for UDDI center to find out the composable services. When the composable services are found, cooperation layer will be responsible for the negotiation process to find out the optimum service for cooperation.

4. AUTOMATIC MATCHMAKING OF THE CROSS-ENTERPRISE SERVICES

4.1 Enhanced UDDI center

In order to support the automatic matchmaking among the cross-enterprise services, UDDI center is enhanced to allow the registration of composability rules, which are described in WSDL using XML.
Figure 2 shows the enhanced UDDI center:

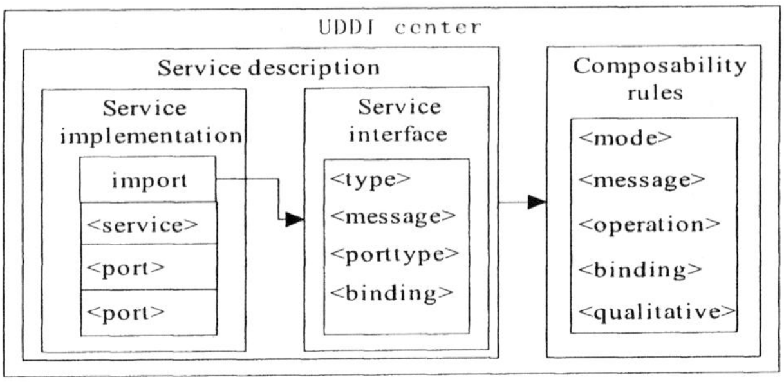

Figure 2. Enhanced UDDI center

Through the enhanced UDDI center, the enhanced WSDL with semantic description is registered for enterprise service to automatically discover the composable cross-enterprise service to achieve the dynamic cooperation. The composability of the cross-enterprise services will be decided according to the synthesis evaluation on the model composability, binding composability, message composability, operation semantics composability and qualitative composability.

4.2 Automatic matchmaking

Figure 3 shows an automatic matchmaking process.

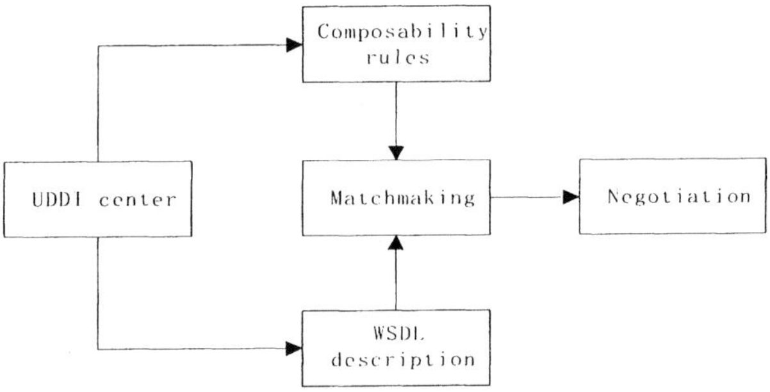

Figure 3. Matchmaking process

There are two ways in the cooperation and composition of cross-enterprise services (Zhou, 2005):

1) Web service internal driven: the cooperation and composition is performed due to the requirement of web service in some enterprise. For example, when the network finance management system in some enterprise wants to implement the functionalities of payoff, tax management and cash management, it should interact with the systems of bank and revenue office to achieve the cooperation.
2) Business requirement driven: the cooperation is established due to the business requirement advanced by some enterprise through the integration of several business systems (web services). For example, when some enterprise wants to invest on some fund, it should obtain the invest

information of the fund such as stock, exchange and bond, etc. So it requires other organizations (i.e. stock exchange, bank) to be integrated to provide the related analysis information on the fund.

For the composition and cooperation of the cross-enterprise services, a composition plan should be built first according to the business logic and composability rules whatever it is web service internal driven or business requirement driven. By composition plan, we mean the list of component services and their interactions with each other (plugging operations, mapping messages, etc.) to form the composite service. The matchmaking algorithm uses as input the composer's specification and UDDI of preexisting service interfaces described in WSDL (extended with semantic constructs). Through the composition plan, a detailed description of the composite service is automatically generated. This description includes the list of outsourced services, mappings between the composite and outsourced services operations and messages, and the control flow of outsourced operations. The control flow refers to the execution order of the operations outsourced by the composite service.

After the component services are automatically matched and selected, the negotiation process will be adopted to achieve the win-win cooperation among the cross-enterprise services.

5. POLICY DRIVEN NEGOTIATION

Some composable services will be chosen out as candidate services to attend the policy driven negotiation process. The policy is described using XML, which are defined in the extended WSDL. Through the multi-steps negotiation process driven by policies, the final cooperation of the cross-enterprise services will be achieved.

5.1. Negotiation Policy in XML

Policies (Yang, 2002; Tang, 2005) are rules governing the choices in the behavior of a system. In our model, policy is adopted to address the adaptive negotiation process.

The negotiation policies are expressed in XML since it has a text-based representation which imposes few restrictions on network technology or protocols and (through the use of XML schema) it has a sufficiently strict syntax to permit automated validation and processing of information in an unambiguous way.

Negotiation policies allow selections from a range of options provided by the developer of the policy-controlled service, comprised of rules in the form of if<condition>then<action> (Tang, 2005). This allows flexibility to be build into a system by supporting a range of different behaviors rather than hard-coding a particular behavior. Furthermore, the use of XML as an intermediate representation still allows the negotiation policies to be developed using any existing approach. Some kind of policy editor GUI can be provided to give a user-friendly and flexible way to manage the service-specific policies.

The following shows the information details, which will be included in the negotiation policy (Figure 4 shows an example of a service policy in XML):

- **Service:** the referenced information of the service such as service id, service name and service type, etc;

- **Info:** the elementary information on the policy itself including policy id, creator, etc. The policy id and service id will be used to identify the service policy;
- **Param-specs:** the available parameters (information) for evaluating the conditions, such as cost, profit ratio, bottom price, QoS, minimal/max adjusted value, etc;
- **EvalFunction:** the evaluation function of synthetic parameters for decision making on actions under different conditions;
- **Trigger:** the relevant policies triggered by different results from the evaluation function to trigger the actions;
- **Action:** the details of actions on the evaluation of the local available parameters.

```
<?xml version = "1.0" encoding = "UTF-8"?>
<policy xmlns = " http://~ tc/policy" xmlns:xsi =
"http:// ~tc /XMLSchema-instance" xsi:schemaLocation
="http:// ~tc /policy file:///home/~tc /policy.xsd">
<service>
      <service-id>SIN001</service-id>
      <name>service001</name>
      <type>Single</type>
</service>
<info>
      <policy-id>Policy1</policy-id>
      <creator>tjf</creator>
      <authority>ALL</authority>
</info>
<param-specs>
      <param-spec variable=var_1>
            <description>descript_1</description>
      </param-spec>
      < param -spec variable=var_2>
            <description>descript_1</description>
      </param -spec>
</param-specs>
<evalFunction value=evalfun_variable>
      EvalFun(param-specs)
</evalFunction>
<trigger variable=evalfun_variable>
      <Branch>
            <Condition value=value_1>
                        <Action>Act_1</Action>
            </Condition>
            <Condition value=value_2>
                        <Action>Act_2</Action>
            </Condition>
      </Branch>
</trigger>
</policy>
```

Figure 4. An example of negotiation policy in XML

5.2. Negotiation process

The specific requirements will be presented by composer for the candidate services, such as QoS, fee, etc. The candidate services will perform evaluation on the requirements according to the negotiation policy and provide the corresponding bids with some contact. During the negotiation process, some adjustments are allowed to be done on the contact, which is also according to the negotiation policy. When the negotiation process ends, the composer will evaluate the final bids and choose the optimum service to establish the cooperation. Figure 5 shows the policy driven negotiation process.

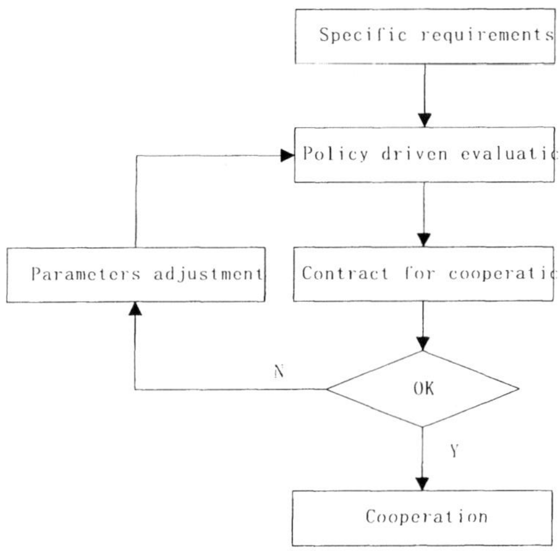

Figure 5. Policy driven negotiation process

Through such policy driven negotiation process, the cross-enterprise services can achieve win-win among the dynamic cooperation process.

6. CONCLUSION

This paper has introduced a service composability model to support dynamic cooperation of cross-enterprise services by applying the semantic rules of composability description. An enhanced UDDI center is provided to support the registration of composability rules by web services. There are two layers defined in web service: semantic layer and cooperation layer. Semantic layer is responsible for the automatic matchmaking process to select the composable services, while cooperation layer is responsible for the negotiation process to establish the final cooperation. Such policy driven negotiation process can achieve win-win among the dynamic cooperation of cross-enterprise services.

7. REFERENCES

1. A. Berfield, P. Chrysanthis, I. Tsamardinos, M. Pollack, S. Banerjee, "A Scheme for Integrating E-Services in Establishing Virtual Enterprises", Proceedings of Research Issues in Data Engineering (RIDE) Workshop, San Jose, CA, February 2002.
2. T. Berners-Lee, J. Hendler, O, Lassila. The Semantic Web. Scientific American 284(5), pp. 34-43. May 2001.
3. B. Medjahed, A. Bouguettaya, A.K. Elmagarmid, "Composing Web Services on the Semantic Web", The VLDB Journal, 12(4), November 2003.
4. C. Bussler, D. Fensel, A. Maedche. A conceptual architecture for Semantic Web enabled Web services. SIGMOD Record 31(4): 24–29, 2002.
5. I. Horrocks. DAML+OIL: A description logic for the Semantic Web. IEEE Data Engineering Bulletin 25(1): 4–9. March 2002.
6. S.A. McIlraith, T.C. Son, H. Zeng. Semantic Web services. IEEE Intelligent Systems. Special Issue on the Semantic Web. 16(2): 46–53, March/April, 2001.
7. J.F. Tang, B. Zhou, Z.J. He, P. Uros, "Adaptive Workflow-Oriented Services Composition and Allocation in Distributed Environment", Proceedings of the third International Conference on Machine Learning and Cybernetics, Shanghai, China, August 2004, pp.599-603.
8. J.F. Tang, B. Zhou, Z.J. He, "Dynamic Economic- and QoS-Based Approach for Workflow-Oriented Distributed Services Allocation", Proceedings of the International Conference on Systems, Man and Cybernetics, Hague, Netherlands, October 2004, pp. 3349-3353.
9. J.F. Tang, B. Zhou, Z.J. He, "Policy driven and multi-agent based fault tolerance for Web services", Journal of Zhejiang University SCIENCE, 2005 6A (7): 676-682.
10. V. Ouzounis, V. Tschammer, "A Framework for Virtual Enterprise Support Services", Proceedings of the 32nd Hawaii International Conference on System Sciences, 1999.
11. G. Weikum (editor). Special issue on Organizing and Discovering the Semantic Web. IEEE Data Engineering Bulletin 25(1): 1–58. March 2002
12. J. Yang, W.J. van den Heuvel and M.P. Papazoglou, "Service Deployment for Virtual Enterprises, Proceedings of the Workshop on Information Technology for VE, 2001.
13. K. Yang, A. Gailis, Todd, C., 2002. Policy-Based Active Grid Management Architecture. Proceedings of 10th IEEE International Conference on Networks (ICON02), p.243-248.
14. B. Zhou, J.F. Tang, Z.J. He, "An adaptive model of virtual enterprise based on dynamic web service composition", Proceedings of the Fifth International Conference on Computer and Information Technology, September 2005, pp. 284-289.

24

ON FORMAL THEORIES AND FORMALISMS FOR VIRTUAL ENTERPRISES

Goran Putnik, Rui Sousa
Production and Systems Engineering Department
University of Minho, School of Engineering
4800-058 Guimarães, PORTUGAL
putnikgd@dps.uminho.pt; rms@dps.uminho.pt

One of the issues in development and implementation of Virtual Enterprise (VE) is a formal theory, or formal theories, of VE. There are serious problems concerning VE formal theory(ies) development. Besides the ones which are commonly known and easily perceived, one of the biggest problems is a general misunderstanding about formalisms and formal theories (FTs) that occurs too often, substituting the formal theory by the formalism. The paper clarifies in an informal way the differences and some implications between formal theory and formalisms, expecting that it will contribute to a better perception of the problem as well as to the directions and approaches concerning VE formal theory development.

1. INTRODUCTION

One of the issues in development and implementation of Virtual Enterprise (VE) is a formal theory, or formal theories, of VE. There is a number of questions concerning a formal theory (FT), or theories, of VE, ranging from its definition, implementation, use, and similar, to the questions like "why a formal theory of VE?". These questions, and respective responses, are similar to the questions and responses on a formal theory, or theories, of other related concepts as organizations, production or manufacturing systems, enterprises, etc.

Formal theories (FTs) have proved their usefulness in "traditional" engineering areas as mechanical, civil and electrical engineering, and (relatively) more recently in computer sciences (e.g. in the 70s the telecommunications area has identified the use of formal approaches as the only mean to deal with the ever-growing complexity of standards and OSI (Open Systems Interconnection) (Turner, 1993)). Consequently, we could say the lack of a FT of VE is a serious obstacle to effective and efficient development and application of a VE concept. However, concerning the "state-of-the-art" of the development of a FT of VE, the authors are not aware of any consistent and rigorous approach towards the FT of VE. This situation could be explained by the fact that there are serious problems concerning VE formal theory(ies) development. Besides the ones which are commonly known and easily

Please use the following format when citing this chapter:

Putnik, G., Sousa, R., 2006, in IFIP International Federation for Information Processing, Volume 220, Information Technology for Balanced Manufacturing Systems, ed. Shen, W., (Boston: Springer), pp. 223–232.

perceived (we will mention some of them later), one of the biggest problems is a general misunderstanding about formalisms and formal theories (FTs) that occurs too often, leading to the substitution of a formal theory by a formalism. This led to the fact that in spite of a number of VE subsystems formal specifications, we do not have, in fact, a FT of VE. Few contributions towards a FT of VE were found by the authors, e.g. (Gruninger and Fox, 1994) and (Janowski et al.; 1998). Within the TOVE (TOronto Virtual Enterprise) project Gruninger and Fox (1994) present a micro-theory for representing "the constraints over the objects and predicates in the ontology".

The objective of this paper is to contribute to the research, development and implementation of VE formal theory, or formal theories (and *not* to present a formal theory of VE – actually, only one example of a formal theory of VE will be presented as a part of discussion), through discussion on some relevant issues. The paper clarifies in an informal way the differences and some implications between Formal Theory and formalisms. In this way, it is expected that the paper will contribute to a better perception of the problem as well as to the directions and approaches concerning VE formal theory.

The results presented in the paper are the results from one of the strategic projects in the Centre for Production Systems Engineering at the University of Minho. The project on Formal Theories of Manufacturing Systems and Enterprises (including the FT of VE) has started on 1997.

2. WHY A VE FORMAL THEORY (OR THEORIES): SOME EXAMPLES OF COMMON MISUNDERSTANDINGS

The need for a FT of VE is mainly due to known reasons as:

- Ambiguities and inaccuracies on the used terminology;
- Inconsistencies and errors on systems specification;
- Conflicting results;
- Implementations not corresponding to specifications;
- The use of a formal language does not imply a formal theory behind;
- The FT allows automation of some phases of VE design process;
- Controlling the growing complexity of standards (Turner, 1993); etc.

Some examples of ambiguities and inaccuracies on the used terminology could be the following. There is not a common interpretation of the terms, i.e. what are we thinking on when we say, "network", "networked (enterprise)", networked organization", or "virtual enterprise", "virtual organization", Figure 1. That is, when we say "network" does it address the networked enterprise or just a networked enterprise domain, i.e. the domain over which the networked enterprise is integrated? The networked enterprise is a subset or a special case of the networked organization or the opposite? Also, could a hierarchy (Figure 2.a) be interpreted as a special case of a network (Figure 2.b)[1] or not. The network is characterized by

[1] The Figures 2.a and 2.b represent hierarchical and network structures, respectively, only under a certain interpretation (implying the interpretation of each arc as a specific relation). As Figures 2.a and 2.b do not have that arc interpretation, somebody could claim that they do not represent hierarchical and network structures. From the graph theory point of view, Figures 2.a and 2.b could be seen as two types of graphs.

relation or by operation? What is the exact meaning of "collaborative"? In an enterprise, even with sequentially performed processes, Figure 3.a, we would say that, anyway, they collaborate. Or, "collaboration" means only that any process is linked to all the other processes and they perform simultaneously, Figure 3.b. Are "hybrid" structures permitted, in the context of "collaborative", Figure 3.c? Is the networked enterprise, or VE, subset, or a special case, of "traditional" enterprise, Figure 4.a, or vice versa, Figure 4.b? What is the (rigorous, of course) criteria? What is the (rigorous, of course) definition of a (VE) Reference Model[2]? Is VE necessarily a networked structure? Etc.

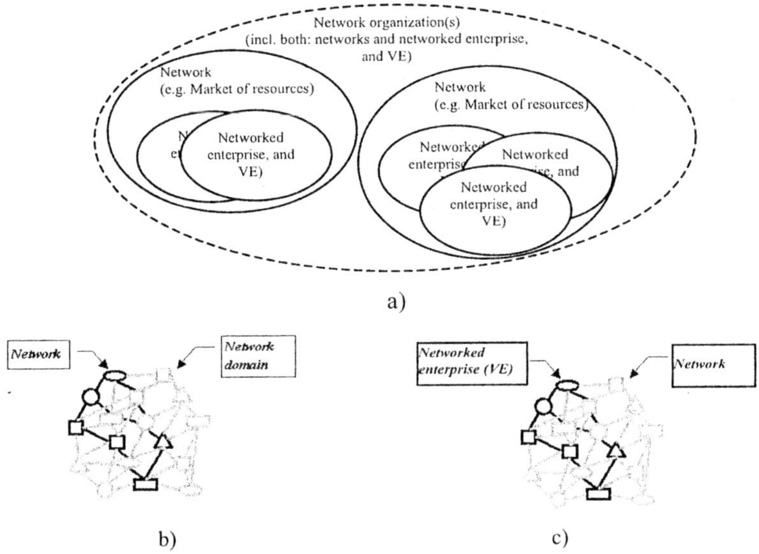

Figure 1 – Different interpretations of the terms "network", "networked enterprise", "networked organization", "virtual enterprise", "virtual organization".

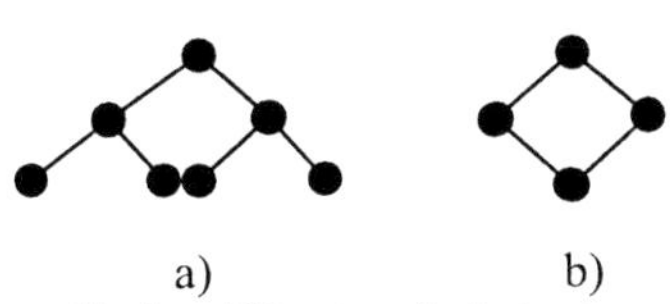

Figure 2 – a) Hierarchical and b) networked structures – usual representation

3. FORMALISMS VS FORMAL THEORIES

It is absolutely necessary to clarify one of the biggest problems for the objective of developing a VE formal theory. That is a general misunderstanding within the engineering community about formalisms and formal theories (FTs), as it is referred above.

[2] E.g., by some authors a condition for a model to be a reference model is its acceptance by the community. By other opinions, that is not a necessary condition. This question deserves a more detailed discussion in some other paper(s).

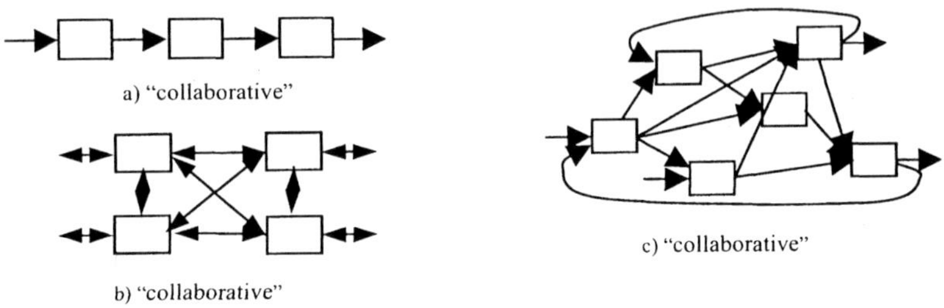

Figure 3 – Different models of the "Collaborative" systems

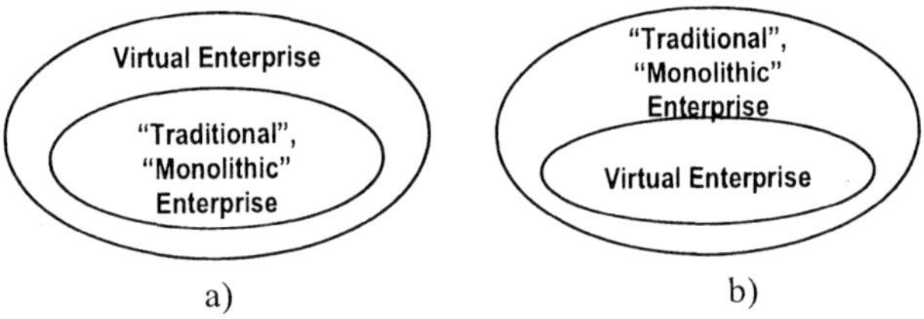

Figure 4 – Relation between VE and "traditional", "Monolithic", Enterprise

Formal Theory is build upon several concepts under well defined conditions[3]. The process of building, or defining, a formal theory could be described (informally) as follows. Based on: (1) some symbols (variables, constant symbols and function symbols) of an **alphabet** and, (2) a set of rules (formation rules, induction rules, or **calculus of terms**), we construct **terms**.

Based on: (1) **terms** and, (2) on the other **alphabet symbols** (including relation symbols) and according to another set of rules (**calculus of formulas**), we construct **Formulas**. The **set of all the formulas**, one can build up from a given **alphabet** using the **calculus of formulas**, is a **language** denoted by L^S (S is the set of alphabet symbols associated to the object concept). Therefore, a **language** is a set of **formulas**. Among all the **formulas** that constitute a **language**, some have a special characteristic – they have no free occurrences of variables (that is, variable occurrences out of $\forall$, $\exists$ quantifiers scope). These **special formulas** are designated as **sentences** (a **formula** without quantifiers and variables is also a **sentence**). A **formula** is **satisfiable** if there is at least one **interpretation** under which that **formula** is true. An **interpretation** J is composed by a **structure** A and an **assignment** β of variables (that is, $J = (A, \beta)$). A **structure** is constituted by a domain A and a map a (that is, $A = (A, a)$). The map a assigns a relation, a function and a constant, from the domain A, to each symbol for relation, function and constant from the **alphabet** symbol set S.

If under a given **interpretation** a **formula** becomes true then that **interpretation** is a **model** of that **formula** (and the **formula** is obviously **satisfiable**). A **sentence** is **satisfiable** if there is at least one **interpretation** under which that sentence is true. When applying an **interpretation** to a **sentence** the **assignment** of variables is

[3] Examples of terms, formulas and other relevant concepts are omitted due to paper length restriction. However, they can be found in (Sousa and Putnik; 2004).

irrelevant (as the **sentence** has no variables occurring free). Thus one can say that a **sentence** is **satisfiable** if exists at least one **structure** making the **sentence** true, that is, if exists at least one **structure** which is a **model** of the **sentence**. As seen before, some of the **formulas** that constitute a **language** are **sentences**. From all those **sentences** some will eventually be **satisfiable**. The subset of **sentences** whose elements are **satisfiable sentences, and closed under consequence**, is a **Theory**. Because **sentences** are **satisfied** by a **structure** *A* the referred subset of **sentences** can be designated by **Theory of** *A*.

Therefore, a **set of formulas** is a **language**. Some of those **formulas** could be **sentences**. Some of those **sentences** could be **satisfiable**. The last ones are a **Theory** (Figure 5).

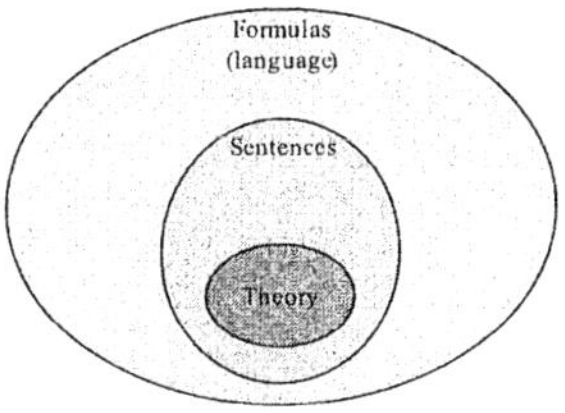

Figure 5 – Formulas, sentences and theory

A **theory** is in fact a **language** which is a subset of an **involving language**.

A **language** could be a **theory** (if exists a **model** for that subset of **sentences**).

A **language** may include none, one or more **theories**.

When applying different **interpretations** to the whole set of **sentences** each one of these **interpretations** could eventually satisfy different subsets of **sentences**. Each one of these subsets is, according to the definition, a **theory**. Thus, the same **language** could include several **theories**, Figure 6. These **theories** can be disjoint, or they can have a common part, or even one **theory** includes another **theory**. (the last situation is referred in [Mendelson, 1987, page 171] via the **finite extension of a theory** concept. In Fig. 6(c), **theory 1** will be a **finite extension of a theory 2**.)

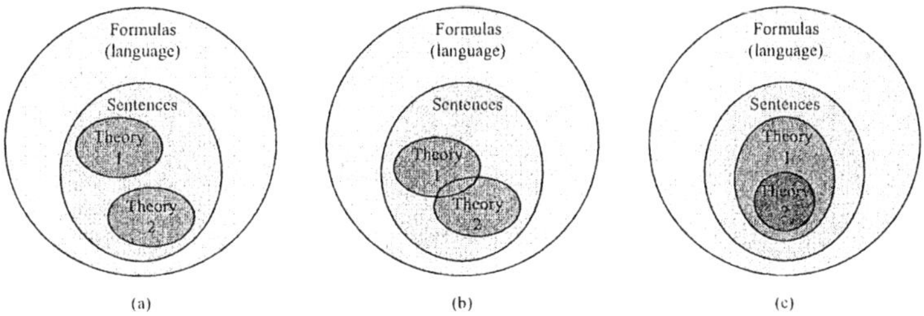

Figure 6 – Theories (a) disjoint (b) intersecting (c) included

The kind of theory described is called **First Order Theories**. Some other characteristics are presented:

- A first order theory *T* is ***consistent*** if and only if there is no *S-sentence* φ such that both φ and $\neg\varphi$ are theorems of *T*.
- A first order theory *T* is ***complete*** if for every *S-sentence* φ, $\varphi \in T$ or $\neg\varphi \in T$.

- A first order theory T is ***axiomatic*** if there is a decidable set Φ of *S-sentence* φ such that $T = \Phi^{|=}$
- A first order theory T is ***decidable*** if there is an **effective procedure** to determine if every *S-sentence* φ is a theorem of T.
- A first order theory T is ***enumerable*** if there is an effective procedure to list all theorems of T.

Therefore, we could say (informally) that a Formal Theory is "a set of statements closed under certain rules of inference." (Wikipedia, 2006). Some of the statements are "initial" statements i.e. axioms, and the statements are composed by a finite set of symbols, i.e. finite "alphabet". We could say now, concerning formalism vs. formal theory of some *object* concept, that formalism means use of any formal language, or even some formal theory (e.g. Set theory, graph theory, etc.) for the concept representation, but, we can not say that we have a (theorem of the) formal theory of the object concept, or that we contribute to the formal theory of the object concept. The sentences from the theory, we can call them a *representational* or *involving* "theory", and correspondent "representational" or "involving" language, which we used for the object concept representation, are the theorems of the *involving* "theory" but they are not, by default, theorems of the object concept formal theory. They could be the theorems of the object concept only by chance or the theorems of an unknown underlying object concept formal theory.

In other words, considering that the formal theory is a language, an object **theory** is in fact a **language** which is a subset of an **involving language** (of the theory used for representation of the object theory). Consequently, we may say, Figure 7.:

Formalism $\neq$ Formal Theory
Formalism $\supseteq$ Formal Theory
Formal Theory $\Rightarrow$ Formalism
but
$\neg$ **(Formalism $\Rightarrow$ Formal Theory)**

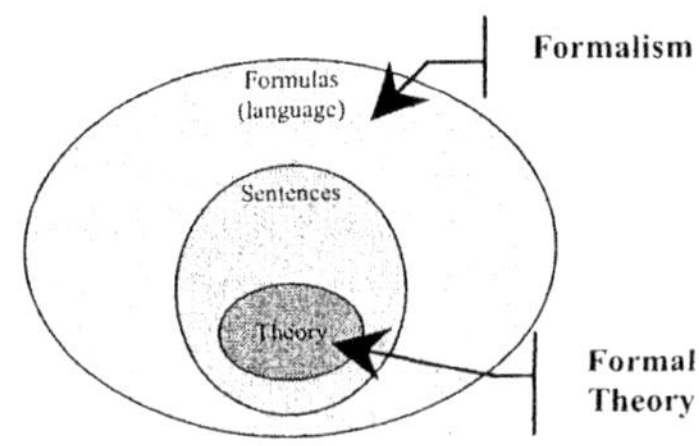

Figure 7 – Formalism and Formal Theory

Thus, the next question is: What does it mean, then, when we use some formalism, i.e. a formal representation for VE (VO – Virtual Organization, NO – Network Organization (e.g. set theory, graph theory, Petri nets, "multi-agents", "metaheuristics", "dynamic programming", C, C++, SDL, RSL, PROLOG, ...) ? This question is already responded above implicitly, but let us be more "object oriented". Regarding VE, or VO, or NO, formalism means use of any formal language, or even some formal theory (e.g. Set theory, logic theory, or graph theory, etc.) for VE, or NO, representation, but, we can not say that we have a (theorem of the) formal theory of VE (VO, NO), or that we contribute to the formal theory of VE. Each formal specification of VE (VO, NO) instance, or "model", that use an involving language of some other theory (e.g. set theory, logic theory, graph theory, Petri nets, "multi-agents", "metaheuristics", "dynamic programming", C, C++, SDL,

RSL, PROLOG, ...), represents, if it is assumed it is a formula, **a "formula" of an unknown underlying formal theory of VE**, Figure 8. The underlying formal theory is "unknown" because we can assume that the formula is a theorem, or even an axiom, but the derivation rules are unknown. In special cases we could assume that the formula is the only theorem (axiom) representing, thus, a trivial case. This case we would call "**a trivial theory**" (this is because **there is no usefulness**, and **there is no science!**, of that case in terms of a VE FT), Figure 9.

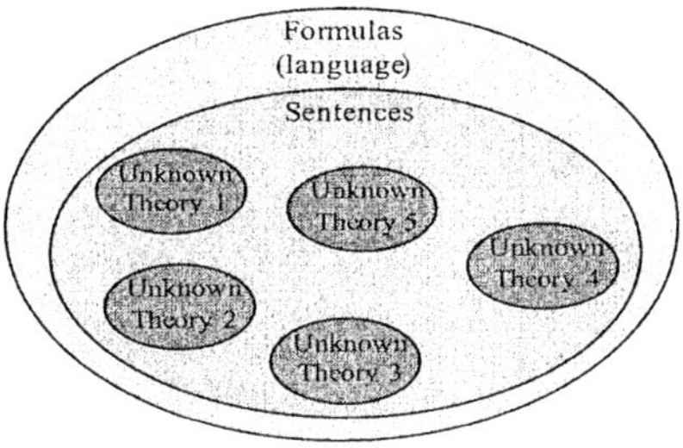

Figure 8 – Sentences as representatives. of "unknown" theories

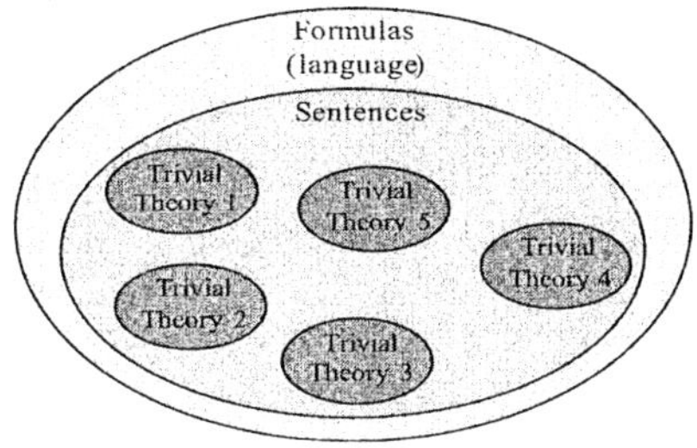

Figure 9 – Sentences as "trivial" theories

The derivation rules of the involving formal language used could lead to inconsistencies and other "errors" (in the sense of the formal theory, but in practical sense as well), regarding the VE model and theory intended. This is because we can be "led" to claim, or to assume, that the theorems of the "involving" theory, e.g., graph theory, are the theorems of our object VE theory. The consequence of this kind of assumption is that we could derivate another theorems using the production rules of the "involving" theory, e.g. graph theory, for which we would believe represents the theorems of the object VE theory while in practice they are not[4]. Otherwise, the "involving" theory, or language, e.g. graph theory, would be equivalent to the object VE theory. In consequence of this eventual equivalency the question is why we talk at all about VE.

4. AN EXAMPLE – THE CASE OF BM_VE FORMALISMS AND FORMAL THEORY

To exemplify the above, we present two instances of BM_Virtual Enterprise[5] (BM_VE) organizational structures, Figure 10.a and Figure 11.a. Their description formulas, or their formal descriptions, are presented on Figure 10.b and Figure 11.b respectively. Presenting these specifications only, even if they are formal and rigorous, we can not say that they are produced rigorously in compliance with eventual BM_VE (formal) theory because we are not sure about the production rules of that BM_VE (formal) theory.

---

[4] See the footnote 1. The comment in footnote 1 illustrates the above.

[5] BM_Virtual Enterprise (BM_VE) is a VE in total or partial compliance with the BM_Virtual Enterprise Architecture Reference Model (BM_VEARM), (Putnik, 2001), (Sousa, 2003), (Putnik et al., 2005).

Without knowing these rules, if we need to specify another BM_VE organizational structure instance we can make an error, i.e. we can specify a structure that does not follow the (derivation) rules by which the structures in Figure 10 and Figure 11 are constructed (actually, although we can claim that both structures in Figure 10 and Figure 11 are constructed by the same rules, rigorously speaking, only apparently they follow the same rules).

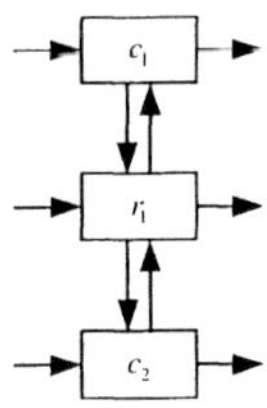

a)

$c_1\left(\downarrow\uparrow r_1 A\left(\right)\downarrow\uparrow \equiv c_2\, s_{eq}\right)$

b)

Figure 10 – An instance of BM_VE organizational structure

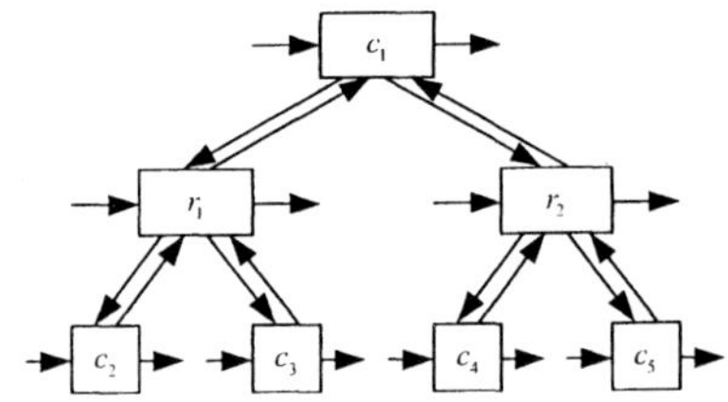

a)

$c_1\left(\downarrow\uparrow r_1\left(\downarrow\uparrow c_2 c_3\right) r_2\left(\downarrow\uparrow c_4 c_5\right)\right)$

b)

Figure 11 – An instance of BM_VE organizational structure

But let us analyze the language defined bellow, represented by the grammar G_{BM}:

$$G_{BM} = (V_T, V_N, S, R) \quad \text{where}$$

$$V_T = \{c_1, \ldots, c_{n_c}, r_1, \ldots, r_{n_r}, s_{eq}, \equiv, \downarrow\uparrow,), (\}$$

$$V_N = \{S, A, B\}$$

$$\begin{aligned} R = \{\, & S \to c_i(\downarrow\uparrow A) \equiv s_{eq}, \\ & A \to r_i(\downarrow\uparrow B), \\ & A \to AA, \\ & B \to c_i(\downarrow\uparrow A), \\ & B \to BB, \\ & B \to c_i \ \} \end{aligned}$$

The BM_VE organizational structure instances in Figure 10 and Figure 11 are generated by the grammar G_{BM}. Moreover, using the grammar G_{BM} we can produce another structure instances, e.g. the structure instance in Figure 12.a, with the correspondent "textual" representation, or "formula", in Figure 12.b.

If we consider all the words generated by the grammar G_{BM} as the instances of the BM_VE organizational structures, then we will say that the grammar G_{BM}[6] is a formal theory of BM_VE organizational structures.

To deal with other aspects of BM_VE, or VE in general, other grammars, as the FT models, would be necessary. At the moment, this work is under development.

[6] Rigorously, the grammar G_{BM} is an attributed grammar. For the purpose of this paper, the grammar attributes presentation is omitted as it does not influence qualitatively the paper's discourse.

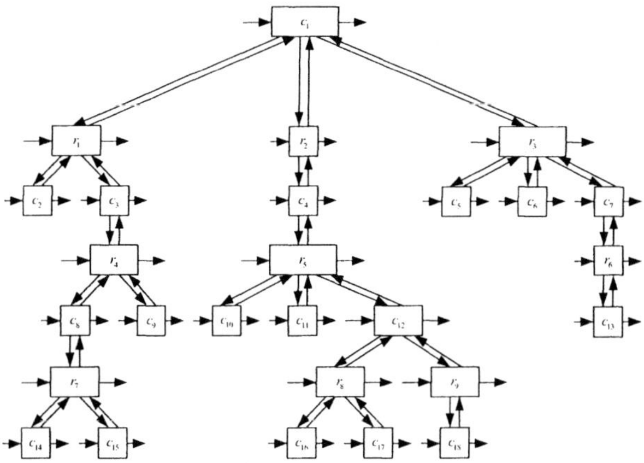

a)

$$c_1\left(\downarrow\uparrow r_1\left(\downarrow\uparrow c_2 c_3\left(\downarrow\uparrow r_4\left(\downarrow\uparrow c_8\left(\downarrow\uparrow r_7\left(\downarrow\uparrow c_{14} c_{15}\right)\right) c_9\right)\right)\right) r_2\left(\downarrow\uparrow c_4\left(\downarrow\uparrow r_5\left(\downarrow\uparrow c_{10} c_{11} c_{12}\left(\downarrow\uparrow r_8\left(\downarrow\uparrow c_{16} c_{17}\right) r_9\left(\downarrow\uparrow c_{18}\right)\right)\right)\right)\right) r_3\left(\downarrow\uparrow c_5 c_6 c_7\left(\downarrow\uparrow r_6\left(\downarrow\uparrow c_{13}\right)\right)\right)\right)$$

b)

Figure 12 – An instance of BM_VE organizational structure

5. WHY NOT A VE FORMAL THEORY (OR THEORIES)

The FT approach has many critics. Four (selected) critics of the FT approach are:

- **It is difficult to understand** (from the cognitive point of view), difficult to learn, difficult to develop;
- **It is difficult** by itself, difficult to develop (objectively; analytical or formal approach - highly complex problem (in terms of complexity theory)).
- Difficult, or impossible, to cover all necessary practical requirements, i.e. user's needs.
- May invalidate practical results if these are not obtained from the theory, or may lead to results without practical importance, etc.

6. CONCLUSIONS

The usefulness of the formal theory, e.g. the formal theory of BM_VE, is obvious. The majority of the problems referred in the section 2 of this paper could be much more easily managed, if not resolved at all. Especially, it is obvious the contribution of a formal theory for engineering tasks, that is, in design, implementation, maintenance, etc., providing the so much desired efficiency and effectiveness (for many engineering tasks). As we have said above, formal theories have proved their usefulness in "traditional" engineering areas. Somebody said: "There is nothing as practical as a good theory".

We would interpret it saying that having a FT of VE means to have **an excellent tool** for VE efficient development, implementation and control.

Naturally, aiming at development of a FT of VE, there is a number of questions to be addressed. Some of them are:

1) Existence of a FT of VE,
2) The theory's "external" problems, i.e. the theory object's nature, i.e. VE nature (which is the main purpose of developing the FT of VE)
3) The theory's "internal" problems, i.e. the properties of the FT of VE itself
4) Relation between FT of VE and FT of PS/MS (Production System/Manufacturing System),
5) Models of FT of VE
6) Development "strategy", etc.

The authors have started the project on FT for VE development, within the research on general issues of formal theories of production systems. Actually, it is developed a FT model for the "canonical" model of BM_VE – the grammar G_{BM} presented in the section 4 (in accordance with the BM_VEARM, under development on the University of Minho, (Putnik, 2001), (Putnik et al., 2005), (Sousa, 2003)).

Concerning the relation between VE formal theory and Production/ Manufacturing Systems (PS/MS) formal theory, the phenomena of the issue is apparently the same (or at least very similar). However, it is an open question if the VE formal theory ia a sub-theory of a PS/MS theory or vice versa, or there is some another relation. This question could have a practical implication in terms of how could we plan development of a VE formal theory, i.e., starting with development of PS/MS formal theory and apply it to VE or starting an independent development, etc. Finally, the lack of the FT of VE doesn't mean that we do not have the Theory of VE. It only means that we do not have a Formal Theory of VE.

7. REFERENCES

1. Gruninger, M., and Fox, M.S. (1994), "An Activity Ontology for Enterprise Modelling", Submitted to: Workshop on Enabling Technologies - Infrastructures for Collaborative Enterprises, West Virginia University (Retrieved from http://www.eil.utoronto.ca/enterprise-modelling/papers/gruninger-wetice94-act.pdf).
2. Janowski T., Giménez L. Zheng H. (1998) Composing Enterprise Models: The Extended and The Virtual Enterprise, in Camarinha-Matos L., Afsarmanesh H., Marik V. (Eds.) *Intelligent Systems for Manufacturing: Multi-Agent Systems and Virtual Organizations*, Kluwer Academic Publishers,
3. Mendelson, E. (1987). *Introduction to Mathematical Logic*, Chapman & Hall.
4. Putnik G. (2001) BM_Vertual Enterprise Architecture Reference Model, in A. Gunasekaran (Ed.) *Agile Maniufacturing: 21st Century Manufacturing Strategy*, Elsevier Science Publ., pp 73-93;
5. Putnik, G. D., Cunha, M. M., Sousa, R., Ávila, P. (2005) BM_ Virtual Enterprise: A Model for Dynamics and Virtuality, in Putnik G., Cunha M. M. (Eds.) *Virtual Enterprise Integration: Technological and Organizational Perspectives*, IDEA Group Publishing, Hershey, PA, USA, pp: 124-143
6. Sousa, R. (2003). Contribuição para uma Teoria Formal de Sistemas de Produção. *Tese PhD*. Departamento de Produção e Sistemas, Universidade do Minho.
7. Sousa, R. M. and Putnik, G. D. (2004). A Formal Theory of BM Virtual Enterprises Structures, in Camarinha-Matos, L. M.; (Eds.), "*Emerging Solutions for Future Manufacturing Systems*", Vienna: Springer, 2004. ISBN 0-387-22828-4, p.315-322.
8. Turner, K. J., Ed. (1993). *Using Formal Description Techniques: An Introduction to ESTELLE, LOTOS and SDL*, John Wiley & Sons.
9. Wikipedia (2006) Theory, Retrieved from http://en.wikipedia.org/wiki/Theory

25 TOWARDS A DISTRIBUTED PROCESS EXECUTION PLATFORM FOR COLLABORATIVE NETWORKS

A. Luis Osório [1], Luis M. Camarinha-Matos [2]
[1] *Instituto Superior de Engenharia de Lisboa, PORTUGAL, aosorio@deetc.isel.ipl.pt*
[2] *New University of Lisbon, PORTUGAL, cam@uninova.pt*

Enterprise collaboration typically occurs in the context of a process. Therefore, in order to support collaboration, there is a need to establish a framework able to handle a diversity of collaborative processes with different computational and communication requirements. This paper discusses the need for such framework, points out key requirements and proposes a strategy for a supporting technological infrastructure. Processes are seen as the central modeling entity in this framework. Portability of process definitions among different technological bindings is also a considered issue.

1. INTRODUCTION

Computer supported collaboration is being pursued in a large diversity of application domains. For instance, there are research groups addressing collaboration in the education area, the computer supported collaborative learning (CSCL) (Dillenbourg et al., 1995), in the collaborative engineering (CE) area, in the virtual enterprises area, and in many other domains. The ECOLEAD project (Camarinha-Matos et al., 2005) is an important effort to understand collaborative ecosystems by defining the foundations and mechanisms for establishing an advanced network-based industry society. There is however a lack of a generic platform to support collaboration activities along their life cycle. From a company's point of view, collaboration acts occur many times in different contexts. They can occur during a phone-call between employees of two companies or they can occur through e-mail exchanges or through other communication forms. In terms of modeling, there is a need to represent different collaboration acts and events independently of them being based on computational transactions or human conversation or through any other channel.

The traditional workflow based approaches to enterprise process automation show various limitations when it comes to model collaboration acts occurring in the enterprise and involving or not business partners. For instance, the concept of work list and work item used to model user interactions requires a more flexible approach. An enterprise user (under some role) might need to contribute to some activity by accessing a dedicated application that is invoked by a process execution kernel

Please use the following format when citing this chapter:

Osório, A. L., Camarinha-Matos, L. M., 2006, in IFIP International Federation for Information Processing, Volume 220, Information Technology for Balanced Manufacturing Systems, ed. Shen, W., (Boston: Springer), pp. 233–240.

(orchestration) following a rather different strategy than the one supported by the worklist/workitem concepts defined by WfMC (Hollingsworth, 1995).

There is a growing awareness among researchers and information and technology practitioners on the need to evolve to scenarios where existing technologies, strategies and tools cope with new challenges for enterprise collaboration. The outlining of key research challenges and reference projects (Camarinha-Matos and Afsarmanesh, 2004) is an important step to the consolidation of the area of collaborative networked organizations (CNOs). CNOs have been addressing different collaboration patterns that can emerge from different breeding environments. A breeding environment offers the necessary preparedness stage for enterprises, organizations or individuals to participate in different collaboration patterns such as virtual enterprise/virtual organization (VE/VO), professional virtual community (PVC), and even e-science and e-government forms (Camarinha-Matos and Afsarmanesh, 2004). The collaboration among members of a CNO needs preparation, trust, and above all, a framework making possible the management of the life cycle of the collaborative activities.

This paper discusses some of these challenges and proposes a strategy for such a collaborative framework. It is assumed that collaboration is modeled by a process that is executed on a distributed platform involving those members that were selected from a breeding environment to form some collaboration pattern. This work is based on practical experience acquired with two application cases, one in the transport systems sector and addressing a holistic approach to the collaborative electronic toll collection (ETC) management system, and another case in the public administration sector aiming to implement an integrated service to citizens in a collaboration scenario involving several public entities. In both cases a service oriented approach is adopted and a key challenge is to contribute to a technological platform and a framework where a number of collaborative processes can be managed.

2. OPEN SERVICE-ORIENTED PLATFORM

Currently there is a lack of a general and adaptable framework able to grasp or integrate the various contributions to inter-enterprise collaboration, mainly those from the electronic business area that have grown up in consequence of the appearance of the XML language and web services.

There are a number of initiatives like those promoted by OASIS (Organization for the Advancement of Structured Information Standards), WS-I (Web Services Interoperability), OAGi (Open Applications Group, Inc.), STAR (Standards for Technology in Automotive Retail), Odette (International Automotive Industry Organisation, initially "Organisation for Data Exchange by Tele-Transmission in Europe"), ITA (Information Technology for the Automotive Industry), AIAG (Automotive Industry Action Group), to mention only a few organisations that are mainly concerned with collaboration in complex scenarios and are led by different industry sectors (potential end-users). Another group of organizations that are more technology-oriented include W3C, IETF, WFMC, BPMI, OMG, WS-I, which are mainly concerned with technological interoperability. Even if there is a trend towards the predominance of a limited number of leading paradigms, mainly when the industry leading companies are involved, most of the developed frameworks are not generic enough to support evolution and are not independent of some vertical industrial sector. As an example, the OAGi organization is defining a set of BODs using accepted languages like XML, XML Schema, and WSDL, to generate message templates to be adopted by cooperating partners. It is however complex to maintain such templates or message rules embedding semantic details, and difficult to evolve and support innovation processes when it is necessary to establish disruptive changes in terms of technology and business levels. Furthermore, the process definitions and the execution platform bindings, an important asset of enterprises, are difficult to move to another execution platform as it requires heavy development efforts (Figure 1).

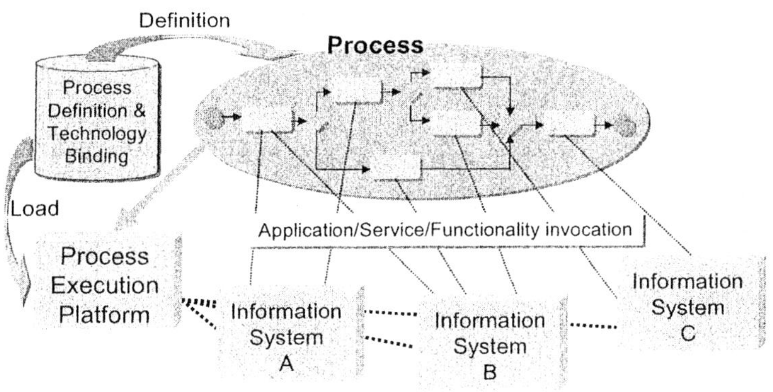

Figure 1 – Portability of process definition and technology bindings among platforms

Existing industrial solutions do not offer a clear strategy on how to move a process definition and technology binding developed for a specific execution platform to another competing platform. The Apache project Beehive (*beehive.apache.org*) is a contribution in this direction by providing a meta-programming model for J2EE, making easier the access to enterprise resources and other model-driven development facilities. The *Controls* concept offered by this project is: "*a lightweight, metadata-driven component framework for building that reduces the complexity of being a client of enterprise resources. Controls provide a unified client abstraction that can be implemented to access a diverse set of enterprise resources using a single configuration model.*" This concept offers an integration layer unifying the invocation from the activities of the process to the enterprise resources (files, services, components). However, this open initiative addresses mainly developments bound to the Java world and does not present a clear "technology agnostic" perspective, or some general enterprise resource invocation framework.

Beyond this lack of portability for process definition and technology bindings among different execution environments, a platform to support collaborative processes among enterprises is a further challenge considering the underlying diversity of cultures (different platforms, programming languages, localization languages, systems, methodologies). Consider the scenario of a company B that has adopted ebXML (Electronic Business using eXtensible Markup Language) for its interactions with another company A. If a company C needs to cooperate with a company B that has adopted ebXML business framework (Figure 2), the solution is not easy and involves an adaptation (effort) investment. This investment might range from the business level to the technological level, by crossing a number of other problem domains like the organizational one.

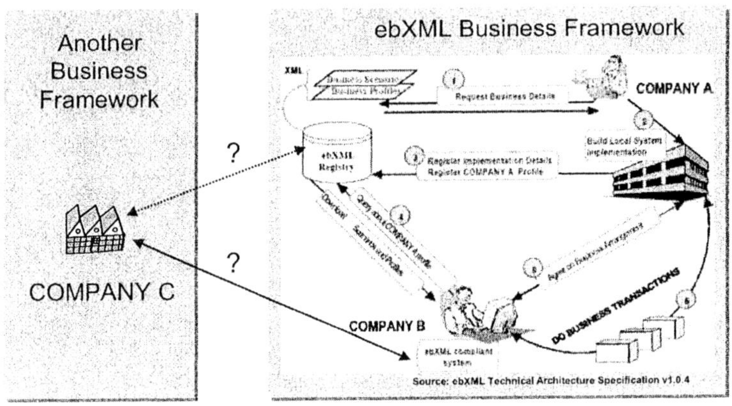

Figure 2 – Challenge for the interoperability between two business frameworks

A strategy grounded on a generalized consensus is not a realistic solution as it happened with the CORBA distributed objects framework from OMG. Even if CORBA continues to be important as one distributed object platform, it coexists with other equivalent technological approaches, like .NET distributed objects. Considering the fast evolution of ICT, the strategy must involve the coexistence of heterogeneous approaches, be them industrial, academic or open community contributions. However, it is important to have some common interoperability facilitators. There are some technologies that are the key to a crescent level of integration namely the XML language and more recently the web services technology with its origin in SOAP (Simple Object Access Protocol), initially developed to overcome firewall problems in distributed applications. The utilization of HTTP and the port 80 usually opened to access enterprise web servers, was envisaged as a prominent tunnelling mechanism to interconnect distributed applications based on remote function calls.

Therefore, in addition to the interoperability problems at technology level, a main concern is the definition of an open service bus based on agreed interfaces and contracts. This bus can be seen as an abstract service layer that is the basis for the functional binding and execution support for processes. It should be based on Web Service standards like WSDL and XML Schema, and supported on a common ontology targeted to each specific application domain. Collaboration scenarios, as described in the next chapter, require the definition of a services' ontology and (interfacing) meta-model entities in order to make possible the dynamic adaptability of services developed by different suppliers.

For most companies, the openness of systems and services regarding multi-vendor solutions is fundamental in order to keep a balanced positioning in the market offer. This requirement establishes a challenging research topic i.e. how to ensure the portability of collaborative process definitions among heterogeneous execution platforms. This portability should be grounded on the diversity of (multi-vendor) execution platforms existing in the various network members, and thus contributing to a paradigm shift on the practice of enterprise collaborative solutions.

3. PROCESS BASED COLABORATIVE ENTERPRISE

In a wide range of application scenarios, processes are the core element of a collaboration framework. Collaboration acts ranging from a simple dialog between two persons to a complex business interaction among a group of partners can be modelled as processes. A (collaborative) process can be the central modelling entity to represent collaboration acts in the context of a virtual organization or any other more traditional collaborative network.

Case 1. As an example, a new generation of electronic toll collection (ETC) systems based on the Intelligent Transport Systems Interoperability bus (ITSIBus) was implemented for the Brisa motorway management company (Osório et al., 2003). The underlying technological infrastructure is based on services with an open interface and contracts. The functional units are implemented as services and some of them are executors of process definitions promoting a shift from a hard-coded business logic to a more flexible and adaptive declarative representation of business logic. In the developed approach, a service layer (Figure 3) establishes an open service bus were a service can advertise its functionalities and events and lookup other services and event based on a discover/advertise mechanism.

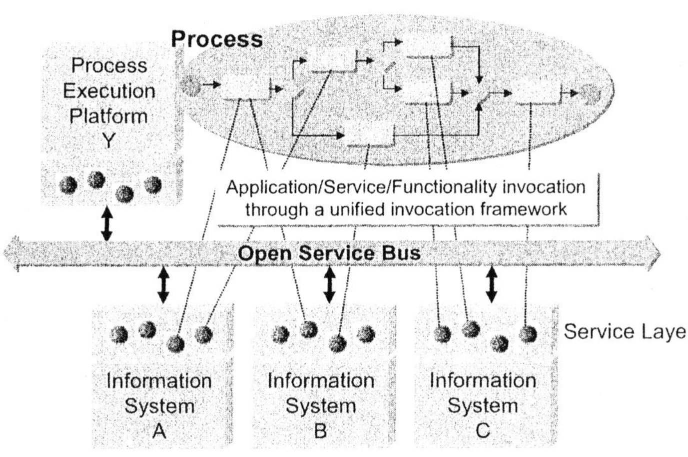

Figure 3 – Unified invocation framework for enterprise resources

In this example, a trend towards a model driven process management system is assumed. But this strategy shows the need for a reference architecture able to establish open meta-models characterizing each contributing service and providing at least three main systemic views: the design view, the operational view and the management view. At design time, there is a need for a new generation of tools able to support a model driven approach to solution construction, test and deployment.

The existing development tools are not prepared to deal with a collaborative process and service-based solution design. There is a need for service interface and contract representation (graphical), collaborative process design and a set of tools to simulate execution and debugging of such complex solutions.

The ITSIBus architecture establishes the concept of system as an execution container running a set of services, as minimal execution entities – computational agents, sharing common environmental facilities like plug-and-play, monitoring and security (Osório et al., 2005). These services can range from simple business logic to intelligent agents, embedding logic to act autonomously in situations where central coordination is not available or decision making is necessary. When playing the role of adapter to legated systems, services can be implemented by intelligent agents as proposed by (Suh et al., 2005) with its collaboration agency architecture. In Brisa's ETC domain processes are well structured, considering that system behaviour is deterministic and defined at process definition phase.

Case 2. However in other application cases the situation can be quite different. As an example, let us consider the case of collaborative networks in e-government, the Citizen Shop case. The goal here is to establish a platform that allows offering integrated services to the citizen on a single attendance point (the "shop"), independently of the governmental organizations that are actually involved in each component of the service. In this case, processes are frequently ill-structured as many situations that could not be fully determined a-priori happen during the interaction with the citizen. In other words, the "process" is composed "on the fly" depending on the specific requests and context. In these situations it might be interesting to explore a meta-model strategy as a basis for cooperative process model definition (Rolland et al., 1998) to cope with decision-oriented products (services). By registering (tracing) collaborative acts not previously defined as a deterministic process can provide a base for future construction of new process definitions by data mining.

At the current stage, the focuses of this work is on supporting processes that can be completely defined a priori. For those classes of collaborative processes, there is a "process owner" representing the network member that defined it. The process participants are services that run autonomously or on behalf of users that belong to one enterprise member of the network (Figure 4).

This "collaborative networked organization – open service bus" (CNO-OSB) platform being developed allows for a flat group of services to be orchestrated according to a number of collaborative processes definitions (service choreographies) and executed. In each enterprise member, there exists a special service with the responsibility of orchestrating distributed processes instances execution. Many instances of this process can be running in an enterprise member each one responsible for the execution of a number of process instances. This distributed and collaborative process execution platform is designed to present the following characteristics:

- Process definition establishes an activity plan where each activity involves a set of operations or a sub-process with associated meta-data to handle invocation of operations or sub-process (local or in another network member);
- Process definition, service interfaces and contracts, and event management for asynchronous operations management following standards like BPEL, WSDL, WS-Events in a tight connection to WS-I interoperability profile standards;

- Transaction management, security, authentication and authoring, execution monitoring (instrumentation) and other functionalities necessary for a dependable distributed process execution.

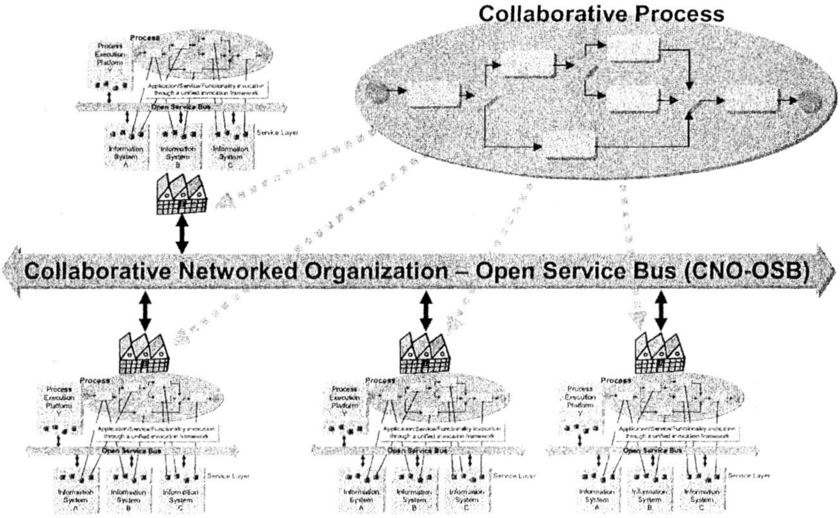

Figure 4 – The collaborative networked organization open service bus (CNO-OSB)

There is another aspect in the proposed approach and that derives from the two application examples, related with the development of a set of tools aiming to offer process experts a simple language, graphical if possible, able for them to contribute for the different life cycles of distributed processes and services based solutions. This aims to cause a shift from a technological cantered development for a process experts centred one. The idea is to work on a new model driven development approach considering developments at model level, deeply based on meta-data, independent from different technological bindings. The challenge is to promote process definition portability among competing technological platforms, creating a competitive market where today proprietary patterns prevail. To some extent, the objective is to follow a strategy established by OMG with the model driven architecture (MDA) initiative where the main objective is to generate a new set of tools developing models for technology independent platforms (PIM – platform independent models) and later bind those generated models to specific platforms (PSM – platform specific models) (Osório et al., 2005).

5. CONCLUSIONS

This paper discussed a distributed process execution platform for collaborative networked organizations based on the requirements from two application projects, one in the private sector and another one involving public institutions. That first case targeted the implementation of an open service-based management infrastructure for toll and traffic management. The second example aims at establishing a platform to offer integrated services to citizens, based on a single attendance point for a number of services executed in different public entities.

In both projects the need for establishing a collaboration culture supported by a flexible ICT infrastructure based on agreed services and contracts was identified. Although from a scientific point of view there is not much innovation in the required platform, there is a need for a pragmatic consideration of interoperability and implementation issues that take into account both existing legacy and the need to move to new business practices. In the proposed system, collaboration acts are modelled as processes, which can range from the support to a simple interaction between two persons to a complex business transaction among a group of partners. In this context there is a need for a number of modelling concepts and mechanisms that allow for the definition of collaborative processes that are able to cope with running exceptions and can easily be bound to different running platforms.

6. ACKNOWLEDGEMENTS

This work is partially supported by BRISA company, through the research and development projects ITSIBus and by the Citizens Shops institute (IGLC - *Instituto para a Gestão das Lojas do Cidadão*) through another research project aiming to develop integrated services involving several public entities.

7. REFERENCES

1. Camarinha-Matos, L. M.; Afsarmanesh, H.; Supporting Infrastructures for New Collaborative Forms, in Collaborative Networked Organizations, pg. 175-192, Kluwer Academic Publishers 2004.
2. Camarinha-Matos, L. M.; Afsarmanesh, H.; Ollus, Martin; Ecolead: A Holistic Approach to Creation and Management of Dynamic Virtual Organizations, IFIP International Federation for Information Processing, Pages: , 3 – 16, Volume 186, Springer 2005.
3. Dillenbourg P., Baker M., Blaye A., O'Malley C. - "The Evolution of Research on Collaborative Learning"', in P. Reimann & H. Spada (Eds). Learning in humans and machines. Towards an interdisciplinary learning science, 189- 211. London: Pergamon, 1995.
4. Hollingsworth, David; Workflow Management Coalition: The Workflow Reference Model, Doc. Nº TC00-1003, Document Status - Issue 1.1, 1995.
5. Osório A. L., Abrantes A. J., Gonçalves J. C., Araújo A.; Miguel J. M., Jacquet, G. C.; Gomes, J. S. - Flexible and Plugged Peer Systems Integration to ITS-IBUS: the case of EFC and LPR Systems, PROVE'03 – 4th IFIP Working Conference on Virtual Enterprises, published by Kluwer Academic Publishers, 2003.
6. Osório, A. Luís; Camarinha-Matos, L.M.; Gomes, J.; A Collaborative Networks case study: The extended "ViaVerde" toll payment system, PRO-VE'05, Valencia – Spain, 26 -28 September 2005.
7. Rolland, C.; Nurcan S.; Grosz, G.; A unified framework for modeling co-operative design processes and co-operative business processes, in the Proceedings of the 31st Annual Int. Conf. on System Sciences, Big Island, Hawaii, USA, 6-9 Jan.98.
8. Suh, Y. H.; Lee K. W.; Kim H.; Collaboration Agency Infrastructure for Virtual Enterprises, Proceedings of the Ninth Int. Conference on Computer Supported Cooperative Work in Design Coventry, pages 56-61, UK, May 24 - 26, 2005.

PART C

INTEGRATED DESIGN AND ASSEMBLY

26

AN INNOVATIVE MAINTENANCE SOLUTION FOR COMPLEX MACHINERY: THE KOBAS PROJECT CASE

Américo Azevedo [1,2], Paula Silva [1], César Toscano[1], Joao Cardoso[1,2]
[1] INESC Porto, Rua Dr. Roberto Frias nº 378,4200-465, Porto, Portugal
[2] Faculdade de Engenharia da Universidade do Porto, Rua Dr. Roberto Frias, 4200-465, Porto, Portugal
ala@fe.up.pt, psilva@inescporto.pt, ctoscano@inescporto.pt, jcard@fe.up.pt

This paper presents an innovative approach in supporting the operation of complex equipment. The concept was developed in the context of the KoBaS project (Knowledge Based Customized Services for Traditional Manufacturing Sectors Provided by a Network of High Tech SMEs) whose main objectives are the development of new knowledge based tools for an intelligent use and management of more sophisticated manufacturing machines; and the creation of an innovative extended network of high-tech SMEs for use, customize, support and make business out of the new development tools. As a concrete example the paper presents a KoBaS solution based on the machine Maintenance. This solution allows one to know the machine condition, to detect and diagnose machine failures, to manage ordinary and extraordinary maintenance plans and maintenance work orders and to provide training support for maintenance interventions.

1. INTRODUCTION

The companies that build, sell and install equipment are increasingly facing the need to develop technologically advanced and consequently more complex equipment, in order to satisfy the demands of high customization of the products processed by them, with low costs, low manufacturing time and high quality levels.

The KoBaS Project (Knowledge Based Customized Services for Traditional Manufacturing Sectors Provided by a Network of High Tech SMEs) is an European project under the Sixth Framework Programme that tends towards a substantial improvement in equipment and industrial machinery, through the development of a set of software components, integrated in their own equipment and functionally directed at the creation of equipment operation programmes (numerical control programmes), analysis and simulation of operation and movement, sequencing of generation operations, configuration of parameters of functioning, maintenance, formation and support to the operation. The project aims also the development of a network business model, made of technology-based SMEs. The main objective is to foster the development and customization of specific solutions required by the

Please use the following format when citing this chapter:

Azevedo, A., Silva, P., Toscano, C., Cardoso, J., 2006, in IFIP International Federation for Information Processing, Volume 220, Information Technology for Balanced Manufacturing Systems, ed. Shen, W., (Boston: Springer), pp. 243–252.

equipment manufacturers according to the requisites demanded by the market they wish to satisfy.

One of the developing components under the KoBaS project is destined to support, in an innovative way, the maintenance activities inherent to the equipment itself. The integration of several black-boxes fully integrated and resulting from the customization of the respective components in the equipment itself, makes it possible to reduce substantially the preparation and operation time and to simultaneously reduce the associated costs.

The manufacturers of equipment goods with the developing technology under KoBaS project, can, therefore, offer more complex products regarding their functionalities, but at the same time easier to operate, which noticeably increases the added value to the client.

The aim of this paper is to address the design of an advanced software solution for maintenance management in the scope of a large on-going European Project (KoBaS Consortium, 2006).

The reminder of this paper is organized as follows. After this introduction, next section presents the project in which the system here described is being developed. The third section presents in particular one of the components in development: the maintenance component. The fourth section addresses the technology and development framework selected for the implementation phase. The fifth and last section concludes the paper.

2. OVERVIEW OF KoBaS PROJECT APPROACH

The concepts and component architecture presented in this paper have been conceived within the European Project KoBaS (Knowledge Based Customized Services for Traditional Manufacturing Sectors Provided by a Network of High Tech SMEs) (KoBaS Consortium, 2006).

The project consortium involves 21 partners from 10 countries, in which 6 partners are European Institutions of R&D: RPK (Germany), INESC Porto (Portugal), ITIA-CNR (Italy), EPFL and ETH Zurich (Switzerland), and Tekniker (Spain).

The project will make use of techniques such as Virtual Reality, 3D & Discrete Events Simulation, Knowledge Based System and Finite Element Analysis, evolved, combined and optimized in their interaction thanks to a new brand of innovative technologies that are developed in the KoBaS project.

2.1 The project vision

The objective of KoBaS project is to create a network of high-tech small and medium enterprises, offering knowledge based services developed within the project to manufacturing companies operating in traditional sectors. Knowledge based services will be a customised integrated solution of KoBaS "components", which will be modules developed for specific purposes. The project vision is to provide a breakthrough in the current practices in the use of manufacturing machines, through the development of a set of generic tools of new conception. This will enable the quick customization of software solutions providing thus new advanced and powerful functionality, machine embedded, and supporting task and process planning, machine maintenance, training, and management. Thanks to this innovative platform, it will be possible for machine-tools and other production equipment to become intelligent, capable to communicate their environment and characteristics, to understand and work with digital models of the parts to produce and to enable an efficient dialog with their operators (KoBaS Consortium, 2004).

These generic tools will be published in a network of High-Tech SMEs. The manufacturing machine builder, confronted with the difficulty of the task and process planning of his equipment and willing to provide a new brand of powerful services to his customer, asks to the Network of High Tech SMEs for a customized software solution for his machine. The KoBaS Network thus provides, thanks to the new instruments and approaches developed within the project, a customized solution that will be embedded in the machine by the Machine Builder allowing him to sell an "intelligent machine" to the end user (KoBaS Consortium, 2004).

In this context, a Component (development tool) is a collection of methodologies, software tools and libraries that will be used by the Network of HT-SMEs to generate a KoBaS Customized Solution. A Kobas Customized Solution is an integrated set of Black-Boxes (software components).

2.2 The KoBaS components

The basic components reflect the main aspects covered by the KoBaS approach in the use of a manufacturing machine, namely:

1. Graphical User Interface Construction Component: with this component the Network will be able to build and promptly configure a customized Man-Machine interface, based on low cost Virtual Reality, in order to make easily and naturally accessible all the functionalities and services to the end-user.

2. Rule-Based Knowledge Core Construction Component: this component is meant to provide the Network with the possibility to customize, for every manufacturing sector/machine confronted, a Rule-based Knowledge Base (process and geometry related), in order to support the machine task programming, configuration, maintenance and training through a knowledge based system.

3. Experience Data Base Construction Component: this component is used by the High Tech SMEs to build a Knowledge DB of past experience related with the Process handled. Once the cases are acquired and formalized, this module provides the methods for knowledge access and reuse. This module operates in strict connection with the Rule-Based Knowledge Core Construction Component.

4. Part Program Creator Component: with this module, the High Tech SMEs Network will be able to quickly provide a customized solution for the offline programming of the machine. For the part program creation, this module relies on the information provided by the knowledge base.

5. 3D Simulation Construction Component: This component is used to create a customized simulation environment. It is meant to move the machine into the virtual environment as it would do in the real world, providing the framework to display the machine tasks or the training and maintenance procedures.

6. Finite Element Analysis (FEM) Component: with this Component, the Network builds a customized FEM analysis. It provides pre and post processing analysis for the process performed by the manufacturing machine under consideration, giving a feedback on the performance of the Production Process.

7. Maintenance and Diagnosis Component: with this component, the Network will be able to build a customized maintenance service for the specific machine. The solution provided is meant to prevent failures, to suggest intelligent maintenance, to demonstrate maintenance procedures. This module should grow wiser (as the rule related modules) as it's used.

8. Training module Construction Component: the Network will use this component to develop a training module for the machine studied, intended to provide training support for the end-user of the machine. The component will also use the other components capabilities (such as simulation) to provide a training virtual environment for the machine end-user

9. Machine Configuration and Mechatronics Construction Component: this component will allow the SMEs Network to develop solutions for the configuration of the machine studied, according to final user needs.

10. Build-in Management Functions Construction Component: this component is meant to provide the network with the ability to develop customized solutions for the management function of the machine. (cost planning, scheduling and optimization)

11. Integration Component: this component is used to build the framework for the solutions proposed, ensuring that all the other modules can communicate, and establishing standards for the methods and classes for every customized solution realized. Within this Component also the data structure and data model for the specific solution will be defined.

Every component is intended to provide the proper instruments and methodology in order to enable an easy customization of machinery according the specific requirements to fulfil in the scope of KoBaS. As an example, the component related to maintenance functionalities, should allow the customization of a specific maintenance solution based on machine monitorization (by sensors), relevant signal analysis, automatic failure detection and diagnosis, alarms and ordinary and extraordinary maintenance plan. The next section describes in particular the Maintenance Customization Solution.

3. THE KOBAS MAINTENANCE SOLUTION

High maintenance costs and production losses due to machine breakdowns are pressing problems for today's manufacturers. In order to increase the reliability and availability of machines it is urgent to shift from scheduled maintenance to the new technology of constantly observing machine condition and predicting its working condition in advance. This is one of the main objectives of the KoBaS Maintenance Solution.

The great majority of the machines produced by European manufacturers do not have built-in monitoring, diagnosis, simulation and analysis abilities that would help the user to identify the proper course of actions nor do they have an intelligent machine capable of assuring its service.

Furthermore, new technologies in sensors recently under development will bring an added dimension to the machine "intelligence" that will need to be dealt with. Also, the openness, the suitability to the manufacturing equipments and the integration with other KoBaS components, will allow an easy adaptation to new machines and will provide a more natural and intuitive interaction with the operator namely, 3D simulation to machine tests and diagnosis, maintenance tasks supported by training, and dynamic and adaptable behaviour through experiences learned from past situations.

This openness will facilitate the management of shorter lifecycles for machine producers and will make these advanced maintenance concepts reachable to smaller machine producers.

3.1 Creation of a Maintenance Customized Solution

As depicted on Figure 1, the Maintenance Construction Component is used by the HT-SMEs to create the Maintenance Black-Box (BB). The main propose of this BB is to implement a customized maintenance management solution supporting the machine's maintenance operations (see above sub-section).

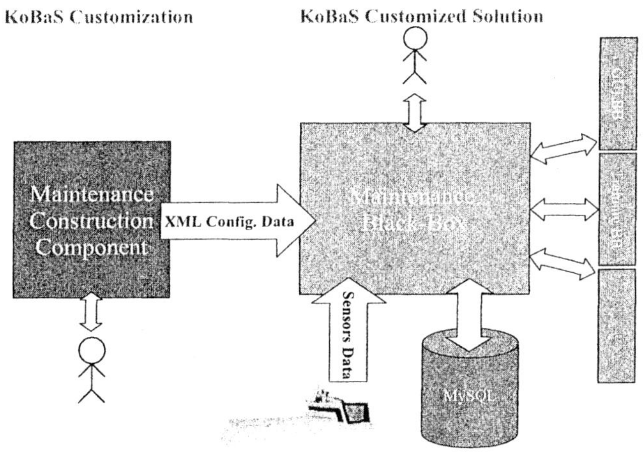

Figure 1: Creation of a Maintenance Customized Solution.

The Maintenance Black-Box is created through a set of XML configuration files and it is integrated with other KoBaS BB as a solution. This solution will be embedded in the machine. The customization phase of the Maintenance BB involves the specification of the several structured information, such as: characterization of the machine, its modules, components and hierarchy; characterization of the machine/components failures; characterization of the preventive maintenance works, including the identification of the maintenance tools to do the work, the spare parts and components to replace and the maintenance training procedures to help the maintenance preventive intervention; characterization of machine parameters; characterization of machine processes and relation with the monitorization parameters.

3.2 The Maintenance BB

Figure 2 presents the architecture overview of the Maintenance BB. This architecture comprises several modules reflecting the set of functionalities considered in the component.

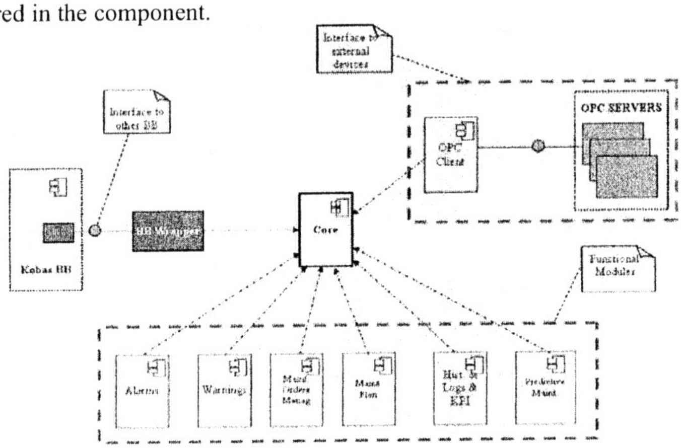

Figure 2: Architecture overview of Maintenance Component.

Interaction with the machine controllers

The Maintenance BB must have a tight interaction with sensors and transducers embedded on the machine in order to receive real-time data and thus provide functionalities related to detection and prediction of machine failures. The approach followed was to design a solution that will act as a client accessing the equipment main controller through a standard interface.

The use of a standard interface enables the interoperability between automation/control applications, field systems/devices and business/office applications. It provides a standard method for exchanging information between the machine controllers and the KoBaS Maintenance BB across a wide range of platforms and involving different technologies. One of the main requirements is the assurance that software is written independently of the type of controller considered.

In factory automation domain, the OPC protocol has been proposed as a standard method for exchanging data between automation elements. This non-proprietary technical specification defines a set of standard interfaces originally based upon Microsoft's OLE/COM/DCOM technology specification (OPC Foundation Website). It is nowadays the preferred connectivity medium into enterprises, and supported by major providers of control systems and industrial instrumentation (Hong and Jianhua, 2006). Currently, OPC XML-DA specification which builds on the OPC Data Access specifications to communicate data in XML, is available. It bases on the web services standards XML, SOAP and WSDL and standardizes the SOAP messages exchanged between clients and server (Zampognaro, 2004). The standardization of messages instead of an application interface allows the implementation on different operating systems and does not base exclusively on Microsoft's COM/DCOM. In this context, the OPC XML-DA standard specification was selected.

Concerning the physical infrastructure, the approach should support two different scenarios:

- OPC Client(s) and Server(s) in the same node – OPC clients and servers run as concurrent processes in the same computer (node);
- OPC Client and Server in different nodes - in this situation OPC clients and servers run in different computer platforms.

The same equipment (machine) can involve more than one OPC server running in the same node or in different nodes (physically located in the machine). Moreover, it is possible to have a direct connection between OPC Server and devices as well as an indirect connection through one or several fieldbus networks.

The Predictive Maintenance module

The Predictive Maintenance Module, which is condition monitoring based, will support monitoring, trigger conditions, and short and long term forecasting for slowly changing sensors. To perform the monitorization the OPC client will make data requests to the Industrial Machine for a specific sensor at specified time intervals. The read value is then appended to a history buffer and applied to the Trigger Condition and Forecast sub-modules.

In the Trigger Condition sub-module the read value is compared with predefined threshold that (if exceeded) can trigger warnings or failures. In this sub-module the values from multiple sensors are also conjunctly evaluated using a Knowledge Base Rules Engine, looking for conditions that can indicate a possible machine or sensor failure, or help diagnose an existing problem. The recent history of the sensor values can also be graphically displayed against the target and threshold limits.

The recent history (minutes) of the sensor values form the basis of a non-linear adaptive ANN-based prediction sub-module, whose function is to predict the sensor value in the very near future and compare the predicted values with predefined thresholds. This module makes it possible to avoid failures originated from punctual and otherwise unexpected situation, such as a suddenly clogged coolant or pressure line.

A similar technique will monitor the long-term trends of a machine component or sub-system monitored by a given sensor. Past sensor values in the range of weeks to months will reveal any long-term trend and, through interaction with the Maintenance scheduler, anticipate or delay a planned Maintenance operation, eliminating unnecessary downtimes and unnecessary parts replacement. Seasonal trends must be detected if possible and not deploy Maintenance operations.

The earlier warnings will be integrated into the Trigger Condition sub-module where, together with other sensor values, it will help to diagnose the problem in advance.

The above described techniques are used for slowly changing data, such as temperature, pressure, etc, but for rapidly changing data sensors, such as accelerometers, instantaneous motor current, etc., further pre-processing steps are needed. This kind of sensors is more effective in the frequency domain, forming the basis of Vibration Analysis Techniques, and for this reason its values will be acquired and process in accordance. Depending on the sensor and techniques to use, the acquired signal will undergo a Fast Fourier or Wavelet Transforms and a Feature

Selection/Extraction step before being applied to the Trigger or Forecast sub-module.

As with slowly changing sensors, rapidly changing sensors will be acquired at specified time intervals, but a burst of hundreds of equally and rapidly sampled values will be read instead.

The Warnings and Alarms modules

The Maintenance BB generates a warning when the monitorization detects one or more parameter value over the warning limits previously configured. The Predictive Maintenance (short or long-term forecast) also generates a warning when it predicts that one or more parameters will come over the warnings limits previously configured. If both monitorization and forecast detect a warning in the same prediction interval only one warning is generated.

The Maintenance BB generates an alarm each time a machine failure occurs (a Corrective MWO or a Predictive MWO is created) or near the planed date of the Systematic Maintenance intervention. Upon the occurrence of an alarm or warning the system waits by the operator acknowledge. The event remains active during the configured time-out. The event will be repeated a number of times according to the configured frequency if there is no the operator acknowledge. When the operator acknowledges the event the acknowledge date is recorded on the system. The system will provide different views over the alarms and warnings data.

The Maintenance Orders Management module

The system allows the management of Corrective MWO, Systematic MWO and Predictive MWO. The machine typical failures should be configured first by the Machine Builder during the KoBaS customization phase and completed later by the End User, according to his experience, during the machine runtime. The same configuration procedure should be made for the maintenance jobs. The Corrective MWO is created by the system when a failure occurs in the machine. The Systematic MWO could be created by the Knowledge Engineer in the KoBaS customization phase or by the End User during the machine operation. The Predictive MWO is created by the system when it predicts the failure occurrence.

When the MWO is created it includes all the information that will be necessary to repair the machine. After the maintenance intervention (according to the End User records and updates) it also includes all the information related with the intervention, namely the maintenance time, the repair time and the components replaced.

The Maintenance Plan module

The Systematic MWO are planned (ordinary maintenance plan) according to the knowledge of the Machine Builder, in the KoBaS customization phase, and in the runtime, according to the experience of the End User. These MWO are periodic and generated automatically by the system according to the maintenance period. The Predictive MWO are planned (extraordinary maintenance plan) by the system according to the failure estimation. The ordinary and the extraordinary maintenance plan are validated by the machine scheduler.

3.3 Interaction with other KoBaS Black-Boxes

The Maintenance BB interacts with other Black-Boxes (BB) using its services (represented by interfaces) in order to obtain essential data necessary to process information to be presented by the graphical user interface. The GUI BB allows the creation of the graphical user interface.

During the maintenance intervention the End User could ask the system about the best procedures to make the intervention. The Training BB is the front end for the end user offering all functionalities to search for maintenance requested training and managing the execution. Before or during the maintenance intervention the End User could access to the Simulation function to test and to underline situations related with the failure, like "put this object in this position", "change the colour of this body" and others. The Simulation BB offers a virtual environment representing the manufacturing machine helping the end user on the failure diagnosis.

The Rule Based BB provides rule based knowledge for the other black-boxes namely the Maintenance BB, in order to allow the detection and diagnosis of machine abnormalities and failures.

The Maintenance BB extracts cases related with the maintenance history in order to help in the failure diagnoses. The Experience Data Base BB is an easy-to-use tool for the creation of a customized case base reasoning expert system.

The Built-in management BB generates production schedules. The Maintenance BB interacts with it to inform and to validate the dates of planned maintenance.

4. DEVELOPMENT AND IMPLEMENTATION

By definition, a KoBaS maintenance solution is an integrated set of black-boxes (software components) targeted to implement maintenance related functionality. Two approaches have been considered to deploy such a KoBaS solution in an equipment machinery: installation of the software in the computer embedded in the equipment machinery itself (and responsible for controlling the machine); installation of the software in a separate computer with a communication channel with the machine's control computer.

Despite the approach taken by the Machine Builder, the software must run in several computational platforms, being the Linux and Windows operating system the minimum requirement. In this context, the Java programming language and the Java 2 Standard Edition constituted the basic ingredients to build a KoBaS solution. A KoBaS solution is then an integrated set of Black Boxes doing their work within a Java virtual machine.

On the other hand, Open Source software constituted another major choice. The development of a KoBaS solution should make use of Open Source software whenever possible. In this context, several tools and frameworks were selected to ease the software development. The MySQL database management system (MySQL, 2005) is being used in the Maintenance solution in order to assure the required data persistency functionality. Access to the data base is implemented by the JDBC API and, on top of it, by the Hybernate framework (Hibernate, 2005). This framework maps Java objects in the object-oriented world to data records in the relational data base world facilitating thus the development of the Maintenance persistency function. The major development tool is the Eclipse IDE (Eclipse Website).

5. CONCLUSIONS

The KoBaS Project (Knowledge Based Customized Services for Traditional Manufacturing Sectors Provided by a Network of High Tech SMEs) is an European project under the Sixth Framework Programme that tends towards a substantial improvement in equipment and industrial machinery.

The customization of specific solutions required by the equipment manufacturers according to the requisites demanded by the market will be supplied by a network business model, made of technology-based SMEs. Each KoBaS customized solution is an integrated set of Black-Boxes generated from software components by the Network of HT-SMEs.

The Maintenance Solution is one of the KoBaS solutions that allow an easy adaptation to new machines and provides a more and intuitive interaction with the operator. It was conceived to provide, machine monitorization, failures detection, failures prediction, alarms, maintenance simulation, training, ordinary and extraordinary maintenance plan and Work Orders management. To fulfil these functions the solution includes the Maintenance BB, based on prediction maintenance condition, and other KoBaS BB, on one hand, to help in the failures diagnostic, and on the other hand, to supply to the end user maintenance training, simulation, and maintenance plan.

The innovative concept presented on this paper allows a greater integration between the machine performances and the related process parameters and promotes the evolution of SMEs operating in traditional industrial sectors, reinforcing and integrating competitiveness, innovation and sustainability.

6. ACKNOWLEDGEMENT

This work has been partly funded by the European Commission through Project KoBaS (No. 505339-2 FP6-2002-NMP-2). The authors wish to acknowledge the Commission for their support and KoBaS project partners for their contribution during the development of various ideas and concepts presented in this paper.

7. REFERENCES

1. Eclipse Website, http://www.eclipse.org
2. Hibernate. Hibernate Open Source Project, http://www.hibernate.org/, accessed 10.1.2006.
3. KoBaS Consortium (2004). KoBaS Description of Work. Technical Report. KoBaS European project 505339-1.
4. KoBaS Consortium (2006). KoBaS Web site, http://www.kobasproject.com, accessed 10.1.2006.
5. MySQL. MySQL Project, http://java.sun.com/jdbc, accessed 10.1.2006.
6. The OPC Foundation Website, www.opcfoundation.org
7. Hong X, Jianhua W. Using standard components in automation industry: A study on OPC Specification. Computer Standards & Interfaces 28 (2006) 386– 395
8. Zampognaro, Francesco. Specifiche ed Implementazione di un Middleware per Applicazioni Real-Time Mediante la Tecnologia Web Service. Master Thesis - Universita degli studi di Roma, 2004

27

REAL-TIME COLLABORATIVE DESIGN SYSTEM FOR PRODUCT ASSEMBLY OVER THE INTERNET

Xiangxu Meng, Weiwei Liu, Yanning Xu
Dept. of Computer Science and Technology, Shandong University, Jinan, P.R. China
mxx@sdu.edu.cn; lww@mail.sdu.edu.cn; xyn@sdu.edu.cn

Product assembly design is a complex activity possible involving collaboration between different designers geographically dispersed. This paper puts forward some significant methodologies and technologies for distributed assembly and presents a collaboration architecture that manages to support working in both synchronous and asynchronous ways. Meanwhile, we adopt a three-level conflicts detection scheme to avoid conflicts effectively and a data streaming technology based on C/P (command/ parameter) to realize the real-time design. Based on the technologies mentioned above, a design system that supports real-time collaborative assembly is developed and we validate it by assembling a mechanical press across network collaboratively.

1. INTRODUCTION

Product assembly plays a critical role in getting products with high quality. However, as to many large and complex products such as mechanical press (Figure 1), designers generally need to assemble and debug all physical parts of products beforehand in the factory, then disassembling all parts for transportation, and at last assemble the parts again in the application place, which results in the great waste of both manpower and material resources. That is why we need a collaborative assembly design system for long-distance guidance and discussion to reduce the time of assembly. Meanwhile it is also propitious to make different domain designers participate in the assembly process planning.

Figure 1 – Mechanical press's pre-assembly in factory

Collaborative assembly design is an important application field of CSCW, however, the current CAD software mainly serves to planar

Please use the following format when citing this chapter:

Meng, X., Liu, W., Xu, Y., 2006, in IFIP International Federation for Information Processing, Volume 220, Information Technology for Balanced Manufacturing Systems, ed. Shen, W., (Boston: Springer), pp. 253–260.

drawing or three-dimensional modeling, and adds strict restrictions to users that there must be only one person manipulating the object at the same time. Such typical software are: SolidWorks eDrawing, Hoops/Net, Actify SpinFire etc, and all of them can't meet the collaborative assembly requirements.

There are many researches about collaborative design up to now: (Li, 2005; Gao, 2004) gave a general summary about the research and development status of collaborative computer-aided design. (Shyamsundar, 2001) presented a new geometric representation called AREP and a prototypical system (cPAD) (Shyamsundar, 2002) based on B/S architecture which, however, couldn't support real-time collaborative work. (Rajarathinam, 2000) studied how to represent a geometric modeling in a multi-modal multi-sensory virtual environment supporting collaborative design, but didn't talk about how to avoid conflicts in the collaborative condition.

To realize a collaborative assembly design in real time, there are still many technologies worth to studying such as: what kind of architectures should be adopted, "Thin server+ strong client" or "Strong server+ thin client"; how to control conflicts during the collaborative manipulations; how to transmit 3D streaming quickly and so on. We studied some key technologies about the real-time assembly on the premise that "if a product can be disassembled, it can be assembled too." and a prototype system is developed supporting remote designers to evaluate assembly planning in real time over the network simultaneously.

2. DISTRIBUTED SYSTEM ARCHITECTURE

At present, there are mainly two categories of distributed system architectures: "Thin server+ strong client" and "Strong server+ thin client". In the first architecture, clients are equipped with whole CAD functions and some communication facilitators. A server plays as an information exchanger to broadcast CAD files or commands generated by a client to other clients. Obviously, easy-deployed is the most straightforward character of this architecture through equipping standalone CAD systems with a communication facility; however, its conflicts are difficult to control without enough information. In the second architecture, the data structures in clients are light-weight and they primarily support visualization and manipulation functions. The main modeling activities are carried out in a common workspace in the server side, which is getting more popular since it brings a new kind of business model—application service provider (ASP). Nevertheless the network bandwidth is heavy-weight, making the delay of communication serious and the server apt to becoming a bottle-neck.

Considering the functional and performance requirement, we extend the functions of both server and clients based on the "Thin server+ strong client" architecture, and put forward a new infrastructure– composite framework (Figure 2). The usage scenario is as follows: when one client loads a product model (module 3) to a new session (module 1, 2), the system will transmit the model with its feasible disassembly sequences which are calculated beforehand (module 4) up to the server automatically. The other clients join the session (module 1, 2) and download specific geometric information (module 3) accordance with specific situations, say network status or different requirements. Then designers can start assembly planning (module 6, 7) in the same virtual environment with various interactive manners (module 5) such as the mouse、keyboard、microphone、Flock of Birds、CyberGlove etc. Meanwhile the assembly route (module 8) reflecting the designer' s interactive operations is recorded automatically. During the above process each client just only transmits those orders that have changed the assembly scene to the order queue (module 9) in the server. And the server takes out one order from the queue ordinally and judges if it will bring a conflict (module 10, 11). After making sure the order will not, the server transmits it forward to the other clients (module 12) where orders are parsed and executed in succession (module 6,7). Different clients can communicate with each other expediently through a chat channel.

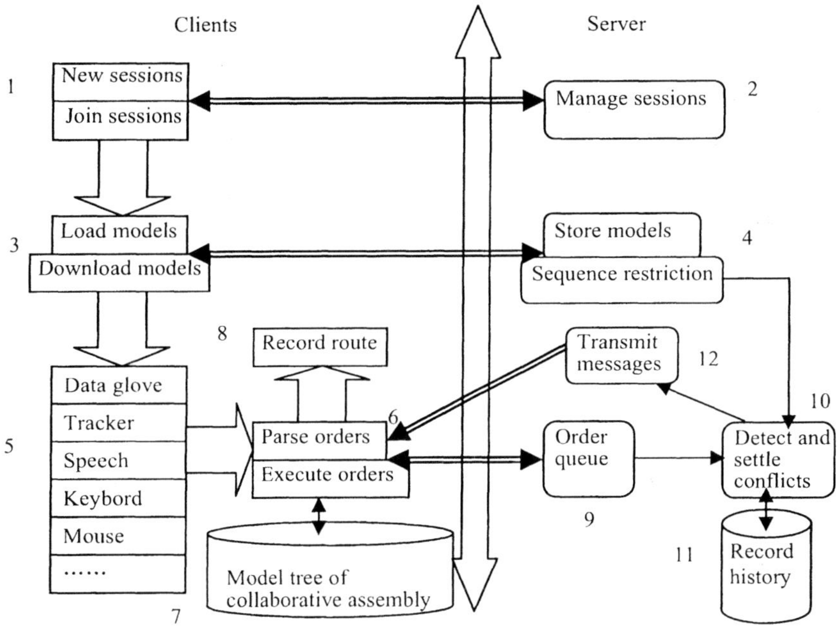

Figure 2 – Architecture of the composite framework

Compared to the typical architecture of "Thin server+ strong client", the composite framework adds several detecting modules to the server which succeeds in controlling conflicts in good time and enhancing the intercurrent ability; meanwhile, it lights the weight of clients by virtual of transferring some functions' execution to server, for instance models storage. In addition, adding the history record module in the server is propitious to working asynchronously or resuming from fault through executing orders from the records in succession. The system is developed from HOOPS 3D and HOOPS/NET, which is owned by the company Tech Soft and supplies many basic collaborative functions and manipulations as object selection, translation, collision detection, view point transform etc. All of those contribute to the implement of our further research and application.

3. CONFLICTS CONTROL MECHANISM

Team management and design coordination functions are the crucial components for establishing a well-organized team and avoiding conflicts on the whole. A working session mechanism is effective for the team management. Each session can be used to organize a collaborative task, and designers in the same session can share the design information dynamically. In a system, different design tasks can be carried out at the same time in different sessions.

Even so, the conflicts can't be avoidable completely because some design tasks must be accomplished by collaboration. There are two primary causes leading to conflicts in the system: incorrect local manipulations and sharing resources. To avoid those conflicts we must pay attention to two aspects: control-class and domain-class (Li, 2002), the former mainly deals with what mechanism is effective to avoid conflicts on the whole, while the latter mainly deals with how to manage the conflicts appeared already. Considering the complexity and diversity of conflicts, a three-level conflicts detection scheme is developed adopting different strategies accordance with different levels, which works effectively via filtrating conflicts level by level (Figure 3).

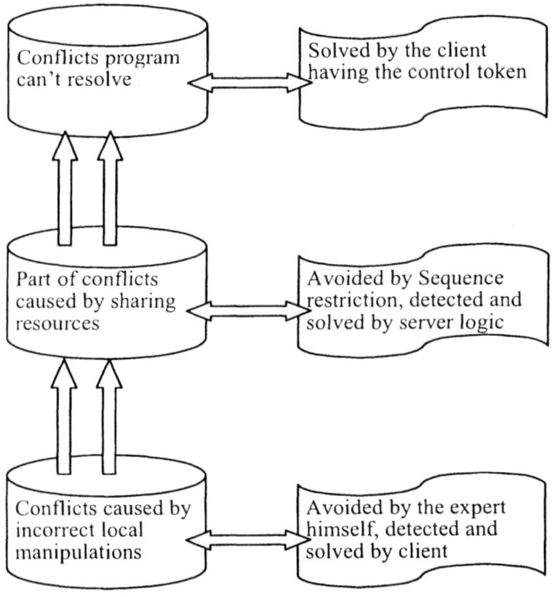

Figure 3 – Three-level conflicts detection scheme

(1) Conflicts caused by incorrect local manipulations. Each part has its own restrictions, for example disassembly direction and tool kits etc, all of which have been stored in the corresponding part restricts table locally. The disassembly actions of an expert are decomposed into many basic manipulations for instance selecting, holding, translating, revolving, releasing and so on. The system automatically

detects conflicts by virtue of the corresponding part restricts table while an expert is operating an object. If any conflict is found, the system will command the "conflicts settling module" to execute an "Undo" order with sending a warning message to the user.

(2) Part of conflicts caused by sharing resources. Due to the large quantity and complex relationships of parts, we can imagine how difficult it is to judge whether the current part can be disassembled or not, that will bring the process of cooperative manipulations into a disordered state and easily induce conflicts. To resolve this problem, We add one "sequence restriction module" to the server, which serves to calculate which parts are knock-down at the present time and change those parts' color into black as hints. So designers now can select and manipulate parts with the assistance of the sequence restriction orderly. In addition, associated with the current disassembly situation and history record, every order likely to cause conflicts must be compared with the conflict classes in the program. If any conflict takes place, program logical module responses to resolve it utilizing correlative domain knowledge and current data.

(3) Conflicts that program can't resolve. The system adopts control token mechanism: there are two classifications in users, one of them with control token and the others not; the client loading models has the control token as default which can be transferred by application. In the situation when there happened those conflicts even the program logic can't resolve, the system will clew users discussing with each other, and finally the client with control token has the most powerful word to resolve the conflicts.

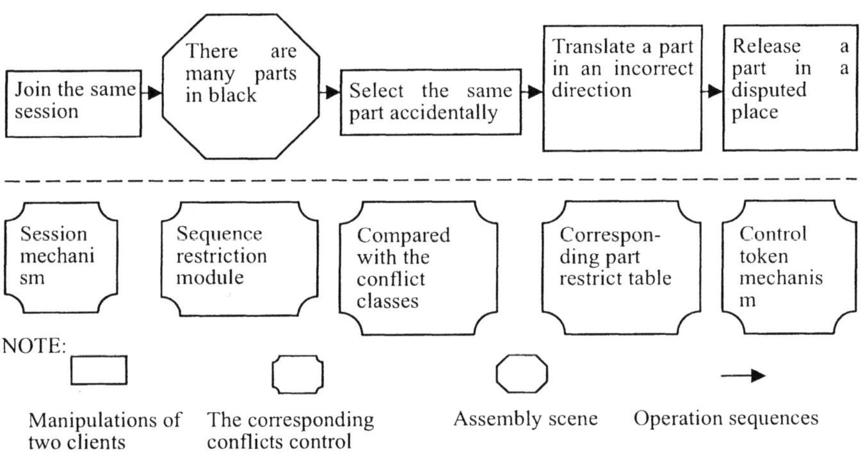

Figure 4 – An example of conflicts control process

Let's take an example to see how conflicts are controlled in the practical collaboration (Figure 4): Suppose that two clients A and B join the same session for carrying out a disassembly task, after joined in there are many parts in black to assist designers operating in the virtual scene. Client A selects one black-color part, written into the history record module immediately, and at one time client B also selects the same part accidentally, which causes a conflict can be easily found out by checking the history record and comparing with the conflict classes in the program. Then program logical module suggests that client B select another part to avoid the conflict. Having selected two different parts, the clients can disassemble objects separately. At the time when one client is translating the part in an incorrect direction, a collision will take place sprung by searching the restricts table of the corresponding part which records the correct disassembly direction of each part, and the program executes an "Undo" order to resume the part's original position. At last, when a part is released in one place where other clients don't agree, the client with control token will make a conclusive decision according to the actual situation.

4. REAL-TIME SYNCHRONOUS COLLABORATION AND DATA TRANSMISSION

Multi-users' assembly platform calls urgently for real-time collaboration, but a significant problem for the above systems is that communication efficiencies are still quite far from satisfactory when large-size feature- and assembly-based models are designed collaboratively. Despite of many studies about effective 3D streaming transmission such as data compression, mesh simplification and incremental transmission, they can't satisfy the real-time requirements.

Representation of models should take into account of practical needs to decide which kind of representation is better. Notice that there are two primary points in our application: collaboration and assembly, deciding that the representation of models should not only care about the speed of network, but also should be convenient for assembly manipulations. In fact there are a mass of data in CAD model not needed in assembly, for example measurement data etc, so our models only keep down those information relevant with assembly, involving: many attributes information such as geometry, lamp-house, view point etc to show how to draw models; and some necessary restrictions information such as disassembly direction, disassembly tools etc to assist assembly simulations. Model tree structure is in favor of assembly manipulation. If someone manipulates a sub-assembly for instance translating or detecting collisions, it will be convenient to deal with by calculating all the sub-nodes of the sub-assembly, changing their correlative attributes, finding out which parts should be redrew, and sending these information forward to the hardware which redraws the scene accordingly.

Although we have predigested models, the amount of data is still large relative to the Internet with limited bandwidth capability. In order to reduce the load of network and increase the speed of transmission, the system adopts a new data streaming technology based on C/P (command/ parameter), i.e. the model data are transmitted only once after clients join a new session, and they will not be transmitted any more in the whole process. The system only transmits commands and parameters changing the disassembly scene to the other clients, and then the clients parse and execute those orders by themselves (Figure 5). Through using this technology we decrease the amount of transmission data and make real-time collaboration a reality.

After clients download model data to their own computers, they can start carrying out disassembly experiments by interaction. There are three kinds of interactive manipulations: the first kind is view point transform genus, for example revolving view points or zooming the scene; the second kind is scene update genus, for example selecting, translating or revolving parts; the third kind is communication genus, for example sending chat messages to one another etc. Because every client manipulates different parts with the different corresponding view points, there is no need to send view point messages to the others; communication messages can be transmitted to each other directly; as to the scene update genus which changes the current disassembly state, it's necessary to send messages in C/P mode to tell other clients changing their own model trees. The corresponding relations are demonstrated in Figure 5.

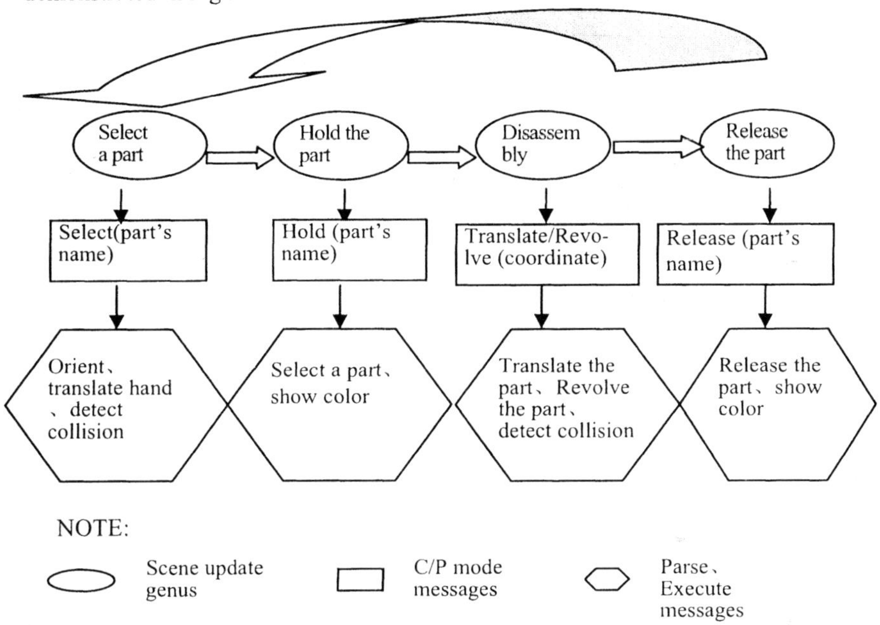

Figure 5 – Corresponding relations between clients' manipulations and transmitted messages

5. CONCLUSIONS

This paper studies some key technologies of collaborative assembly, involving framework of system, representation of models, transmission of data, avoidance of conflicts etc, and produces a real-time collaborative working platform proved availability by disassembling a mechanical press in the different places simultaneously (Figure6, 7). Further work would include how to translate different quantity of information to users according to their specific requirements and how to supply more convenient channels to communicate with each other effectively.

Figure 6 – Collaborative work by the designers from different domains with multimodal interaction(speech,data glove and tracker)

Figure 7 – Scene of collaborative disassembly (every hand stands for one client, the parts in black are knock-down for the moment.)

6. REFERENCES

1. Shuming Gao, Fazhi He. Survey of distributed and collaborative design. Journal of Computer-Aided Design & Computer Graphics, 2004; 12:149-157.1.
2. Yuemei Li, Shouqian Sun, Yunhe Pan. Multi-agent conflict resolution in cooperative design. Journal of Computer-Aided Design & Computer Graphics, 2002; 2:158-162.
3. W.D. Li, W.F. Lu etc. Collaborative computer-aided design – research and development status. Computer-Aided Design, 2005;37:931-940.
4. Rajarathinam Arangarasan, Rajit Gadh. Geometric modeling and collaborative design in a multi-modal multi-sensory virtual environment. Proceedings of DETC'00. ASME 2000 Design Engineering Technical Conference and Computers and Information in Engineering Conference, Baltimore, Maryland, September, 2000;10-13.
5. N.Shyamsundar, R.Gadh. Internet-based collaborative product design with assembly features and virtual design spaces. Computer-Aided Design, 2001; 33:637-651.
6. N.Shyamsundar, R.Gadh. Collaborative virtual prototyping of product assemblies over the Internet. Computer-Aided Design, 2002;34:755-768.

28

AN INTEGRATED APPROACH TO A MULTI-ATTRIBUTE DECISION PROBLEM THROUGH A CASE STUDY

Ashish Ganguli[1], Yong Zeng[1], Akif A. Bulgak[2]
[1]*Concordia Institute for Information Systems Engineering*
[2]*Department of Mechanical and Industrial Engineering,*
Faculty of Engineering and Computer Science, Concordia University,
1455 de Maisonneuve West, Montreal, QC H3G 1M8, Canada
ashi_gan@alcor.concordia.ca; zeng@ciise.concordia.ca; bulgak@vax2.concordia.ca

This article presents an integrated approach to reaching a robust decision in a full-scale multi-attribute decision problem inspired from a real-word case. The Hypothetical Equivalents and In-Equivalents Method (HEIM) is first validated using the full-scale decision problem. The major drawbacks of HEIM have been identified. In an attempt to eliminate the drawbacks of the HEIM, the Game Relative Importance Method (GRIM) is integrated into the method for a robust solution. The results obtained from the application of this integrated approach to the real-life decision problem have been presented. We conclude that the new approach presented yields good and efficient solutions with less computational time in comparison to applying only the HEIM.

1. INTRODUCTION

One of the initial stages in a product development process is the generation of design concepts. Once several alternative design concepts are generated, a decision has to be made to select the best alternative for manufacturing. This decision is usually based on the utility of both the customer and the producer. Multiple conflicting criteria are often involved, where some alternatives may be superior in one aspect but inferior in others (Shah et al., 2000). This type of problem is known as the Multiple Selection Criteria Problem (MSCP) or the Multiple Criteria Decision Making (MCDM) (Fernandez et al., 2005). Various approaches and tools have been developed to solve the MSCP problems. One of the approaches proposed to address the problem is the Utility Theory in which a utility function is defined for modeling the uncertain values of decision maker's preference (Hazelrigg, 1998) to provide the preference structure (Fernandez et al., 2005). However, the utility function cannot be used to formulate the problem when precise values are available. As an alternative, the Hypothetical Equivalent and In-equivalent Method (HEIM) is one of the most advanced methods recently developed. It is successful in accommodating multiple attributes and formulating the MSCP problem to find a robust solution by calculating the weights and equating the value functions but it is unable to identify

Please use the following format when citing this chapter:

Ganguli, A., Zeng, Y., Bulgak, A. A., 2006, in IFIP International Federation for Information Processing, Volume 220, Information Technology for Balanced Manufacturing Systems, ed. Shen, W., (Boston: Springer), pp. 261–270.

relevant attributes (Shah et al., 2000). The number of attributes incorporated into the problem is done arbitrarily. Methods such as factor analysis (Urban and Hauser, 1993) and weighted factor analysis (Hessami and Hunter, 2002) can be used to list important attributes. However, such analysis requires extensive efforts from a group of experts to predict the positive and negative effects of each factor and making a hierarchical decomposition of those factors (Hessami and Hunter, 2002). Such a process is time consuming and expensive. A need for a quick and efficient process is hence evident. The proposed Game Relative Importance method (GRIM) takes into account the contribution of each attribute into the problem and finds such a hierarchical order. To establish an order a comparison is done based on differences in attribute level within the alternatives. In this article, an integrated approach combining HEIM and GRIM is proposed to overcome the existing problems of current methods.

2. INTEGRATION OF HEIM AND GRIM METHODS

HEIM method solves for robust weights by re-constraining the problem (See and Lewis, 2003). The utility function can be given as a multi-linear, multiplicative or an additive function. In the first case there is a set of attributes $Y=\{A_1, A_2,\ldots,A_n\}$ needed to characterize the design and the complement of A is Y^{l}, where Y is independent of Y^{l}. For the second case each attribute should be independent of other and so is in the third case. For this to happen, any important attribute not selected for solving the MSCP problem will independently act as an inhibitor to the design decision. Hence, the major drawbacks of HEIM are its inability to identify the significant attributes for MSCP problem (See and Lewis, 2003) as well as the absence of constraint analysis. The absence of constraint analysis may result in lengthening the problem unnecessarily or may overconstrain the problem. The general formulation for the weights is given by:

$$Min\ F(w) = (1 - \sum_{j=1}^{n} w_j)^2 \tag{1}$$

$$s.\ t.\ \ h(x) = 0 \text{ and } g(x) \leq 0$$

where x is the attribute weight vector, n is the number of attributes, w_j is the weight of jth attribute, and both h(x) and g(x) define constraints. For example, h(x)=0 means that the decision maker prefers alternative A and B equally. This gives V(A) = V(B) and the inequality constraints are formed when decision maker prefer alternative A over B, i.e. V(A) > V(B) or V(A)–V(B)-α = 0, where

$$V(A) = \sum_{i=1}^{n} w_i a_i . \tag{2}$$

α is the penalty function, a small positive number to get the equality. After getting a solution set $S_n \in \{S_1, S_2,\ldots,S_n\}$ for multiple set of weights, the value functions are equated:

$$\sum_{i=1}^{n} w_i a_i = \sum_{j=1}^{n} w_j a_j \tag{3}$$

Upon rearrangement,

$$V(h_1)=\sum_{i-1}^{n} h_{1i}\, w_i \quad \text{and } V(h2)=\sum_{j-1}^{n} h_{1j}\, w_j \tag{4}$$

where h_1 and h_2 are new hypothetical alternatives.

Further preference structure should be given by stating either $h_1>h_2$ or $h1<h2$ to form g(x) and the process is repeated until a robust solution has been reached (Gurnani et al., 2003). From the complete list of the design attributes, one can often observe that while some attributes are very important, some others may not be irrelevant. The constraint analysis done in the case study, presented in section 3, with 13 design variables out of the 42 availables, based upon 3 level 3^{13-10} factorial , resulted in an over constraint problem with no solution. After the constraint analysis it was observed that when $N(G_i) \rightarrow 10$ & $N(f) \rightarrow 3$:

$$S_n \in (S_1, S_2, S_3, \ldots, S_m) \rightarrow Q(S_1, S_2 \ldots S_m) \tag{5}$$

where N(f) is the number of factor level, $N(G_i)$ is the number of constraints, and $Q(S_i)$ is a set of "no solution" or "solution" based on constraints rearranged by changing the preference structure. Human judgment is required for selection of the correct list of attributes, which often contains bias. A new Game Relative Importance Method (GRIM) is proposed to identify those attributes, which make significant contribution. After the decision matrix is prepared relative importance is provided on a scale of 5 to 1. A baseline alternative is selected marking "S". Alternatives if having different attribute values are marked S1, S2 and so on without an increasing or decreasing order. The game value and justified importance is calculated and the relevant attributes are hence selected based upon a preset criterion. The method will be clearer through the case study. In the GRIM, the final selected design efficiency can be represented as:

$$\eta \oplus A(_{ijk\ldots\ldots n})^{JI} \tag{6}$$

where $A(_{ijk\ldots\ldots n})^{JI}$ is the set of last offspring attribute with the highest justified importance, and $\oplus$ is the structure.

These two techniques are integrated in our approach in order to find a robust selection. This integrated approach makes the initial relative importance change according to the game value. The effect of such change is prominent in the decision-making process. In the beginning, a particular attribute may seem to be unimportant for selection but its sub-attribute may have a direct impact on the final efficiency of the product depending upon its comparison with other such criteria for all the alternatives. If A_i, B_i, C_i are set as different parent attributes, then for a product:

$$A_{ij} \in (A_i)^P \tag{7}$$

where A_{ij} are the offspring attribute and $(A_i)^P$ is the parent attribute. Similarly the next generation of attributes $A_{ijk} \in A_{ij}$ and the last offspring attribute $A_{ijk\ldots n}$ will be the set of the last offspring attribute. Any one of these sets of offspring attributes having a higher justified importance affects the product efficiency inspite of having a low importance of its parent attribute. GRIM identifies the number of alternatives in the decision matrix to be n and m being the number of alternatives with the same value for each criterion. It solves for the number of true comparisons within an attribute for all the alternatives. The game value and justified importance of the attributes are given as:

$$G = [\ \{n(n-1)/2\} - \{\ m(m-1)/2\}] \tag{8}$$

$$JI = RI\ + [\ n - G] \tag{9}$$

where n = number of alternatives, m = number of alternatives with same attribute value, RI = Initial relative importance, and JI= Justified importance.

3. CASE STUDY

An example case study is given in this section to illustrate the integrated approach. Five hypothetical aircrafts belonging to the super-light business jet category were selected for the case study: Jet Mod 04, Jet Mod 04 XR, Jet Mod 06 super-light, Jet Mod 06 XR super-light, and Jet Mod 08 midsize. These small aircrafts generally have a seating capacity between 7 and 9 and mainly used for business travel. A general specification chart is given in following subsequent tables categorized into general, performance and dimensions. A complete list of data with 42 criteria for selection was acquired to analyze the problem. A relative importance was given according to a scale from 5 to 1, where 5 is deemed to be the highest important parameter and 1 to be the lowest one. The parent attribute sets are identified as general, performance, dimension, and rated as shown in Table 1.

Table 1. Relative importance table

	General = 3			Performance = 5					Dimension = 4		
Importance	Cap	Avionics	Engine	Range	Speed	Airfield Performance	Ceiling	Noise	External	Internal	Weights
	2	5	4	5	4	5	4	4	4	5	5

After the initial relative importance rating was given and the analysis of each offspring attribute was done and rated, the Jet Mod 04 (JM04) was selected as the baseline and marked as "S". Depending on the actual attribute value of each aircraft (i.e. S1, S2,...Sn), different attribute values are marked without necessarily being in an increasing or decreasing order. These S values were used to find the number of games occurring within the problem. Combining equation 8 and 9, the equation 10 was developed:

$$JI = IR + [n- [\{n (n-1)/2\}-\{m (m-1)/2\}]] \quad \textbf{(10)}$$

The justified importance rating was given at the bottom row of each table. If one of the most important attributes having a rating of 5 in each alternative is present with a game value of zero, then its contribution to the overall problem is null since it gives no valuable input to reject its competitive alternative. Thus only the attributes with relative importance≥3 and JI=5 are selected and marked as "5S" and if JI=5 and relative importance≤3 then it is marked as "5E" and not selected.

Based on the results of general, performance (Tables 2, 3) and dimension criteria 7 parameters were selected as an input to the MSCP problem. A modified limit calculation for normalization was done to prevent absolute 0 and 1 values. The calculation is represented in equation (11). The selected attribute normalized data table or decision matrix (Table 4, 5) is prepared showing actual data as well as the modified normalized scores according to the equation 11. The normalized score is obtained based on the utility preference structures for each attribute. The value functions are calculated and shown in Table 6.

$$\text{Delta} = 10\ \%\ \text{of (Max value} - \text{Min value)} \quad \textbf{(11)}$$

$$\text{Lower limit} = \text{Min value} - \text{Delta}$$

$$\text{Upper limit} = \text{Max value} + \text{Delta}$$

Table 2. Comparison & Relative importance for General Criteria

Alt.	General = 3											
	Cap.=2		Avionics=5							Engine=4		
	Crew=1	NOP=2	F1=3	F2=3	F3=5	F4=4	F5=5	F6=5	F7=5	F1=3	F2=4	F3=3
JM04	S	S	S	S	S	S	S	S	S	S	S	S
JM04XR	S	S	S	S	S	S	S	S	S	S	S	S
JM06	S	S1	S	S	S	S	S	S	S	S	S	S
JM06XR	S	S1	S	S	S	S	S1	S1	S1	S	S	S
JM08	S	S	S	S1	S	S	S	S	S	S	S1	S
JI	1	3	0	2	0	0	4	4	4	0	3	0

Engine: F1= Honeywell TFE731-20-AR, F2= Thrust, F3= Flat rating ISA+16 °C. ***Avionics:*** F1= Honeywell primus 1000, 4 tube EFIS, F2= Crew advisor system, F3= Honeywell primus 660 weather radar, F4=Dual primus nav/comm. System, F5=Traffic collision avoidance system (TCAS 2000), F6=Enhanced ground proximity warning system (EGPWS) with wind shear, F7= Emergency locator transmitter (ELT).

Table 3. Comparison and Relative importance for Performance Criteria

Alt.	Performance = 5										
	Rng.=5	Speed=4			AP=5		Ceiling=4		Noise=4		
	M.R. =5	HSC=4	TCS=5	LRCS=4	BFL=2	LD=5	MOA.=4	CCA=5	TONL=4	App.=4	SDL.=4
JM04	S	S	S	S	S	S	S	S	S	S	S
JM04XR	S	S	S	S	S2	S	S	S2	S2	S	S
JM06	S1	S	S	S	S1	S	S	S1	S1	S	S
JM06XR	S2	S	S	S	S2	S	S	S3	S2	S	S
JM08	S3	S	S	S	S2	S	S	S4	S3	S1	S1
JI	5S	0	0	4	5E	5E	0	5S	5S	3	3

Rng.=Range, M.R.= Maximum Range, HSC= High speed cruise, TCS= Typical cruise speed, LRCS=Long range cruise speed, BFL= Balanced field length, LD=Landing distance, MOA=Maximum operating altitude, CCA=Climb to cruise altitude, TO=Take off, App=Approach, SDL=Sideline

Table 4. Selected attribute normalized data table

Alt.	Maximum range (Km)		Climb to cruise altitude (sec)		Take off Noise (EPNdb)		Floor area (ft square)	
	W1		W2		W3		W4	
JM04	3378	12	80	12	73.7	95	55.7	12
JM04XR	3378	12	85	12	75.5	90	55.7	12
JM06	3763	30	82	30	74.5	93	62	50
JM06XR	3865	35	95	35	75.5	93	62	50
JM08	4617	88	38	88	78.9	60	68.9	90

Table 5. Selected attribute normalized data table

Alt.	Total Vol. (Cubic ft.)		Basic operating wt.(lb)		Max. fuel weight (lb)	
	W5		W6		W7	
JM04	363	20	13,718	12	5375	5
JM04XR	363	20	13,718	12	5375	5
JM06	410	86	13,888	25	6062	25
JM06XR	410	86	13,888	25	6062	25
JM08	453	95	14,772	88	7910	88

A 2^{7-4} factorial design was used to generate hypothetical alternatives with -1 and +1 as low and high-level attribute value respectively. Stating the decision maker's preference from the hypothetical alternatives constraints g1, g2...g4 are formed. Re-writing equation (1) a minimization problem is formed. The objective function ensures the sum of weights being equal to unity. The minimization problem is represented is shown in equation (12).

$$\text{Min } F(w)=(1-\sum_{j=1}^{n} w_j)^2 \quad \textbf{(12)}$$

subject to:

$$g1= -2w2+2w4-2w6+2w7-\alpha\leq 0$$
$$g2= -2w3-2w5+2w6+2w7-\alpha\leq 0$$
$$g3= 2w2+2w4-2w6-2w7-\alpha\leq 0$$
$$g4= -2w1-2w3-2w4-2w6-\alpha\leq 0$$

Side Constraints: $0 < w_j < 1$

Table 6. Selected attribute value function.

Alt.	Maximum range (Km	Climb to cruise altitude (sec)	Take off Noise level (EPN db)	Floor area (ft square)	Total Volume (Cubic feet)	Basic operating (wt. lb)	Max. fuel weight(lb)	Total value
	W1	W4	W3	W4	W5	W6	W7	
JM04	12	80	95	12	20	12	5	.12w1+.8w2+.95w3+.12w4+.20w5+.12w6+.05w7
JM04XR	12	85	90	12	20	12	5	.12w1+.85w2+.9w3+.12w4+.20w5+.12w6+.05w7
JM06	30	82	93	50	86	25	25	.3w1+.82w2+.93w3+.5w4+.86w5+.25w6+.25w7
JM06XR	35	95	93	50	86	25	25	.35w1+.95w2+.93w3+.5w4+.86w5+.25w6+.25w7
JM08	88	38	60	90	95	88	88	.88w1+.38w2+.6w3+.90w4+.95w5+.88w6+.88w7

The entire two step process of first evaluating the selection criteria (Table 2, 3) and dimension criteria through the GRIM process to generate the final set of attributes can be understood through the flowchart (see Figure 1). It explains the process of selecting or rejecting any particular attribute.

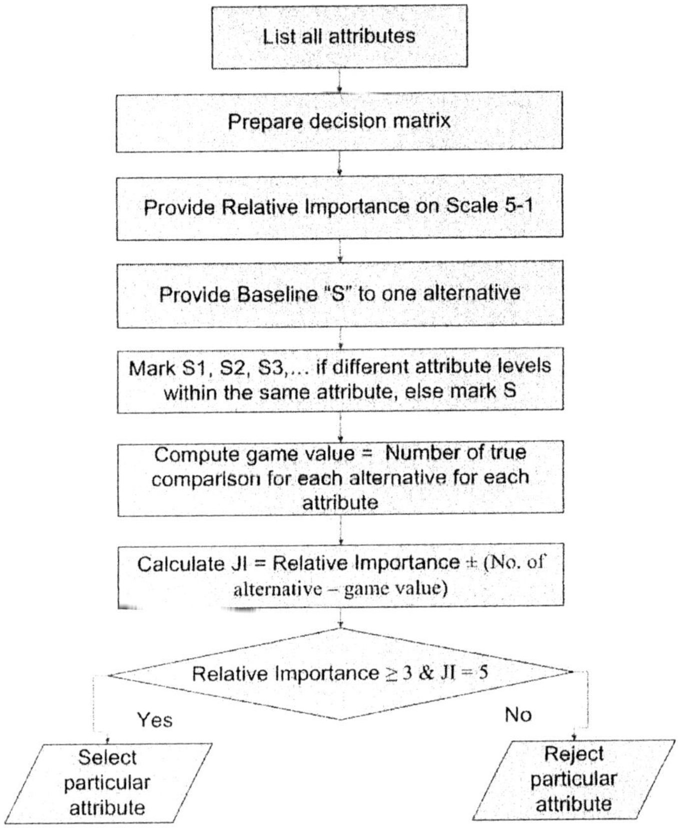

Figure 1- Flowchart of game relative importance method

This set then serves as an input to the HEIM process. A graphical illustration of the HEIM process is showed (see Figure 2). The HEIM process, as shown in figure 2, acts on the generated final set of attributes to give our final robust solution.

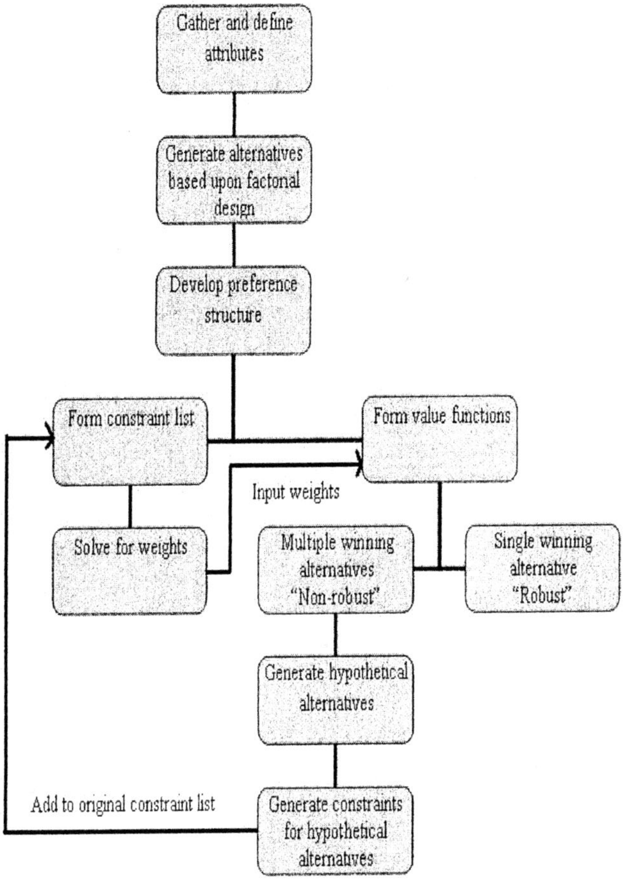

Figure 2 - Flowchart of the HEIM

A "C" program was generated to solve the minimization problem. The problem was first solved by using the HEIM alone, according to which the JM06XR is the winner. On equating the value functions and generating further hypothetical constraints two cases were reached with JM06XR as winner in case A and JM08 as winner in case B. Further, the problem was re-constrained and a robust solution was found. Multiple solutions were reached with JM08 as winner in all three cases.

4. CONCLUSIONS AND FUTURE WORK

The major disadvantage of the HEIM is its inability to identify the attributes which makes a significant contribution to the MSCP problem. The integrated approach taken to solve the MSCP problem with identification of maximum contributing attribute by GRIM and then solving by HEIM resulted in robust solution with comparatively less computation time. In this study, we additionally validated the HEIM on a full-scale problem with real time data. The robust solution indicated the Jet Mod 08 as the best choice. Our future work will focus on other applications of the approach to test its efficiency on a larger domain of decision-making problems involving multiple selection criteria. One important aspect is to effectively state the preference structures in the MSCP problem.

5. REFERENCES

1. Fernandez, M.G et al. "Decision support in concurrent engineering- The utility – based selection decision support problem", Concurrent Engineering: Research and Application", 2005
2. Gurnani, A., See, T.K., Lewis, K., "An approach to robust multiattribute concept selection" ASME Design technical conferences, Design Automation conference, DETC03/DAC-48707.
3. Hazelrigg, G.A.(1998). A framework for decision based engineering design, Journal of Mechanical Design, 120:653-658.
4. Hessami, A., Hunter, A., 2002, "Formalization of Weighted Factors Analysis", Knowledge-Based Systems, Volume 15, Issue 7, Pages 377-390
5. Shah, J. J., Kulkarni, S.V.,Vargas-Hernandez, N., 2000, "Guidelines for Experimental Evaluation of Idea Generation Methods in Conceptual Design," *Journal of Mechanical Design*, **122**(4) pp. 377-384.
6. See, T.K., Lewis, K., "Multiattribute decision making using hypothetical equivalents" ASME Design technical conferences, Design Automation conference, DETC03/DAC-34079.
7. Urban, G. L., and Hauser J. R., 1993, Design and Marketing Of New Products, 2nd Edition, Prentice Hall, pp.1-16.

29

MAKING MANUFACTURING CHANGES LESS DISRUPTIVE: AGENT-DRIVEN INTEGRATION

Joseph Neelamkavil,
Weiming Shen, Qi Hao, Helen Xie
Integrated Manufacturing Technologies Institute
National Research Council Canada
joseph.neelamkavil@nrc.gc.ca;
weiming.shen nrc.gc.ca; qi.hao nrc.gc.ca; helen.xie nrc.gc.ca

This paper presents some results of our recent investigation on how to address changes in manufacturing environments. The results presented here are based on our recent visits to manufacturing plants (in order to understand current industrial practice) and a comprehensive literature review. It has become apparent that agents and multi-agents based technologies have the power and flexibility to deal with the shop dynamics. This paper also discusses research opportunities and challenges in this area, presents our recent research work in developing agent-based technologies to streamline and coordinate design and production activities within a manufacturing enterprise, and between the enterprise and its suppliers.

1. INTRODUCTION

In recent years manufacturing companies have gone through a rapid paradigm change. This change is primarily due to a shift from vendor's market to that of customer's. Customers want products with lower prices, higher quality and faster delivery; yet they also demand products customized to match their unique needs. The customer demands also fluctuate over short periods of time; yet if not served promptly, they are lost to the competition; products also become obsolescent in a short time for a variety of reasons (out of fashion, new products in market, energy cost, and so on). To meet with many of these challenges, manufacturers will require a paradigm shift to adapt their business model to be compatible with agility and mass customization. It combines mass production's economies of scale with custom manufacturing's flexibility. Requirement for lower product cost has dictated reductions in production cost also, which further causes a ripple effect on the complete production process, and demand enhanced productivity and flexibility. But the productivity targets get bypassed when unexpected changes (described as disturbances often) occur at the shop level (e.g., new production/process introduced,

Please use the following format when citing this chapter:

Neelamkavil, J., Shen, W., Hao, Q., Xie, H., 2006, in IFIP International Federation for Information Processing, Volume 220, Information Technology for Balanced Manufacturing Systems, ed. Shen, W., (Boston: Springer), pp. 271–280.

machine breaks down, parts missing, etc.). For most manufacturers, it is a challenge to deal with the frequent product changes and customer demand fluctuations.

It should be stated that, the changes attributable to order fluctuations, cancellations and occasional wrong decisions are unavoidable in an industry. For example, to realize an undesired design decision would trigger changes to products data. The impact of the changes needs to be assessed properly, and the focus in this research is to study how best the impact of both planned and unplanned changes can be addressed at the manufacturing shop and resource level, to optimize production via downtimes reduction and optimum utilization of the plant resources, within the constraints imposed by the factory dynamics. Rapid reconfigurations of the manufacturing, flexible planning, and fast response control systems have been echoed as the business models to deal with these scenarios. To crystallize our ideas above, we visited several manufacturing plants, recorded and analyzed findings, classified and summarized the results, and arrived at approaches that could deal with the changes and disturbances in production – the changes – in general.

This paper presents some results of our recent investigation on how to address changes in manufacturing environments. The results presented here are based on our recent visits to manufacturing plants (in order to understand current industrial practice) and a comprehensive literature review. It has become apparent that agents and multi-agents based technologies have the power and flexibility to deal with the shop dynamics. This paper also discusses research opportunities and challenges in this area, and briefly presents our recent research work in developing agent-based technologies to streamline and coordinate design and production activities within a manufacturing enterprise, and between the enterprise and its suppliers.

2. EFFECT OF CHANGES ON PRODUCTION

In a manufacturing shop, the occurrence of changes is a daily event. The term "change" here is used to mean those, which occurs primarily at the production level. It is somewhat different from the management of engineering changes (often termed as Engineering Change Management - ECM), which is a well-established area. The ECM is meant to cater for the changing needs of products and product functionalities; and more details of this topic can be found in (Pikosz and Malmqvist, 1998). In the manufacturing context, the major reasons why changes are bound to occur are the following:

- Customers change their minds and come up with new requirements
- Suppliers stumbles on their delivery status
- Customer service collects recalls and expects quick fixes
- Design engineers improve functionality and introduce new product variants
- Marketing & Sales adds extra features and likes introducing newer products
- Production implements continuous improvement and new product lines
- Unexpected breakdown creates chaos on the shop floor
- Manufacturing adds new production techniques
- Regulation forces change both at the product and process levels.

The MASCADA report (Valckenaers, 1998) has come up with distinct definitions for the terms production "changes" and production "disturbances", namely: "A production change is an alteration to the production conditions, which is intentionally performed by the plant. A production disturbance is an unanticipated change to production conditions with a negative effect on the (manufacturing) process performance". The important characteristic that distinguishes changes from disturbances is pivoted on the question whether the production facility intends the change. Bear in mind that, a disturbance in one area of production can cause a change in another area, and vice versa. Looking differently, a production change is a planned event, while a disturbance is something that is not at all planned, and hence can only be reacted to.

From a research viewpoint, the impact of changes and disturbances is often assessed from a qualitative perspective. Note that, a quantitative assessment is difficult to attain because, the real impact can vary from plant to plant, and location to location. The nature of disturbance also plays an important role – whether these are caused by external factors (e.g., customers and suppliers) or internal matters (e.g., machine breakdown). The effect of disturbance based on internal matters also depends on criticality of the resources involved (e.g., a key machine), or whether it is localized or global within the plant (e.g., a power outage in the plant).

To conduct research on impact of changes, firstly one has to collect data pertaining to the changes and disturbances; the reliability and the source of the data also will become critical. For example, a senior executive in a plant may not even be aware of the machine break down of a manufacturing cell. The noted MASCADA study (Valckenaers, 1998) based their findings on 19 plants from 17 European companies (13 being discrete manufacturing facilities) categorized the disturbances into 4 categories, based on who/what caused them: a) customers, b) suppliers, c) internal (e.g., resource, quality or co-ordination breakdown) and d) regulatory/environmental. Also, the disturbances may be abrupt vs. gradual (e.g., wear & tear), random vs. systematic (e.g., avoid production during peak electrical hours), time (e.g., late delivery) vs. poor quality based. The above study has categorized the changes into 4 categories also: a) introduction of new products or product variants, b) increase/decrease in production capacity, c) introduction of new production technology, and d) changes in work force.

In most plants studied by the MASCADA team, it was discovered that most planning and control systems treated disturbances (sometimes the changes also) as exceptions that were undesirable, and were therefore ignored mostly during regular planning activities. The planning systems were based on a top-down approach (MRP, for example), which didn't allow any feed back from the lower levels. This means that if production gets deviated from the plan, the planning needs to be restarted again. The study concluded with a number of guidelines on how to handle the changes and disturbances:

- Manufacturing systems need to be easily adaptable to new requirements.
- Manufacturing systems need to be scaleable at small steps.
- Workers and engineers need to be integrated seamlessly into the production process that optimizes the usage of technical and human resources.
- Shop systems need to be able to flexibly react to varying production orders.
- Products and production processes need to be structured such that the product distinguishing features are built-in as late as possible.
- Shop processes should be organized in such a way that if disturbances do occur, production can continue with reduced capacity instead of just collapsing it.

- If disturbances are bound to occur, resources need to be re-assigned efficiently. The re-assignment however, should not cause any more disturbances.
- Planning and control systems should make global consequences of local decisions explicit to user, and support the co-ordination of all related decisions.

3. INDUSTRY PRACTICE

We visited several manufacturing plants (truck manufacturing plant, medical instrument manufacturer, landing gear assembly plant, mold making and injection molding shops, etc.) to discuss the impact of changes in production, and to find out issues and problems that typically exist between the design and production departments – like integration issues, data handling issues, change management issues, communication issues, and so on. Some of the findings are the following:

The need to deal with change is woven all over onto the truck manufacturing operations. The visited plant assembles over 100 trucks daily, and operates 3 shifts/day, having variable production rates for each shift. It is a challenge to address the schedule change issues, and to do proper load balancing since the chassis line, engine line, cab line, etc. need to converge to proper entry points just in time for assembly to take place without hampering the production line. In addition, every second or third truck is of a totally different kind, and the changes and variations due to different parts combination become large. The plant needs to deal with over 10,000 different kinds of parts; and the occurrences of missing (or wrong) parts or parts not at right place at right time become a common event. In some manner, the assembly line and the material management system have to absorb such changes, or come up with provision to complete the assembly offline at some other time. In trucking industry, they have a relatively long (3-5 year time frame) product development time.

The medical instrument manufacturing plant that we visited focuses on the design and assembly of the instrument, while the actual manufacturing is outsourced. Their product development time is relatively short (1 to 1.5 year time frame). Requirement changes, missing and/or incomplete requirements have been problem areas. The product development cycle needs to cater for prototyping and piloting prior to full production. Integration and traceability issues throughout the complete product life cycle are seen very important, and are brought under control via frequent but formalized stage gating procedures. This is a time consuming process, but it works for them, as there are not many players manufacturing the same kind of apparatus. Material management is well under control as they have limited number of parts to deal with, while at the same time they follow a strict lean manufacturing and part replenishment procedures.

The issues associated with the changes in production and integration problems in other plants fall typically somewhere between the two cases described above, and they are not elaborated here.

4. SITUATION ANALYSIS

Based on the industry visits that we have had so far, the two most important issues that became noticeable were 'embracing the change' and the 'need for integration'. In this context and throughout this research study, the phrase 'change' is used loosely to mean both the 'changes' and the 'disturbances'. We observed that, the effects of changes in industries have been addressed mostly at the business and information technology level only. Not much attention is given to the changes of the manufacturing system itself; yet the shop floor suffers from continuous evolution of changes that are imposed on it.

It appears that the reasons why industries find it hard to cope with a change is due to inefficient communication procedures, ineffective shop floor control systems, and the difficulty of achieving integration arising out of several technology components and islands of automation. For example, a capacity analysis software determines a master production schedule that sets long-term production targets; a scheduling system determines the sequence in which manufacturing resources are used in producing the products; a manufacturing execution system tracks the real-time status of work in progress, enforces process plans, and reports labor/material claims. In addition, these applications are often legacy systems developed over a number of years. Each system may perform well for its designated tasks; they need to respond very quickly to changes, but often are not equipped well to handle complex business decisions that involve the coordination of several applications. Rapid but timely solutions to these scenarios are crucial to agile manufacturing, especially in the current era of globalization, automation, and telecommunication.

Recalling what we mentioned earlier, shorter product life cycles and global competition have forced many manufacturing companies to become agile. Industry needs to be able to respond very quickly to deal with this and to minimize the impact of changes on production performance. In this environment, configuring the shops for production and establishing effective control is crucial. Valckenaers et al. (1999) have summarized the requirements associated with this scenario into 5 categories namely: proactive-ness, reactivity, flexibility, reconfigure-ability and connectivity.

By being proactive, a shop control system can plan for the future. It anticipates certain events, and is guided to those states in which such events are best handled, if they do occur. Proactive-ness also enables a shop control system to look ahead into the future. By being reactive, the control system is able to react rapidly to certain events and adapt to such situations. Alternative shop resources, transportation routes, process plans and trade-offs between yields and speeds are examples of process flexibility. The user interaction flexibility addresses how flexible a system supports a specific user to perform several tasks, or how many users can do certain tasks. Reconfigure-ability is a feature that allows for easy adaptation of the system for new productions, product variants, plant layouts, and so on. How tightly a shop control is connected to an actual production environment and its equipment controllers are representative of the connectivity. Included here is the ability to collect data from factory floor to perform the required shop floor control. The control system maintains and has access to the latest available information about shop resources, and is able assign/reassign orders to any available resources.

According to Wyns (1999), a shop floor control is that group of activities directly responsible for managing the transformation of a planned order into a set of outputs, and those indirect activities that needs to be done to cover the changes or adaptations of the control solution in an easy way. During the operative phase, it governs short-term detailed planning, execution, and monitoring of process preparation and resource allocation activities that are needed to control the flow of work order from the time it is released by planning system for the execution, until the order is completed. Two well-recognized shop floor reference architectures that are much contextual here are the 'holonic' manufacturing and multi-agent manufacturing systems (Deen, 2003; Marik et al., 2003; Shen et al., 2001; van Brussel et al., 1998).

5. CHALLENGES AND OPPORTUNITIES

Based on the industry practice analysis and research literature review, a number of research challenges as well as opportunities have been identified.

System Architecture

To create a system that supports agile shop floor control is a hard and complex task. The challenge here is to come up with an architecture that supports agility, and that facilitates the design of control systems that accommodates heterogeneous manufacturing resources, yet without expending too much programming effort. It requires the formulation of an instant manufacturing cell (Barata de Oliveira, 2003) (with various rules and procedures to work in harmony) composed of various resources represented by multi-agents, based on new and/or changed capacity. Issues related to the integration of several controllers, as well as their co-ordination need to be supported.

Modeling and Presentation of Agents in Action

A major thrust here is to formulate a modeling paradigm that can address both the modeling and controlling aspects. Creation of agent models that are representative of all shop resources with pertinent capability features and control aspects is always a challenge. It should be dynamic in the sense that any modifications and/or capability changes to the physical resource should reflect in its agent model immediately. At the same time one has to ensure that it doesn't overreact to minor changes (like spending more time changing its mind rather than acting, or triggering excessive set-ups and material handling activities in lieu of production). For easy communication with the manufacturing personnel, an agent embedded system should also provide good presentation and visualization capabilities, perhaps with the aid of virtual models.

Ontology-based Communication

At a high level, ontology aims at sharing and reusing knowledge by providing an explicit specification of conceptualization. It can support agent communications in two ways. First, it can provide common vocabulary for communication and shared understanding. Second, it can support data interoperability among agents or computer systems by providing semantic foundations for translators. Consider the scenario in which a planning agent wants to communicate with a resource agent regarding various machine tools. In particular, a terminology they may need to communicate is the name of machine tool manufacturer. In a new design of agents, one can take advantage of ontology by specifying a common vocabulary, i.e. "Made by" for the machine tool manufacturer. And, agents can use this common term for communication. In a legacy design, the planning agent and resource agent may regard the machine tool manufacturer as "Manufactured by" and "Maker is", respectively. Ontology can provide semantic foundations by mapping "Maker is" and "Manufactured by" to "Made by", enabling agent interactions in natural manner.

Data Collection & Filtering

The systems environment should be developed based on a data-centric approach in which sensors couple with databases that ensures quality, integrity, reliability, redundancy and security. In a dynamic shop environment, on-line and automatic data collection from the physical shop floor will become almost mandatory. A form of data mining and filtering is necessary to screen out noise and other unnecessary data. If decisions are taken based on real-time simulations, then to ensure selection and extraction of correct data towards on-line simulation always poses a challenge. Data required for real-time decisions, and the amount of details (hence the data) that need to be modeled will be very critical in a simulation based control approach.

Quality of Decisions and Real-time Adaptation

In an agile environment, the shop dynamics will necessitate a manufacturer to deal with frequent changes, and real-time simulation will be helpful to try out different scenarios, and to make decisions based on an optimized shop. Artificial intelligence and distributed simulation technologies can be applied to address such real-time decision processes, which can also help to reduce the computation time involved in making those decisions (Yoon and Shen, 2005). The quality of decision itself is affected by factors such as the simulation frequency, embedded details, incorporated assumptions, and so on.

Conflict Resolution

In the shop environment the resources always face the dilemma of local vs. global optimization. Reconciling different viewpoints and conflicting objectives among resources (hence agents trying to coordinate their actions) is a real shop floor challenge. The agents need to act intelligently in making decisions and taking actions while at the same time accommodating the global effects of local decisions.

6. AGENT-BASED SOLUTIONS

In the change scenario, a manufacturer needs to be able to implement, execute, update, and maintain the network representing the launch of new products and/or processes, and also addresses those issues pertaining to changes that are forced upon on the shop floor. Delays and failures may affect the production processes and transportation system, which may force re-grouping or re-routing of the orders. When re-routing or re-scheduling is needed, one has to ensure that the material handling aspects are also taken care of. A tool built to tackle such issues must support the user in coordinating and controlling most of these activities through a set of small, distributed, autonomous, configurable, intelligent, and communicating systems designed to satisfy their specific goals, while globally achieving optimum performance at the enterprise level (Paolucci and Sacile, 2005). It is becoming somewhat obvious that multi-agent systems (Peng et al., 1999; Brückner et al., 1998; Shen and Norrie, 1999; Parunak, 1999) provide promising technologies for rendering the modern shop floor to be agile.

The manufacturing resources, typically under the control by various controllers, are the basic set based on which the complete factory production is planned. Each resource (representative of an agent) has its own basic core capabilities or core skills, and through cooperation can build additional capabilities. And when they interact and cooperate, they can generate aggregated functionalities that are compositions of their individual capabilities. Many resources need to work together in harmony to deal with the changes and to achieve a common goal. For the manufacturing system to work well within the dynamics of production, an agent-based system needs to know what constitutes a change upfront. With help from a human, or instruction from an analysis support module, a decision module can decide on what constitutes a change. A control module can decide on what data is to be monitored to detect those changes. Based on messages from the control module on the occurrence of a change, the decision module can make contextual decisions, in liaison with the support module. Here, the breakdown of a critical machine can be considered as a big change, whereas failure of other machines may be considered a small one. Proper coordination and evaluation of the resource capabilities are needed to deal with the change scenario. Essentially, one can think of the change scenario in terms of a manufacturing resource configuration and allocation problem in real-time.

Based on our understanding of industrial needs and our experience on agent-based collaborative design and manufacturing, particularly agent-based manufacturing scheduling and as described in (Shen et al., 2005; Hao et al., 2005; Shen and Norrie, 2001), we are currently working in collaboration with industrial partners on following research directions:

- Implementing just-in-time materials handling using intelligent software agents and optimization techniques. Some results of this research work are reported through a separate paper submitted to the same conference by Hao and Shen (2006).
- Investigating schedule repair and buffer size optimization through agent-based simulation.
- Integrating RFID (Radio Frequency Identification) technology with agent-based real-time shop floor scheduling and materials handling.
- Developing software tools by integrating software agents and workflow technology for coordinating/managing between design and production departments.

7. CONCLUDING REMARKS

Shorter product life cycles and global competition have forced manufacturing companies to become very dynamic and agile. Modern manufacturing environments are subject to frequent and rapid changes (e.g. new products or product variants, machine breakdowns, demand fluctuations, etc.). Manufacturers need to respond quickly to deal with this situation in order to minimize the impact of changes on their production performance, and for the survival of the organizations. Agent technology and multi-agent systems have the power and flexibility to provide rapid assessments on the need for production changes and factory dynamics, recommend and/or trigger real-time actions, and adjust manufacturing plans and production.

Our broad vision is to develop agent-based technologies to streamline and coordinate design and production activities within a manufacturing enterprise, and between the enterprise and its suppliers. We believe that, by integrating with other advanced technologies including cooperative workflow, simulation, and optimization, intelligent software agents can be successfully applied to address dynamic changes in the manufacturing shop floor locally without disruptions to the overall production – a problem that cannot be solved by traditional planning and scheduling systems because these changes are not predictable in advance.

8. REFERENCES

1. Barata de Oliveira, JA. "Coalition Based Approach for Shop Floor Agility – A Multi-agent Approach", Ph D Thesis, University of Lisbon, Portugal, 2003.
2. Brückner, S, Wyns, J, Peeters, P, Kollingbaum., M. "Designing Agents for Manufacturing Control", 2nd AI & Manufacturing Research Planning Workshop. Albuquerque, NM. August 1998.
3. Deen, SM (Ed.) Agent-Based Manufacturing - Advances in the Holonic Approach. Springer-Verlag, Heidelberg, Germany, 2003.
4. Hao, Q, Shen, W. "An Agent-Based Simulation of a JIT Material Handling System". Submitted to BASYS'06, 2006.
5. Hao, Q, Shen, W, Zhang, Z. An Autonomous Agent Development Environment for Engineering Applications, International Journal of Advanced Engineering Informatics, 2005, 19(2):123-134.
6. Marik, V, McFarlane, D, Valckenaers, P. (Eds.) Holonic and Multi-Agent Systems for Manufacturing. Springer-Verlag, Heidelberg, Germany, 2003.
7. Paolucci, M, Sacile, R. Agent-Based Manufacturing and Control Systems, CRC Press, New York, 2005.
8. Parunak, HVD. "Industrial and Practical Applications of DAI". In Multi-agent System: a Modern Approach to Distributed Artificial Intelligence, G. Weiss (Ed.), Cambridge, Massachusetts: MIT Press, 1999, pp. 377-421.
9. Peng, Y, Finin, T, Labrou, Y, Cost, RS, Chu, B, Long, J, Tolone, WJ, Boughannam, A. "An Agent-Based Approach for Manufacturing Integration - The CIIMPLEX Experience", International Journal of Applied AI, 1999, 13(1-2):39-44.
10. Pikosz, P, Malmqvist, J. "A comparative study of engineering change Management in three Swedish engineering Companies", DETC'98, Paper No DETC98/EIM-5684, Atlanta, GA, USA, 1998.
11. Shen, W, Norrie, DH. "Agent-Based Systems for Intelligent Manufacturing: A State-of-the-Art Survey", Knowledge and Information Systems, an International Journal, 1999, 1(2):129-156.
12. Shen, W, Norrie, DH. "Dynamic Manufacturing Scheduling Using both Functional and Resource Related Agents". Integrated Computer-Aided Engineering, 2001, 8(1):17-30.
13. Shen, W, Norrie, DH, Barthes, J-P. Multi-Agent Systems for Concurrent Intelligent Design and Manufacturing, Taylor and Francis, London, UK, 2001.
14. Shen, W, Lang, S, and Wang, L. iShopFloor: An Internet-Enabled Agent-Based Intelligent Shop, IEEE Transactions on Systems, Man, and Cybernetics, Part C, 2005, 35(3):371-381.

15. Valckenaers, P. "WP1 Dissemination Report: Analysis and evaluation of change and disturbances in industrial plants (MASCADA ESPRIT LTR 22728)," tech. rep., K.U. Leuven PMA, A.I. Systems, Daimler Benz Research, University of Cambridge, VTT Automation, 1998.
16. Valckenaers, P, Heikkilä, T, Baumgaertel, H, McFarlane, D, Courtois, J-P. "Towards a novel manufacturing control principle", Second International Workshop on Intelligent Manufacturing Systems, Katholieke Universiteit Leuven, 1999, pp. 871-875
17. van Brussel, H, Wyns, J, Valckenaers, P, Bongaerts, L, Peeters, P. "Reference Architecture for holonic Manufacturing Systems: PROSA". Computers in Industry, 1998, 37(3):255-274.
18. Wyns, J. "Reference Architecture for Holonic Manufacture - the key to support evolution and Reconfiguration". Unpublished PhD thesis, Katholieke Universiteit Leuven, Leuven, 1999.
19. Yoon, HJ, Shen, W. "Simulation-Based Real-Time Decision Making for Manufacturing Automation Systems: A Review", Int. J. Manufacturing Technology and Management, 2005.

30

A VISION-BASED APPROACH OF BARE-HAND INTERFACE DESIGN IN VIRTUAL ASSEMBLY

Xiaobu Yuan and Jiangnan Lu
University of Windsor, Windsor, Ontario, Canada N9B 3P4
{xyuan|lu12}@uwindsor.ca

This paper presents the continuous work on a previous project on vision-based hand interaction design. The original work developed a vision-based approach of bare-hand-based posture recognition and motion tracking. This paper further investigates its application in virtual assembly. The accuracy and robustness of the developed approach enable product engineers to perform assembly operations while manipulating virtual objects directly in virtual environments with bare hands. The direct human involvement creates a user-defined assembly sequence, which contains the human knowledge of mechanical assembly. By extracting the precedence relationship of machinery parts, the system becomes capable of generating alternative assembly sequences for robot reprogramming. Operation of the presented approach is illustrated and analyzed with experiments.

1. INTRODUCTION

Virtual assembly replaces physical objects with the virtual representation of machinery parts, and provides advanced user interfaces for human operators to design and generate product prototypes, to analyze and optimize manufacturing processes, and to verify and control work-floor implements directly in computer-synthesized environments. The elimination of physical prototyping and on-site verification makes virtual assembly a powerful tool to reduce the life-cycle of manufacturing and to adapt changes or introduce new products.

With the support of new devices, virtual assembly develops new means for human operators to directly manipulate virtual objects with their hands. Hand-based interface in virtual assembly or other manufacturing systems can be considered as an upgrade to the conventional interface design, and has become the key to improve the naturalness of human-computer interaction in virtual environments. Consequently, the tracking and recognition of hand shapes and movements are the main issues in interaction design for virtual assembly (Rivière, 2004).

In the past, special virtual reality devices are employed to recognize hand shapes and to track hand movements (Fels, 1993). Data gloves, for example, are such a type of device that produces sensory inputs to determine the spatial positions of hands and to monitor the temporal parameters of hand shapes. However, data gloves are

Please use the following format when citing this chapter:

Yuan, X., Lu, J., 2006, in IFIP International Federation for Information Processing, Volume 220, Information Technology for Balanced Manufacturing Systems, ed. Shen, W., (Boston: Springer), pp. 281–290.

expensive and fragile. Cables connecting to the computer also impose limits to the movement of human operators (Baudel, 1993). The problems of glove-based devices have motivated active research to search for better solutions, and a vision-based approach of bare-hand-based posture recognition and motion tracking was developed (Yuan and Lu, 2005).

This paper further applies the developed approach in virtual assembly. The structure of this paper is as follows. After providing a brief survey of related work in the next section, this paper first highlights the vision-based approach of posture recognition and motion tracking in Section 3, and then discusses interaction design with bare hands in Section 4. Afterwards, Section 5 explores virtual assembly with bare-hand interaction. The discussions of this paper finish with experiment results and conclusions at the end.

2. RELATED WORK

Virtual assembly involves two processes, i.e., robot programming to teach robotic manipulators carrying out assembly tasks and sequence planning to determine the order of assembly tasks for optimal mechanical assembly. The old-fashion online approach suffers from the down-time of robot operations, the danger imposed upon human operators, and the difficulty of making adjustments for new products. In comparison, virtual assembly works inside the computer. It promotes the development of automated manufacturing tools and allows for the integration of different technologies from a wide range of originally separated areas.

However, the lack of flexible and user-friendly human-computer interaction has been a major bottleneck. Text-based programming, low-end graphics interfaces, and advanced graphics systems have been the common approaches to program robots accomplishing assembly tasks. While robot programming languages suffer from high complexity and inaccurate calibration, graphics interfaces with traditional input devices, such as keyboards and mouse, inherently limit the speed and naturalness. In addition, high-level robot programming without motion planning creates problems in implementation as objects run freely in virtual environments but not in the physical world.

In fact, the paradigm of robot programming has not changed until the introduction of hand functioning in human-computer interaction. Data glove devices have been successfully used to track hand movements and to measure hand configuration parameters (Ma, 1998). One typical example is programming-by-demonstration (PbD), which uses data gloves to convert the movement of human operators into assembly operations (Friedrich, 1996). Unfortunately, virtual reality devices are commercially expensive and cumbersome to use. Moreover, glove-based interaction is powerless for remote control when the user at the far side is not equipped with such devices.

Recent advancement in computer vision stimulated studies of vision-based hand interfaces (Ionescu, 2005). Without the need of mechanical devices, vision-based interfaces provide a more convenient platform that visually interprets hand postures and tracks hand positions. While glove-based interfaces quantify human hands with electronic signals, vision-based interfaces represent the model of bare hands with visual features that reside in image frames (Comport, 2003). Uniquely colored gloves or markers, for example, belong to such vision-based approaches. They present distinguishable color patches in images for hand tracking. Colored gloves and marks are cheaper than data gloves, but are still inconvenient to use. Besides, changes to marker colors or lighting conditions require system modification and re-initialization.

Ideal hand-based interaction neither relies on data-glove devices nor needs any artificial markers. It uses bare hands as human beings do in their daily lives. Different models of bare hands have been developed for the localization and recognition of hand shapes (Rivièrel, 2005). They use a variety of hand features, including skin colors, contours, and fingertip and binary silhouette. Most of them help to segment and identify bare hands for posture classification, but fail to preserve depth information. Nevertheless, depth information is essential for the tracking of hand rotations and movements, and plays an important role in virtual programming as robot operations take place in 3D work cells.

3. BARE-HAND PROCESSING

Hand segmentation and posture recognition are two major parts of bare-hand interaction. While the former deals with the detection of low-level hand features, the latter uses hand features to reconstruct 3D hand trajectories.

3.1 Hand Localization and Segmentation

Color-based approaches of hand processing are robust to dynamic and changing light environments, but work only with monochromic backgrounds. They are also computationally intensive, and therefore unable to implement in real-time. Motion-based mechanisms, in comparison, are fast in computation and work well with complex backgrounds. Nonetheless, motion-based algorithms are highly susceptible to noise and require static background. The drawbacks make techniques in the second category inapplicable in real environments.

A hybrid model is developed by combining skin color detection with motion differencing. This hybrid model maximizes the system's ability to separate skin class from non-skin class while minimizing negative effect of illumination. Experimental results reveal that HSV color space yields 72% correct classification rate, and is therefore the choice for skin color detection. Furthermore, by extending skin color detection from individual pixels to a window of 3x3 pixels, not only the candidate pixel but the neighboring pixels are involved in classification.

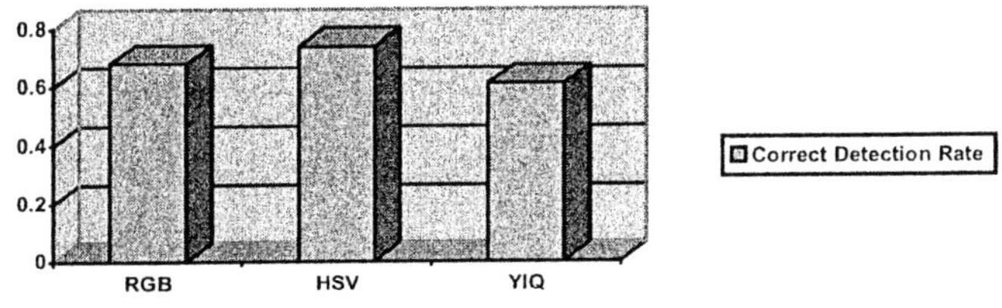

Figure 1 – A comparison of different color spaces

The selection of an HSV color and the use of neighboring pixels help to remove most constraints and to alleviate the computational burden. However, noise still remains due to changes in lighting condition and the presence of other parts of human body. To eliminate the impacts of noise, a skin tone filter is also designed. It is a gradient pattern matching for the hue component of skin color in HSV color space, and the hue pattern of user skin is obtained from the image differencing algorithm given in Equation 1 (Yuan and Lu, 2005). Analysis shows that the hue values vary from 0.8 to 1.0 for the same bare hand, but never reach 0.65 for noisy hands or the background.

$$F_H = F_u / F_d \qquad (1)$$

where,

$$F_u == \sum_{(u,v)\in W} H_{user}(u,v) \bullet H_{noise}(x+u, y+v)$$

$$F_d = \sqrt{\sum_{(u,v)\in W} (H_{user}(u,v) - \overline{H}_{user})^2 \bullet \sum_{(u,v)\in W} (H_{noise}(x+u, y+v) - \overline{H}_{noise})^2}$$

3.2 Fingertip Detection and Tracking

Feature selection is critical in bare-hand-based interaction, and fingertips are one of the frequently used features in hand-based interaction design. Among other fingertip detection techniques, such as template matching and texture mapping, contour-based detection has demonstrated its potentials for its simplicity, flexibility, and efficiency. Contour-based detection relies on the characteristic properties of fingertips in an image. For instance, curvature of a fingertip outline follows a characteristic pattern (low-high-low), which can be cues for feature detection. Other benefits for contour-based detection come from its robustness to noise and tolerance to image scaling.

The developed approach uses K-curvature measurements for fingertip detection. Once the hand is segmented from the original images, the system extracts boundary pixels in the hand contour image. The pixels that reside in the outline of the hand form a list in which each pixel is denoted by $P_i=(x_i, y_i)$. The K-curvature is the angle C_i between two vectors $[P_{i-k}, P_i]$ and $[P_i, P_{i+k}]$, where k is a constant. When curvatures are greater than a fixed threshold with positive or negative magnitudes respectively, the curvatures of pixels along the boundary reach a local maximum/minimum. They are the "local features" in forms of "peaks" or "valleys".

Fingertip tracking needs to find corresponding fingertips in consecutive frames of an image stream, and to record hand movements. It is more challenging than rigid object tracking as it has many degrees of freedom to deal with but few features to use. To eliminate motion blurring caused by fast fingertip movements, the developed approach identifies fingertips by re-running the fingertip-finding algorithm for each frame. Suppose $p_t=(x_1, y_1\ t)^T$ and $q_t=(x_2, y_2, t)^T$ are two coordinates of a fingertip in a pair of stereo-vision frames at moment t. The 3D position of the target fingertip can be calculated with projective geometry. Fingertip trajectory is then reconstructed and stored in a sequence of matrices.

4. BARE-HAND INTERACTION

A bare hand interface allows operators to specify and evaluate assembly operations directly in the virtual environment with their bare hands. In particular, hand postures trigger control commands for object manipulation, and hand gestures map to 3D trajectories of collision-free paths.

4.1 Identification of Hand Gesture and Postures

Each assembly task involves both the global and fine motions of objects. The bare-hand-based interaction uses four controlling commands to quit an action and to point to, hold on, or free objects. Accordingly, a set of four postures defines the vocabulary for the four commands respectively as shown in Figure 2. This interface uses the number of fingertips to define controlling commands. In particular, the "point" posture uses only one finger, and "hold" uses two fingers. When the posture shows five fingers, it means "free"; and a no-finger hand posture means "quit".

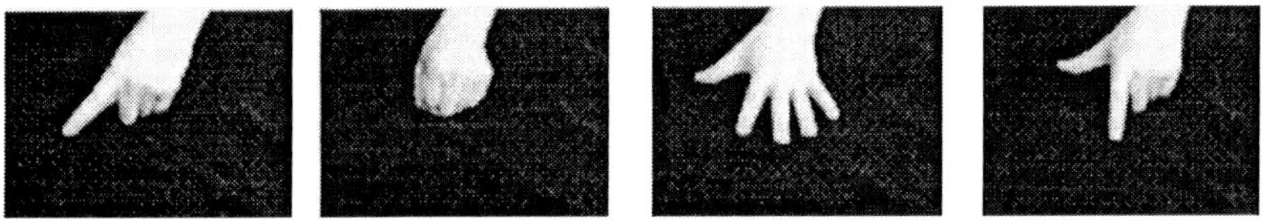

Figure 2 – Hand postures for object manipulation

Assembly operations translate and/or rotate machinery parts in the 3D space. Fingertip tracking in bare-hand-based interaction makes translation relatively easy to achieve, but the measurement of rotations needs three fingertips to determine. Due to the fact that the fingertips of the thumb, the index finger, and the little finger can hardly be collinear, they are the fingertips to define rotations. In addition, the difference in shape between the thumb and the little fingers makes them easy to distinguish. Hand morphology and biomechanics further suggest that the index finger is always in the middle of the two fingers in natural gesticulations. The index fingertip is then identified by examining their relative positions.

4.2 Bare-Hand-Based Object Manipulation

Shown in Figure 3 is a state transition diagram that illustrates hand functioning in the bare-hand-based interface. The posture and gesture start as two concurrent states. They operate in the super state of 'continuous modes', generating control commands and motions. In particular, a closed hand triggers the *hold* command to make a selection or uphold an action; an open hand implies *free*, which releases an item in the possession of the virtual hand or frees the hand from constraints; a point sign invokes items for selection; and an "quit" sign terminates a running process.

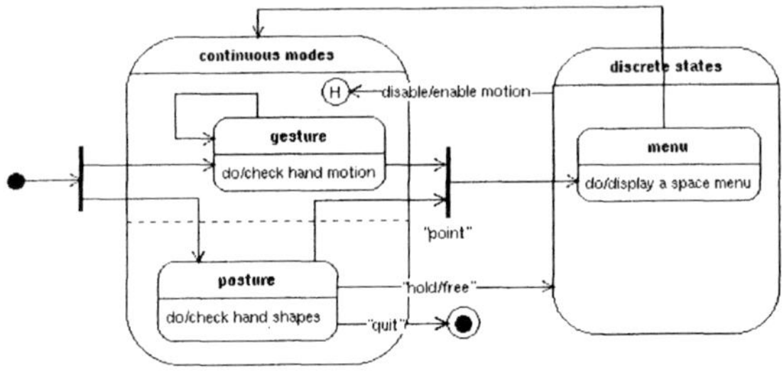

Figure 3 – The state transition diagram of "hand"

Before the hand picks up an object, the *hold* command works primarily with *free* for hand adjustment. It releases the hand from the continuous control mode. The motion of the hand in the physical world no longer contributes to any motion in the virtual environment. In such a way, the hand is able to prepare itself to a desired position or orientation. Object manipulation begins with a *point* command, which prompts up a space menu in the virtual environment and activates menu selection. A casting ray from the index finger then highlights the menu item it hits, which becomes selected when followed immediately by *hold.*

If the selected item is labeled as "reference", the control goes into a remote selection mode, in which a user selects graphics features to define reference coordinate systems. Alternatively, the selection of menu item "motion" starts object manipulation. At this time, a hold posture grabs the object under selection, and makes it to move together with the tip of the index finger until a free posture releases it. In either the "reference"' or "motion" mode, a *quit* command returns the control to the menu state. One of them remains active until the selection of menu item "done".

5. VIRTUAL ASSEMBLY WITH BARE-HAND INTERACTION

In a virtual assembly, the positions of objects are visually presented as part of the virtual environment. The user's manipulation of virtual objects through a human/computer interface transforms into the definition of assembly tasks and a user-defined assembly sequence. An assembly task starts when the hand grabs an object and finishes until the object reaches to its destination. An assembly is a product put together from a set of parts P by a virtual robot manipulator accomplishing a sequence of assembly tasks T in the three-dimensional workspace. In the assembly, $P = P_0+P$, where P_0 denotes a stable workstation and $P = \{P_1, P_2, \ldots, P_n\}$ is the set of machinery parts, and $T = t_1\ t_2 \ldots t_n$ is an ordered assembly tasks.

The goal of virtual assembly is to produce optimized assembly sequences that are ready to direct robot manipulators implementing assembly tasks and putting together machinery parts into products. Given a user-defined assembly sequence, virtual assembly applies geometric and symbolic reasoning to determine the precedence constraints of individual objects, and converts the constraints into an assembly tree for the generation of alternative sequences (Yuan and Gu, 1999).

Optimization comes from the evaluation of alternative sequences against certain selected measures, and calibration takes the characteristics of each individual robot manipulator into consideration.

Precedence constraints are the ordering constraints that guarantee the validity of assembly task sequences. A sequence is feasible if it satisfies common requirements such as geometric, stable, and orientation constraints. Following the order of object manipulation in the user-defined assembly sequence, constraints deduction checks one by one if the assembly of a machinery part must wait for the assembly of any others. This information then leads to the construction of an assembly tree, in which the root stands for the stable workstation P_0 and each path that links from the root to a leaf node represents an assembly sequence.

In addition to the generation of alternative sequences, the virtual assembly system is also responsible to assist human operators with their work of programming assembly tasks and evaluating assembly sequences. The scope of assistance may range from robotics simulation to automatic reprogramming. The former uses computer graphics to present assembly tasks by animating object movements and robot operation with inverse kinematics and dynamics. In comparison, the latter allows the import of human expertise through virtual programming, but still reduces human involvement to the least possible level.

6. EXPERIMENTS AND ANALYSIS

Experiments have been conducted in two groups to verify the operation of virtual assembly with bare-hand interaction, and to evaluate the performance in hand posture recognition and hand movement tracking.

6.1 The Assembly of a Die-Set

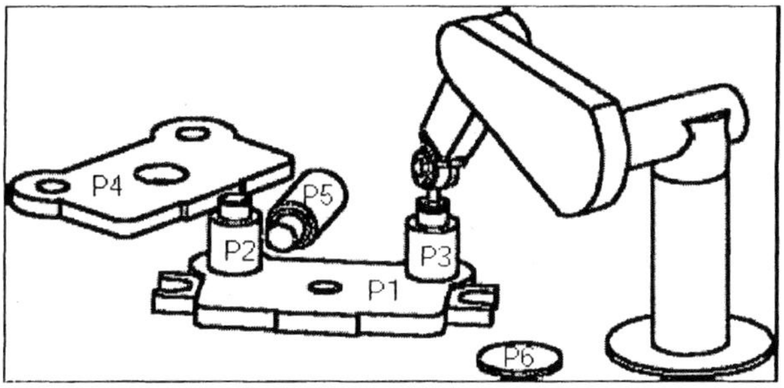

Figure 4 – A Virtual Environment for the assembly of a die-set

Figure 4 illustrates a typical die-set. To put together the die-set with five components and to place the metal plate for stamping, a user-define feasible sequence of assembling the objects can simply be in the order of P_1,P_2, P_3, P_4, P_5, and P_6. Correspondingly, the assembly tasks to place the objects are t_1, t_2, t_3, t_4, t_5, and t_6. Each of the tasks t_i as listed below, contains the information to instruct a robot manipulator reaching to, moving around, and releasing an object P_i.

t_1: place P_1 with its half-open hole facing up;
t_2: stand up P_2 by inserting it into a guide-hole;
t_3: stand up P_3 at the other guide-hole of P_1;
t_4: sit the two corner-holes of P_4 on the guideposts;
t_5: insert P_5 half-way through the bigger hole of P_4;
t_6: slide P_6 on top of the half-open hole of P_1.

Right after the user defined assembly sequence, the virtual assembly system determines the precedence relationship between objects. Sequence generation then begins to construct the assembly tree. Figure 5 shows ten possible sequences to assemble the die-set in the constructed assembly tree, including the one defined by the user. Taking a sequence out of the tree, a replay moves each object in the sequence one after another by applying the corresponding assembly task recorded during programming. For the first sequence in the tree, as an example, operations t_1, t_2, t_3, t_6, t_4, and t_5 are applied to objects P_1, P_2, P_3, P_6, P_4, and P_5. Experiment shows that there is no problem applying the original tasks in this new sequence.

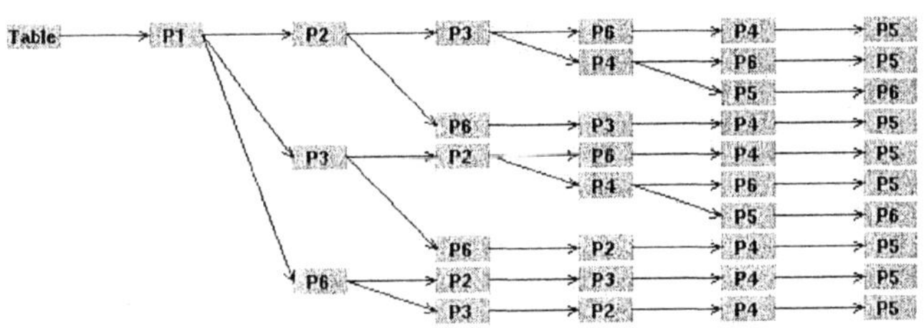

Figure 5 – An automatically constructed assembly tree of the die-set

6.2 Robustness Analysis

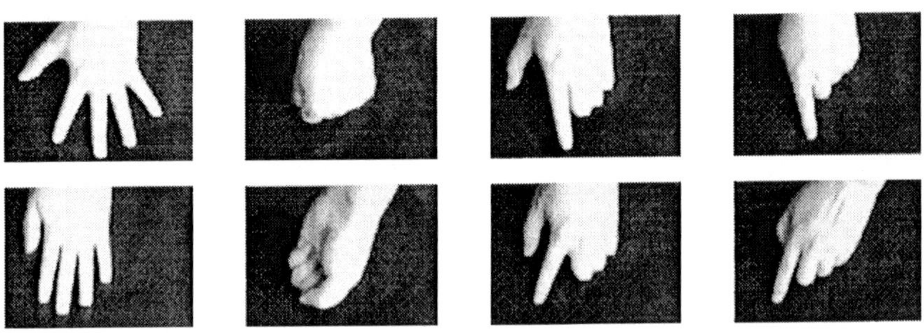

Figure 6 – Hand postures recognition

Experiments in the second group evaluate the performance of hand posture recognition and hand movement tracking with bare-hand interaction. Hand posture recognition plays a major role in bare-hand-based interaction design, and its robustness in virtual assembly was a focus of the evaluation. As shown in Figure 6, the shape of the four defined hand postures was intentionally changed or rotated in experiment, and results showed that all of the distorted postures were recognized correctly.

Bare-hand-based interface is primarily designed to eliminate the dependency of virtual assembly on specialized devices. Due to the fact that no one can repeat the same hand movements in the 3D space, it is impossible to evaluate a bare-hand-based interaction by comparing the tracking results with the measurements from data glove devices. In comparison, marker-based hand tracking is similar to bare-hand-based hand tracking in the way that it tracks hand movements in image streams. The use of extra color marks on hands is expected to produce better results than bare hands.

Accuracy evaluation is therefore conducted by comparing results obtained from the two algorithms on the same hand movements. Three test cases were built in experiment, each of which contains a pair of hand trajectories reconstructed by the two tracking algorithms. In all three test cases, ten sampling points were checked. A direct comparison of the 3D coordinates with the sampling points indicates the accurate rate was at 87.3%, 91.2%, and 89.1% for the three test cases respectively. The result is very encouraging as 90% is considered good accuracy for hand tracking with data glove devices.

7. CONCLUSIONS

Human-computer interaction based on bare hands has undeniable advantages in virtual manufacturing. This paper investigates the application of bare-hand-based interaction in virtual assembly. By visually tracking hand motions and interpreting controlling commands, the presented approach enables human operators to perform assembly operations in an intuitive and natural way. The user-defined assembly sequence embodies human knowledge of mechanical assembly, which in turn helps to generate alternative assembly sequences for robot reprogramming. Experiments demonstrated the robustness and efficiency of the presented approach. Active research is being conducted to further investigate its application in practice.

8. ACKNOWLEDGEMENTS

This work was funded by Natural Sciences and Engineering Research Council of Canada (NSERC).

9. REFERENCES

1. Lakatta EG, Cohen JD, Fleg JL, Frohlich ED, Gradman AH. Hypertension in the elderly: age- and disease-related complications and therapeutic implications. Card Drugs Ther 1993; 7: 643-54.
2. Smith, Adam. "An Inquiry into the Nature and Causes of the Wealth of Nations". In Classics of Economics, Charles W. Needy, ed. Oak Park, IL: Moore Publishing, 1989.
3. Ward, Benjamin. What's Wrong with Economics. New York: Basic Books, 1972.
4. Baudel T, Charade M. Remote control of objects using free-hand gestures. Comm. ACM. 1993; 36: 28-35.

5. Comport A, Marchand É, Chaumette F. "A real-time tracker for markerless augmented reality." In Proc. 2nd IEEE and ACM International Symposium on Mixed and Augmented Reality. 2003 ; 36-45.
6. Rivière J, Guitton, P. Hand postures recognition in large-display VR environments. Lecture Notes in Computer Science. 2004; 2915: 259-268.
7. Rivière J, Guitton, P. Model-based video tracking for gestural interaction. Virtual Reality. 2005; 8: 213-221.
8. Fels S, Hinton G. A neural network interface between a data-glove and a speech synthesizer. IEEE Trans. Neural Networks. 1993; 4: 2-8.
9. Friedrich H, Munch S, Dillmann R, Bocionek S, Sassin M. Robot programming by demonstration (rpd) - supporting the induction by human interaction. Machine Learning. 1996, 23: 163-189.
10. Ionescu B, Coquin D, Lambert P, Buzuloiu V. Dynamic hand gesture recognition using the skeleton of the hand. Journal on Applied Signal Processing. 2005; 13: 2101-2109.
11. Ma L, Lau R, Feng J, Ng C. Shape modifying and controlling with sensor glove in virtual-reality. Progress in Natural Science, 1998; 8: 488-497.
12. Yuan X, Gu Y. "An integration of robot programming and sequence planning". In Proc. the 1999 IEEE International Conference on Robotics and Automation. 1999; 102-107.
13. Yuan X and Lu J. "Virtual programming with bare-hand-based interaction". In Proc. of 2005 IEEE International Conference on Mechatronics and Automation. 2005; 896-900.

31 ON THE ASSEMBLY LINE: A VIEW FROM INDUSTRIAL SOCIOLOGY

Paula Urze[1], Tiago Machado[2]
[1] *FCT/UNL – Faculdade de Ciências e Tecnologia, Universidade Nova de Lisboa*
[2] *UNINOVA – Instituto de Desenvolvimento de Novas Tecnologias*
pcu@fct.unl.pt; tgm@uninova.pt

This paper presents the scientific concerns and some results of the LiMITE – on the Assembly Line: Innovation, Work and Employment – project. Even considering the vast number of experimented industry production technologies, the classic assembly line has proved to be a structuring element within the frame of mass production, whereas known the difficulty to find practical alternative solutions. In this sense, the constraints brought up by the assembly line in order to develop innovative working practices remain relevant. For this reason, it is also important to look for the possibility to develop innovative working practices under the dominant model, i.e. the assembly line inherited from Ford. The paper's empirical section is based on two case studies pointed to two companies from the automotive components industry.

1. INTRODUCTION

Even thought the design of production systems has always been considered a crucial problem to industrial engineers, the global competition and the fast evolution of production technologies have contributed to increase its importance. Rekiek *et al.* (2002) point out that production systems are nowadays characterized by shorter cycle times, higher automation levels and by the emergence of new production equipments.

Naturally, the configuration of assembly lines has been evolving since Henry Ford's era. Besides the traditional line, based on an unidirectional sequential workflow, and designed to assemble just one model, there is nowadays a multiplicity of new and more flexible configurations like mixed-model and multi-model lines, asynchronous lines interleaved by buffers, U-shaped lines, among other variants.

In 1973, the European Economic Community announced that the assembly line would have to be abolished from the European car industry (cf. Emery, 1975: 4). But after more than three decades, and considering the vast number of production technologies available today, the classic assembly line still seems to be an attractive industrial device for mass production. That is, even considering the vast number of experimented industry production technologies, the classic assembly line has proved to be a structuring element within the frame of mass production, whereas known the difficulty to find practical alternative solutions. In this sense, the constraints brought up by the assembly line in order to develop innovative working practices remain relevant.

For this reason, we shall keep on researching for the possibility to develop innovative working practices – surpassing assembly line repetitive work and the degradation of human labor – under the dominant model, *i.e.* the assembly line

Please use the following format when citing this chapter:

Urze, P., Machado, T., 2006, in IFIP International Federation for Information Processing, Volume 220, Information Technology for Balanced Manufacturing Systems, ed. Shen, W., (Boston: Springer), pp. 291–300.

inherited from Ford. Are we going to experiment a return back to the old conflicts related to the repetitive work and to the work degradation or are there alternatives able to guarantee sustainability and attractiveness of the work on line?, asks Jürgens (1997). So we add a couple more questions: can mass production effectively discard the assembly line without lost of competitiveness? To what extent it continues to limit the existing options considering work organisation and its contents? Bringing to nowadays a classical expression (Trist *et al.*, 1963) is there room for a genuine "organisational choice"?

2. ORGANISATIONAL CHOICE (S)

In a simplified assertion, the assembly line can be seen as a set of workplaces distributed sequentially and interconnected through a material transportation system. For each workplace a set of tasks are executed following a pre-defined assembly process which involves: (1) the time needed to execute the task; (2) the precedence relation between tasks establishing the execution sequence; (3) association zone restrictions in order to avoid the assignment of different tasks to the same workplace.

Assembly lines are synchronized in order to regulate the flow of components. Each workplace has a pre-defined time period to accomplish all associated tasks (cycle time). The transportation system is designed in such a way that when the cycle time is finished the subassembly that has been done in a workplace is transferred for the next workplace resulting in a new subassembly. When the time variability for tasks execution is high asynchronous lines can be used. In these lines, each workplace incorporates the assigned tasks at a certain rate. In this case the component is transferred to the next station only when all associated tasks are concluded. The decisions about the tasks to assign to each workplace in order to minimize the costs, satisfying market demand and simultaneously coping with precedence and zone restrictions, are the main challenges when designing an assembly line (assembly line balancing).

With more or less a century of existence, and although the recognition of some problems, the classic assembly line is still largely adopted in the worldwide industry. Actually, Jürgens reminds us (1997: 255-256) that the assembly line is since its inception a controversial socio-technique innovation. Before the World War II, the opinions about the effects of this technological instrument on the work quality and the workers satisfaction were divided. Later, in the 50s and 60s, other voices criticizing the assembly line raised. In 1952, Walker e Guest, in *The Man on the Assembly Line*, wrote that "Of all occupations in modern industry none has attracted such controversial comment as that of the assembly line worker, and especially of the auto assembly worker on the 'final line'." The authors point out the negative effects of the assembly line for the automotive industry workers – mainly, the repetitive nature of the tasks and the "tight" dependency on the assembly line pace.

In 1964, Blauner, in *Alienation and Freedom*, emphasized the importance of the technological structure, namely of the assembly line, on the nature and organisation of work, in the following terms: "The assembly line's inexorable control over the pace and rhythm of work is most critical; it is largely responsible for the high degree of pressure, the inability to control the quantity of work, and the lack of free movement." Two years after, Goldthorpe concluded that "Assembly line production in the automobile industry is now generally regarded as the locus classicus of worker alienation." (Cf. Jürgens, 1997). Later, in the 70s, Coriat, in *L'Atelier et le Chronomètre*, argued in the same direction : "A l'origine de la chaine, violence calculée, systématiquement appliquée contre le travail des hommes, ce rêve originel du capital à la recherche du 'mouvement perpétuel' de la fabrique. La production à flux continu, 'clé de voûte' de tous les systèmes d'organisation de travail (...)." (Coriat, 1994: 67).

We emphasise the Coriat approach due to the fact that, as Berggren (1992) underlines, ever since the edition of *Organizational Choice* in the 60s (Trist *et al.*, 1963), a group of social scientists relied on the thesis that work organisation could be modeled almost independently from technological constraints – there would always be room for an "organizational choice". It is remarkable, adds the Swedish author that in the Trist *et al.* (1963) there's no support for this thesis, a perspective that can only be supported by a short term diagnostics but harder to sustain within the framework of longer studies, covering periods of economic expansion and contraction, plain employment and unemployment phases. Berggren's case studies clearly reveal the mutual interaction and interdependency between technical production design and work organisation.

Shimokawa *et al.* (1997) studied the alternatives to the traditional mass production. The authors begin to explain that the options related to the assembly systems adopted by the enterprises are influenced by their strategies and options and also by factors linked to the context and environment. Considering that the enterprises or regions deal with different objectives and constraints, assembly systems change. Within this framework, different assembly systems have been identified in the automobile industry: neo-fordist, Uddevalla, neo-craft and lean system. Revising the different models, the authors concluded that none of the existent alternative models has shown sustainable advantages comparing to traditional mass production system.

In Bergreen (1992), the alternatives to the traditional production system have been analysed. But in a more focused way, considering the configuration of the assembly systems as central dimension of analysis side by side with the work organisation dimension. It is important to underline the methodology developed by this work, due to the fact that, by articulating the technical and organizational dimensions, it emphasized the conditioning role of the technical subsystem in relation to the social system. This is, of course, the privileged focus of our research project.

At the end of this decade, Engström, Jonsson e Medbo (1998) tried to demonstrate the superiority (efficiency) of the assembly systems supported by an "organic" flow over the conventional assembly systems. Taking as a starting point the typology developed by Wild (1975), three type of losses were identified: (1) it is not possible to divide uniformly the work among the different workstations – losses associated to the balancing; (2) a large amount of work in the workstation is required for materials handling; (3) the assembler can not change his natural work rhythm – loss associated to the system. Based on empirical information provided by the Swedish automobile industry, the authors conclude that the assembly system supported by an "organic" flow was more efficient than the conventional one. In fact, the sociological research related to alternatives to the assembly line leads mainly to the solutions experimented by Volvo. First in Kalmar's plant, and latter in the Udevalla plant. The interest in Volvo-ism has been intensified by the fact that the Uddevalla's unit, more innovative than its precedent, was closed in 1993 to reopen two years later under the name of Autonova (a result from the collaboration between Tom Wilkinshaw Racing and Volvo).

Udevalla is, among all the known cases (Toyota Kyûshû, Honda Flexifactiories, Volvo Kalmar, Uddevalla e Autonova, etc.), the most emblematic one. Because it is an example of the "reflexive production" paradigm – 42 parallel product workshop assembly (instead of a sequential assembly line), 480 minutes cycle time, working teams assembling the whole car and being responsible for planning, quality control, development of methods, etc. Uddevalla had exceptional conditions to promote socio-organisational innovation - "This plant was in all its aspects a humanistic production system." (Granath, 1998).

As already pointed out, the shutdown of Uddevalla unit was followed by the opening of Autonova some years later. But, in June 2001, the TT News Agency reported that Volvo Cars, subsidiary of Ford Motor Company, announced the decision of reintroducing the conventional assembly line in this plant. This decision from Volvo seemed to confirm the end of the "Volvo-ist" model. The executive management board and representatives from the Metallurgical Union agreed that this was the only way to ensure the future of the plant. On this matter, Volvo's Metallurgical Union President, Mattias Jonsson, declared that the abandon of "reflexive production" was a regrettable decision. Nevertheless, with the maintenance of jobs in stake, he also stated that it was not possible to argue in favor of such a production system. Basically, Jonsson's words synthesized the idea that the conventional assembly line was a matter of survival for the plant.

3. TWO CASES IN THE AUTOMOBILE INDUSTRY

This paper's empirical section is based on two case studies pointed to two companies from the automotive components industry. These companies – Huf and Iber-Oleff – have 391 and 320 (Dec. 2004) employees, respectively. The first company, located in the north of Portugal, belongs to an European group. The second one, located in the centre of Portugal, has resulted from a joint-venture between a Portuguese and a German group. Both companies combine different types of assembly lines (regarding shape, automation levels, etc.) and therefore they seem to be a privileged fieldwork considering the LiMITE project' focus. The interviews address different hierarchical levels, always including operators that work within different assembly lines systems.

Huf Portugal manufactures electronic and mechanic lock systems for the automobile industry. Amongst their products, for example, are the locks for the Opel Corsa and Astra, Ford Focus and Peugeot 307. Furthermore, it also manufactures blocking systems for the steering and external handles for doors. The company started operating on 1991. Iber-Oleff started production in 1992 in the area of plastic technical components, namely the air diffusion system for the Multi Purpose Vehicle produced by AutoEuropa. This company produces plastic functional components, ashtrays, air diffusers, speaker grills, ventilation grills, command systems for ventilation and air conditioning, parts and covers for car radios.

3.1. From the layout ...

At Huf Portugal we can distinguish, even though if only roughly, three periods corresponding to three assembly line concepts. Naturally, within each of these periods co-exist assembly lines that are the answer to different concepts. In any case, at present, there are assembly lines in the factory layout which have very different formats and levels of technological sophistication.

In any case, during the launch period (phase 1), between 1991 and 1992, the lines were installed following mainly a traditional model – straight-line format, sequential logic, little technological sophistication and manual work. The assembly lines of the following period (phase 2), already with a vertical development, present a U shape and started to include elevators (buffers) near each workplace – "They introduced the elevators and the boxes to accumulate material between workplaces and the material stopped accumulating, and thus breaking, between workplaces", explains one of the operators in the PSA (Peugeot-Citroen) line. In 2002 (phase 3), a new concept started to be planned. But even nowadays, the phase 2 system is kept for some products such as some of GM's (General Motors) products (steering blocking). There are also hybrid lines, such as the Ford line, where they produce sequentially and without large intermediary warehouses (the warehouses are the spaces between the equipment). The most advanced system is that of the new PSA line, that already incorporates more functions and also has the peculiarity that it combines the two products typical of Huf Portugal in the same line – on the one hand the locks, on the other, the blocking systems.

The new PSA line, phase 3, emerged following a partnership between Huf Portugal and Huf Germany. The former does not have a commercial department, meaning it is the Germans who agree the contracts with the clients. A new market opportunity arose and the possibility to increase the PSA market quota with this contract, from 20% to 60%, hence the investment to create this new infrastructure. As the production manager noted, "It is important to be competitive amongst the different Huf companies. We design our layouts to absorb a larger chunk of the market, that is so that Huf Germany gives us more market. We never say no." In a hypothetical situation, he adds, "Let us suppose that a line was designed to produce 1400 sets. But if the client asks us for 3200 we have to set up another line, another structure. We have to invest."

These new lines, in installation and testing phase, are characterised by their serpentine format, high level of automation and the fact that in the same infrastructure, not one, but several different products are manufactured. In this model, the intermediary levels (buffers) are again suppressed. Which does not mean that between two equipments, due to some problem that makes a workplace slower than the others, there may not in effect exist a small number of stocks – it generally does not happen, but, when it does, that situation is due, mostly, to some problem in the equipment or to a problem in the quality of some component. In any case, the process is now supported by pallets (and no longer in boxes). In the lines introduced in phase 2, as we mentioned, the sets move inside boxes that, in turn, are accumulated in elevators. In this new model, "There are no boxes, there are pallets, and it all follows in sequence.", summarises the line head of the new PSA. It is a very recent system, which means that, in the words of the same speaker, "The machines are not even in the right places yet. It is the experimental phase. We started last week. But we are improving. Today, I am producing more and tomorrow it will be better."

One of the advantages of these new lines is their flexibility, allowing the assembly of different types of products at the same time. In the previous PSA line, in a U shape, it was possible to assemble different products, but it was a different situation. According to the production manager, "Product A went in and product A came out. In this line [new PSA] it is possible to introduce product A and, immediately afterwards products B and C. Then you branch out depending on the needs, such and such an operation. We have a common trunk [shared components] for the composition of the locks. The ignition is common to all the projects in the PSA line. Then the keys change, with the logotypes for Fiat, Citroen, Peugeot. (...) Previously, when the line was U shaped, we changed the machine. Even though they were machines that engaged quickly, there was always a certain loss of time."

In fact, perfectly sequential systems are more demanding from the point of view of equipment maintenance and availability of the equipment itself. By resorting to the concept of intermediary buffers, present in the lines introduced in phase 2, small stops and breakdowns can be solved in 5 or 10 minutes with no effect on the line output. These buffers are present within the framework of the philosophy of "n-1" people with relation to the number of workplaces. Therefore, there is always a workplace that does not have the human resources to man it, which means that people must be rotated. In the systems introduced in phase 3, any breakdown causes a stop in production, which also means that the support processes, such as maintenance, have to be rethought, so that everything works efficiently and within the predicted rhythm. With reference to that, as opposed to the U-shaped lines that incorporate a support for components lasting about two hours (meaning that if the equipment stops, production carries on), in totally sequential lines, tolerance is limited to a few boxes, so that, if there is a breakdown, it is more probable that the production flow has to be interrupted.

At Iber-Oleff the VW240 (Volkswagen) assembly line (ventilation grills, command systems for ventilation and air conditioning) was installed about 5 years ago. The layout was conceived by the German company, an expert in line manufacture and assembly. There were several studies around the assembly line, until they reached the configuration that is essentially maintained to date. "Because I don't know if you noticed, we have a conveyor from behind that picks up the parts for painting down here and surely was taken into account for the layout." The industrial engineer also adds "We agreed with and criticised their work in a positive way. We will question as to why they are here and not there. Because we think that if it were somewhere else we would have these and these advantages. We face each other this way." The VW 240 assembly line has separate lines where the parts circulate, in different stages of product insertion. As a consequence, the same part is placed on different lines depending on the number of products that have been incorporated into it. The conveyor belt rotates continuously and depending on the task, the operator gets the product from the corresponding line.

From the time the line was installed, some suggestions have been made for alterations, but which did not compromise the structure of the initial project. "This type of alterations takes place a lot here. We constantly work on it. We change the layout very often.", mentions the production manager. In the same line, says the area head that "[the layouts are altered] within a matter of months." The proposals usually arise when the line is already working, as it is then that some alterations seem advantageous so that it is optimised. Concluding that in the case of the automobile industry the lines are usually connected to a product

There are adjustments that have been made a little at a time and that, from the point of view of the engineer that at the time oversaw the installation, should be reported to the Germans so that the information can be used, and when new lines are created they can incorporate that knowledge. It is a capital that is worth accumulating, re-using it in the assembly lines. Speaking of this, this engineer explains that "We can make those alterations, sometimes we report them, other times not, depending on how important we decide they are. If it is very important, and to collaborate with our experience, we report it. We evaluate everything, from ergonomics, the chair itself, and the height of the table. All aspects, may be one or so may escape us at first sight, but we always pay attention so that we don't forget anything."

The proposals that come from the Portuguese factory, in general, do not face great obstacles from the company responsible for the layouts. On this subject, the industrial engineer recounts another experience: "Just today I had a meeting to define a new assembly line, where it had been predicted that a certain piece of equipment should be in a certain place and here, we thought not, and suggested a change and they are going to accept the change because it is the most appropriate for that specific piece of equipment."

The discourses agree with regards to the advantages of the layout of the VW240 line: "We have advantages in terms of flexibility and balancing. The advantage is the flexibility and a better balancing, i.e. I can see if the work phases (process) are balanced or not, and they give the operator flexibility in his/her work. In terms of productivity, we do not impose but let the operator do the managing. The result is obvious, no need to call attention to it", says the production manager.

When we asked to the industrial engineer about the disadvantages of this line, the answer is conclusive: "There are no disadvantages." As for the extension of this configuration in the industry, the importance of the product is stressed. There are products for which "We cannot even think of that layout model. The basis for this part is this back part. On top of this back part which we call the housing, the lamellae are mounted, as well as a valve, a sponge, an axis, we assemble a link and a wheel (...). We assemble all this. It makes sense, I have the part on the first line and as I add operations I change the part to other lines until we reach the end where the last operator fits in the last part and then places it in the box." This type of configuration is shown as advantageous when there is a basic product and operations are added to the same part.

3.2. ... to the work organisation

It is the time to go into the question that is at the origin of the project LiMITE. In other words, to evaluate the conditioning role technology plays on the organisational work model. On that subject, the engineering manager tells us the following: "There is always the possibility of organizing the work differently. Usually, the person who designs the assembly line thinks one way, and this is the way we use when work starts. Everybody assumes that it is the best way, but that is not always true. With the experience that the line heads have, their technical knowledge, which is far superior to ours, and that of other people, there are always a few adjustments to be made. The alterations are discussed, and are discussed amongst everybody." This optimisation of the layout, if it takes place, is the responsibility of the engineering and the production managers. Usually, it is always the person responsible for production who participates in the project development phase. When launching the project, it is also the production (in this case, the manager and line heads) that suggests alterations to be discussed with the engineering manager. Picking up the initial question again, how does the technical aspect rule the other aspects? It varies. But, for example, it has been concluded that in the new PSA line, the way the equipment was conceived does not make it possible to advance to another workplace, to have access to the components.

In any case, and since the technical and organisational subsystems do not coexist independently from each other, together with the advances in the assembly infrastructure, the management philosophy has also evolved somewhat. Quoting the human resources manager, "The first one was of the type *Chaplin's Modern Times*, where each operator worked at a single post [one person for a workplace], there was no movement, no rotation. Therefore, a traditional assembly line. Then, we introduced working groups, continuously improving, and started to implement a more modern and more flexible assembly line philosophy." Following this reasoning, she explains, "The rhythm of the assembly line is set by the first workplace. What cannot be recovered in one workplace can be recovered in another. Automatically, our flexibility started in the middle stations of the assembly lines." However, in the opinion of the engineering manager, the introduction (on its own) of the U line does not alter the working model significantly. "Working in a U or in a line, for me, is the same concept. The only reason for working in U is that there isn't enough space to place twelve pieces of equipment. The line was just bended to gain space."

In the new assembly lines, still in the development phase, but where in reality work is already being carried out, the alterations introduced in terms of work organisation are not substantial. In the words of an operator in the new PSA line, it turns out to be "Better, because we work facing the machine, we do not need to move so much. I think it is more cost-effective, small parts, we work with pallets, each one takes a set, all one after the other, always moving, with no elevators. [Do you then prefer to work without elevators?] That way the machine is in front of us, all we have to do is take it from the machine and place it on the pallet, whereas on the other line [with elevators], we used much bigger boxes and had to work with them on the machine – remove, turn, place – which meant more work." According to the engineering manager, in the transition from phase 2 to phase 3, "What happens is that the workplace as we understood it in concept 2 is more restricted in concept 3." That is, the tasks are more dispersed, which indicates some automation of the work.

But, still according to the engineering manager, "The organisation itself, the day to day of a person who works with this system, will not vary a lot. There is some specificity of one or other pieces of equipment, but not of the way people work." This idea is confirmed by an operator in the new PSA line: "The work is the same. It's just that in the old line everything was very manual." Very similar, the testimony of a colleague: "It is different because there are several robots (with regards to that we have evolved a bit) which now mount several things, may be it is not so hard, but, essentially, it is the same thing." That is, if it is true that the investment on electronic equipment in phase 3 brought in more flexibility, better technical knowledge and better sensibility, in what is essential, the work model ends up by not being that different from the previous situation. In terms of tiredness, the new technical infrastructure will not bring many changes, according to the declaration of the engineering manager himself: "It does not lighten the work load. We still take into account the same technical factors of tiredness in the time measuring studies."

At Iber-Oleff, the configuration of the line conditions a great deal the organisation choices. The structural options take place at the time that the layout is defined, there are options to choose from. Even though the content of the workplace remains, in general, very poor, there is a group logic that this type of configuration facilitates. Sharing the conveyors from which the operators remove the circulating parts, even though in a sequential logic, creates a team dynamic that usually escapes the work in a simple sequential line. "When they are in a sequential line, people are near each other and anything could happen. Whereas, there, we are sharing. There is an opening to the work team." The rhythm imposed by the conveyor belts and the limited content of the operations brings us near the traditional concept that we usually associate to the work in the assembly line. "The line itself helps people to gain the appropriate rhythm, that is, it almost creates a common rhythm.", says the production manager. However, there is a different dynamic imported from the variation of the sequential configuration of the line, which seems to restrict the limitations due to different rhythms and diminish the isolation of the workplaces and the dependence on each other.

It is worth stressing the tasks that are difficult to substitute by technical equipment, at least in an affordable way. Automation is emphasised, there is no doubt that the company is moving in that direction. Maintenance, or not, of the workplaces depends greatly on the technical advantages of the equipment and the return on the necessary investment for its acquisition. There are advantages in human work that seem to favour certain workplaces, just as there are technological investment costs that in some cases do not seem to compensate.

Automation usually means fewer jobs, even though those tasks are very repetitive. The area head tells us that "They have fewer workplaces. But, in any case, they are usually repetitive tasks. And sooner or later they would bring trouble. If we did not rotate people, surely tendonitis and such would emerge..."

The VW240 line has not suffered significant alterations in terms of the technologies used. The adjustments made to the line are connected, mainly, to the displacement of some workplaces, or the introduction of others. It is worth mentioning that production quantities take more people to that line. That is, when production increases, some workplaces are activated, similar to others that already exist in the line, in order to increase the resources available and that way to guarantee a response to production demands. But this organization counts on the *subassys* (lamellae assembly) operators who in these cases are allocated to the VW240 line. In other cases, they move operators from other areas to work in this line. There is a combination of efforts to fulfil the objectives established.

4. FINAL REMARKS

In relation to the question that encompasses the whole of the LiMITE project – within the framework of this technology, is there margin for new organisational choices? – and to which we have been answering throughout the case studies, the

Huf's human resources manager answered in these terms: "Always, processes, ways of doing things, can be improved."

On the one hand, the different assembly lines are the basis of the technical infrastructure of these factories, which means that, objectively, they define restrictions in terms of execution times and gestures for the operators, which is, to start with, a limitation. However, the two companies have been able to find relatively innovative ways to qualitatively optimise their work practices - and, amongst the different strategies they have initiated, it is worth mentioning the task rotation system (well organised) and management (in the hands of the operators themselves). Underneath this structure it was possible, in spite of everything, to find some elements that seem to contradict some painful characteristics usually associated to this type of work. It is precisely this cross-over between technological restriction and qualitative innovation that make Huf Portugal and Iber-Oleff interesting case studies within the conceptual framework of the LiMITE project.

5. ACKNOWLEDGMENTS

This work has been supported by the LiMITE project. A project funded by the Operational Program for Employment, Training and Social Development, sponsored by the Portuguese Ministry for Labour and Social Solidarity.

6. REFERENCES

1. Bergreen, C., *Alternatives to Lean Production: Work Organization in the Swedish Auto Industry*, New York: ILR Press, Capítulo 5 – Organizational and Technical Design of Swedish Automotive Industry, 1992, pp. 90-100.
2. Blauner, R. *Alienation & Freedom – the Factory Worker and his Industry*, Chicago: The University of Chicago Press, (1964).
3. Coriat, B. *L'Atelier et le Chronomètre: Essai sur le Taylorisme, le Fordisme et la Production de Masse*, Paris: Christian Bourgois Éditeur, Cap. 14 – La Chaine, 1994, pp. 67-85.
4. Durand, J-P.; Stewart, P.; Castillo, J-J, *Teamwork in the Automobile Industry* – Radical Change or Passing Fashion?, Macmillan, GERPISA,1999.
5. Emery, F., *The Assembly Line: Its Logic and Our Future*, Occasional Papers in Continuing Education, No. 7, 1975.
6. Engström, T.; Jonsson, D.; Medbo, L., "The Uddevalla plant and interpretations of industrial design process", *Integrated Manufacturing Systems*, Vol. 9, 5, MCB University Press, 1998, pp. 279-295.
7. IMTI, *Manufacturing processes & Equipment Roadmap*, Section 5 – Assembly, 24 July 2000.
8. Jürgens, U., Rolling Back Cycle Times: the Renaissance of the Classic Assembly Line in Final Assembly, *in*: Shimokawa, K.; Jürgens, U.; Fujimoto, T. (eds), *Transforming the Automobile Assembly: Experience in Automation and Work Organization*, Berlin: Springer, 1997, pp. 255-273.
9. Kahaner D K, *Assembly Line Automation Activities in Japan*, Asian Technology Information Program (ATIP), Japan, 1993.
10. Lopes, J. M., *O Problema do Layout: uma Visão Global*, dissertação de doutoramento, Braga: Universidade do Minho, 1990.
11. Rekiek, *et al.*, "State of the art of optmization for assembly line design", *Annual Reviews in Control*, 26 2002, pp. 163-174.
12. Shimokawa, K.; Jürgens, U.; Fujimoto, T., *Transforming the Automobile Assembly: Experience in Automation and Work Organization*, Berlin: Springer, Introdução, 1997, pp. 1-16.
13. The Sky Group, *The Product Layout – a facility layout concept*, Production/Operations Management, *on-line*.
14. Trist *et al.*, *Organizational Choice*, London: Tavistock Institute, 1963.
15. Walker, C. R.; Guest, R. H., *The Man on the Assembly Line*, Massachussets: Harvard University Press, 1952.
16. Wild, R., "On the selection of mass production systems", *International Journal of Production Research*, Vol. 13, No. 5, 1975, pp. 463-61.

32 AN OBJECT BASED VIRTUAL REALITY SIMULATION TOOL FOR DESIGN VALIDATION OF A NEW PARADIGM CLOTH MANUFACTURING FACILITY

Fabio Bonsignorio
PMARlab DIMEC University of Genova
bonsignorio@dimec.unige.it

Rezia Molfino
PMARlab DIMEC University of Genova
molfino@dimec.unige.it

Within Leapfrog project, the general task is to design an effective and efficient manufacturing solution coping with the needs of the mass customization competitive scenario, overcoming the limitations which have so far prevented the automation of the garment assembly, through a comprehensive and general ('holistic') redesign of the cloth manufacturing process properly exploiting flexible automation and robotic technologies. Virtual Reality simulation is a core part of the interactive validation of design choices. The garment assembly line simulator has various objectives: (a) Validation of the garment assembly line 'holistic concept'; (b) Layout definition against realistic production data; (c) Prediction of realistic production metrics after line deployment. The simulator has been developed paying a special attention to easiness of maintenance: adding new resources or reports or changing the statistic of the production representative measures must require a limited effort; VR representation of what-if scenarios make easier comprehensive evaluation of alternative layout solutions.

1. INTRODUCTION

The Leapfrog (Leadership for European Apparel Production From Research along Original Guidelines) project is a major European project aiming at a complete textile/clothes supply chain and manufacturing process redesign in order to reestablish European leadership in the textile industry.

Although the textile industry represents perhaps the oldest example of modern industrialization (the first cotton mill at Cromford, Derbyshire, UK, is usually considered the first example of a modern factory) the level of automation reached in this field of manufacturing engineering is far to be completed.

The process leading from raw materials to fabric rolls is already highly automated, on the contrary the final assembly of garments after the cutting of the fabric to the final sewing and finishing of the garment are currently performed by human workers: this is one of the main causes of the ongoing outsourcing process in the textile industry.

Please use the following format when citing this chapter:

Bonsignorio, F., Molfino, R., 2006, in IFIP International Federation for Information Processing, Volume 220, Information Technology for Balanced Manufacturing Systems, ed. Shen, W., (Boston: Springer), pp. 301–308.

A holistic radically new textile/clothes production paradigm is needed.

The system to be designed is a complex non linear manufacturing system: the relation between the design parameters and the constraint cannot be expressed in a closed form.

This is a typical case where the simulation of the proposed alternative design solutions is the main quantitative support for the design of the production facility [1,3,6,7,8].

As what it is needed is a disruptively new layout of a complex mechatronic manufacturing system.

The new system is based on strong innovation of state-of-the-art solutions and new solutions:

1. Fast and automated single-ply cutting through multiple cutting heads: multiple plies cutting needs human, manually performed, separation of cut parts.
   Instead of a single cutting head and multiple fabric plies we have single ply and multiple cutting heads, but we eliminate the need for human intervention
2. Fusing of all cut parts
3. Automatic robotic pick up of cut parts on the single ply cut table (see point 1) and transfer to an automatic hanging transport system
4. Automated 3D manipulation of the parts to prepare for the 3D sewing
5. Sewing process with:
   - new 3-D-robot-sewing-technology and
   - traditional sewing technique (different automation levels)

This is a key issue as sewing of the garment is currently done operating manually 2D sewing machines quite similar to the original century old Singer patent.

This system is thought as a first step toward achieving a high degree of mass customization. This relies on deploying a system that is flexible enough to lead to varied batches of mass production.

This is possible through the introduction of robotic technologies in several steps of the, reshaped, production process: these technologies allow the automation of operations that were not so far possible to automate, and, on the other hand, being the robots programmable machines, it is possible to have the benefits of mass production in a very versatile easily reprogrammable environment, enabling the so called 'mass customization'.

The system is devised to overcome most of the drawbacks of the current state of the art production methods by the creation of more advanced and rather breakthrough components or by devising methodologies that totally bypass those restrictions.

A simulation tool, named GALsim (Garment Assembly Line simulator), has been developed in order to support the design and assess the technologies to be used [2,4,10,11].

The development of a new dedicated simulation tool was needed since the resource involved and the problems addressed are quite new, as GALsim has the objective of modeling a new paradigm cloth manufacturing facility.
It is an innovative simulation tool for an innovative production system design.

This simulation tool, based on Delmia Quest software, is a modular and scalable event-driven object based simulator and allows realistic VR representation of what-if scenarios.

1.1 Intended use of the simulator

The simulation technology plays a key role in the solution design and in the technology assessment allowing the quantitative, and also 'visually animated' in VR, verification of the design choices and allowing to test in advance if the different proposed technologies can fit in the overall framework.

As the overall garment assembly process is being radically redesigned , there are many design variables, with a wide spectrum of new resource kinds, based on new robotics and mechatronic technologies and on new physical-chemical processes.

The new garment assembly line design is greatly helped, and perhaps made possible by the possibility of on the fly evaluation of alternative scenarios and solution frameworks based on different levels and kinds of process, and product, innovation.

A costly trial and error process can be shortened working on virtual models of the different mechatronic system alternative layouts.

In summary the key objectives pursued by means of the simulator tool are:

- Validation of the garment assembly line 'holistic concept': first of all the viability of the proposed technologies and solutions must be checked roughly on the basis of their guessed characteristics (timings, capacities etc...)
- Layout definition against realistic production data: after that the core choices have been done, the design specification are checked quantitatively against realistic production scenarios drawn from current state of the art manufacturing practices
- Progressive development of the model adopting what-if approach: the possible design choices are evaluated in advance quantitatively against the desired performance indexes
- Prediction of realistic production metrics after line deployment: assumed and defined the design parameters the future production scenarios can be checked
- Validation of 'operation procedures' (how production is managed): in the varied operation condition during the life of the manufacturing facility many scenario can occur: from a scheduled maintenance to a subsystem failure. 'Operation procedures' have to be defined in order to properly manage any foreseeable occurrence. A realistic simulation model allows to test and compare in advance the operation practices.
- Ergonomics and labor effort analyses: the tools allow to verify some labor efforts and to know in advance excessive requests on the workers.
- Visualization of the model as a virtual reality environment: it also provides a quick 'at glance' preview of how an envisioned solution could work

It is also devised that the GALsim simulator will be used as a decision support tool in order to improve the utilization of the line [5,7,10,11].

This also needs paying a special attention to the easiness of maintenance in order to be able to evaluate in a reasonable time the different production scheduling alternatives.

2. GARMENT ASSEMBLY LINE

As told above the garment assembly line is conceived as a complex mechatronic system conceived in order to cope with the flexibility needs of the emerging mass customization scenario. In the figure below (fig. 1) we can show the layout of the garment assembly line as it is represented by the GALsim system.

The 'holistic' MES (Manufacturing Execution System) is conceived as a multiagent intelligent modular system.

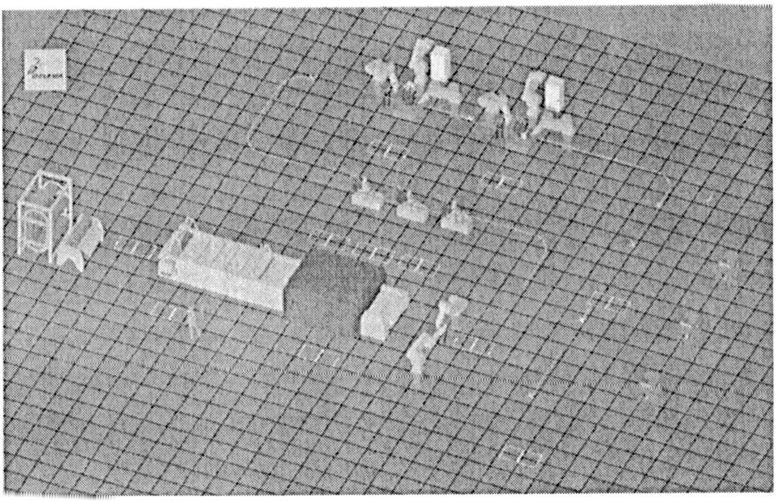

Figure 1 - Garment Assembly Line Layout

The entities constituting the line are:

- Cutting table: a digital single-ply multiple head cutting table which cut the fabric rolls in order to obtain the 'cut parts' , the pieces of fabrics which are sewed together
- Fusing
- Unloading of Cutting Table: it is performed by robots through particular advanced grippers
- 2D Automatic Sewing Machines they are dedicated to the sewing of some specific subassemblies like sleeves
- Garment Assembly Cell (GAC): this cell is constituted by a reconfigurable mould, by dedicated positioning equipment
- 2D Manual Sewing: these are quite classical sewing machines like those currently used for most sewing operation of the cloth manufacturing cycle.

In Figure 2 is summarized the multiagent structure of the MES system.

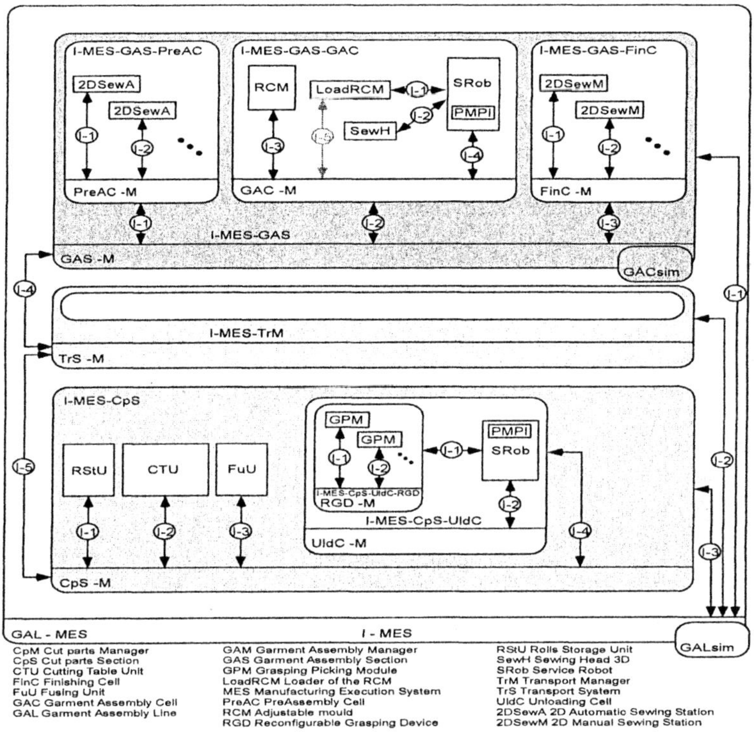

Figure 2 - MES Multiagent structure

3. SIMULATION MODEL

The modeling of the system requires that we properly define the part model, the resource model and the process model.

The part model is the modeling of raw materials, semi-finished, finished goods that flow through the plant.

The resource model is the modeling of machines, transport systems, buffers.

The process model is the modeling of the manufacturing operations defining how operations are performed in the machines, how transportations are managed in the transport systems, how buffers are managed, in general all the rules governing the evolution of the garment assembly line.

3.1 Part Model

The part model requires that are defined the items ('parts') flowing through the line, as number and kind.

The reference finished good item considered is the formal men jacket.

The main kinds of items we have considered are: Fabric rolls, Cut parts, Formal jacket sub assemblies, and Formal jackets.

3.2 Resource Model

The resource model requires that we define the resources by which the items are transformed and moved from a place to another in the plant.

The kinds of resources include various types of machines (cutting table, loading/unloading robots, sewing robots, reconfigurable moulds, 2D sewing automated machines), human operators, transportation systems.

For any resource a number of parameters must be defined, regarding: set up/operation times; logics, capacity, etc.; input type (push/pull); output type (push/pull); cycle process (times of operation/operation-steps, ...); setup process: set up (lot/n of pieces), timings; unloading process; part routing (depending on layout topology); MTTR, MTBF; Shifts

The statuses that can be assumed by the resources are: Idle, Busy, Blocked, Failed, Off shifts.

3.3 Process Model

The modeling of the manufacturing facility requires that the process model is defined, i.e. the definition of logic rules stating the reciprocal interaction of elements in the model.

The flow of production is simulated with an event driven approach, defined by means of resource attributes.

It is necessary to define:

- Dispatching rules: e.g., first available machine, minimum set up time, priority
- Connections between the resources
- Priorities

The dispatching rules and connections are between involved resources.

3.4 Simulation Outputs

As we have seen in section 1.1, GALsim, as a core design tool for a complex system that cannot be represented in closed form, allows to compare the effects of different design choices and to assess the alternative technologies (for example for the reconfigurable mould).

The reporting capabilities allow to identify: bottlenecks, responsiveness to failures, throughput, resource saturation, labor effort, queue lengths/times, work in process.

It is possible to show VR animation of garment assembly process evolution during the simulated time, updating the reports and diagrams online.

In the figures below some examples of output: two VR animation snapshot pictures, in fig 3 (left), the line before the start of the shift of the workers at the manual sewing cells, in fig 3 (right) the line during a cut table failure; in fig. 4 (up), an example of utilization pie chart of the cut table around a failure, in fig. 4 (down) a snapshot of a body jacket subassemblies present in the Garment Assembly Cell at a given time.

It must be noticed that the snapshots in fig. 3 are VR animation snapshots, the view point can be moved quite freely allowing to see the scenes from different perspectives, this is particularly useful when analysing ergonomics (for the workers) or mobility issues (for the robots).

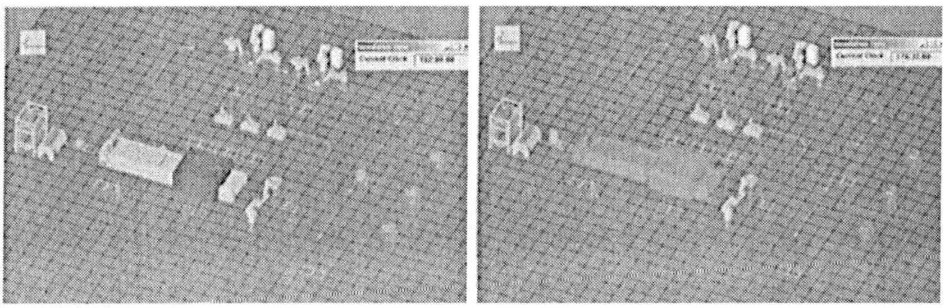

Figure 3 - Line snapshot before worker shift begin (left), line snapshot in failure condition (right)

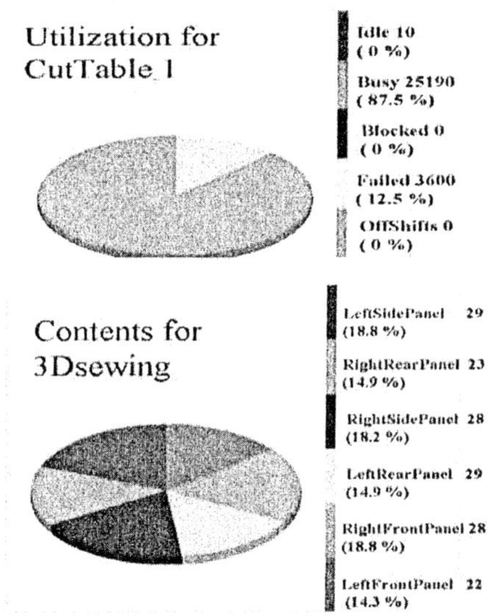

Figure 4 - Utilization for Cut Table (up) , 3Dsewing part content snapshot (down)

4. FURTHER DEVELOPMENTS AND CONCLUSIONS

Current version of GALsim allows to simulate production management of the overall system in the different conditions envisioned at design time and to provide an operational decision support system after the facility deployment.

A detailed mechanical simulator of the most critical activities of handling and grasping from the cutting table, of loading unloading on the transport system and of the body jacket 3D sewing will be developed. It will be integrated seamlessly with the general simulator of garment assembly line.

A new concept textile manufacturing facility design substantially reshapes the current cloth manufacturing methodologies.

The resulting multiagent holistic mechatronic production facility represents an example of complex non linear system.

A new paradigm garment assembly line design requires new resource and production management models, thus requiring a new dedicated simulation tool.

An advanced Virtual Reality simulation tool provides the necessary quantitative support for the analysis of design alternatives and for the assessment of proposed technologies.

5. ACKNOWLEDGMENTS

The Leapfrog (Leadership for European Apparel Production From Research along Original Guidelines) project is funded by the EC under the FP6-2003-NMP-NI-3, Proposal n° 515810-2.

6. REFERENCES

1. Aguiar, M.W.C., Murgatroyed, I.S., Edwards, J.E. 1996. "Object-oriented resource models: their role in specfying components in integrated manufacturing systems", Computer Aided Manufacturing Systems, Vol9, N°1.
2. Chalmeta, R., Williams, T.J., Lario, F. and Ros, L. 1997. "Developing an Object-Oriented Reference Model for Manufacturing", in Manufacturing Systems: modelling, management and control, P. Kopacek, Ed., Preprints of the IFAC Workshop on Manufacturing Systems: modelling, management and control (Vienna, Austria, Feb. 3-5). 379-384.
3. Leondes, T. (Ed.) Computer Aided and Integrated Manufacturing Systems Techniques and Applications, Vol. II: Computer-Integrated Manufacturing (Boca Raton, Florida: CRC Press LLC, 2001).
4. Moss, S., Davidsson, P. Multi-Agent-Based Simulation, Springer, Mar 1, 2001
   Appa Iyer Sivakumar, Chin Soon Chong, Development of an Object Oriented Simulation Engine for On-Line Simulation and Optimization, 33rd Annual Simulation Symposium pp. 291-299
5. McNally, P., Heavey, C. Developing simulation as a desktop resource. International Journal of Computer Integrated Manufacturing, Volume 17, Number 5 / July-August 2004, pp. 435 - 450
6. Acaccia, Chiavacci, Michelini, Callegari, Benchmarking clothing industry: effectiveness by computer simulation," Proc. of 11th European Symposium: Simulation in Industry ESS-99, Erlangen, October 26-28, 1999, pp. 519-524.
7. Acaccia, Conte, Maina, Michelini, "Computer aids for intelligent manufacture of quality clothes," Computer in Industry 50 (1) 2003, pp. 71-84.
8. Acaccia, Conte, Maina, Michelini, Molfino, "Integrated manifacture of high standing dresses for customized satisfaction," Globalisation of Manifacturing in the digital communication era, KLUWER, Boston, 1999, pp.511-523.
9. Bruzzone, L., Molfino, R.M., Zoppi, M. A discrete event simulation package for modular and adaptive assembly plants. In M.H. Hamza, editor, 22nd Int. Conf. Modelling, Identification and Control MIC2003, pages 280-282, Innsbruck, Austria, February 10-13 2003. IASTED/ACTA Press.
10. Bonsignorio F., Ferrari R., Busatti M., Bottero M. "A simulation tool integrated in a short term planning system", in PROCEEDINGS OF THE 4TH INTL. CONF. ON SIMULATION IN MANUFACTURING, London 1988, IFS
11. Bonsignorio F., Ferrari R., Trebino O., "Practical integration of a short term planning system in a production management information system", in Proceedings of the 1989 Intl. Workshop on Decisional Structures in Automated Manufacturing, Genova 1989, IFAC-CIRP-IFIP- IFORS

33 ON THE RE-CONFIGURABILITY DESIGN OF PARALLEL MACHINE TOOLS

Dan Zhang
Faculty of Engineering and Applied Science,
University of Ontario Institute of Technology
Oshawa, Ontario, L1H 7K4, Canada
Dan.Zhang@uoit.ca

The ongoing globalisation and international competition of manufacturing is requiring manufacturers to become more agile to efficiently use production resources. The ability to rapidly reconfigure production processes allows rapid ramp-up and robustness to adapt to unexpected events. In this paper, the reconfigurable machine tools and its characteristics is discussed in detail. Firstly, the basic modules of reconfigurable parallel machine tools are introduced. Then the criteria to form the reconfigurable machine are discussed. Some conceptual examples are illustrated and one case study is conducted.

1. INTRODUCTION

As the main component of reconfigurable manufacturing systems, the reconfigurable machine tools are machine tools that are built from machine modules [1]. A machine module could be an actuator, a joint, a link, a tool holder or a spindle. These modules are designed to be easily reconfigured to accommodate new machining requirements. Therefore, the reconfigurable machine tools have the characteristics of modularity, convertibility, integrate-ability, customisation, flexibility and cost-effectiveness.

The main objective of reconfigurable machine tool design is reconfigurable components that are reliable and cost-effective. It is noted that machine design will be conducted at the component level.

Some researchers already focused on the RMT study. Koren and Kota [2] received a patent for an RMT where the type, location, and number of spindles could be adjusted in response to changing product requirements. Landers et al.[3,4] presented an overview of reconfigurable machine tools and their characteristics,. Zhang et al. [5-7] developed a generic kinetostatic model for RMT stiffness analysis and design optimisation. Yigit and Ulsoy [8] discussed the vibration isolation of RMTs and proposed different isolation strategies depending on the RMT requirement.

As described by Koren et al. [9], the modular construction of Parallel Kinematic Machines (PKMs), allow them to be used as a special class of reconfigurable machine tools, consist of simple and identical modules that can be configured into different machines. Parallel kinematic machines can be also used as particular modules for machining lines.

Please use the following format when citing this chapter:

Zhang, D., 2006, in IFIP International Federation for Information Processing, Volume 220, Information Technology for Balanced Manufacturing Systems, ed. Shen, W., (Boston: Springer), pp. 309–316.

According to Jovane [10], reconfigurable parallel kinematic machines should have:

- Modular and reconfigurable structure with the possibility to change the total DOF,
- Light structure and high dynamic performances, as a standard PKM,
- Possibility to use innovative materials in the machine design
- Possibility to use fast actuators such as linear motors
- Due to the intrinsic modularity, a DOF decoupling is recommended
- Low complexity in the kinematic chains.

2. BASIC COMPONENTS

There are 4 basic types of joints, they are: revolute joint, Hook (universal) joint, Spherical joint and prismatic joint (Figure 1). One can develop the reconfigurable parallel machine by combination of those joints and links.

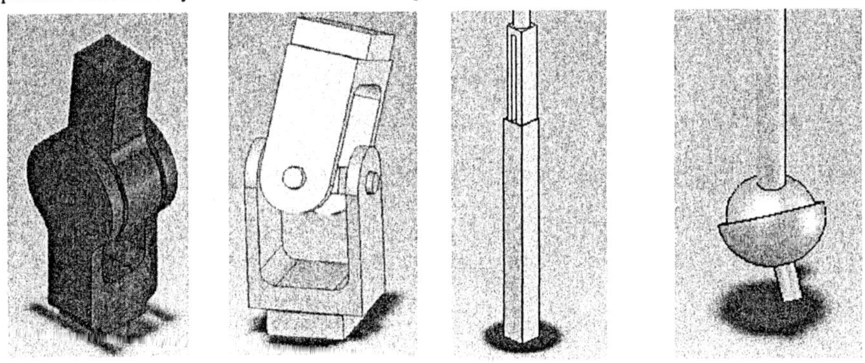

Figure 1 - Basic modular components

3. DESIGN CRITERIA

In order to design the reconfigurable parallel kinematic machines properly, there are several criteria that need to be followed with, here are some major issues: the degree of freedom of the architecture; design constraints used for most promising typology selection and kinetostatic model for stiffness analysis and design optimisation.

3.1 Mobility

A preliminary evaluation of the mobility of a kinematic chain can be found from the Chebychev-Grübler-Kutzbach formula.

$$M = d(n - g - 1) + \sum_{i=1}^{g} f_i \tag{1}$$

where M denotes the mobility or the system DOF, d is the order of the system ($d = 3$ for planar motion, and $d = 6$ for spatial motion), n is the number of the links including the frames, g is the number of joints, and f_i is the number of DOFs for the ith joint.

3.2 Constraints

In order to ensure the required motions (i.e., 5-DOF between the machine tool and the workpiece), the DOFs distribution numbers and the type of motions for each leg should be properly selected. Each leg can be facilitated with spherical, prismatic, Hooke and revolute joints.

The legs used in machine tools must be simple and practical to implement. For the sake of the simplicity and dexterity of mechanism, we prefer to use "spherical" pairs as the joints between link and the moving platform for those legs with more than 3 DOFs. Since the serially connected revolute joints easily lead to singularities and the manufacturability is difficult, so we forgo their use for cases where more than 2 revolute joints connected in series.

The study is based on fully-parallel mechanisms, but one can add legs (with 6-dof) to keep the structure symmetric. For the shape of the platforms, one should avoid the use of regular polygon, since it may lead to singularities.

4. SOME EXAMPLES

In Figures 2 – 4, we proposed a series of reconfigurable parallel mechanisms which consist of n identical actuated legs with six degrees of freedom and one passive leg with n degrees of freedom connecting the platform and the base. The degree of freedom of the mechanism is dependent on the passive leg's degree of freedom. One can improve the rigidity of this type of mechanism through optimization of the link rigidities to reach a maximized global stiffness and precision. Finally, this series of mechanisms have the characteristics of reproduction since they have identical actuated legs, thus, the entire mechanism essentially consists of repeated parts, offering price benefits for manufacturing, assembling and purchasing. It acts as a kind of reconfigurable parallel machines, based on a number of identical active kinematic chains plus a passive central leg, to compensate torsion. The reconfiguration process consists on the possibility to change the total DOF by adding or removing active struts, and, consequently constraining the passive leg.

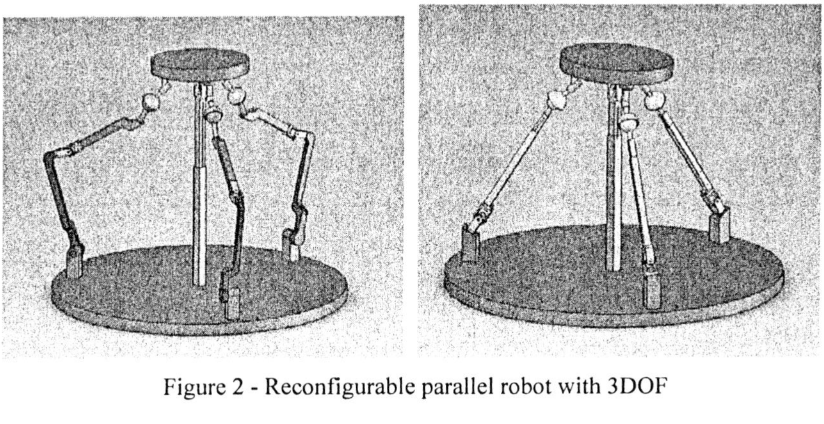

Figure 2 - Reconfigurable parallel robot with 3DOF

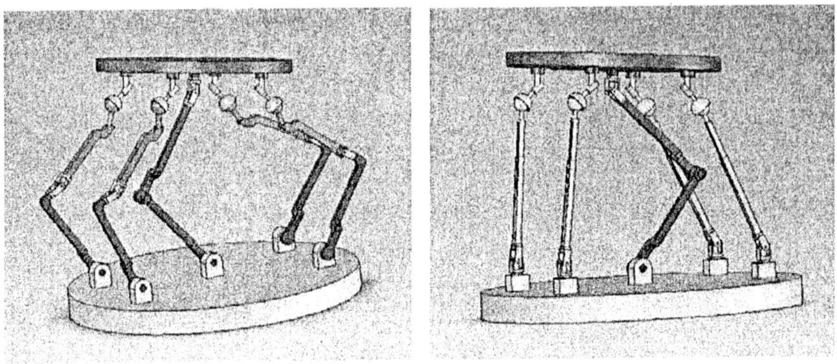

Figure 3 - Reconfigurable Parallel Robot with 4DOF

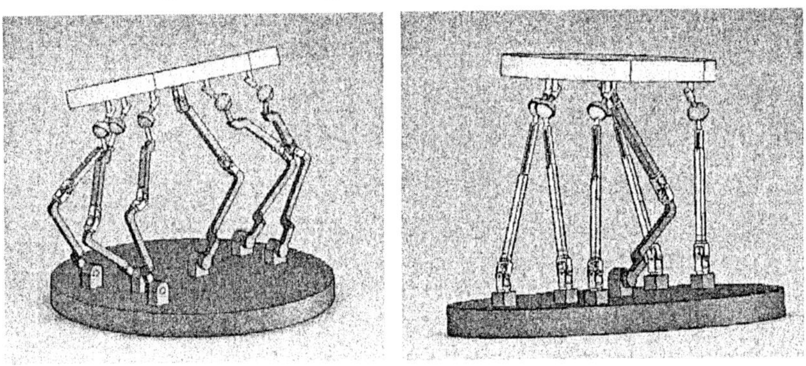

Figure 4 - Reconfigurable Parallel Robot with 5DOF

5. CASE STUDY

We take a 3DOF reconfigurable parallel robot (Figure 2) with prismatic actuators as an example. The Cartesian coordinates of the platform are given by the position of point P with respect to the fixed frame, and the orientation of the platform (orientation of frame P-X'Y'Z' with respect to the fixed frame), represented by the joint angles of the fourth leg $\theta_{42}.\theta_{43}$ (central leg) or by the matrix **Q.**

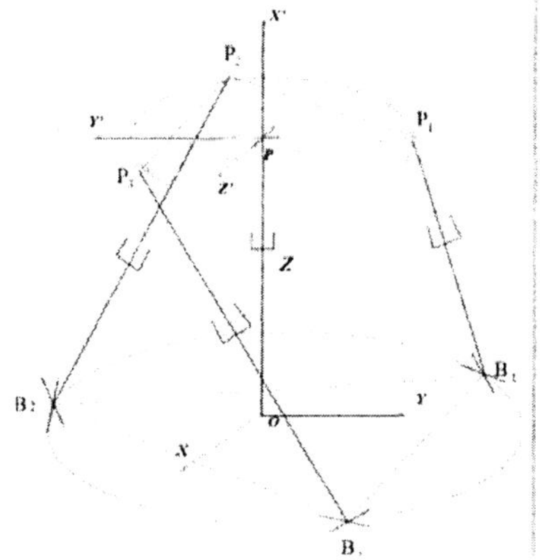

Figure 5 - Schematic Representation of Reconfigurable Parallel Robot with 3DOF

If the coordinates of the point P_i in the moving reference frame are represented with (x'_i, y'_i, z'_i) and the coordinates of the point B_i in the fixed frame are represented by vector $\mathbf{b}_i$, then for i=1,...,4, we have

$$\mathbf{p}_i = \begin{bmatrix} x_i \\ y_i \\ z_i \end{bmatrix} \mathbf{r}'_i = \begin{bmatrix} x'_i \\ y'_i \\ z'_i \end{bmatrix} \mathbf{p} = \begin{bmatrix} x \\ y \\ z \end{bmatrix} \mathbf{b}_i = \begin{bmatrix} b_{ix} \\ b_{iy} \\ b_{iz} \end{bmatrix} \tag{2}$$

where $\mathbf{p}_i$ is the position vector of point P_i expressed in the fixed coordinate frame whose coordinates are defined as x_i, y_i, z_i, and $\mathbf{r}'_i$ is the position vector of point P_i expressed in the moving coordinate frame, and $\mathbf{p}$ is the position vector of point P expressed in the fixed frame as defined above,

We can then write

$$\mathbf{p}_i = \mathbf{p} + \mathbf{Q}\mathbf{r}'_i \tag{3}$$

where $\mathbf{Q}$ is the rotation matrix corresponding to the orientation of the platform of the manipulator with respect to the base coordinate frame.

Subtracting vector $\mathbf{b}_i$ from both sides of eq.(3), we obtain

$$\mathbf{p}_i - \mathbf{b}_i = \mathbf{p} + \mathbf{Q}\mathbf{r}'_I - \mathbf{b}_i \tag{4}$$

then taking the Euclidean norm of both sides of eq.(4), we derive

$$\rho_i^2 = (\mathbf{p}_i - \mathbf{b}_i)^T(\mathbf{p}_i - \mathbf{b}_i) \tag{5}$$

where ρ is the length of the ith leg, i.e. the value of the ith joint coordinate. The solution of the 3-dof platform is therefore completed.

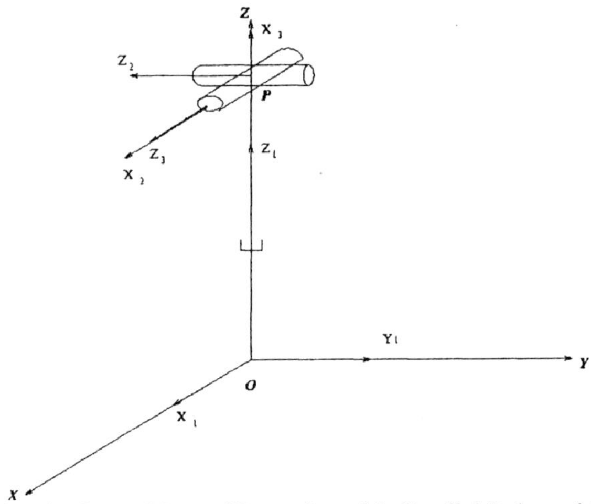

Figure 6 - The Central Leg of Reconfigurable Parallel Robot with 3DOF

Figure 6 illustrates the configuration of the passive leg. From the Figure 6, we can obtain the Denavit-Hartenberg parameters as in Table 1,

Table1: The DH parameters for the passive constraining leg with rigid links

i	a_i	b_i	α_i	θ_i
0	0	0	0	0
1	0	Z	90	0
2	0	0	90	θ_{42}
3	0	0	0	θ_{43}

with the Denavit-Hartenberg parameters above, we can obtain the kinematic model.

The kinematic chain can be taken as a serial manipulator, the kinematics of the serial manipulators comprise the study of the relations between joint variables and Cartesian variables. A Hooke joint can be replaced by two orthogonal revolute joints in this case.

For the passive kinematic chain, we have the velocity equation

$$\mathbf{J}_4 \theta_4 = \mathbf{t} \tag{6}$$

where

$$\dot{\boldsymbol{\theta}}_4 = \begin{bmatrix} \dot{\rho} & \dot{\theta}_{42} & \dot{\theta}_{43} \end{bmatrix}^T \tag{7}$$

and the Jacobian matrix of the 4th leg of the mechanism $\mathbf{J}_4$ can be expressed as

$$\mathbf{J}_4 = \begin{bmatrix} 0 & \mathbf{e}_{42} & \mathbf{e}_{43} \\ \mathbf{e}_{41} & \mathbf{e}_{42} \times \mathbf{r}_{42} & \mathbf{e}_{43} \times \mathbf{r}_{43} \end{bmatrix} \tag{8}$$

Differentiating eq.(5) with respect to time, we have the velocity equation as

$$\mathbf{At} = \mathbf{B}\dot{\boldsymbol{\rho}} \tag{9}$$

where vectors $\dot{\boldsymbol{\rho}}$ and $\mathbf{t}$ are the joint velocity and the twist of the platform defined as

$$\dot{\boldsymbol{\rho}} = \begin{bmatrix}\dot{\rho}_1 & \dot{\rho}_2 & \dot{\rho}_3\end{bmatrix}^T \tag{10}$$

$$\mathbf{t} = \begin{bmatrix}\boldsymbol{\omega}^T & \dot{\mathbf{p}}^T\end{bmatrix}^T \tag{11}$$

According the principle of virtual work, we have

$$\boldsymbol{\tau}^T\dot{\boldsymbol{\rho}} = \mathbf{w}^T\mathbf{t} \tag{12}$$

where $\boldsymbol{\tau}$ is a vector of the actuator forces applied at each actuated joint and w is the wrench (torque and force) applied to the platform and where it is assumed that no gravitational forces act on any of the intermediate links.

We have $\mathbf{w} = \begin{bmatrix}\mathbf{n}^T & \mathbf{f}^T\end{bmatrix}^T$ where **n** and **f** are respectively the external torque and force applied to the platform.

Regarding eq.(9) and substituting it in eq. (12), we obtain

$$\boldsymbol{\tau}^T\mathbf{B}^{-1}\mathbf{At} = \mathbf{w}^T\mathbf{t} \tag{13}$$

Now, substituting eq.(6) into eq.(13), we have

$$\boldsymbol{\tau}^T\mathbf{B}^{-1}\mathbf{AJ}_4\dot{\boldsymbol{\theta}}_4 = \mathbf{w}^T\mathbf{J}_4\dot{\boldsymbol{\theta}}_4 \tag{14}$$

The latter equation must be satisfied for arbitrary values of $\dot{\boldsymbol{\theta}}_4$ and hence we can write

$$(\mathbf{AJ}_4)^T\mathbf{B}^{-T}\tau = \mathbf{J}^T{}_4\mathbf{w} \tag{15}$$

The latter equation relates the actuator forces to the Cartesian wrench, **w**, applied at the end-effector in static mode. Since all links are assumed rigid, the compliance of the mechanism will be induced solely by the compliance of the actuators. An actuator compliance matrix **C** is therefore defined as

$$\mathbf{C}\boldsymbol{\tau} = \Delta\boldsymbol{\rho} \tag{16}$$

where $\boldsymbol{\tau}$ is the vector of actuated joint forces and $\Delta\boldsymbol{\rho}$ is the induced joint displacement. Matrix C is a (3x3) diagonal matrix whose ith diagonal entry is the compliance of the ith actuator.

and finally, we obtain

$$\Delta\mathbf{c} = \mathbf{J}_4(\mathbf{AJ}_4)^{-1}\mathbf{BCB}^T(\mathbf{AJ}_4)^{-T}\mathbf{J}^T{}_4\mathbf{w} \tag{17}$$

Hence, we obtain the Cartesian compliance matrix as

$$\mathbf{C}_c = \mathbf{J}_4(\mathbf{AJ}_4)^{-1}\mathbf{BCB}^T(\mathbf{AJ}_4)^{-T}\mathbf{J}^T{}_4 \tag{18}$$

with

$$\Delta\mathbf{c} = \mathbf{C}_c\mathbf{w} \tag{19}$$

where $\mathbf{C}_c$ is a symmetric positive semi-definite (6x6) matrix, as expected. With this model, we can do design optimization, performance evaluation and integrated control and trajectory control.

6. CONCLUSIONS

The design of reconfigurable machine tools has been addressed in the paper. Design methodology for this kind of machine tool, especially for reconfigurable parallel kinematic machine tool is discussed in detail. A case study is given to show the procedures of generating the reconfigurable parallel mechanism, and models for further design and analysis.

7. ACKNOWLEDGEMENTS

The author acknowledges the support for this work provided by the Natural Sciences and Engineering Research Council of Canada (NSERC) and Kayla Viegas for the CAD models.

8. REFERENCES

1. D. Fedewa, M. Mehrabi, S. Kota, and V. Gopalakrishnan, Design of a parallel structure ficture for reconfigurable machining systems, Proceedings of the 2000 Japan-USA Flexible Automation Conference, Ann Arbor, USA, 2000
2. Y. Koren and S. Kota, 1999, Reconfigurable machine Tools, US patent 5,943,750
3. R.G. Landers, A New Paradigm in Machine Tools: Reconfigurable Machine Tools, Japan-USA Symposium on Flexible Automation, Ann Arbor, USA, 2000
4. R.G. Landers, B.K. Min and Y. Koren, Reconfigurable Machine Tools, Annals of the CIRP, Vol.50/1, pp. 269-274, 2001
5. D. Zhang, C. Mechsfske and F. Xi, Stiffness Analysis of Reconfigurable Parallel Mechanisms with Prismatic Actuators, CIRP 1st International Conference on Reconfigurable Manufacturing, Ann Arbor, Michigan, 2001
6. D. Zhang, C. Mechsfske and F. Xi, Optimization of Reconfigurable Parallel Mechanisms with Revolute Actuators, CIRP 1st International Conference on Reconfigurable Manufacturing, Ann Arbor, Michigan, 2001
7. D. Zhang, Kinetostatic Analysis and Optimization of Parallel and Hybrid Architectures for Machine Tools, Ph.D. Thesis, Laval University, Quebec, 2000.
8. A.S. Yigit, A.G. Ulsoy, Design of Vibration Isolation Systems for Reconfigurable Precision Equipment, Japan-USA Symposium on Flexible Automation, Ann Arbor, USA, 2000
9. Y. Koren, U. Heisel, F. Jovane, T. Moriwaki, G. Pritschow, G. Ulsoy, H. Van Brussel, Reconfigurable Manufacturing Systems, Annals of the CIRP, Vol. 48/2, 1999
10. F. Jovane, S. P. Negri, I. Fassi, L. Molinari Tosatti, Design Issues for Reconfigurable PKMs, 3th Chemnitz Parallel Kinematics Seminar: Development Methods and Application Experience of Parallel Kinematics. Chemnitz 2002

34 EVOLVABLE ASSEMBLY SYSTEMS BASIC PRINCIPLES

Mauro Onori
Royal Institute of Technology, onori@iip.kth.se
José Barata
Universidade Nova de Lisboa /UNINOVA, jab@uninova.pt
Regina Frei
Royal Institute of Technology, onori@iip.kth.se

This paper addresses the underlying principles of Evolvable Assembly Systems. This paradigm was recently proposed as an answer to the requirements faced by assembly companies in the current world of business and technological changes. The basis for this new approach lies in a multi-disciplinary study of the needs and requirements, and shifts the technological focus from complex, flexible, multi-purpose systems to simpler, process-oriented, dedicated swarms of machine modules.

1. INTRODUCTION

Notwithstanding recent technological advances, the social and economic situation for assembly-intensive companies in Europe is facing considerable challenges. Recent studies quantify European outsourcing at 21% of total assembly activities, and have forecasted a rise to over 40% by 2007 (Kirsten 2002). The hidden issue behind such facts is that assembly has not been rendered cost-effective yet, and this is particularly true for novel products and markets, such as the micro-assembly domain. In order to counter this lack of adequate automation, and to rise to the challenge, the Evolvable Assembly Systems (EAS) paradigm was launched a few years ago (Alsterman et al. 2004; Barata et al. 2005; Onori 2002). The basis for this new approach lies in a multi-disciplinary study of the needs and requirements, and shifts the technological focus from complex, flexible, multi-purpose systems to simpler, process-oriented, dedicated swarms of machine modules. From a paradigm point of view, the EAS approach rejects machine flexibility (sub-optimal at any specific task) for application agility. This paper will attempt to detail the recent developments within Evolvable Assembly Systems, including ontological, methodological, and application developments.

2. BASIC FOUNDATIONS

Evolvable Assembly Systems (EAS) is commonly understood to be just another interpretation of Reconfigurable Assembly Systems (RAS). Since this is not the

Please use the following format when citing this chapter:

Onori, M., Barata, J., Frei, R., 2006, in IFIP International Federation for Information Processing, Volume 220, Information Technology for Balanced Manufacturing Systems, ed. Shen, W., (Boston: Springer), pp. 317–328.

case, it is of some importance to clarify how EAS differs, and which new aspects it proposes. Re-configurable systems take a system perspective and start with, in most cases, the current product and user requirements. In fact, re-configurability is a further development of flexibility. Flexible systems provided multi-purpose machines that were fairly good at many tasks but optimal in none. And the cost was high. Re-configurability has sub-divided these machines into smaller units and focussed on the system interfaces, but the focus is far too limited in time: current products and company aspects! EAS differs radically, and the following clarifies some aspects:

- Main focus – RAS focuses on re-configurability (geometric setup) of system components; not necessarily automatic. EAS focuses on adaptability of system components through capture of emergent properties; semi or fully automatic. This may raise safety issues.
- Development trigger issue – RAS uses current (existing product features as a development trigger, whereas EAS focuses on the re-engineering needs of the assembly system (product shift).
- Modularity level – RAS applies conventional subdivision of assembly into "transport-handling-assembly-finalisation" blocks, which results in coarse granularity. EAS applies lower-level modularity based on process-level characteristics: the "operational" level becomes subject to modularity, hence resulting in fine granularity.

In truth, the differences are many, and EAS provides, through its tight process-oriented modularity, a much closer link between product design and assembly system development. In order to further clarify how EAS operates, and upon which premises it builds its theoretical and technological foundations, a brief description of the main aspects will follow.

2.1 System Concept

EAS proposes, basically, a radical new way of thinking: in terms of assembly systems, what is required is not a complex solution which tries to accomplish all of the envisaged assembly needs within a closed unit (Flexible Assembly systems) but, rather, a solution which, being based on many simple, re-configurable, task-specific elements (system modules), allows for a continuous evolution of the assembly system. This, in many ways, is an interpretation of swarm strategies. In fact, it is the aggregation of many small, simple entities that will enable a given functionality, and this functionality can disappear by the removal of some of these entities. New functionalities are simply created by forming different formations, or coalitions of entities.

EAS also goes a step further by proposing a totally new way a considering the products and assembly systems: the design cycle of the products will be influenced by which modules are available. In other words, many simple, strictly task-oriented components with standard interfaces are better than few, very flexible but extremely expensive solutions that cannot be integrated within existing systems: in effect, there are no defined systems within EAS, only process-oriented modules. This, however, brings us to a discussion of what a system is. In fact, a system's capability is NOT the sum of the capabilities of its parts. It could be a completely different scenario altogether. The issue being raised here is that EAS will consist of a vast range of inter-connectable modules. When a system will be created according to the EAS principles, the resulting capability of the sum of the modules will not be so easily predicted. What happens, in fact, when a multitude of small entities is brought together, is that new, unthought-of capabilities may emerge from this coalition. Hence the need to start studying the principles of *emergence*, since the capabilities being brought together may also be viewed as particular functionalities (skills) being offered by each module. A very important point to highlight is the fact that the properties that emerge out of the interaction of the modules that compose the assembly system represent just the complex functionalities (complex skills) of the system we want to create (Barata 2005). The interesting conclusion is that we can generate any functionality we want as long as we provide the control architecture able to accommodate the interaction of modules.

2.2 Control Concept

Starting with the basic assumptions enunciated in the previous section, it becomes clear that traditional control paradigms cannot be used to support EAS. In effect the control solution for EAS must cope with the following aspects:

1) Support the integration of modular components that might include their own controllers with different levels of intelligence,
2) product changes,
3) fluctuations in demand,
4) support the addition and removal of components during normal production,
5) provide support for the operative phase.

Support the Integration of Modular Components

In terms of integration of modular components, then the centralised approach is immediately at a loss since the basic assumption under which EAS is grounded is the idea of independent modules that may be reused and put together at will, according to the system that is being created (coalition). With such an objective the main goal is about plugability or the possibility that coalitions of modules need to be created in a very fast way without any or just minor programming changes. This requirement cannot be achieved with a centralised approach which may be optimal for static systems but copes poorly with dynamic life cycle systems. The hierarchical approach is also very weak because hierarchies deal poorly with structural changes. Even if each module has its own controller, and therefore becomes distributed, it is not convenient that each module controller has functionalities that allow essentially the creation of very static hierarchies.

Hence a different approach should be followed in which each module should have a level of intelligence that enables it to participate in societies of modules (coalitions) following different types of hierarchical structures. This means a completely new approach on how module controllers should be designed because modules do not know á priori (statically) what kind of requests and from whom these requests will come. These new extra functionalities for participating in dynamic coalitions with different types of hierarchical structures requires new challenges on the module controller architecture.

The multiagent paradigm is a good choice for two reasons. The first reason is related to modelling and abstraction. In fact it is quite simple to make the connection between a module and a modular component. The second reason is related to the need for plugability. In effect if a modular system is being considered they must be plugged or unplugged at will. Therefore a multiagent system seems to be ideal because, by definition, a multiagent system is adequate to plug and unplug agents (in this case component modules). Another aspect connected to a modular system is the fact that a particular assembly or production system is a composition of modules, which can then be also considered as a coalition of modules. This is another connection between production modules and the agent world. A very well known and established work is in fact the work on agent coalitions (Castelfranchi et al. 1992; Klusch and Gerber 2002; Pechoucek et al. 2002; Shehory and Kraus 1995), which can be applied also here (Barata 2005).

It is important to emphasise that modularity does not immediately imply modules with very intelligent controllers. However, if it is considered that modules can be plugged and unplugged in a very simple way (none or minor programming effort) then at least some functionalities related to plugability are needed at module level.

Product changes

Product changes may imply alterations in the structure of the assembly system. The system evolves by changing, rapidly, the modules that compose the coalition. This change must be done with no or minor programming. The aspects of plugability referred to in the previous point are fundamental to ensure a smooth change. However, more than just simple plugability is needed because one must consider an entity that will be responsible for the changes in the coalition or society, and the controller modules must be able to cope with this entity by providing them with relevant information (such as the skills that each one is able to perform) in addition to information about their past experience.

Therefore, an entity that should be called upon whenever a major change is required to the coalition is being implied. This entity (tool) should be able to guide a systems integrator in the process of creating a new coalition fitted for a product family. The most important aspect is that this tool should be able to guide the user by suggesting that the coalition is not created based only on product features but also on the skills made available by the modules.

In otner words, each module should be able to supply information of its own skills or functionalities to be used by the external tool or configurator. In addition to this information, the user should also get information about the quality of the module he/she is choosing. Consequently, each module controller should include functionalities to store relevant information such as the number of faults, number of working days, etc and also to supply them whenever requested by the configurator. This means extra computational capability (or intelligence) for each module controller.

It is relevant to consider that whenever the coalition is created the set of module controllers (agents) start to interact and a new set of behaviours (skills) emerges. This means that whenever a coalition is created it is not necessary to program the interactions between its members. The only thing the configurator has to do is assist in the creation of the coalition.

The relation with the multiagent paradigm for this requirement is that an agent supports various behaviours. In fact, in the illustrated situation, the agent in addition to its normal work of controlling its module should also perform behaviours related to answering requests about its own skills and monitoring and storing relevant information about its physical entity (module). On top of these aspects the configurator can itself be an agent, in this case a configurator agent that interacts with a user and with the society of modules to get information.

If the change on the product only implies alterations of the flow within the system then the change does not imply any changes to the coalition but only changes in the way the product is processed. If it is considered that the product is an entity that know its process steps (process sequence) and considering that the transportation system is intelligent enough to know what kind of modules and which skills they publish, then whenever there is a product change that can be coped within the system nothing is required to change since it is all about negotiation between the product and the transportation system. It must be clarified that what is being considered here is a situation in which each traveling product (piece) is represented within the control system as an intelligent entity that can interact with the other entities. If, for instance, it is considered that each product is transported by an autonomous entity (kind of AGV) that can move to the locals where the other modules might do some operation on the product, and in addition to that the transporter can make broadcasts to know which modules are available and which are their skills, then it becomes easier to understand why product changes of this type do not require any reprogramming.

Taking into consideration the previous requirements it is quite easy to make the connection between product changes and the need for a controller based on the multiagent paradigm. The autonomy makes the agent responsible for its own acts, which means that each module is individually responsible for the actions over the product. Adding or removing one of these modules does not have a big impact on the overall system structure since each module is a confined individual entity. It must be noted that being confined does not mean any interactions with the other modules. In fact the social ability that characterises the multiagent world is fundamental to ensure that the modules interact among themselves.

Fluctuations in the demand

This aspect is mostly related with the ability to plug or unplug modules, as well as with the capability of the modules to adapt themselves to the new requirements. If the demand increases a very obvious solution is adding new modules to increase throughput or in case of lower demand modules can be removed. But a less obvious solution of system adaptation to the new requisites may take place. For instance lets suppose the system is composed of intelligent modules that always tries to save energy. On a situation of low throughput, each module understands that it can do the work at slower speed, therefore saving energy. It can even be the case that parts of the system are shutdown, which means that certain modules are shutdown; only part of the controller is active, looking for changes in the demand. If a demand increases the modules detects this new situation and the ones already working start to produce faster and the ones that have shutdown themselves start working. The reverse happens in a situation of reduced demand.

This capacity flexibility can only be achieved if each module is intelligent enough to be able to execute these actions. It is very important to consider that the agents needs to interact among themselves in order to be able to sense that a demand is increasing or decreasing. It would be ideal that the agents could figure out what is going on based on their local knowledge.

Addition or removal of modules during normal operation

This aspect is about minor changes that might be made to a system during normal operation. These minor changes might happen due to minor changes in the product that do not require the intervention of the configuration tool and/or to support maintenance interventions without having to stop the system. For instance, if the system is operating and at certain level a feeder or even a gripper is required to be removed for short period maintenance then the system should be able to still operating, even if at reduced capability. It might be the case that the removal of the gripper inhibits the capability of the cell to do some operations but the other might still be supported.

These requirements are again best suited for the multiagent paradigm since only with intelligent modules that support emergence in the sense that the new capabilities of the system are automatically derived whenever a new module is added or removed, it is possible to solve this problem.

An interesting side effect of this requirement is that the system can be composed manually without the need for a configurator. Because of the mergence that is supported by each module controller the behaviour of the system emerges naturally out of the coalition.

Operative phase

The need for better operational support means all the tasks that are required to support the shop floor while it is operating. This include aspects such as online reconfiguration of production parameters, maintenance support functionalities, advanced diagnostic systems, advanced user interfaces, etc. All these requirements are also better solved if the multiagent paradigm is considered, because among other aspects it is important to model each manufacturing component or module as an entity that includes all relevant information (maintenance parameters, process related variables, etc) that can be used in supporting the aspects referred before. This shows how the connection between modular component and agents is done. Another important aspect at this level is the autonomy that each agent (module) must possess. In fact each agent must support simultaneously different behaviours or actions. For instance an agent representing a certain module participating in a coalition (system) must simultaneously do the following tasks: 1) answer requirements for executing some work (for instance, the request to transport one piece from one point to another received by a gantry robot), 2) keeping monitoring the internal sensors of the module, 3) answering requests from external users for internal parameters values, 4) interacting with other modules for advanced diagnosis functionalities, etc.

2.3 Emergence

The EAS paradigm, considers the optimisation under conditions of change. As a *system*, an assembly system actually consists of many different, small, but cooperating, entities, or if we like, organisms. There will be higher organisms (the system as a whole, or modules) and lesser ones (grippers, fixtures). As this approach is adopted by EAS, with many simple, dedicated process units, special conditions arise. This brings us to how systems should be interpreted, and a branch of studies called "Systems Thinking".

Systems Thinking enables one to progress beyond simply seeing events to seeing patterns of interaction and the underlying structures which are responsible for the patterns. A system is something that maintains its existence and functions as a whole through the interactions of its parts. It consists of many different parts and organs, each acting separately yet all working together and each affecting the others (analogy with EAS: The modules are the parts and the assembly system is the system).

The essential difference, from a EAS point of view, is that in a true system, when one part is added, the additional functionality is not an obvious addition of the added parts' functionality. This is what is called emergent properties, and is the basis of EAS. *Systems have emergent properties that are not found in their parts. You cannot predict the properties of a complete system by taking it to pieces and analysing its parts.*

Similarly, the smaller the constituents of the system, the easier it is to define, structure and coordinate the skills being brought in and out of it.

Note: the constituents have to be classified in terms of the overall system demands, changing external conditions, and particular tasks to de carried out.

It is suggested, due to the potentially large level of complexity, and liability issues, to subdivide emergence into 3 levels: primitive, communal, and full emergence.

Primitive emergence:
This refers mainly to very simple self-organisation. The main prerequisite is the ability of the system modules to self-recognition. This is to be viewed as a fundamental objective of plugability. There must be an open information exchange between the modules and the organiser of assembly events, whether this be at configurator level or performed autonomously by more "intelligent" (agent-based) modules. This type of emergence has a smaller effect on systems based on coarse modules: i,e.- the finer the granularity, the higher the need for primitive emergence to satisfy the reconfigurability & evolvability demands.

Communal emergence:
This refers mainly to self-organisation with adaptability. In this case the system is not only fully capable to self-program and setup itself, but it also enables some forms of adaptability. This is not only self-calibrations but, primarily, the ability of the system to adapt if changes are brought to the "world perimeter" its comprises, such as the product or fluctuation scenarios described earlier.

At this point, the emergent solution may detect possibilities and inform the user/controller of such emergent properties. An example of such Community-adaptive emergence could be the creation of a Pick&Place capability out of a gripper and one or two linear or rotational movements.

Full/Evolutionary emergence:
This refers to full adaptability and evolvability. This level enters the realm of classical emergence, in which the emergent property itself is often unpredictable and unprecedented, and may represent a new level of the system's evolution. Full

emergence means that EAS learns how to perform "tricky" tasks, similar to the way a human operator gets better when he gains more experience. This means that the systems has to be taught how to use its acquired knowledge to draw further conclusion. Of great importance is the idea that certainty/uncertainty aspects surface. The emergent properties have never before been tested and exploited, are the result of previously unknown logical/social or other patterns, all of which raises liability and safety issues. *It is for this reason alone that it is of utmost importance to understand the value of an emulator:* it will enable the user to verify the emerging skills/properties before one builds the actual system! If this is available, the liability issues would also become more treatable.

2.4 Evolvability

When considering evolvability, or EAS, one often makes the terrible mistake of considering qualitative and quantitative features on the same plane of thought. This has led the EAS developments into quite some confusion, and some preliminary clarification is required.

A qualitative feature describes a given EAS characteristic from a general viewpoint, and cannot easily be assigned any set metric for quantification, unless this metric is clearly defined.

A quantitative feature describes a performance characteristic and can be measured. Hence, when discussing evolvability one should first ask whether such concepts may be measured at all. In an attempt to clarify this, the article proposes a tentative set of definitions.

First of all, the evolvability resides within the system characteristics (we will always assemble, etc.- functionality not evolvable, but the qualitative attributes), which must be assumed to be mechatronic in nature. The assembly system may then be defined as an evolvable system if:

- It is a fully "reconfigurable" mechatronic system platform that exhibits an emergent behaviour.
- The assembly units and modules are mechatronically integratable.
- The evolvable & reconfigurable system is composed of process-oriented components.
- The system can automatically determine its functionality based on the components' skills (when components are plugged together to form it).
- There is no (or minimal) investment in the programming & coding, but, rather, in how to establish and exploit relations.
- Maintenance, documentation and the ability to store information in support of operational stability.

However, evolvability is not attained by a single giant leap forward in technology. It is attained in carefully measured steps, that start with specified modularity. From a process-oriented set of modules, one must then assure plugability, followed by reconfigurability and, finally, evolvability. A full set of definitions and clarifications will be given later.

3. EAS PREREQUISITES

Several partners have, for the past few years, elaborated on the EAS paradigm. Some of the more recent developments, such as the ABAS platform, clearly attempt to fathom out the prerequisites and move forward. The EUPASS Integrated Project has been the largest endeavour to date to follow the EAS principles, and has, together with more pinpointed efforts between KTH, UNINOVA, and EPFL, elaborated a set of foundations for EAS.

3.1 Definitions and metrics

As stated earlier, it became quite clear, at a very early stage, that means to validate the concept were needed. The first step was to define the way certain aspects were being interpreted, and then move forward to try and set up adequate metrics for exploitable validation. The basic definitions are given below, starting with the basic EAS prerequisite, which is a process-oriented set of assembly modules.

Module – Any unit that can perform an operation and integrates a specific interface. Granularity level needs to be defined (the lower the level, the higher the degree of emergence)

Granularity – The lowest level of device being considered within a reference architecture. The lower the level of building block (tool, gripper), the higher the emergent behaviour: if a gripper can "communicate" with a robot, new operational characteristics may emerge (flip product/part in motion, fine positioning...). However, this implies that an overload of definitions and information management may arise, and a minimum level needs to be clearly set.

Plugability – The ability to rearrange and integrate system components within the framework of a given system architecture. The resulting new layout does not preclude efficient performance, one simply and physically plugs together a new arrangement.

Reconfigurability (interoperability) – The ability to reaarrange available system components to perform new, but pre-defined operations (plugability plus characteristics that ensures the efficient performance of the resulting new layout).

Evolvability – It is a fully "reconfigurable" mechatronic system platform that exhibits an emergent behaviour which introduces new or refined levels of functionality. It requires a stringently defined reference architecture to enable the correct application of the relevant characteristics.

These are **qualitative** features. Therefore, an attempt at setting quantitative features to the EAS objectives was carried out, resulting in the table below:

EAS Qualitative Features	EAS Quantitative Features
Evolvability-conformity	Skills repository & Management
Plugability- control specifications	Module description/blueprint
Plugability- user requirements	Application guidelines
Safety conformity	CE & safety certification procedures
Evolvability & Safety	Rules related to emergent behaviour
Plugability-practical implementation	EAS "wrapper" solution: hardware
Evolvability-practical impl.	EAS "wrapper" solution: software

At the highest level the EAS application will require some form of virtual repository that stores the protocols, guidelines, and other support structures for users to be able to comply with its specifications. The Reference Architecture will be available through this level. Plugability is not only the required control structures and approaches applied to exchange and adapt data, but it must also indicate for external users what may be required in order to comply. Emergence, defined earlier, also implies that unwanted characteristics may emerge, and raise safety or liability issues. This will have to be considered carefully, and specific rules will need to be enforced. One of the major keys to a successful EAS application resides in how the modules will interact with one another, which implies that much more than traditional interfaces will be required. These are, for the time being, defined as "wrappers", which also implies that legacy components may be adapted to the EAS format. Finally, but also fundamentally, EAS requires an extremely well defined Reference Architecture. This architecture will be a mirror of the ontology developed for a particular class of products. This Reference Architecture is currently being developed out of the ontology definition that will be given in the next section.

3.2 Ontologies

Ontology is a concept widely used today in Knowledge Engineering, Artificial Intelligence and Computer Science in any application that involves knowledge management and information management. Its importance in other engineering domains such as mechanical engineering and assembly, in particular, is quite natural since this domain requires computer based supporting tools to help in the design and operation of agile assembly systems such as Evolvable Assembly Systems. The full development of the EAS concept cannot be done without design supporting tools and advanced control solutions whose main requirement is the ability to quickly change and adapt. However, this requires the assembly domain to be fully understood and modeled in a way that can be used by the different computerized tools involved within assembly. Only if the main concepts behind the assembly domain and how they interrelate are fully understood and conveniently represented using some kind of formalism it will be possible to create computer programs that support the development and operation of EAS. In addition to the central aspect of creating computer representations that model the concepts and their relationships it is fundamental that these representations are agreed by the assembly community because it must be taken into account that different tools will be used and developed across the domain and therefore it will be of great advantage if the tools could somehow reuse the models that are developed. These computer representations (models) can be defined using ontologies and this is the reason why they are being addressed here.

Ontologies are much more than simple taxonomies since they allow appending semantics to relations among entities. So, we may envision relations that describe dynamics, social interactions, etc. However, much work needs to be done in knowledge modelling and knowledge management, which should not be a surprise since ontologies are about representing knowledge. For knowledge management existent methodologies, like CommonKADS (Schreiber et al. 1994) may be exploited. But the great challenge is about knowledge modelling, since the approach for ontology creation needs always knowledge modelling both for the situations in which the ontology is created using a top-down or bottom-up approach. One challenge is modelling operational knowledge, which is not clearly stated in products' data sheets. For instance, two modules can be mechanically interconnected but practical experience may indicate that such a connection is usually ineffective as mechanical stress is imposed, which in turn it is application dependent. The problems are increased if the knowledge of several experts needs to be merged, if the knowledge is contradictory, etc. The ontology research work about creating ontologies tries to provide answers to this question (Gómez-Pérez et al. 2004).

Some steps are being done in the direction of structuring the assembly domain in the European project EUPASS, in which some preliminary ontologies about the assembly process are being done using OWL. The most important concepts that are necessary in defining a precision assembly ontology to implement successful EAS must include the following sub-ontologies: (1) ontology of modules, (2) ontology of processes, (3) ontology of skills, and (4) ontology of products.

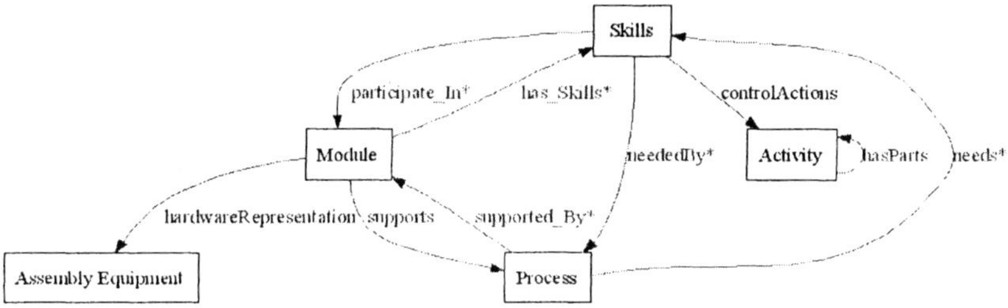

Figure 1 – Most important concepts and how they interrelate

An ontology of modules should identify the modules available to be used in coalitions (assembly systems). These modules can range from simple entities such as a gripper to a very complex entity such as an assembly cell. The ontology of processes should identify all the classes of processes that are used in assembly such as gluing, pick&place, joining, etc. The ontology of skills needs to formalise the skills that may be found in assembly. Not only is it necessary to define the individual skills associated to each module but also the skills that emerge out of the basic ones. The ontology of products should identify families of products that share certain assembly processes. These concepts grouped as sub ontologies are related according to figure 1. In figure 2 it is shown just part of the taxonomy of the ongoing ontology work that describes the assembly process.

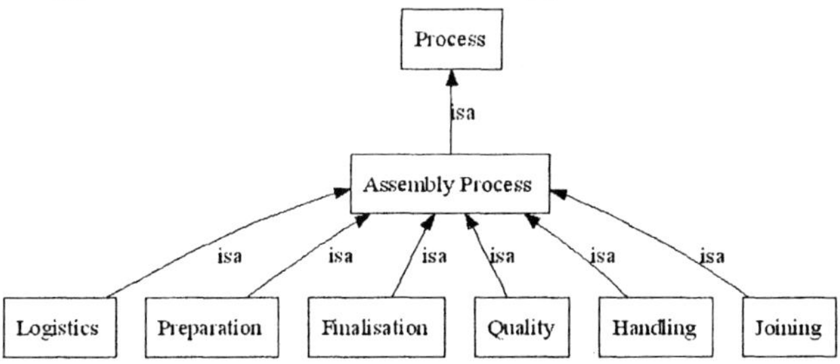

Figure 2 – Part of the taxonomy of the assembly process

4. CONCLUSION

Results are currently being published (Barata et al. 2006), and include methodologies for selecting modules, ontologies, and Reference Architectures. This article has focussed on the underlying principles. This is important since a new paradigm is taking form. As stated earlier, the underlying foundation of EAS is evolvability. Evolvability is, in rough terms, the most advanced form of adaptability. Hence it implies that the systems must adapt to changes. Changes may be either predictable or unpredictable in nature. Changes may occur at internal system level or external gobal level. Unpredictable behaviours that may affect EAS systems are termed as "emergent" and can be either great opportunities or extremely disruptive. A solution that cannot even attempt to tackle emergent events cannot be deemed as Evolvable. Therefore, the work is now focusing on Reference Architectures and validation cases.

5. ACKNOWLEDGEMENTS

This work has been partially done with the support of the European Commission through the Integrated Project EUPASS. The authors would also like to thank all their colleagues in the EUPASS project.

6. REFERENCES

1. Alsterman, H., Barata, J., and Onori, M. (2004). "Evolvable Assembly Systems Platforms: Opportunities and Requirements." Intelligent Manipulation and Grasping, R. Molfino, ed., IMG'2004, Genova, 18-23.
2. Barata, J. (2005). *Coalition Based Approach For ShopFloor Agility*, Edições Orion, Amadora - Lisboa.
3. Barata, J., Camarinha-Matos, L. M., and Onori, M. "A Multiagent Based Control Approach for Evolvable Assembly Systems." *INDIN 05 - 3rd International IEEE Conference on Industrial Informatics*, Perth - Australia.
4. Barata, J., Onori, M., and Frei, R. "Applying Evolvable Assembly Systems." *ISIE'06 - IEEE International Symposium on Industrial Electronics*, Montreal - Canada.
5. Castelfranchi, C., Micelli, M., and Cesta, A. (1992). "Dependence Relations Among Autonomous Agents." Decentralized A.I. 3, E. Werner and Y. Demazeau, eds., Elsevier Science Publishers B. V, Amsterdam, NL, 215-227.
6. Gómez-Pérez, A., López, M. F., and Corcho, O. (2004). *Ontological Engineering*, Springer-Verlag, London.
7. Kirsten, M. (2002). "Present: Trends in the Western European Electronics ", Reed Electronics Research.
8. Klusch, M., and Gerber, A. (2002). "Dynamic Coalition Formation Among Rational Agents." *IEEE Intelligent Systems*, 17(3), 42-47.
9. Onori, M. "Evolvable Assembly Systems - A New Paradigm?" *International Symposium on Robotics*, Stockholm, Sweden.
10. Pechoucek, M., Marik, V., and Bárta, J. (2002). "A Knowledge-based Approach to Coalition Formation." *IEEE Intelligent Systems*, 17(3), 17-25.
11. Schreiber, A. T., Wielinga, B. J., De Hoog, R., Akkermans, J. M., and Van de Velde, W. (1994). "CommonKADS: A Comprehensive Methodology for KBS Development." *Ieee Intelligent Systems & Their Applications*, 9(6), 28-37.
12. Shehory, O., and Kraus, S. (1995). "Coalition Formation among Autonomous Agents: Strategies and Complexity." From Reaction to Cognition, C. Castelfranchi and J. P. Muller, eds., Springer-Verlag, Heidelberg, 57-72.

35 A CELLULAR NEURAL NETWORK FOR DEFORMABLE OBJECT MODELLING

Y. Zhong[1], B. Shirinzadeh[1], X. Yuan[2], G. Alici[3] and J. Smith[4]
[1]Department of Mechanical Engineering, Monash University, Australia
[2]School of Computer Science, University of Windsor, Canada
[3]School of Mechanical, Materials, and Mechatronics Engineering
University of Wollongong, Australia
[4]Monash Medical Centre, Department of Surgery, Monash University, Australia

This paper presents a new methodology for the deformation of soft objects by drawing an analogy between cellular neural network (CNN) and elastic deformation. An improved CNN model is developed to simulate the deformation of soft objects. A finite volume based method is presented to derive the discrete differential operators over irregular nets for obtaining the internal elastic forces. The proposed methodology not only models the deformation dynamics in continuum mechanics, but it also simplifies the complex deformation problem with simple setting CNN templates.

1. INTRODUCTION

Deformable object modelling is essential for many industrial and medical applications such as assembly and disassembly of flexible parts, and surgery simulation. To this end, a significant amount of research efforts have been dedicated to simulating the behaviours of deformable objects. These research efforts can be divided into two classes. The first class of studies is focused on real-time simulation such as mass-spring models [Choi et al, 2003] and spline surfaces used for deformation simulation and visualization [Bockholt et al, 1999]. The advantage of these methods is that the computation is less time consuming and the algorithm is easier to be implemented. However, the method does not allow accurate modelling of material properties, and more importantly, increasing the number of springs leads to a stiffer system. The other class of investigations focuses on deformation modelling using techniques such as Finite Element Method (FEM) [Basdogan et al, 2004] and Boundary Element Method (BEM) [Monserrat et al, 2001]. In FEM or BEM, rigorous mathematical analysis based on continuum mechanics is applied to accurately model the mechanical behaviours of deformable objects. However, these methods are computationally expensive and are typically simulated off-line. The pre-calculation [Monserrat et al, 2001], matrix condensation [Bro-Nielsen, 1998], the space and time adaptive level-of-detail [Debunne et al, 2001] and explicit finite element [Cotin et al, 2000] techniques are used to enhance the computational performance.

Please use the following format when citing this chapter:

Zhong, Y., Shirinzadeh, B., Yuan, X., Alici, G., Smith, J., 2006, in IFIP International Federation for Information Processing, Volume 220, Information Technology for Balanced Manufacturing Systems, ed. Shen, W., (Boston: Springer), pp. 329–336.

There are several investigations that combine neural network with deformable modelling [Nurnberger et al, 2001; Duysak et al, 2003]. However, in these methods, neural networks are mainly used to determine the parameters of mass-spring models. To the best of our knowledge, this study is the first to directly use neural network techniques to mimic the behaviours of deformable objects under externally applied loads.

This paper presents a new methodology for accurate modelling deformable objects by drawing an analogy between cellular neural network (CNN) and elastic deformation. The deformation is formulated as a dynamic CNN. An improved autonomous CNN model is developed to mimic the deformation of soft objects through the CNN activity. The internal elastic forces are incorporated in the CNN model as the local connectivity of cells and the external applied force as the independent current source to model the deformation dynamics. A finite volume based method is presented to derive the discrete differential operators over irregular nets. The finite volume method enforces the conservation of energy in a discrete sense and provides an intuitively geometric discretization rather than an interpolating function in the finite element method to calculate internal elastic forces. The proposed methodology not only models the physical dynamics of soft object deformation, but it also simplifies the complex deformation problem with simple setting CNN templates.

2. DESIGN OF CNN MODEL

2.1 CNN Analogy

A CNN model can be applied to different grid types. Without loss of generality, we consider a CNN on a rectangular grid with M rows and N columns. Each node on the grid is occupied by a cell. The dynamics of the array of M×N cells is described by the following equation and conditions [Chua and Yang, 1988]:

$$C\frac{dv_{xij}(t)}{dt} = -\frac{1}{R_x}v_{xij}(t) + \sum_{(k,l)\in N_r(i,j)} A(i,j;k,l)v_{ykl}(t) + I_{ij} \tag{1}$$

where *(i, j)* refers to the cell associated with the node under consideration, *(k,l)* to a cell in the neighborhood of the cell *(i, j)*, namely $N_r(i, j)$, within a radius *r* of the cell *(i, j)* (r=1 for simplicity). *C* is the capacitance of a linear capacitor, R_x is the resistance of a linear resistor, I_{ij} is the current of the independent linear/nonlinear current source, and *A* is the feedback template. $v_{uij}(t)$, $v_{xij}(t)$, and $v_{yij}(t)$ denote the input, state and output of the cell *(i,j)* at the time *t*, respectively. $v_{yij}(t)$ is a non-linear function of $v_{xij}(t)$.

One significant feature of CNN, as well as the basic difference from other neural networks, is the local connectivity of cells [Chua and Yang, 1988], i.e. any cell in CNN is connected only to its neighbouring cells. Adjacent cells directly interact with each other. Cells not directly connected to each other have indirect effect because of the propagation effects of the continuous-time dynamics of CNN.

Therefore, given the initial state and the external environment, CNN activity is only determined by the local connectivity of cells. The local connectivity of cells is similar to the internal force since the deformation is only determined by the internal force under the given external force and the initial state.

Another significant feature of CNN is that the individual cells are non-linear dynamical systems, but that the coupling between them, i.e. the local connectivity of cells, is linear [Slavova, 2003]. The feature makes CNN very suitable for modelling non-linear materials since CNN conserves the physical properties of the continuous structure.

Further, CNN offers an incomparable computation speed due to the collective and simultaneous activity of all cells [Roska et al, 1995; Kozek et al, 1995]. The computation advantage of CNN is very suitable for real-time the computation requirement of deformable object simulation.

In the proposed CNN analogy, the deformation of soft objects is treated as the activity of a CNN. The object surface is treated as a CNN by using a number of locally connected cells. The external force is treated as the current source of the contact cell. The local interactions generated by the local connectivity of cells are treated as internal forces, and the CNN dynamic activity is treated as the dynamics of deformation. Therefore, such a CNN with the current source, the local connectivity and the dynamic activity can be seen as a communication medium among an external force, internal forces and deformation.

2.2 CNN Model for Deformation

The dynamics of soft object deformation is usually described by the Lagranage dynamics. Therefore, the CNN model is formulated as the Lagrange dynamics to describe the dynamics of soft object deformation.

When an external force is applied to a soft object, the contact point of the external force is replaced with a new position. As a result, the other points not influenced by the external force are in an unstable state. The external force is propagated among mass points to establish a new equilibrium state by generating the corresponding internal forces. Based on the equilibrium state, the new position of each point is obtained. The dynamic deformation process is governed by the Lagrangian equation, which is to balance the externally applied force with the internal forces due to the deformation of the soft object. The Lagrange equation that governs the motions of each node is [Goldsteln, 1980]:

$$m_i \frac{d^2\mathbf{U}_i}{dt^2} + \gamma_i \frac{d\mathbf{U}_i}{dt} + \mathbf{G}_i = \mathbf{F}_i \tag{2}$$

where $\mathbf{U}_i$ is the position vector of node i at time t, and m_i and γ_i are the mass and damping constants of node i, respectively, $\mathbf{F}_i$ is the external force applied to node i at time t, and $\mathbf{G}_i$ is the net internal force applied to node i at time t. In the case of linear elasticity, the linear strain makes the internal force linear with respect to the position vector, i.e.

$$\mathbf{G}_i = K_i \mathbf{U}_i \tag{3}$$

where K_i is a coefficient at node i, which is related with the stiffness of a material.

The second-order differential equations can be divided into a system of first order differential equations by introducing a velocity function $\mathbf{V}_i$.

$$m_i \frac{d\mathbf{V}_i}{dt} = \mathbf{F}_i - K_i \mathbf{U}_i - \gamma_i \mathbf{V}_i \qquad \frac{d\mathbf{U}_i}{dt} = \mathbf{V}_i \tag{4}$$

According to Eq. (4), the CNN model for dynamic deformations is designed as a second-order autonomous CNN:

$$C\frac{dv_{xij}(t)}{dt} = -\frac{1}{R_x}v_{xij}(t) + \sum_{(k,l)\in N_r(i,j)} A(i,j;k,l)v_{ykl}(t) + I_{ij} \qquad \frac{dv_{yij}}{dt} = v_{xij} \tag{5}$$

where

$$v_{yij} \to \mathbf{U}_{ij} \quad v_{xij} \to \mathbf{V}_{ij} \quad I_{ij} \to \mathbf{F}_{ij} \quad C = m_i \quad R = \frac{1}{\gamma_i} \tag{6}$$

In most of CNN applications, the independent current source of each cell has the same value. For the purpose of soft object deformation, the independent current source is only set by the external force. Therefore, the current source value at the contact cell is directly obtained from the external force, and the current source values of other cells are set to zero since there is no external force applied to them.

Due to the fact that any CNN has only one output, a three-lay CNN is constructed to compute the X, Y and Z coordinates of displacement vectors.

The solution of the Lagrangian equation needs initial values and boundary conditions. The initial values and boundary conditions should also be incorporated in the CNN model. The initial values can be directly associated with the initial state of the CNN. The simplest boundary condition is the Dirichlet boundary condition, i.e. the given boundary values. The Dirichlet boundary condition is realized by using fixed-state cells.

3. FORMULATION OF LOCAL CONNECTIVITY OF CELLS

From the above section, it is obvious that the local connectivity of cells is related to the internally elastic forces. A straightforward approach to derive the internal elastic force is the Hooke's law, and the relationship between the displacement u and the applied internal elastic force F is [Timoshenko and Goodier, 1970]:

$$\mathbf{F} = S[\mu\Delta\mathbf{u} + (\lambda + \mu)\nabla(\nabla \bullet \mathbf{u})] \tag{7}$$

where Δ is the Laplace operator, ∇ is the gradient operator, $\nabla \bullet$ is the divergence operator and S is the measure of the area the force applied on.

To get the internal elastic force at each node for setting the local connectivity of cells, Laplace operator and the gradient-of-divergence operator has to be discretized on the object surface. The discretization of these operators on a regular net is straightforward, and can be easily done by the techniques of finite difference. The difficulty is how to discretize these operators on an irregular net.

3.1 Finite Volume Method for Discrete Operators

For the discretization of the operators on an irregular net, a finite volume method [Versteeg and Malalasekera, 1995] is used to aid establishing the discretization at each node. The finite volume method has the strength of dealing with unstructured grids and has been widely used in computational fluid and heat transfer. One straightforward finite volume method is Voronoi diagram [Barth, 1992], which derives the discretized equation at each node from the energy conservation law. It subdivides the domain into a finite number of non-overlapping cells or control volumes, over which the conservation of energy is enforced in a discrete sense. Fig. 1(a) shows the finite volume constructed by the Voronoi scheme. The finite volume of point $\mathbf{P}_0$ is constructed by connecting each intersection points between the perpendicular centerlines of each edges adjacent to point $\mathbf{P}_0$. Fig. 1(b) shows a segment of the Voronoi finite volume.

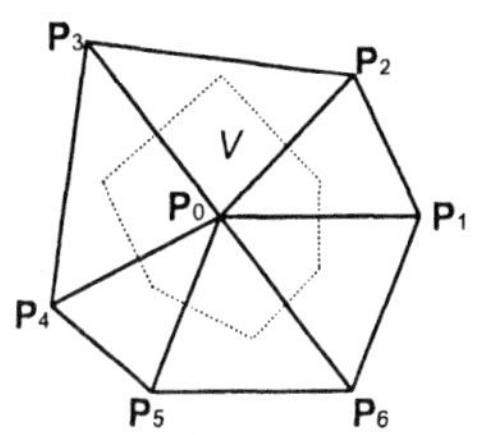

(a) A Voronoi finite volume

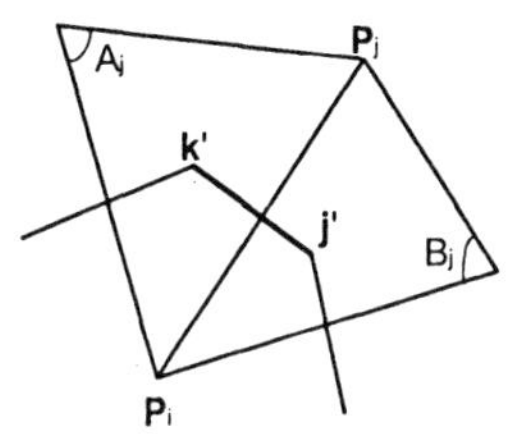

(b) A segment of the finite volume

Figure 1 – Voronoi finite volume on a triangular net

By derivations, Laplace operator at $\mathbf{P}_i$ may be written as:

$$(\Delta \mathbf{u})_{\mathbf{P}_i} = \frac{1}{2S} \sum_{j \in N_i} (\cot A_j + \cot B_j)(\mathbf{P}_j - \mathbf{P}_i) \tag{8}$$

where S is the measure of the finite volume, and N_i is the set of the neighbour points of point $\mathbf{P}_i$ (see Fig. 1(b)).

By derivations, the gradient-of-divergence operator at point $\mathbf{P}_i$ may be written as:

$$\nabla(\nabla \bullet \mathbf{u}) = \frac{1}{4S^2} \sum_{j \in N_i} \Big(\sum_{k \in N_j \cap k \notin N_i} (\mathbf{u}_j + \mathbf{u}_k) \bullet \mathbf{n}_{\overrightarrow{\mathbf{P}_j\mathbf{P}_k}} \Big) \mathbf{n}_{\overrightarrow{\mathbf{P}_i\mathbf{P}_j}} \tag{9}$$

where $\mathbf{n}_{\overrightarrow{\mathbf{P}_i\mathbf{P}_j}} = \dfrac{\overrightarrow{\mathbf{P}_i\mathbf{P}_j}}{\left\| \overrightarrow{\mathbf{P}_i\mathbf{P}_j} \right\|}$.

3.2 Formulation for the Local Connectivity of Cells

The local connectivity of cells is set by the internally elastic forces. Replacing Eqs. (8) and (9) in Eq. (7), the internal force at $\mathbf{P}_i$ is:

$$\mathbf{F}_i = \frac{1}{2}\sum_{j\in N_i}\left(\mu(\cot A_j + \cot B_j) + \frac{\lambda+\mu}{2S\left\|\overrightarrow{\mathbf{P}_i\mathbf{P}_j}\right\|}\sum_{k\in N_j \cap k\notin N_i}(\mathbf{P}_j + \mathbf{P}_k)\bullet \mathbf{n}_{\overline{\mathbf{P}_j\mathbf{P}_k}}\right)(\mathbf{P}_j - \mathbf{P}_i) \qquad (10)$$

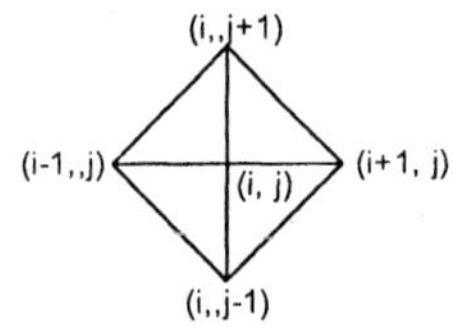

Figure 2 – A triangular net

From Eq. (10), the local connectivity of the CNN model, i.e. the template A can be obtained. For example, assuming the object surface is a net shown in Fig. 2, the template A is:

$$A_{i,j;i,j} = -\frac{1}{2}\sum_{m,n\in N_{i,j}}\left(\mu(\cot A_{m,n} + \cot B_{m,n}) + \frac{\lambda+\mu}{2S\left\|\mathbf{P}_{m,n} - \mathbf{P}_{i,j}\right\|}\sum_{k,l\in N_{m,n}\cap k,l\notin N_{i,j}}(\mathbf{P}_{m,n} + \mathbf{P}_{k,l})\bullet\frac{\mathbf{P}_{k,l} - \mathbf{P}_{m,n}}{\left\|\mathbf{P}_{k,l} - \mathbf{P}_{m,n}\right\|}\right)$$

$$A_{i,j;i-1,j} = \frac{1}{2}\left(\mu(\cot A_{i-1,j} + \cot B_{i-1,j}) + \frac{\lambda+\mu}{2S\left\|\mathbf{P}_{i-1,j} - \mathbf{P}_{i,j}\right\|}\sum_{k,l\in N_{i-1,j}\cap k,l\notin N_{i,j}}(\mathbf{P}_{i-1,j} + \mathbf{P}_{k,l})\bullet\frac{\mathbf{P}_{k,l} - \mathbf{P}_{i-1,j}}{\left\|\mathbf{P}_{k,l} - \mathbf{P}_{i-1,j}\right\|}\right)$$

$$A_{i,j;i+1,j} = \frac{1}{2}\left(\mu(\cot A_{i+1,j} + \cot B_{i+1,j}) + \frac{\lambda+\mu}{2S\left\|\mathbf{P}_{i+1,j} - \mathbf{P}_{i,j}\right\|}\sum_{k,l\in N_{i+1,j}\cap k,l\notin N_{i,j}}(\mathbf{P}_{i+1,j} + \mathbf{P}_{k,l})\bullet\frac{\mathbf{P}_{k,l} - \mathbf{P}_{i+1,j}}{\left\|\mathbf{P}_{k,l} - \mathbf{P}_{i+1,j}\right\|}\right) \qquad (11)$$

$$A_{i,j;i,j-1} = \frac{1}{2}\left(\mu(\cot A_{i,j-1} + \cot B_{i,j-1}) + \frac{\lambda+\mu}{2S\left\|\mathbf{P}_{i,j-1} - \mathbf{P}_{i,j}\right\|}\sum_{k,l\in N_{i,j-1}\cap k,l\notin N_{i,j}}(\mathbf{P}_{i,j-1} + \mathbf{P}_{k,l})\bullet\frac{\mathbf{P}_{k,l} - \mathbf{P}_{i,j-1}}{\left\|\mathbf{P}_{k,l} - \mathbf{P}_{i,j-1}\right\|}\right)$$

$$A_{i,j;i,j+1} = \frac{1}{2}\left(\mu(\cot A_{i,j+1} + \cot B_{i,j+1}) + \frac{\lambda+\mu}{2S\left\|\mathbf{P}_{i,j+1} - \mathbf{P}_{i,j}\right\|}\sum_{k,l\in N_{i,j+1}\cap k,l\notin N_{i,j}}(\mathbf{P}_{i,j+1} + \mathbf{P}_{k,l})\bullet\frac{\mathbf{P}_{k,l} - \mathbf{P}_{i,j+1}}{\left\|\mathbf{P}_{k,l} - \mathbf{P}_{i,j+1}\right\|}\right)$$

$$A_{i,j;i-1,j-1} = 0 \quad A_{i,j;i-1,j+1} = 0 \quad A_{i,j;i+1,j-1} = 0 \quad A_{i,j;i+1,j+1} = 0$$

4. IMPLEMENTATION RESULTS AND DISCUSSIONS

A prototype system has been implemented for interactive deformable object modelling with force feedback. A PHANToM haptic device is configured with the system to carry out the deformations of deformable objects with force feedback. The PHANToM is a six degree-of-freedom haptic device from Sensable Technologies, and provides force feedback to a user. A user can use the device to touch the soft object and deform it by pushing, pulling and dragging the object surface in a natural 3D environment.

Fig. 3 illustrates the deformations of an elliptic sphere with 400 mass points. Fig. 3(a) shows the undeformed elliptic sphere. Fig. 3(b) and (c) show different views of the object deformed under a tensile force. Fig. 4 shows the deformation modelling of a shoes-shaped object under a compressive force.

Compared to the mass-spring method, the proposed methodology is more accurate than mass-spring due to the formulation of the local connectivity of cells based on continuum mechanics. Compared to BEM and linear FEM, the proposed methodology has the computational advantage due to the collective and simultaneous activity of all cells. In addition, the CNN model is easier to be formulated than the complex BEM and linear FEM models. Only surface mass points are involved in computation and rendering without any inside points, while the interior must be meshed and calculated in FEM.

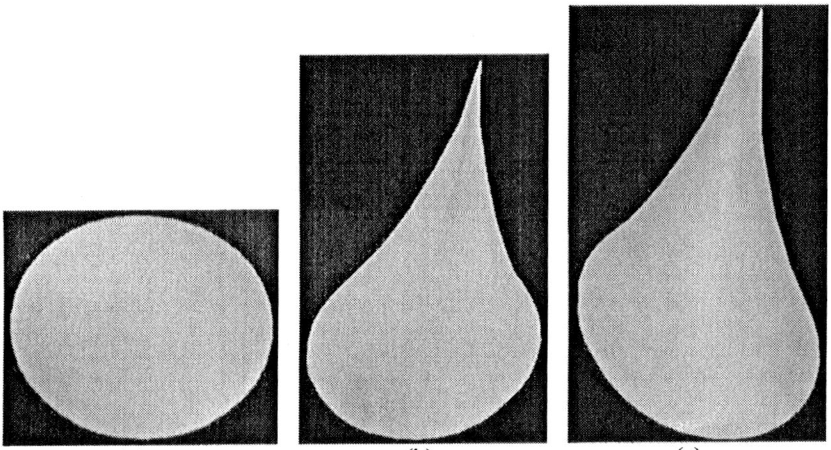

(a) (b) (c)

Figure 3 – Deformations of an elliptic sphere

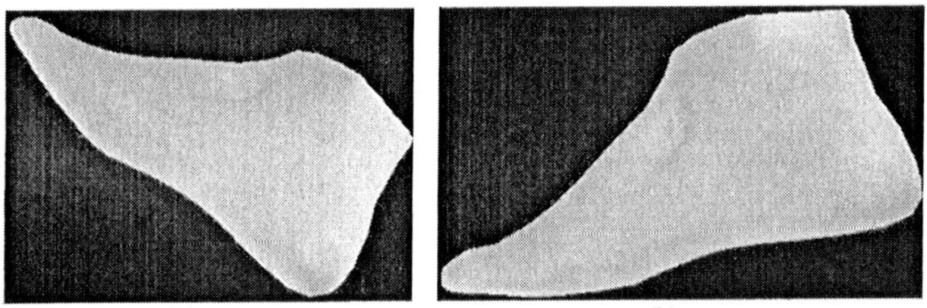

(a) (b)

Figure 4 – Deformations of a shoes-shaped object

5. CONCLUSIONS

Presented in this paper is a new methodology to mimic the deformation of soft objects by drawing an analogy between CNN and elastic deformation. The contribution of this paper is that the CNN techniques are used to describe the deformation dynamics through incorporating the externally applied force and the internally elastic forces, and new discrete differential operators are derived to ensure the conservation of energy and provide the geometric intuitiveness. An improved CNN model is developed to describe the dynamics of soft object deformation. A finite volume based method is presented to derive the discrete differential operators over irregular nets for obtaining the internal elastic forces. The finite volume method enforces the conservation of energy in a discrete sense and also provides an intuitively geometric discretization rather than an interpolating function in finite element methods to calculate internal elastic forces. The proposed methodology not only models the physical dynamics of soft object deformation, but it also simplifies the complex deformation problem with simple setting CNN templates.

6. ACKNOWLEDGEMENTS

This paper is supported by Australian Research Council (ARC) and NSERC of Canada.

7. REFERENCES

1. Choi K., Sun H. and Heng P. A. Interactive deformation of soft tissues with haptic feedback for medical learning, IEEE Transaction on Information Technology in Biomedicine, vol. 7, no. 4, 2003, pp358-363
2. Bockholt U., Müller W. and Voss G. (etc.). Real-time simulation of tissue deformation for the nasal endoscopy simulator (NES), Computer aided surgery, vol. 4, no. 5, 1999, pp281-285
3. Basdogan C., De S. and Kim J. (etc.). Haptics in minimally invasive surgical simulation and training, IEEE Computer Graphics and Applications, vol. 24, no. 2, 2004, pp56-64
4. Monserrat C., Meier U. and Alcaniz M. (etc.). A new approach for the real-time simulation of tissue deformations in surgery simulation, Computer Methods and Programs in Biomedicine, vol. 64, 2001, pp77-85
5. Bro-Nielsen M. Finite element modeling in surgery simulation, Proceedings of The IEEE, vol. 86, no. 3, 1998, pp490-503
6. Debunne G., Desbrun M., Cani M. P. and Barr A. H. Dynamic real-time deformations using space & time adaptive sampling, Proceedings of the 28th annual conference on Computer graphics and interactive techniques, Los Angeles, USA, 2001, pp31-36
7. Cotin S., Delingette H. and Ayache N. A hybrid elastic model allowing real-time cutting, deformations and force-feedback for surgery training and simulation, The Visual Computer, vol. 16, no. 8, 2000, pp437-452
8. Nurnberger A., Radetzky A. and Kruse R. Using recurrent neuro-fuzzy techniques for the identification and simulation of dynamic systems, Neurocomputing, vol. 36, 2001, pp123-147
9. Duysak A., Zhang J. J. and Ilankovan V. Efficient modelling and simulation of soft tissue deformation using mass-spring systems, International Congress Series, vol. 1256, 2003, pp337-342
10. Chua L. O. and Yang L., Cellular neural network: Theory, IEEE Transactions on Circuits and Systems, vol. 35, no. 10, 1988, pp1257-1272
11. Slavova A. Cellular Neural Networks: Dynamics and Modelling, Dordrecht, London: Kluwer Academic Publishers, 2003
12. Roska T., Chua L. O., Wolf D., Kozek T., Tetzlaff R. and Puffer F. Simulation nonlinear waves and partial differential equations via CNN—Part I: Basic Techniques, IEEE Transactions on Circuits and Systems, Vol. 42, No. 10, 1995, pp. 807-815
13. Kozek T., Chua L. O., Roska T., Wolf D., Tetzlaff R., Puffer F. and Lotz K. Simulating nonlinear waves and partial differential equations via CNN—Part II: typical examples, IEEE Transactions on Circuits and Systems, vol. 42, no. 10, 1995, pp. 816-820
14. Goldsteln H. Classical Mechanics, Second Edition, Reading, MA: Addison-Wesley, 1980.
15. Timoshenko S. P. and Goodier J. N. Theory of Elasticity. McGraw-Hill, 1970
16. Versteeg H. K. and Malalasekera W. An introduction to computational fluid dynamics: the finite volume method, Harlow, Essex, England: Longman Scientific & Technical, 1995
17. Barth T. J. Aspects of unstructured grids and finite-volume solvers for the Euler and Navier-Stokes equations, AGARD Report 787, 1992, pp6.1-6.61

PART D

MONITORING AND CONTROL

36 A LOAD BALANCING METHOD FOR DEDICATED PHOTOLITHOGRAPHY MACHINE CONSTRAINT

Arthur Shr[1], Alan Liu[1], Peter P. Chen[2]

[1]*Department of Electrical Engineering, National Chung Cheng University Chia-Yi 621, Taiwan, R.O.C.*

arthurshr@gmail.com, aliu@ee.ccu.edu.tw

[2]*Department of Computer Science, 298 Coates Hall, Louisiana State University Baton Rouge, LA 70803, U.S.A.*

pchen@lsu.edu

The dedicated photolithography machine constraint in semiconductor manufacturing is one of the new issues of photolithography machinery due to natural bias. In this paper, we propose the heuristic Load Balancing (LB) scheduling approach based on a Resource Schedule and Execution Matrix (RSEM) to tackle this constraint. The LB method is to schedule each wafer lot at the first photolithography stage to a suitable machine, according to the load balancing factors among machines. We describe the proposed LB scheduling method and present an example to demonstrate the proposed method and the result of the simulations to validate the approach.

1. INTRODUCTION

Semiconductor manufacturing systems are different from the traditional manufacturing operations, such as a flow-shops manufacturing system in assembly lines or a job-shops manufacturing system. In a semiconductor factory, one wafer lot passes through hundreds of operations, and the processing procedure takes a few months to complete. The operations of semiconductor manufacturing incrementally develop an IC product layer by layer.

To solve the complex and challenging scheduling problems in the semiconductor manufacturing, many queuing network scheduling polices or methods have been published to formulate the complexity of semiconductor manufacturing problems. These scheduling policies deal with the buffer competing problem in the re-entrant production line (Kumar, 1993), wherein they pick up the next wafer lot in the queue buffers when machines are becoming idle. Two scheduling policies have been proposed to reduce the mean and variance of product cycle time (Kumar, 1994) (Lu, 1994). Wein's research used a Brownian queuing network model to approximate a multi-class queuing network model with dynamic control to the process in the semiconductor factory (Wein, 1988). SDA-F, a special family-based scheduling rule, uses a rule-based algorithm with threshold control and the least slack principle

---

Please use the following format when citing this chapter:

Shr, A., Liu, A., Chen, P. P., 2006, in IFIP International Federation for Information Processing, Volume 220, Information Technology for Balanced Manufacturing Systems, ed. Shen, W., (Boston: Springer), pp. 339–348.

to dispatch wafer lots in photolithography stages (Chern, 2003). A study proposed a stochastic dynamic programming model for scheduling new wafer lot release and bottleneck processing by stage in the semiconductor factory (Shen, 2003). One other research used the Petri Net approach to modeling, analysis, simulation, scheduling and the control of the semiconductor manufacturing system (Zhou, 1998).

Recently, a study concerning the load balancing issue developed a load balance allocation function by applying a dynamic programming method to the machine constraint in the photolithography machines (Miwa, 2005). Two approaches were reported to use simulations to model the photolithography process, and one of them proposed a Neural Network approach to the photolithography scheduling problem (Arisha, 2004). The approach followed a qualifying matrix and the lot scheduling criteria to improve the performance of the photolithography machines. The other is to decide the wafer lots assignment of the photolithography machines when the wafer lots were released to the manufacturing system to improve the load balancing problem in the photolithography area (Mönch, 2001).

Although these scheduling polices or methods have been developed and applied in the semiconductor factories, they did not concern the dedicated photolithography machine constraint, which will cause load unbalancing among the photolithography machines in the semiconductor factory. In fact, the wafer lots in a load unbalancing semiconductor factory usually need to be switched from the highly congested machines to the idle machines. This takes much time and relies on experienced engineers to manually handle alignment problems of the wafer lots with a different situation off-line. It is inefficient to determine, one lot at a time, which wafer lot and machine need to be switched. This method cannot meet the fast-changing market of the semiconductor industry.

In this paper, motivated by the issue described above, we propose a Load Balancing (LB) scheduling approach based on the Resource Schedule and Execution Matrix (RSEM) developed in our previous research work (Shr, 2006a) (Shr, 2006b) (Shr, 2006c) to tackle the dedicated machine constraint and load balancing issue.

The paper is organized as follows: Section 2 describes the dedicated photolithography machine constraint in detail. Section 3 presents our proposed LB scheduling method based on RSEM to apply to the dedicated constraint issues of semiconductor factory. Section 4 shows the simulation results that validated our approach. Section 5 discuses the conclusion.

2. DEDICATED PHOTOLITHOGRAPHY MACHINE CONSTRAINT

One of the challenges in the semiconductor manufacturing systems is the dedicated photolithography machine constraint which happens because of the natural bias of the photolithography machine. Natural bias will impact the alignment of patterns between different layers. The smaller the dimension of the IC products (wafers), the more difficult they will be to align between different layers, especially when we move on to a smaller dimension IC for high technology products. The wafer lots passing through each photolithography stage process have to be processed on the same machine. The purpose of the limitation is to prevent the impact of natural bias and to keep a good yield of the IC product. A research considered different process control policies including the machine dedication policy in their simulation study for semiconductor manufacturing has reported that the machine dedication policy had the worst performance of photolithography process (Akcalt, 2001). The machine dedication policy is similar to the dedicated machine constraint we are discussing here. Figure 1 describes the circumstance of the dedicated machine constraint.

With this dedicated photolithography machine constraint, when wafer lots enter each photolithography operation stage and if the wafer lots have been dedicated to the machine ***X***, they need to be processed and wait for ***X***. They cannot be processed by other machines, e.g., the machine ***Y***, even if ***Y*** is idle. On the other hand, when wafer lots enter into other stages, without any machine constraints, the wafer lots can be scheduled to any machine of ***A***, ***B*** or ***C***.

The photolithography process is the most important process in semiconductor manufacturing since the yield of IC products is always dependent on a good photolithography process, while at the same time the process can also cause defects. Not surprisingly, the performance of the factory will rely on the photolithography machines. Therefore, the dedicated photolithography machine constraint is the most important challenge to improve productivity and fulfill the request for customers as well as the main contributor to the complexity and uncertainty of semiconductor manufacturing.

This load balancing issue is derived mainly from the dedicated photolithography machine constraint. This is because once the wafer lots have been scheduled to one of the machines at the first photolithography stage, they must be assigned to the same machine in the subsequent photolithography stages until they have passed the last photolithography stage. Therefore, any unexpected abnormal events or breakdown of one machine will cause a pile-up of many wafer lots waiting for the machine and make it critical to the factory. Some of the photolithography machines will become idle and remain so for a while, due to the fact that no wafer lots can be processed, and the other will always be busy while many wafer lots bound to this machine are awaiting processing. As a result, some wafer lots will never be delivered to the customer on time, and the performance of the factory will have been decreased and impacted.

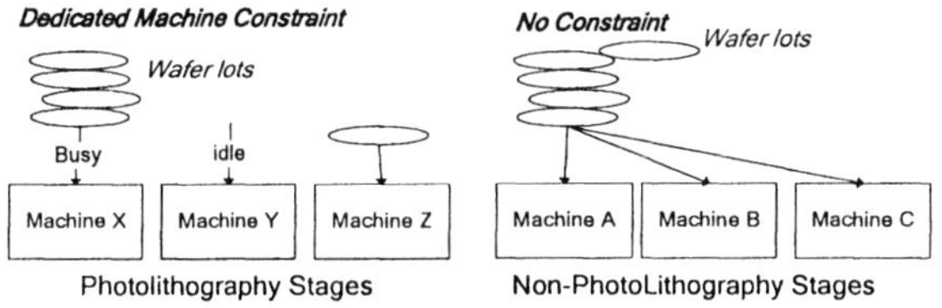

Figure 1–Dedicated photolithography machine constraint

3. LOAD BALANCING SCHEDULING APPROACH

In this section, we apply the Load Balancing (LB) scheduling approach to the dedicated machine constraint of the photolithography machine in semiconductor manufacturing. The LB approach uses the RSEM as a tool to represent the temporal relationship between the wafer lots and machines during each scheduling step. We use an example to demonstrate the LB approach in the following.

The RSEM construction process consists of three modules including *Task Generation, Resource Calculation,* and *Resource Allocation* module. The first module is to model the tasks for the scheduling system and it is represented in a two-dimensional matrix. We generate the two-dimension matrix for the tasks that are going be processed by machines. One dimension is reserved for the tasks $t_1, t_2, \ldots, t_n$, the other is to represent the periodical time events (or steps), $s_1, s_2, \ldots, s_m$. Each task has a sequential *Process Pattern* to represent the resources it needed during the process sequence from a raw material to a product and we put the process pattern in an array. We define each type resource as r_k, $k=1$ *to* o, e.g., the process pattern, $r_1, r_2, \ldots, r_o$, means that a particular task needs the resources in the sequence of r_1 first and r_2 following that until r_o is gained. Therefore, the matrix looks as follows:

	s_1	s_2	.	.	.	.	.	s_j	.	.	s_m
t_1	r_1	r_2	r_3	..	..	..	r_k.	..	..	r_o	..
t_2		r_3	r_4	..	..	r_k	..	..	..	..	..
.							..	..			
t_i					r_3	r_4	..	..	r_k	..	
.							.	.			
t_n					..	..	r_k	..	..	..	

The symbol r_k in the matrix entry $[t_i, s_j]$ is to represent the task t_i needs the resource r_k at the time s_j. If t_i starts to the process at s_j in the system and the total numbers of steps needed for t_i is p, we will fill its process pattern into the matrix from $[t_i, s_j]$ to $[t_i, s_{j+p-1}]$ with r_k. All the tasks, $t_1, \ldots t_n$, follow the illustration above to form a task matrix in the task generation module. To represent the dedicated machine constraint, the symbol r_k^x, a replacement of r_k, is to represent that t_i has been dedicated to a particular instance x of a resource type r_k at s_j. One more symbol w_k represents the wait situation when the r_k cannot serve t_i at s_j. We will insert this symbol in the *Resource Allocation* module later.

The *Resource Calculation* module is to summarize the value of each dimension as the factors for the scheduling rules of the *Resource Allocation.* For example, we can determine how many steps t_i is processed by counting task pattern of the row, t_i row in the matrix. We can also realize how many wait steps t_i has had by counting w_k from the starting step to the current step in that row of the matrix.

We need to generate the task matrix, obtain all the factors for the scheduling rules, and build up the scheduling rules before starting the execution of the *Resource Allocation* module. The module is to schedule the tasks to the suitable resource according to the factors and predefined rules. To represent the situation of waiting for r_k, i.e., when the resource of r_k is not available for t_i at s_j, then we will not only insert w_k in $[t_i, s_j]$ of the matrix, but also shift one step for the process pattern following of t_i.

After obtaining the process flow for customer product from the database of semiconductor manufacturing, we can use a simple program to transform the process flow into the process pattern and task matrix representation. For a typical factory, there are thousands of wafer lots and hundreds of process steps. There is an example to transform the process pattern of wafer lots into a task matrix. We let r_2 represent the photolithography machine and the others are to represent non-photolithography machines. The symbol r_2^x in the matrix entry $[i,j]$ is to represent t_i need of the photolithography machine x at s_j with dedicated machine constraint, while the r_k $(k \neq 2)$ in $[i,j]$ is to represent t_i needing the machine type k at s_j without dedicated machine constraint. There is no assigned machine number for the photolithography machine before the wafer lot has passed first photolithography stage.

Suppose that $r_1r_3r_2r_4r_5r_6r_7r_2r_4r_5r_6r_7r_8r_9r_1r_3r_2r_4r_5r_6r_7r_3r_2r_8r_9$ is the process pattern of t_1 and it starts to release to the factory at s_1, we will fill its process pattern into the task matrix from $[t_1,s_1]$ to $[t_1,s_{25}]$, which indicates that t_1 needs the resource r_1 at the first step, resource r_3 at the second step, and r_9 at the last step. The photolithography process, r_2, in this process pattern has not dedicated to any machine yet and total of the steps for t_1 is 25. The wafer lot t_2 in the following task matrix has the same process pattern as t_1 has but it starts at s_3. The wafer lot t_i in the task matrix starts from s_8 and it requires the photolithography machine, but the machine is different from the machine t_2 needed at s_{10}, i.e., t_2 needs the machine m_1, while t_i has not been dedicated to any machine yet. Moreover, at s_{11}, t_2 and t_i might compete for the same resource r_4 if r_4 is not enough for them at s_{11}.

	s_1	s_2	s_3	s_4	s_5	s_6	s_7	s_8	s_9	s_{10}	s_{11}	s_{12}	s_{13}	s_{14}	s_{15}	s_{16}	s_{17}	s_{18}	s_{19}	s_{20}	s_{21}	s_{22}	s_{23}	s_{24}	s_{25}	..	..	..	s_m
t_1	r_1	r_3	$\underline{r_2}$	r_4	r_5	r_6	r_7	$\underline{r_2}$	r_4	r_5	r_6	r_7	r_8	r_9	r_1	r_3	$\underline{r_2}$	r_4	r_5	r_6	r_7	r_3	$\underline{r_2}$	r_8	r_9				.
t_2			r_1	r_3	$\underline{r_2}$	r_4	r_5	r_6	r_7	$\underline{r_2}^1$	r_4	r_5	r_6	r_7	r_8	r_9	r_1	r_3	$\underline{r_2}$	r_4	r_5	r_6	r_7	r_3	$\underline{r_2}$	r_8	r_9		
t_i								r_1	r_3	$\underline{r_2}$	r_4	r_6	r_5	..	..	..	..	..	..	..	..	..	..	..	..	..	..		
..												..	..																

We need to obtain some factors for our proposed LB scheduling approach. The definitions and formulae of these factors are as follows:

W: wafer lots in process,
P: numbers of photolithography machines,
O: types of machine (resource)

(1) How many wafer lots will need the k type machine (photolithography machine, k = 2) at s_j:

$$RR(r_k^x, s_j) = \sum_{t_i \in W} \begin{cases} 1 \text{ if } [t_i, s_j] = r_k^x \\ 0 \text{ otherwise} \end{cases}, \quad 1 \leq x \leq P$$

(2) Step Count:
(2.1) How many wait steps t_i has had before s_j:

$$WaitStep(t_i) = \sum_{j=start}^{current\ step} \begin{cases} 1 \text{ if } [t_i, s_j] = w_k \\ 0 \text{ otherwise} \end{cases}, \quad 1 \leq k \leq O$$

(2.2) How many steps t_i will have.

$$Steps(t_i) = \sum_{j=start}^{end\ step} \begin{cases} 1 \text{ if } [t_i, s_j] \neq null \\ 0 \text{ otherwise} \end{cases}$$

(3) The load factor of the machine m_x, wafer lots × remanding photolithography stages

$$Load\,(m_x, s_j) = \sum_{t_i \in W} \{t_i * R(t_i) \mid pm(t_i) = m_x\}$$

(3.1) How many remaining photolithography stages of t_i:

$$R(t_i) = \sum_{j=current}^{end\ step} \begin{cases} 1 \text{ if } [t_i, s_j] = r_2^x \\ 0 \text{ otherwise} \end{cases}, \quad 1 \le x \le P$$

(3.2) $pm(t_i)$: the photolithography machine number that t_i is dedicated to.

We will use these factors in the third module. In this example, load is defined as the wafer lots limited to the machine m multiply their remaining photolithography layers stage. The larger load factor means that the more required service from the wafer lots which have been limited to this machine. The LB uses these factors to schedule the wafer lots to a suitable machine at the first, unconstrained, photolithography stage.

Suppose we are currently at s_j, and the LB scheduling system will start from the photolithography machine. First, we check if there is any wafer lot which requires the resource of the photolithography machine at the first stage. LB will assign the m_x with smallest Load(m_x, s_j) (formula (3)) for them one by one. After that, these wafer lots have been dedicated to a photolithography machine. For each m_x, the LB will select one of the wafer lots dedicated to m_x which has the largest WaitStep(t_i) for it. Load(m_x, s_j) of m_x will be updated after these two processes. The other wafer lots dedicated to each m_x which can not be allocated to the m_x at this step s_j will insert a w_2 for them in their process pattern. For example, at s_{10}, t_i has been assigned to m_1, therefore, t_{i+1} will have a w_2 inserted into at s_{10}, and then all the following required resource of t_{i+1} will shift one step. The following matrix shows the situation. All the other types of machines will have the same process, without the need of being concerned with the dedicated machine constraint. The LB will schedule one of the wafer lots with the largest WaitStep(t_i), then the second largest one, and so on, for each machine. Similarly, the LB will insert a w_k for the wafer lots not assigned to machines r_k.

	s_9	s_{10}	s_{11}	s_{12}	s_{13}	s_{14}	..	..	s_j	..	s_m
					..	..					
t_i	..	r_2^1	r_4	r_5	r_6	r_7	..	..			
t_{i+1}	..	w_2	r_2^1	r_4	r_6	r_5	..	..	..	..	
..		↑	→	→	→	→					

To better understand our proposed scheduling process, the flowchart of the RSEM is shown in Figure 2.

The process of using the RSEM starts from the *Task Generation* module, and it will copy the predefined task patterns of tasks into the matrix. Entering the *Resource Calculation* module, the factors for the tasks and resources will be brought out at the current step. This module will update these factors again at each scheduling step. The execution of the scheduling process is in the *Resource Allocation* module.

When we have done the schedule for all the tasks for the current step, we will return to check for new tasks and repeat the whole process again by following the flowchart. We will exit the scheduling process when we reach the final step of the last task if there is still no new task appended to the matrix. After that, the scheduling process will restart immediately when the new tasks arrive in the system.

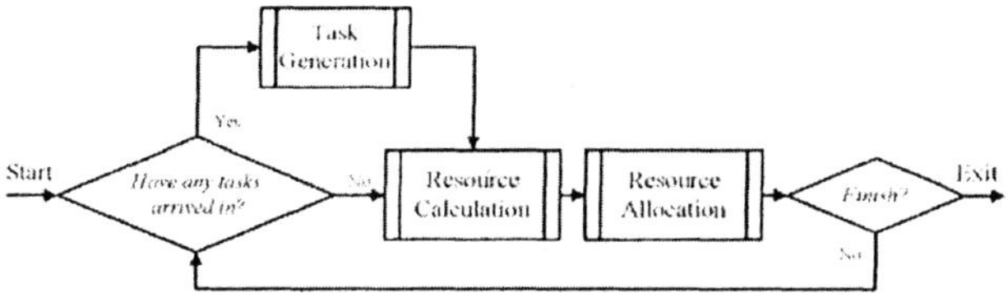

Figure 2–Flowchart of the RSEM

We assume that all the resource types for the wafer lot will have the same process time in this example, i.e. all the steps will have the same time duration. The assumption simplifies the real semiconductor manufacturing system, but could be focused on the issue of the dedicated machine constraint. However, it is not difficult to approach the real world on a smaller scale time step. Most scheduling polices or methods can provide neither the exact allocation in accepted time, nor a robust and systematic resource allocation strategy. We use the RSEM to represent the complex tasks and to allocate resources by the simple matrix calculation. This reduces much of the computation time for the complex problem.

4. SIMULATION RESULT

We have done two types of simulations for a Least Slack (LS) time scheduling policy and our LB scheduling approach. The LS policy has been developed in the research, Fluctuation Smoothing Policy for Mean Cycle Time (FSMCT) (Kumar, 2001). FSMCT scheduling policy is for re-entrant production lines. The LS scheduling policy sets the highest priority to a wafer lot whose slack time is the smallest in the queue buffer of one machine. When the machine is going to idle, it will select the highest priority wafer lot in the queue buffer to service next. The entire class of LS policies has been proven stable in a deterministic setting (Kumar, 1994) (Lu, 1991) without the dedicated constraint issue. However, the simulation result shows that our proposed LB is better than the LS approach. For simplifying the simulation to easily represent the scheduling approaches, we have made the following assumptions:

1. *Each wafer lot has the same process steps and quantity.*
2. *All photolithography and other stages have the same process time.*
3. *There is no breakdown event in the simulations.*
4. *There is unlimited capacity for non-photolithography machines.*

We have implemented a simulation program in Java and run the simulations on the NetBeans IDE 4.1. Figure 2 depicts our simulations result. Our simulations are to set up six to eleven photolithography machines and 1000 wafer lots. Each wafer lot in the simulations has 85 steps, and 20 of them are photolithography stages. For example, the task pattern is as follows:

$r, r, r_2, r, r, r, r, r, r_2, r, r, r, r, r_2, r, r, r, r, r, r, r, r, r, r, r_2, r, r, r, r, r, r_2, r, r, r, r, r, r, r_2, r, r, r, r, r, r_2, r, r, r_2, r, r,$
$r_2, r, r, r_2, r, r, r_2, r, r, r_2, r, r, r_2, r, r, r_2, r, r, r_2, r, r, r_2, r, r, r_2, r, r, r_2, r, r, r_2, r, r, r_2, r, r$

We let the symbol r represent the non-photolithography stage and r_2 the photolithography stage. In the following task matrix, the wafer lot t_2 starts to process in the simulation when t_1 has passed two steps (s_3). The wafer lot t_3 starts when t_1 has passed three steps (s_4).

$s_1, \ldots \ldots \ldots \ldots \ldots \ldots \ldots \ldots \ldots \ldots \ldots \ldots s_m$
t_1: $r, r, r_2, r, r, r, r, r, r_2, r, r, r, r, r_2, r, r, r, r, r, r, r, r, r, r, r_2, r, r, r, r, r, r_2, r, r, r, r, r, r, r_2, r, r, r, r, r, r_2, \ldots \ldots$
t_2: $r, r, r_2, r, r, r, r, r, r_2, r, r, r, r, r_2, r, r, r, r, r, r, r, r, r, r, r_2, r, r, r, r, r, r_2, r, r, r, r, r, r, r_2, r, r, r, r, r, r_2, \ldots$
:
t_i $r, r, r_2, r, r, r, r, r, r_2, r, r, r, r, r_2, r, r, r, r, r, r, r, r, r, r, r_2 ..$
:
t_{1000}: :

In the simulations, the wafer arrival rate between two wafer lots is a Poisson distribution. We ran ten-iteration for each simulation and the simulations result is shown in (a) and (b) of Figure 3.

We applied the LS and LB methods to select the next wafer lot to process in the simulations. When the wafer lot needs to wait for its dedicated photolithography machine, we will insert a w_r in the task pattern of the wafer lot to represent the situation. After finishing the simulations, we count the pattern of wafer lots to know how much time they have used. Although the simulations are simplified, they reflect the real situation we have met in the factory. After applying the LS method to the above simulations and counting the required resource (the formula $RR(r_k^x, s_j)$ in Section 3, $k = 2, x = 6$ *to* 11) for the photolithography machines at each step we can realize that the load of these machines becomes unbalanced during the simulations.

It is not difficult to extend the simulation with more wafer lots and stages of photolithography or non-photolithography. Moreover, we can use different numbers of r_2, e.g., r_2, r_2r_2, or $r_2r_2r_2r_2$, together for the task patterns to represent different process time for different photolithography stage. Both LB and LS scheduling approaches are applied to the same task matrix during each simulation generated by the Task Generation module described in Section 3. The result of simulation in Figure 3(a) shows that in terms of the outperformed steps and percentage, the LB method is better than the LS method set with 10 machines. While Figure 3(b) shows that in terms of the outperformed steps and percentage, the LB method is better than the LS method set with different number of machines.

By comparing the mean of cycle time, the LS has an average 39.45 steps more than the LB and that is the LB is better than the LS 30.93% on average in the simulation using 10 machines. The simulations result of different photolithography machines indicates that when the capacity of the photolithography area in the manufacturing system is overloaded (the number of the photolithography machine for these simulations is less than or equal to 10), the more the photolithography machines, the better the LB method performs than LS method dose. On the other hand, when the capacity of the photolithography area in the manufacturing system is under loaded, the advantage of LB will decrease, but LB is still better than LS 22.37 steps and 20.10% in the simulations.

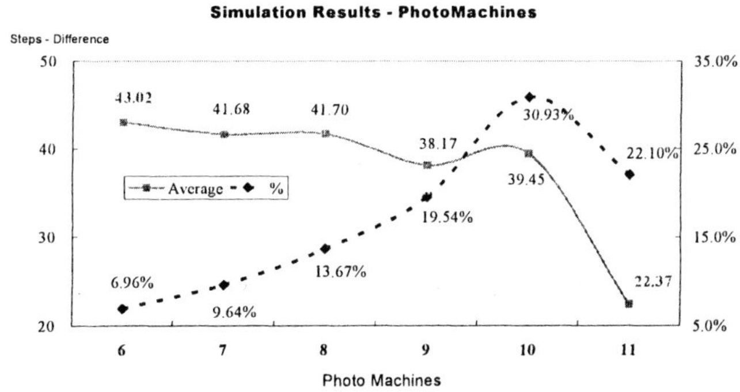

(a) 10 iterative simulations

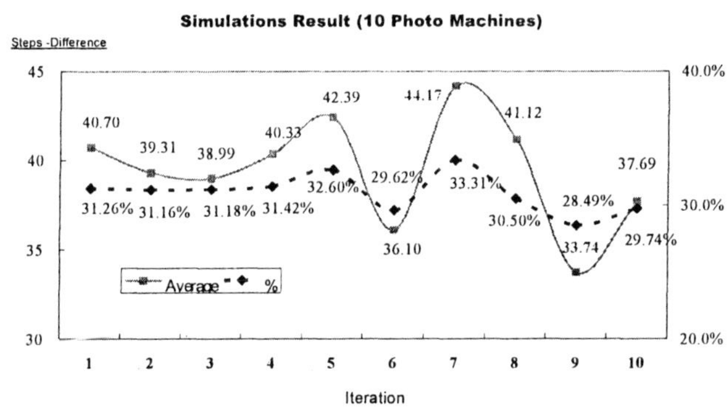

(b) Simulations of different photolithography machines

Figure 3–Simulations results of LB and LS scheduling approaches

5. CONCLUSION

To provide the solution to the issue of dedicated photolithography machine constraint, the proposed Load Balancing (LB) scheduling method has been presented. Along with providing the LB scheduling method to the dedicated machine constraint, we have also presented a novel model–the representation and manipulation method for the task patterns. In addition, the simulations have also shown that the proposed LB scheduling method was better than the LS method. The advantage of the LB method is that we could easily schedule the wafer lots by simple calculation on a two-dimensional matrix, RSEM.

6. ACKNOWLEDGMENTS

The authors would like to thank the anonymous reviewers for their valuable comments. This research was partially supported by the U.S. National Science Foundation grant No. IIS-0326387. This research was also supported in part by the Ministry of Education under grant EX-91-E-FA06-4-4 and the National Science Council under grant NSC-94-2213-E-194-010 and NSC-92-2917-I-194-005. One of us, A. Shr, is grateful to Ms. Victoria Tangi for English proofreading.

7. REFERENCES

1. Akcalt, E, Nemoto, K, Uzsoy, R. Cycle-time improvements for photolithography process in semiconductor manufacturing. IEEE Transactions on Semiconductor Manufacturing 2001; 14, 1: 48-56.
2. Arisha, A, Young, P. "Intelligent Simulation-based Lot Scheduling of Photolithography Toolsets in a Wafer Fabrication Facility". 2004 Winter Simulation Conference 2004; 1935-1942.
3. Chern, C, Liu, Y. Family-Based Scheduling Rules of a Sequence-Dependent Wafer Fabrication System. In IEEE Transactions on Semiconductor Manufacturing 2003; 16, 1: 15-25.
4. Kumar, PR. Re-entrant Lines. In Queuing Systems: Theory and Applications, Special Issue on Queuing Networks 1993; 13, 1-3: 87-110.
5. Kumar, PR. Scheduling Manufacturing Systems of Re-Entrant Lines. Stochastic Modeling and Analysis of Manufacturing Systems, David D. Yao (ed.), Springer-Verlag, New York 1994: 325-360.
6. Kumar, S, Kumar, PR. Queuing Network Models in the Design and Analysis of Semiconductor Wafer Fabs. In IEEE Transactions on Robotics and Automation 2001; 17, 5: 548-561.
7. Lu, SCH, Ramaswamy, D., Kumar PR. Efficient Scheduling Policies to Reduce Mean and Variance of Cycle-time in Semiconductor Manufacturing Plants. In IEEE Transactions on Semiconductor Manufacturing 1994; 7, 3: 374-385.
8. Lu, SH, Kumar, PR. Distributed Scheduling Based on Due Dates and Buffer Priorities. In IEEE Transactions on Automatic Control 1991; 36, 12: 1406-1416.
9. Miwa, T, Nishihara, N, Yamamoto, K. Automated Stepper Load Balance Allocation System. IEEE Transactions on Semiconductor Manufacturing 2005; 18, 4: 510- 516.
10. Mönch, L, Prause, M, Schmalfuss, V. "Simulation-Based Solution of Load-Balancing Problems in the Photolithography Area of a Semiconductor Wafer Fabrication Facility". 2001 Winter Simulation Conference 2001: 1170-1177.
11. Shen, Y., Leachman, R.C. Stochastic Wafer Fabrication Scheduling. In IEEE Transactions on Semiconductor Manufacturing 2003; 16, 1: 2-14 .
12. Shr, AMD, Liu, A, Chen, PP. "A Load Balancing Scheduling Approach for Dedicated Machine Constraint". 8th International Conference on Enterprise Information Systems, Paphos, Cyprus, May 2006a (to appear).
13. Shr, AMD, Liu, A, Chen, PP. "A Heuristic Load Balancing Scheduling Method for Dedicated Machine Constraint". 19th International Conference on Industrial, Engineering & Other Applications of Applied Intelligent Systems (IEA/AIE'06), Annecy, France, June 2006b (to appear).
14. Shr, AMD, Liu, A, Chen, PP. "Load Balancing among Photolithography Machines in Semiconductor Manufacturing". 3rd International Conference on Intelligent Systems (IEEE IS'06), London, England, September 2006c (to appear).
15. Wein, L. M. Scheduling Semiconductor Wafer Fabrication. In IEEE Transactions on Semiconductor Manufacturing 1988; l, 3: 115-130.
16. Zhou, M. and Jeng, MD. Modeling, Analysis, Simulation, Scheduling, and Control of Semiconductor Manufacturing System: A Petri Net Approach. In IEEE Transactions on Semiconductor Manufacturing 1998; 11, 3: 333-357.

37 ITERATIVE HEURISTICS FOR PERMUTATION FLOW SHOPS WITH TOTAL FLOWTIME MINIMIZATION

Xiaoping Li, Qian Wang
School of Computer Science &Engineering, Southeast University, Nanjing, P.R .China, 210096
xpli@seu.edu.cn, qwang@seu.edu.cn

In this paper, flow shop scheduling problems with total flowtime minimization is considered. IRZ (iterative RZ, presented by Rajendran and Ziegler, EJOR, 1997) is found to be effective to improve solutions and LR (developed by Liu & Reeves, EJOR, 2001) is suitable for initial solution developing. By integrating FPE (forward pair wise exchange) and FPE-R(forward pair wise exchange-restart) with IRZ, two efficient composite heuristics, ECH1 and ECH2, are proposed. Computational results show that the proposed three outperform three best existing ones in performance and ECH1 is best. IRZ is the fastest heuristic. ECH2 is a trade-off between IRZ and ECH1 both in effectiveness and efficiency.

1. INTRODUCTION

Flow shop scheduling is an important manufacturing system widely existing in industrial environments. A flow shop can be described as n jobs being processed on m machines and each job having the same machine-order [2]. Total flowtime (or equivalently mean flowtime if all machines are available at time zero) is an important performance measure in flow shop scheduling which can lead to stable or even utilization of resources, rapid turn-around of jobs and minimizing in-process inventory [6][15].

A flow shop with respect to total flowtime minimization is NP-complete [7]. For decades, heuristics have been developed for the problems considered. Heuristics presented by Gupta[8], Rajendran & Chaudhuri[12], Rajendran[16], Ho[9] and Wang et al[18] were efficient algorithms for these problems before NEH [11] seems to be the best heuristic for makespan minimization, while it is not the best for total/mean flowtime minimization [9,12,16]. NEH is based on job-insertion, in which an unscheduled job of the seed generated by some rule is respectively inserted into every possible slot of the current solution (a schedule/partial schedule) and the best generated one is selected as the new current solution. For total/mean flowtime minimization flow shop scheduling, it seems that FL (proposed by Framinan and Leisten [5]), WY (presented by Woo and Yim [19]) and RZ (developed by

---

Please use the following format when citing this chapter:

Li, X., Wang, Q., 2006, in IFIP International Federation for Information Processing, Volume 220, Information Technology for Balanced Manufacturing Systems, ed. Shen, W., (Boston: Springer), pp. 349–356.

Rajendran and Ziegler [13]) are efficient constructive heuristics. FL is similar to NEH but every job-insertion is followed by a pair-wise exchange to improve the current solution. A pair-wise exchange is to generate solutions by exchanging positions of every pair of jobs in the current solution. If the current solution is worse than the best of the generated, it is replaced with the best. WY is also derived from NEH but no seed is predetermined and all unscheduled jobs should perform job-insertion. RZ is based on a different job-insertion from NEH, in which a seed is resulted by sorting jobs with weighted processing times and it is set as the current solution. Every job of the seed performs job-insertion to the subsequence of the current solution without the inserted job. From the comparisons of WY against RZ in [19][13], it is found that RZ outperforms WY for small instances but the relative performance of WY improves with the number of jobs and finally WY outperforms RZ. Computational results of Framinan and Leisten [5] show that FL outperforms both WY and RZ for majority of the randomly generated instances. The temporal complexities of FL and WY are $O(n^4m)$ while that of RZ is $O(n^3m)$.

Recently, many composite heuristics are proposed, such as IH1~IH7 (described by Allahverdi and Aldowaisan [1]), IH7-FL (given by Framinan and Leisten [5]), FLR1 and FLR2 (presented by Framinan, Leisten and Ruiz-Usano [4]). Of these heuristics, FLR2, IH7_FL and FLR1 seem to be the most efficient ones, which adopt FL, the most efficient constructive heuristic, to construct a solution or improve the current solution. Most existing heuristics improve their solutions only by some one-pass method. However, most solutions can be greatly improved by an iterative way, which will be shown by IRZ, ECH1 and ECH2 proposed in this paper.

The paper is organized as follows. Iterative methods and IRZ are described in section 2. In section 3, initial solution development is introduced. ECH1 and ECH2 are proposed in section 4. Computational results are described in section 5, followed by conclusions in section 6

2. ITERATIVE METHODS

Framinan, Leisten and Ruiz-Usano [4] gave a clear framework to divide a heuristic into three phases: index development, solution construction and solution improvement. Each heuristic can include one or more of these phases. A heuristic is regarded as composite if it employs another heuristic for one or more of the three above-mentioned phases. Consequently, a heuristic is regarded as simple if it does not contain another heuristic within any of the three phases.

Procedure of most heuristics, such as constructive ones FL[5], WY[19] and RZ [13], is actually searching an optimal solution among the neighbor solutions generated from an initial solution (which may be produced by combination of several algorithms) by some rule. For example, sequences generated during RZ can be regarded as neighbor solutions from the ASC(w sum(pt)) produced seed by RZ-insertion rule. The best is selected from the neighbor solutions as the final solution of RZ. The solution of FL is also selected from neighbor solutions of the seed generated by DESC(sum(pt)), which are constructed by NEH-insertion and PE (pair-wise exchange) rule.

However, such heuristics may be considerably improved by iterating the rule, i.e. if the solution of a heuristic is better than its initial solution then it is set as the new initial solution and performs the same search procedure again. This process repeats until the solution of some iteration (the best of the generated neighbor ones) is not better than its initial solution, which can be illustrated in Figure 1.

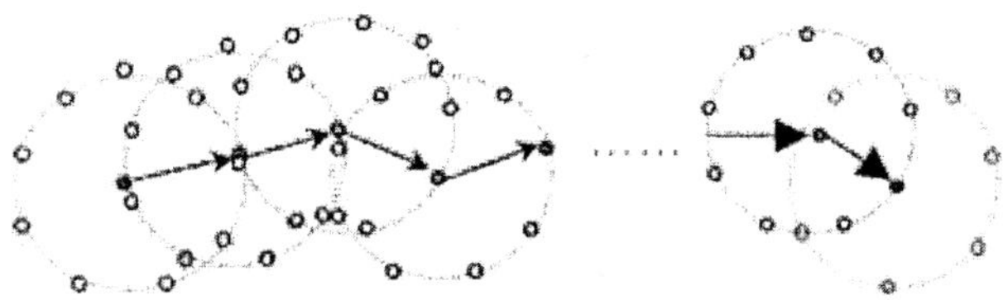

Figure 1 – Procedure of an Iterative Heuristic

Different heuristic has different improvement by iterative method. As for RZ, WY and FL, the corresponding iterative ones are denoted by IRZ (iterative RZ), IWY (iterative RZ) and IFL (iterative RZ) respectively. The six heuristics are tested by 110 benchmark instances generated by Tailard [17]. To compare effectiveness of the six heuristics, ARPD (Average Relative Percentage Deviation) is used which is similar to mean error (used by Allahverdi & Aldowaisan [1], Woo& Yim [19]) and Relative percentage increase in total flowtime (adopted by Rajendran &Ziegler [14]). In this paper, ARPD is slightly different from the one defined by Framinan &Leisten [5], Framinan, Leisten and Ruiz-Usano [4], which is defined as follows.

$$ARPD = \sum_{j=1}^{N} \left(F_i(H)/Best\_known_i - 1\right)/ N \times 100 \quad (1)$$

in which $F_i(H)$ is the total flowtime of instance i obtained by some heuristic H, $Best\_known_i$ is the best total flowtime among the compared heuristics for instance i and N is the number of instances for the same size of combination of jobs and machines. So N is 10 for Tailard's benchmark instances. ARPD of the six heuristics is shown in Table 1.

Table 1 – ARPD comparison of the six heuristics

Prob.	n	m	RZ	IRZ	WY	IWY	FL	IFL
T01	20	5	1.702	0.381	2.066	2.066	1.524	0.642
T02	20	10	1.231	0.164	2.121	2.106	1.762	0.981
T03	20	20	1.055	0.233	2.044	2.030	1.403	0.872
T04	50	5	1.559	0.364	2.418	2.211	0.920	0.302
T05	50	10	1.628	0.358	1.888	1.743	0.778	0.573
T06	50	20	1.394	0.197	1.678	1.661	0.739	0.438
T07	100	5	1.548	0.414	1.400	1.363	0.311	0.089
T08	100	10	1.792	0.180	1.382	1.190	0.939	0.156
T09	100	20	1.691	0.000	2.102	1.761	1.019	0.686
T10	200	10	1.685	0.388	1.484	1.264	0.399	0.075
T11	200	20	1.754	0.036	1.700	1.608	1.149	0.668
Average			1.549	0.247	1.844	1.728	0.995	0.498
Improved times				6.27		1.07		1.99

Note: Improved times is a ratio of ARPD of some heuristic to that of its iterative one.

Table 1 shows that iterative method increases RZ about 5.27 times, FL nearly 1 times but WY very little. Though RZ is far worse than FL in ARPD on average, IRZ (with ARPD 0.247) is even much better than IFL (with ARPD 0.498). WY and FL are based on NEH-insertion which is different from RZ-insertion. The above results indicate that iterative method can improve RZ-insertion much more than NEH-insertion.

Average iterations of RZ, FL and WY are about 8.4, 3.2 and 2.4 respectively. Also, because the temporal complexities of FL and WY are $O(n^4m)$ while that of RZ is $O(n^3m)$, CPU-time of IRZ should be much less that that of IWY or IFL for limited loops. For the above experiment, IRZ needs only 37.63s while IWY requires 194.18 and IFL spends 481.53 on average for class T11.

From above results, it can be seen that the iterative method is desirable for RZ-insertion based heuristics in both effectiveness and efficiency.

3. INITIAL SOLUTION DEVELOPMENT

There are several rules or algorithms (or their combination) for initial solution development, such as ASC(w sum(pt)), DESC(sum(pt)), ASC(sum(pt)), DESC(ABS(dif pt)) and LR(x) (developed by Liu & Reeves [10]). For IRZ, seven initial solution developing methods are performed on the instances in section 4.1. Experimental results indicate that LR can always obtain good results. On average, the best two are LR(n) and LR(n/m) while RANDOM is the worst. LR(1) and FLR1 can make IRZ obtain similar performance, which are only better than RANDOM. Though FLR1 (which is the combination of LR(1) and FL) outperforms FL, it always deteriorates performance of IRZ when it is used to generates initial solution instead of FL. In other words, good initial solution cannot ensure good result for IRZ. As for LR(x), performance of IRZ increases with x on average. However, the larger x is, the more CPU-time needs. For T11, CPU-times of IRZ with LR(n), LR(n/m) and LR(2) generating initial solution are 516.69s, 54.17s and 43.87s respectively.

From above, we can see that LR(n/m) is similar to LR(n) in performance while it only needs little more CPU-time than LR(2) does. Therefore, LR(n/m) is reasonably selected to develop initial solution.

4. NEW COMPOSITE HEURISTICS

In this section, three composite heuristics, IRZ, ECH1 and ECH2, are presented for flowtime minimization flow shop scheduling problems. For convenience, we combine index development and solution construction into one phase, initial solution development. So a heuristic consists of initial solution development and solution improvement phases. For the problem considered in this paper, iterative RZ-insertion is rather efficient for solution improvement and LR(n/m) is desirable for initial solution development, which can be illustrated in the following.

According to literature [4], pair-wise exchange strategies, such as FPE, FPE-R, BPE and BPE-R, can always efficiently improve solutions. In this subsection, FPE-

R and FPE are respectively integrated with RZ-insertion during IRZ procedure and ECH1 and ECH2 are proposed. In other words, ECH1 improves sequence of LR(n/m) by repeating the combination of RZ-insertion with FPE-R until no improvement can be made. ECH1 conducts in a similar way.

As well, the maximum iteration number of IRZ in the experiment of section 4.1 is 14 and the average is 8.4, which means that the improvement is very slight when the iteration-number exceeds some constant. So we choose the stop criterion as either improvement can be made or iteration-times is greater than a constant (20 in this paper), which is also applied to IRZ in the following experiments. Because of the similarity of ECH1 and ECH2, we just give the formal description and the execution procedure of ECH2 as follows, in which GFC is adopted.

1. Call LR(n/m) to generate seed π^s.
2. $k \leftarrow 1, \pi^c \leftarrow \pi^s$.
3. Repeat
   3.1 $\pi^s \leftarrow \pi^c$, $F(\pi^s) \leftarrow F(\pi^c)$.
   3.2 Call RZ-insertion method to obtain current solution π^c.
   3.3 Call FPE and keep the best solution among the procedure in π^b.
   3.4 If $F(\pi^b) < F(\pi^c)$, then $\pi^c \leftarrow \pi^b$ and $F(\pi^c) \leftarrow F(\pi^b)$.
   3.5 $k \leftarrow k+1$.
4. Until $F(\pi^s) \le F(\pi^c)$ or $k \ge 20$.
5. Stop. π^s is the final solution.

The time complexity in step 1 is $O(n^2)$. The number of loops in step 3 cannot exceed 20, which means that its time complexity depends on that of RZ and that of FPE. Time complexity of RZ is $O(mn^3)$ and it is obvious that time complexity of FPE is also $O(mn^3)$. Hence, time complexity of both IRZ and ECH2 is $O(mn^3)$. However, the worst computation effort of FPE-R is about cmn^4 in which c is a constant, i.e. the worst time complexity of FPE-R is $O(mn^4)$. So the time complexity of ECH1 is $O(mn^4)$.

To illustrate the computational procedure of ECH2, an 8-jobs 6-machines problem as shown in Table 2 is considered.

Table 2 – The 8-jobs 6-machines problem

	J_1	J_2	J_3	J_4	J_5	J_6	J_7	J_8
M_1	22	36	9	89	16	23	66	82
M_2	57	61	87	34	80	78	95	54
M_3	14	23	52	39	23	28	63	50
M_4	82	1	15	12	72	36	28	42
M_5	93	55	43	4	20	72	26	75
M_6	64	80	4	62	74	67	19	14

The seed sequence obtained by LR(n/m) in step 1 is (2,4,1,5,8,3,6,7) with total flowtime 4171. Six iterations are performed in step 3 and 4, i.e. $k = 6$ when the heuristic stops, the corresponding results are shown in Table 3.

Table 3 – Solutions in ECH2 for the instance

No.	Operation	Result sequence	Flowtime
$k = 1$	RZ	(5,2,1,3,4,8,7,6)	4079
	FPE	(5,2,8,3,4,1,7,6)	4022
$k = 2$	RZ	(5,3,4,2,8,7,6,1)	3979
	FPE	(5,3,4,2,8,7,6,1)	3979
$k = 3$	RZ	(3,4,2,5,8,7,6,1)	3887
	FPE	(3,4,2,1,8,7,6,5)	3870
$k = 4$	RZ	(3,4,2,1,8,7,5,6)	3864
	FPE	(3,4,2,1,8,7,5,6)	3864
$k = 5$	RZ	(3,4,2,1,8,5,6,7)	3854
	FPE	(3,4,2,1,8,5,6,7)	3854
$k = 6$	RZ	(3,4,2,1,8,5,6,7)	3854
	FPE	(3,4,2,1,8,5,6,7)	3854

ECH2 stops when neither RZ nor FPE can improve the current solution and (3,4,2,1,8,5,6,7) is the final solution with total flowtime 3854.

5. EXPERIMENTAL RESULTS

To compare the proposed three heuristics with the best existing composite ones (FLR1 and FLR2, which have been proposed by Framinan, Leisten and Ruiz-Usano [4] and IH7_FL [5]) both in effectiveness and efficiency, all the 120 benchmark instances generated by Tailard [17] are tested. ARPD defined in subsection 4.1 is also adopted to evaluate effectiveness for each group, in which every current best solution $Best\_known_i$ is the minimum total flowtime among Rajendran & Ziegler [14], Liu & Reeves [10] and the six heuristics considered in this section. Efficiency is measured by average CPU-time (in seconds) spent on 10 instances of each combination of m and n.

All the heuristics are implemented by Visual Basic 6.0 and performed on IBM PC 2.0GHz with 256M RAM. Experimental results are given in Table 4 and Table 5.

Table 4 shows that the average performance of the proposed three is better than that of the existing three (IH7-FL, FLR1 and FLR2). ARPD of ECH1 is the least among the six composite heuristics, i.e. ECH1 is the most effective heuristic except two cases (IRZ is the best for T03 and ECH2 is the best for T04). ECH2 is worse than ECH1 but it outperforms the other four for majority cases. IRZ outperforms the existing three except that it is outperformed by FLR2 for T01, T02 and T03. As for the existing three, though ARPD of IH7-FL is less than that FLR2 for three cases (T04, T05, T10), FLR2 is better than IH7-FL on average. FLR1 is the worst for all cases among the compared heuristics.

Table 4 – Effectiveness comparison of the six heuristic

Prob	IH7-FL	FLR1	FLR2	IRZ	ECH1	ECH2
T01	1.621	1.702	1.011	1.016	0.680	0.711
T02	1.702	1.989	1.163	1.392	1.161	1.394
T03	1.406	2.439	1.383	1.398	1.414	1.448
T04	1.176	1.820	1.254	0.692	0.586	0.565
T05	1.600	2.740	1.813	1.142	0.861	1.095
T06	1.807	2.656	1.749	1.282	0.937	1.128
T07	0.856	0.864	0.504	0.353	0.165	0.193
T08	1.420	1.987	1.201	0.623	0.408	0.535
T09	1.724	2.420	1.645	1.311	1.148	1.244
T10	0.901	1.302	0.952	0.289	0.096	0.151
T11	1.112	1.406	0.910	0.172	0.121	0.139
T12	0.889	1.061	0.729	0.108	0.040	0.121
Aver.	1.351	1.866	1.193	0.815	0.635	0.727

Table 5 – Efficiency comparison of the six heuristics

Prob	IH7-FL	FLR1	FLR2	IRZ	ECH1	ECH2
T01	0.02	0.01	0.02	0.02	0.03	0.02
T02	0.02	0.02	0.03	0.02	0.03	0.03
T03	0.04	0.03	0.05	0.03	0.04	0.04
T04	0.28	0.24	0.32	0.32	0.41	0.41
T05	0.47	0.43	0.60	0.45	0.65	0.66
T06	0.89	0.80	1.08	0.59	1.14	0.88
T07	4.03	3.22	4.90	3.00	4.29	4.56
T08	7.65	5.85	9.35	3.92	9.02	6.04
T09	15.61	11.14	18.03	5.69	13.98	8.73
T10	112.14	103.57	146.94	43.08	122.00	71.48
T11	221.07	213.16	298.38	58.94	201.06	95.62
T12	6058.22	5804.27	8808.07	1521.61	8297.74	1920.65

Table 5 shows that all the compared composite heuristics have a similar time increasing tendency which is in accordance with their same time complexity. Though time complexity of ECH1 and the existing three is $O(mn^4)$, FLR2 and ECH1 are more time-consuming than the other two (FRL1 and IH7-FL) and they needs more than 8000s for T12 (500×20 instances) on average. IRZ and ECH2, with time complexity $O(mn^3)$, need much less CPU-time than the other four do (only 1521.61s and 1920.65s respectively for T12).

So among the six heuristics considered in this paper, ECH1 is the best in performance while it requires CPU-time nearly as much as FLR2 does. IRZ is the fastest, its overall performance is much better than FLR1, IH7-FL and FLR2 but worse than ECH1 and ECH2. ECH2 is a trade-off between IRZ and ECH1 in performance with CPU-time consuming similar to IRZ

6. CONCLUSIONS

In this paper, flow shop scheduling problems with flowtime minimization was considered. Based on iterative method and LR(n/m) initial solution development, three composite heuristics, IRZ, ECH1 and ECH2, were proposed and compared with the best existing composite ones, FLR1, FLR2 and IH7_FL. Computational results showed that the proposed three outperform the best existing three. ECH1 is the best among the six heuristics in effectiveness while nearly needs as much CPU-time as FLR2 does. IRZ is the fastest but its overall performance is worse than ECH1 and ECH2. ECH2 is a trade-off between IRZ and ECH1 both in effectiveness and efficiency.

7. ACKNOWLEDGEMENT

This work is supported by National Natural Science Foundation of China under Grants (No. 60504029 and No.90412014).

8. REFERENCES

1. Allahverdi A, Aldowaisan T. New heuristics to minimize total completion time in m-machine flowshops. International Journal of Production Economics 2002; 77(1): 71-83.
2. Baker KR. Introduction to sequencing and scheduling. New York: Wiley, 1974.
3. Framinan JM, Leisten R, Rajendran C. Different initial sequences for the heuristic of Nawaz, Enscore and Ham to minimize makespan, idletime or flowtime in the static permutation flowshop sequencing problem. International Journal of Production Research 2003; 41(1): 121-148.
4. Framinan JM, Leisten R, Ruiz-Usano R. Comparison of heuristics for flowtime minimisation in permutation flowshops. Computers & Operations Research 2005; 32(5): 1237-1254.
5. Framinan JM, Leisten R. An efficient constructive heuristic for flowtime minimisation in permutation flow shops. OMEGA 2003; 31(4): 311-317.
6. French S. Sequencing and scheduling: an introduction to the mathematics of the job-shop. Chichester: Ellis Horwood, 1982.
7. Gray MR, Johnson DS, Sethi R. The complexity of flowshop and jobshop scheduling. Mathematics of Operations Research 1976; 1(2): 117-129.
8. Gupta JND. Heuristic algorithms for multistage flow shop problem. AIIE Transaction 1972; 4(1): 11-18.
9. Ho JC. Flowshop sequencing with mean flowtime objective. European Journal of Operational Research 1995; 81(3): 571-578.
10. Liu J, Reeves CR. Constructive and composite heuristic solutions to the P//ΣCi scheduling problem. European Journal of Operational Research 2001; 132(2): 439-452.
11. Nawaz M, Enscore EE, Ham I. A heuristic algorithm for the m-machine n-job flow-shop sequencing problem. OMEGA 1983; 11(1): 91-95.
12 Rajendran C, Chaudhuri D. An efficient heuristic approach to the scheduling of jobs in a flowshop. European Journal of Operational Research 1991; 61(2): 318-325.
13. Rajendran C, Ziegler H. An efficient heuristic for scheduling in a flowshop to minimize total weighted flowtime of jobs. European Journal of Operational Research 1997; 103(1): 129-138.
14. Rajendran C, Ziegler H. Ant-colony algorithms for permutation flowshop scheduling to minimize makespan total flowtime of jobs. European Journal of Operational Research 2004; 155(2): 426-438.
15. Rajendran C. A heuristic for scheduling in flowshop and flowline-based manufacturing cell with multi-criteria. International Journal of Production Research 1994; 32(11): 2541-2558.
16. Rajendran C. Heuristic algorithm for scheduling in a flowshop to minimise total flowtime. International Journal of Production Economics 1993; 29(1): 65-73.
17. Tailard E. Benchmarks for basic scheduling problems. European Journal of Operational Research 1993; 64(1): 278-285.
18. Wang C, Chu C, Proth JM. Heuristic approaches for $n/m/F/\sum C_i$ scheduling problems. European Journal of Operational Research 1997; 96(3): 636-644.
19. Woo DS, Yim HS. A heuristic algorithm for mean flowtime objective in flowshop scheduling. Computers & Operations Research 1998; 25(3): 175-182.

38

PARAMETER ESTIMATION FOR TIME-VARYING SYSTEM BASED ON COMBINATORIAL PSO

Weixing Lin[1, 2], Peter X. Liu[2]
[1)] Faculty of Information Science and Technology, Ningbo University, Ningbo, Zhejiang Province, China 315211
[2)] Department of System and Computer Engineering, Carleton University, Ottawa, Ontario, Canada K1S 5B6
wlin@sce.carleton.ca

In this paper, a novel Particle Swarm Optimization (PSO) identification algorithm for time-varying systems with a colored noise is presented. Presented criterion function can show not only outside system output error but also inside parameters error in order to explain more difference between actual and estimative system. Identification algorithm may consist of many different PSO algorithms that are named the combinatorial PSO. The estimating and tracking of parameters make use of characteristics of different PSO algorithms. The simulation and result show that the identification algorithm for time-varying systems with noise was indeed more efficient and robust in combinatorial PSO comparing with the original particle swarm optimization.

1. INTRODUCTION

For time-varying system identification and parameter tracking is still a very active research field in the recent years [1]. Most of identification algorithms are still the Recursive Least Squares (RLS) or gradient algorithms, which search along the direction of parameters change slowly. If the search space is not differentiable or the parameters are nonlinear, usually the optimal global solutions would not be found. In the algorithms the selection of the gain or forgetting factor becomes very important. In general, a large gain or a small forgetting factor makes the RLS or gradient algorithms to have a better ability for tracking the variation of parameters, but also makes them sensitive to white noise. On the other hand, a small gain or a big forgetting factor makes the algorithms less sensitive to white noise, but at the same time results in a poor tracking ability for slowly time-varying systems. There are many new identification algorithms such as identification using Genetic Algorithms (GA) in order to have good tracking ability and be not sensitive to white noise [2] [3]. GA can define search direction and scopes only based on the fitness function converted from the object function and doesn't need to know the differential of object function and other auxiliary information. This will be very convenient for the

Please use the following format when citing this chapter:

Lin, W., Liu, P. X., 2006, in IFIP International Federation for Information Processing, Volume 220, Information Technology for Balanced Manufacturing Systems, ed. Shen, W., (Boston: Springer), pp. 357–368.

functions whose differentials are difficult to find or are not existent. Above algorithms use all models with white noise to study.

The particle swarm optimization (PSO) is similar to GA where the system is initialized with a population of random solutions. It is unlike GA, however, where every potential solution is also assigned a randomized velocity and the potential solutions called particles are "flown" through the search space. In this paper a novel PSO identification algorithm for time-varying systems with a colored noise is presented. Because PSO algorithm is a stochastic optimization algorithm as same as GA, it should be above advantages of GA.

2. OVER VIEW OF PARTICLES SWARM OPTIMIZATION

2.1 Features of Particle Swarm Algorithm

In 1995, Kennedy and Eberhart first introduced PSO method [4]. The PSO is a stochastic optimization technique that can be likened to the behavior of a flock of birds and that is derived from the social-psychological theory. The method has been found to be robust in solving problems featuring nonlinearity and no differentiability, multiple optimization, and high dimensionality through adaptation.

The PSO is a swarm based optimization technique. A simple explanation of the PSO's operation is as follows. Every particle represents a possible solution to the optimization task at hand. During iterations every particle accelerates in the direction of its own personal best solution found so far, as well as in the direction of the global best position discovered by the particles in swarm so far. This means that if a particle discovers a promising new solution, all the other particles will move closer to it, exploring the region more thoroughly in the process.

In PSO every particle has the following attributes and is treated as a volume-less point in the D-dimensional search space. The *ith* particle can be described by a vector $X_i = [x_{i1}(t), x_{i2}(t) \Lambda\ x_{iD}(\text{t})]^T$. The personal best position of the *ith* particle in the search space is recorded and represented as $P_i = [p_{i1}(t), p_{i2}(t) ... p_{iD}(t)]^T$. The global best position found by any particle during previous steps is represented by the symbol P_g, i.e. $P_g = [p_{g1}(t), p_{g2}(t) ... p_{gD}(t)]^T$. The flying velocity of particle i is represented as $V_i = [v_{i1}(t), v_{i2}(t) ... v_{iD}(t)]^T$. The maximum velocity of particles is $V_{\max} = (v_{\max 1}, v_{\max 2} ... v_{\max D})^T$, and the range of the particles is $\text{X}_{\max} = (x_{\max 1}, x_{\max 2} ... x_{\max D})^T$. Assuming that the function J_h is to be minimized, the velocity and position of every particle can be modified respectively by the following equations

$$\begin{aligned} v_{id}'(t+1) = {} & w \times v_{id}(t) + C_1 \times rand_1 \times [p_{id}(t) \\ & - x_{id}(\text{t})] + C_2 \times rand_2 \times [p_{gd}(t) - x_{id}(\text{t})] \end{aligned} \tag{1}$$

Where $i = 1,2\Lambda\ N_1$, $d = 1,2\Lambda\ D$. N_1 is the number of particles. C_1 and C_2 are acceleration coefficients that are positive constants. Acceleration coefficients control how far a particle will move in a single iteration. Typically, these are both set to 2.0, although assigning different values to and sometimes leads to improved performance [5]. $rand_1 \sim U(0,1)$ and $rand_2 \sim U(0,1)$ are two uniform random sequences in the range (0,1). The component value in every vector V_i can be clamped to the range $[-v_{\max d}, v_{\max d}]$ to reduce the likelihood of particles leaving the search space. The value of $v_{max\,d}$ is usually chosen to be $k \times X_{\max}$, with $0.1 \le k \le 1.0$ [6]. The formula of clamped velocity may show as

$$v_{id}(t+1) = \begin{cases} v_{\max d} & v_{id}'(t+1) > v_{\max d} \\ v_{id}'(t+1) & -v_{\max d} \le v_{id}'(t+1) \le v_{\max d} \\ -v_{\max d} & v_{id}'(t+1) < -v_{\max d} \end{cases} \qquad (2)$$

The new position of a particle is calculated using

$$x_{id}(t+1) = x_{id}(t) + v_{id}(t+1) \qquad (3)$$

The personal best position of each particle is updated using

$$P_i(t+1) = \begin{cases} P_i(t), & J_h(X_i(t+1) \ge J_h(P_i(t)) \\ X_i(t+1), & J_h(X_i(t+1) < J_h(P_i(t)) \end{cases} \qquad (4)$$

and the global best position found by any particle during previous steps, is defined as

$$P_g(t+1) = \arg \min_{P_i} J_h(P_i(t+1)) \qquad 1 \le i \le N_1 \qquad (5)$$

The first item of (1) is the momentum term and w is its inertia weight.

The inertia weight w is employed to control the impact of previous velocity on the current velocity in order to influence global and local exploration abilities of the "flying points". A larger inertia weight facilitates global exploration while a smaller inertia weight tends to facilitate local exploration to the current search area. Suitable selection of the inertia weight can provide a balance between global and local exploration abilities and require less iteration on average to find the optimum. There is a kind of simple way the value w is linearly decreased with the iteration going. By means of the mathematics, the description is [8]

$$w = w_{\max} - \frac{w_{\max} - w_{\min}}{iter_{\max}} \cdot iter \qquad (6)$$

Where $w_{\max}$ is the initial weight, $w_{\min}$ is terminative weight, $iter_{\max}$ is the biggest iteration times, $iter$ is the current iteration times.

Usually, we define the inertia weight w to linearly decrease from 0.9 to 0.4 during the iterations. Every particle is updated according to its own flying experience and the group's flying experience. From simulation we know the particle maybe loss its optimum when the inertia weight w too large, thus the algorithm will be failed to converge. So far we do not know the best and accurate changing regulation of the inertia weight. To overcome this shortage, we induce fuzzy logic rules algorithm into PSO. The algorithm is the PSO with fuzzy self-adapting inertia weights (FSPSO).

Considering fuzzy logic rules and the last results of particle's searching, it modifies dynamically the values of inertia weight in order to promote the convergent capability of the particle swarm [7].

If we establish the variable J_h of the square sum of output residual reflected directly, the opposite quantity $\Delta\delta$ related directly with the square sum of the output residual is described. The mathematics formula may show as $\Delta\delta = \frac{J_h - J_{h\min}}{J_{h\max} - J_{h\min}}$, where J_{hmax} and J_{hmin} is the top and bottom limit that J_h may take the value respectively. In the formula we get the value $\Delta\delta$ placed in [0, 1] interval, which is more fit than J_h to be the input of the fuzzy reasoning machine. Two inputs of the fuzzy reasoning are defined as three fuzzy sets respectively. It is named as small, medium and big. Nine kinds of different combinations of inputs correspond with nine outputs to decide the next iteration's value w. We set the fuzzy self-adapting rule and mainly consider two points. One is that when J_h is big, the increment of the next iteration w is big so that particles can be in the big range searching, vice versa. The other one is that when the value of the current iteration w is big, the minus quantity of the next iteration w is big so that particles can accelerate the convergence. The rule table of fuzzy self-adapting algorithm is showed in Table 1. The simulation result shows that the strategy adjusting w can obtain the more satisfied result.

Recently, work by Clerc [9]-[11] indicated that a constriction factor may help to ensure convergence. Application of the constriction factor results in (7).

$$\begin{aligned} v_{id}'(t+1) = K[v_{id}(t) + C_1' \times rand_1 \times [p_{id}(t) \\ - x_{id}(t)] + C_2' \times rand_2 \times [p_{gd}(t) - x_{id}(t)]] \end{aligned} \tag{7}$$

where $K = \frac{2}{\left|2 - \varphi - \sqrt{\varphi^2 - 4\varphi}\right|}$ and $\varphi = C_1' + C_2'$, $\varphi \geq 4$.

If you compare (7) with (1) you will find that $w = K, C_1 = KC_1'$ and $C_2 = KC_2'$. For example: if $C_1' = C_2' = 2$, (7) is the same as (1) in w=1. It is seen that the values of w, C_1 and C_2 are important in PSO.

2.2 Combinatorial PSO

Eberhart and Shi [12] have shown that the constriction factor alone does not necessarily result in the best performance. Combining more approaches could result in the fastest convergence overall. These improvements appear to be effective on a large collection of problems.

Kennedy has taken this LBEST version of the particle swarm and applied to it a technique referred to as "social stereotyping" [13] [14]. A clustering algorithm is used to group individual particles into "stereotypical groups". The cluster center $G_i(t)$ is computed for every group and then substituted into (1), yielding three strategies to

Table 1 - The Rule table of The Fuzzy Self-adapting Algorithm

Inertial weight of the next iteration [0.4,0.9]		The opposite quantity directly related with the square sum of the output residual $\Delta\delta$, [0, 1]		
		Small [0,0.35]	Medium [0.35,0.7]	Big [0.7,1]
Inertial weight of the current iteration w	Small, [0.4,0.6)	0.4	w +0.08;	w +0.15
	Medium, [0.6,0.75)	w -0.05	w	w +0.10
	Big, [0.75.0.9]	w -0.10	w -0.08	w +0.05

calculate the new velocity

$$V_i'(t+1) = w \times V_i(t) + C_1 \times rand_1 \times [G_i(t) - X_i(\mathrm{t})] + C_2 \times rand_2 \times [P_g(t) - X_i(\mathrm{t})] \tag{8}$$

$$V_i'(t+1) = w \times V_i(t) + C_1 \times rand_1 \times [P_i(t) - X_i(\mathrm{t})] + C_2 \times rand_2 \times [G_i(t) - X_i(\mathrm{t})] \tag{9}$$

$$V_i'(t+1) = w \times V_i(t) + C_1 \times rand_1 \times [G_i(t) - X_i(\mathrm{t})] + C_2 \times rand_2 \times [G_i(t) - X_i(\mathrm{t})] \tag{10}$$

The results presented indicate that only the method in (8) performs better than the standard PSO of (1). This improvement comes at increased processing cost, as the clustering algorithm needs a nonnegligible amount of time to form the stereotypical groups. In a time-varying system we define:

$$G_i(t) = \frac{1}{h}\sum_{l=0}^{h} P_i(t-l) \tag{11}$$

Where h is a width of window.

Moreover, following simulations have shown. We are able to make use of the advantage of more approaches in the time-varying system. For example we take two approaches of PSO. We can divide particles of swarm into two types crossly. The first type is used for FSPSO. Another type is used for the PSO of inertia weights in (1). The particles of swarm are two times more than number of parameters in identification. The particles of each type are more than parameters in identification.

3. IDENTIFICATION FOR ARMAX MODEL WITH TIME-VARYING PARAMETERS

The considered stochastic ARMAX model with time-varying parameters is given by[15]

$$A(q^{-1},k)y(k) = q^{-d(k)}B(q^{-1},k)u(k) + \frac{1}{C(q^{-1},k)}e(k) \tag{12}$$

where

$$A(q^{-1},k)=1+a_1(k)q^{-1}+a_2(k)q^{-2}+\Lambda\ +a_{na}(k)q^{-na} \quad ,$$
$$B(q^{-1},k)=b_1(k)q^{-1}+b_2(k)q^{-2}+\Lambda\ +b_{nb}(k)q^{-nb} \quad ,$$
$$C(q^{-1},k)=1+c_1(k)q^{-1}+c_2(k)q^{-2}+\Lambda\ +c_{nc}(k)q^{-nc}\ .$$

$a_1(k), a_2(k),\Lambda\ , a_{na}(k), b_1(k), b_2(k),\Lambda\ ,$ $b_{nb}(k), c_1(k), c_2(k),\Lambda\ , c_{nc}(k)$ are the time-varying parameters of the system. $d_{\min} \le d(k) \le d_{\max}$ is the time-varying delay. *u (k)*, *y(k)* and *e(k)* are system input, output and white noise serial respectively. If $C(q^{-1},k) \ne 1, \dfrac{e(k)}{C(q^{-1},k)}$ shows a type of colored noise in (12). Suppose the model of the system is

$$\hat{A}(q^{-1},k)\hat{y}(k)=q^{-\hat{d}(k)}\hat{B}(q^{-1},k)u(k)+\frac{1}{\hat{C}(q^{-1},k)}e(k) \tag{13}$$

where

$$\hat{A}(q^{-1},k)=1+\hat{a}_1(k)q^{-1}+\hat{a}_2(k)q^{-2}+\Lambda\ +\hat{a}_n(k)q^{-na} \quad ,$$
$$\hat{B}(q^{-1},k)=\hat{b}_1(k)q^{-1}+\hat{b}_2(k)q^{-2}+\Lambda\ +\hat{b}_{nb}(k)q^{-nb} \quad ,$$
$$\hat{C}(q^{-1},k)=1+\hat{c}_1(k)q^{-1}+\hat{c}_2(k)q^{-2}+\Lambda\ +\hat{c}_{nc}(k)q^{-nc}$$

$\hat{\theta}^T(k)=[\hat{d}(k),\hat{a}_1(k),\hat{a}_2(k),\Lambda\ ,\hat{a}_{na}(k),\hat{b}_1(k),\hat{b}_2(k),\Lambda\ ,\hat{b}_{nb}(k),\hat{c}_1(k),\Lambda\ ,\hat{c}_{nc}(k)]$
is
$\theta^T(k)=[d(k),a_1(k),a_2(k),\Lambda\ ,a_{na}(k),b_1(k),b_2(k),\Lambda\ ,b_{nb}(k),c_1(k),\Lambda\ ,c_{nc}(k)]$
estimation of the parameters at *k* time. The error between the actual and the estimated system output is defined by

$$\begin{aligned}\varepsilon = y(k)-\hat{y}(k) = {} & [\frac{B(q^{-1},k)}{A(q^{-1},k)}q^{-d(k)}-\frac{\hat{B}(q^{-1},k)}{\hat{A}(q^{-1},k)}q^{-\hat{d}(k)}]u(k) \\ & +[\frac{1}{A(q^{-1},k)C(q^{-1},k)}-\frac{1}{\hat{A}(q^{-1},k)\hat{C}(q^{-1},k)}]e(k)\end{aligned} \tag{14}$$

When the identification model is different from the actual system, $\varepsilon \ne 0$. We define the performance criterion function is as follows

$$\begin{aligned}J_h(\theta(k)) = {} & \sum_{i=0}^{h}\lambda^i\{[y(k-i)-\hat{y}(k-i)]^2 + \\ & \mu[\theta_{best}(k-i)-\hat{\theta}(k-i)]^T[\theta_{best}(k-i)-\hat{\theta}(k-i)]\}\end{aligned} \tag{15}$$

Where h is a width of window. The faster parameters of time-varying change, the smaller choice h is to have batter result. $\hat{y}(k)$ is the estimated output in system. λ is the forgetting factor. Typically, $0 < \lambda \le 1$ is the range of λ .Actually we use a value of λ from 0.90 to 0.98. The much smaller λ^i is, the more *i* increase in order to track the dynamic system and forget older data. μ shows coefficient in square error of parameters. Its value may balance the ratio of error of parameters and system output.

Actually we use a value of μ from 0.3 to 0.5. In (15) the first part shows error of system output and second part shows error of parameters in order to explain more difference between actual and estimative system.

A flow chart for such an algorithm, referred to here as combinatorial PSO, is given in Figure 1. This algorithm is capable of providing desirable performance and convergence properties in most any context.

In (12) all parameters of model are given: $a_1(k)=-1.5$, $a_2(k)=0.7$, $b_2(k)=0.5$, $c_1(k)=1.0$, $c_2(k)=0.41$, $d(k)=2$. Where the time-varying parameter is

$$b_1(k) = \begin{cases} 1.0 & (k < 200) \\ 1 + 0.4 * \sin[0.2(k - 200)] & (k \geq 200) \end{cases}$$

The noise e(k) is the white noise whose mean is null and $\sigma^2 = 0.1$. The input signal $u(k)$ is the white noise whose mean is null and amplitude is 1.

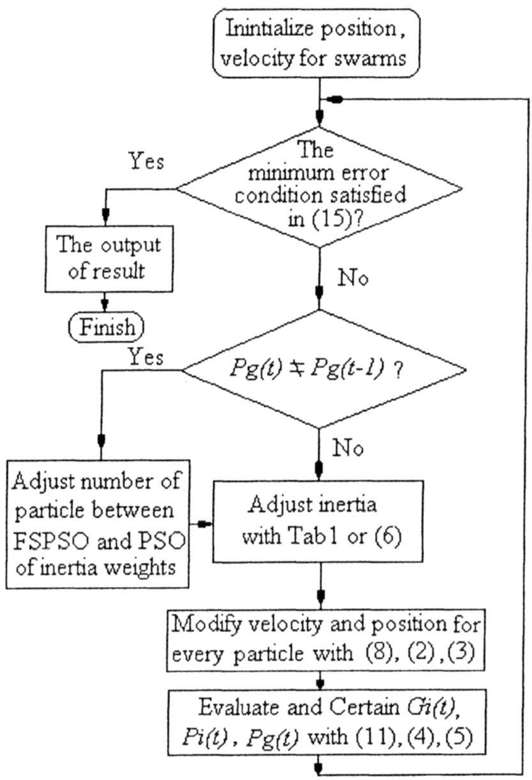

Figure 1 - A flow chart for such an algorithm based on combinatorial PSO

In following figures horizontal coordinate is iteration times and vertical coordinate is value of parameters. In figures green line is actual value of parameters and red stippling is estimated value of parameters. In order to show clearly in figures we only give out the result of $a_1(k)$ and $b_1(k)$. $a_1(k)$ represents time invariant parameter. b_l *(k)* represents time-varying parameter. In order to avoid bad convenience and velocity too fast to control range of velocity we set the maximum velocity into 1 ($V_{d\max} = 1$). The number of particles N_1 takes 30. Acceleration coefficients C_1 and C_2 take 2 equally. The forgetting factor λ takes 0.95. μ takes 0.4.

4. DIGITAL SIMULATION

4.1 Identification with PSO of Inertia Weights

Figure 2 shows the result in the PSO of inertia weights when h is 3. Its inertia weights changes from 0.9 to 0.4 with (6). Similarly Figure 3 and Figure 4 are results when h is 5 and 10 respectively. From results we may find that the better result of tracking parameter is, the smaller h is. The later tracking of parameters is batter than forward convenience in PSO of inertia weights. Before 200 iteration times result is not good.

4.2 Identification with FSPSO

Figure5 shows the result in FSPSO when h is 3. Its inertia weights changes from 0.9 to 0.4 with fuzzy logic rules in Table 1. Similarly Figure 6 and Figure 7 are results of h=5 and 10 respectively. From results we may find that forward convenience of parameters is batter than the later tracking of parameters in FSPSO. Its convenience (about 70 iteration times) is faster than the PSO of inertia weights (about 200 iteration times). The result with h=3 (Figure 5) is batter than other (Figure 6 or Figure 7).

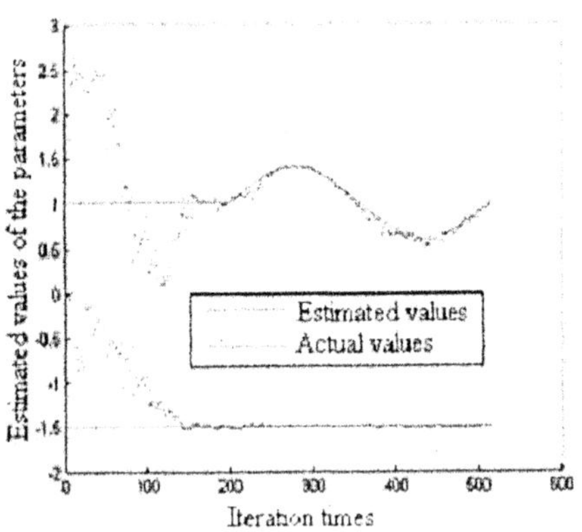

Figure 2 - PSO of inertia weights with h=3

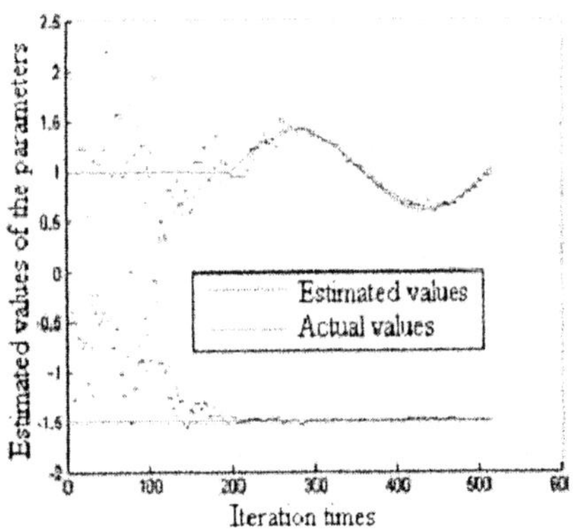

Figure3 - PSO of inertia weights with h=5

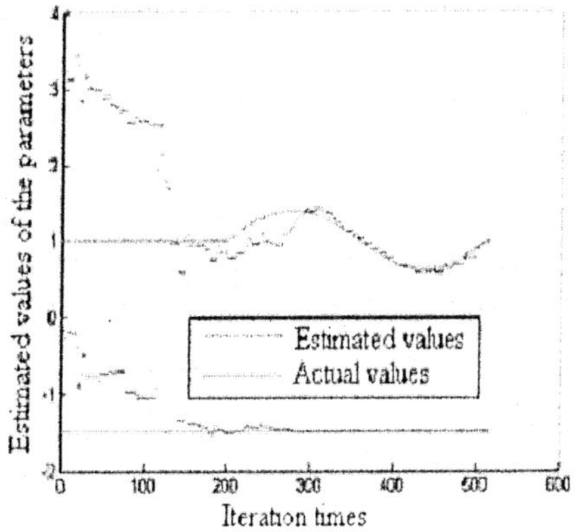

Figure 4 - PSO of inertia weights with h=10

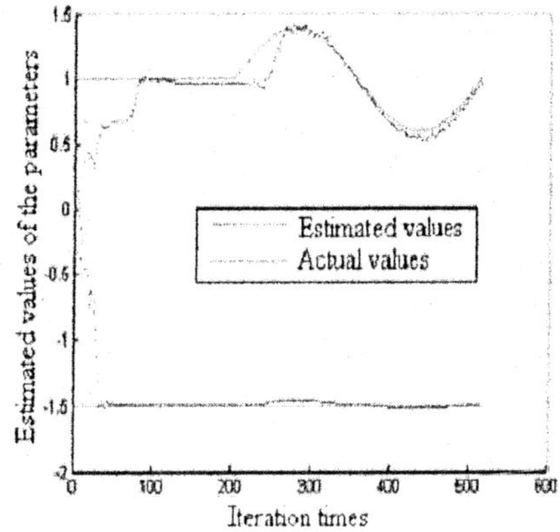

Figure 5 - FSPSO with h=3

We may compare FSPSO and the PSO of inertia weights with above results. Then we can discover that FSPSO combines the combinatorial PSO with the PSO of inertia a weights to take thire advantages and to make up another degradation.

4.3 Identification with combinatorial PSO

If we use combinatorial PSO we meet the allocation of swarms. According to lots of simulation we select an half particles (15 particles) for FSPSO and other particles (15 particles) for PSO of inertia weights. Figure 8 shows the result in combinatorial PSO when h is 3. Similarly Figure 9 and Figure 10 are results when h is 5 and 10 respectively. It is seen that in the combinatorial PSO tracking of parameters and convenience is much batter than that one in FSPSO or the PSO of inertia weights. The better result of tracking parameter is, the smaller h is when h is more than 2. Or the convenience is imperfect.

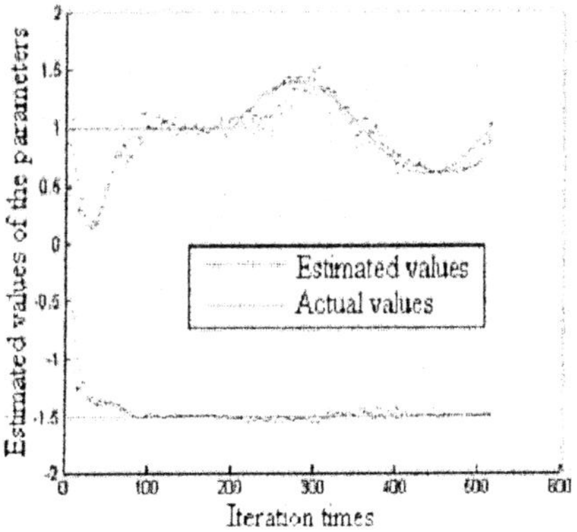

Figure 6 - FSPSO with h=5

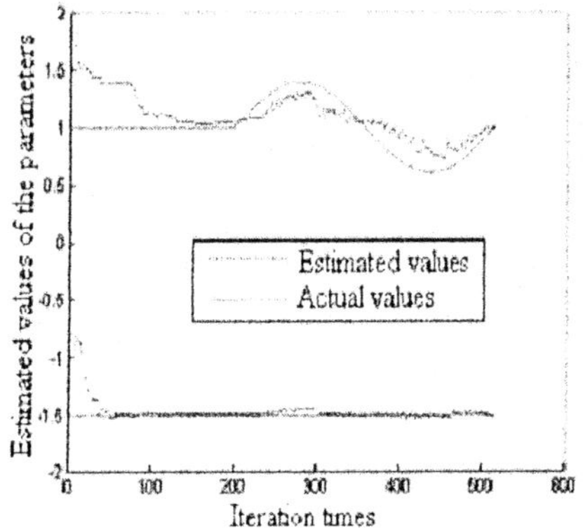

Figure 7 - FSPSO with h=10

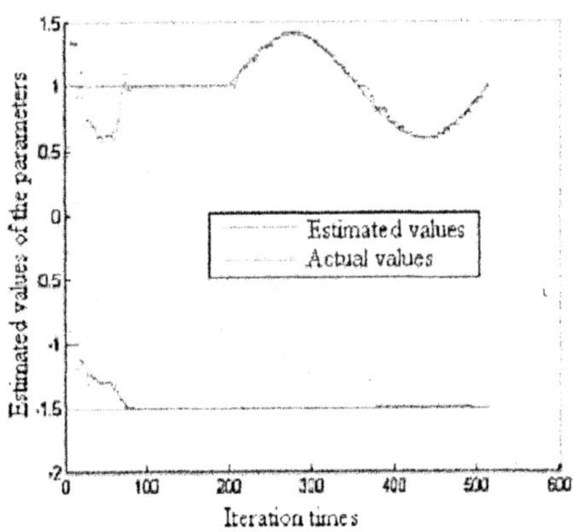

Figure 8 - Combinatorial PSO with h=3

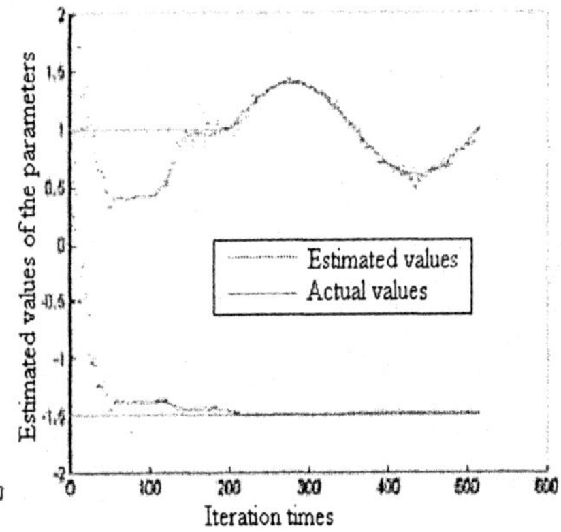

Figure 9 - Combinatorial PSO with h=5

A lots of simulations show that the more particles for the PSO of inertia weights are, the batter dynamic tracing is and that the more particles for FSPSO are the batter the convergence of time invariant systems is.

4.4 The robustness with colored noise in system

In the above results h=3 is good selection. When the ratio of colored noise to signal is 10 percentages simulation result is shown in Figure 11. Similarly Figure 12 and Figure 13 are shown respectively when the ratios of colored noise to signal are 20 and 30 percentages. Results of tracking and robustness are more satisfied.

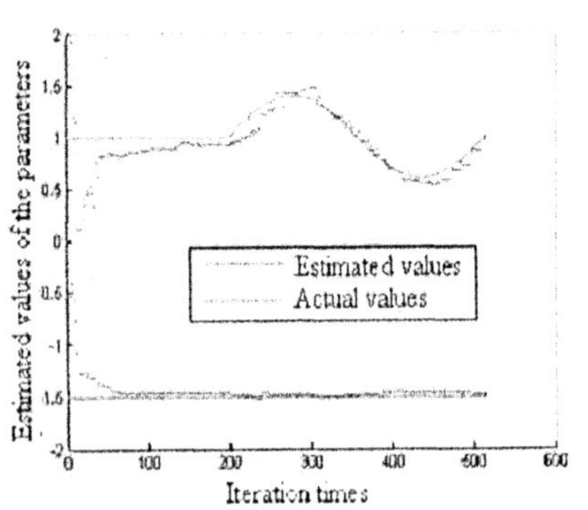

Figure 10 - Combinatorial PSO with h=10

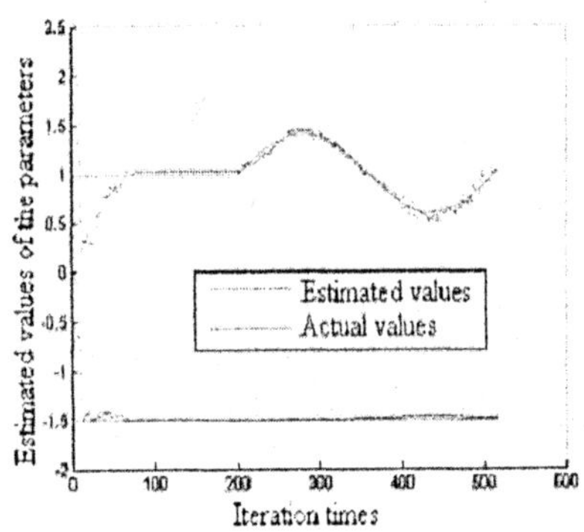

Figure 11 - Combinatorial PSO with noise 10%

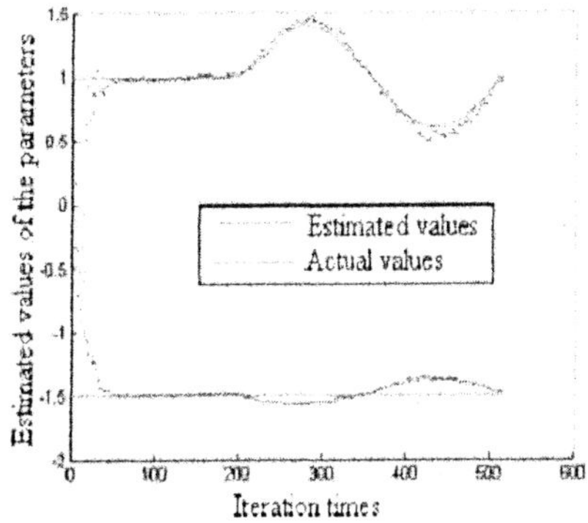

Figure 12 - Combinatorial PSO with noise 20%

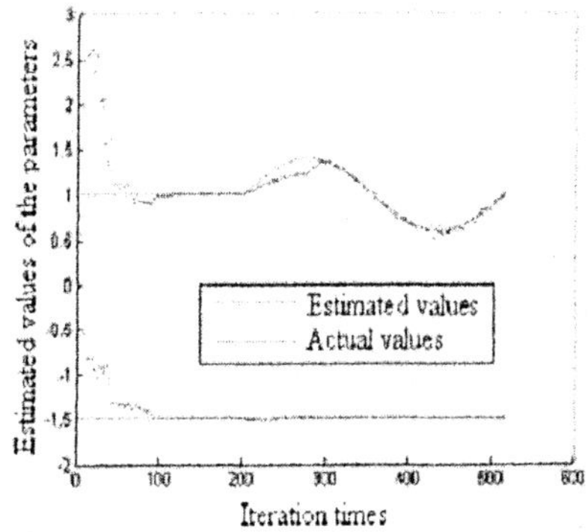

Figure 13 - Combinatorial PSO with noise 30%

5. CONCLUSIONS

In this paper, parameter estimation of the time varying for process models is converted to an optimization problem. Presented (15) can show not only outside system output error but also inside parameters error in order to explain more difference between actual and estimative system. We are able to make use of advantages of more approaches in the time-varying system. We take the combinatorial PSO that FSPSO combines PSO of inertia weights in simulation.

The identification algorithm for time-varying systems with colored noise was indeed more efficient and robust in combinatorial PSO comparing with FSPSO or PSO of inertia weights.

6. ACKNOWLEDGEMENT

This work is supported by the K. C. Wong Education Foundation

7. REFERENCES

1. Feng Ding, Tongwen Chen, "Performance Bounds of Forgetting Factor Least-Squares Algorithms for Time-Varying Systems with Finite Measurement Data", *IEEE Trans. Circuits and Systems,* 52(3): 555-566, 2005.
2. R. Salomon, "Evolutionary algorithms and gradient search similarities and differences," *IEEE Trans. Evolutionary Computation*: 2(2), 45-55, 1998.
3. Yuncan Xue, Qiwen Yang, Jixin Qian, "Parameter estimation for time-varying system based on improved genetic algorithm", *Proc. the 28 Annual Conference of the IEEE Industrial Electronics Society*, Sevilla, Spain: 2007-2010, 2002.
4. R. C. Eberhart and J. Kennedy, "A new optimizer using particle swarm theory," *Proc. 6th Industrial Symp. Micro Machine and Human Science*, Nagoya, Japan: 39–43, 1995.
5. P. N. Suganthan, "Particle swarm optimizer with neighborhood operator," *Proc. Conference Evolutionary computation*, Washington, DC: 1958–1961, 1999.

6. R. C. Eberhart, P. Simpson, and R. Dobbins, *Computational Intelligence PC Tools*: Academic, ch. 6: 212–226, 1996.
7. Weixing Lin, Chongguang Jiang, Jixin Qian, "The Identification of Hammerstein Model Based on PSO with Fuzzy Adaptive Inertia Weight", *Journal of Systems Science and Information*, 3(2): 381-391, 2005.
8. Y. Shi, R. C. Eberhart, "Empirical Study of particle swarm optimization", *Proc. IEEE International Conference. Evolutionary Computation*, 3: 101-106, 1999.
9. M. Clerc, "The swarm and the queen: toward a deterministic and adaptive particle swarm optimization," *Proc. ICEC'99*, Washington, DC: 1951–1957, 1999.
10. D. Corne, M. Dorigo, F. Glover, Eds., *New Ideas in Optimization*. New York: McGraw-Hill, ch. 25: 379–387, 1999.
11. M. Clerc, J. Kennedy, "The particle swarm: explosion, stability, and convergence in a multi-dimensional complex space," *IEEE Trans. Evolutionary Computation*, 6(1): 58–73, 2002.
12. R. C. Eberhart, Y. Shi, "Comparing inertia weights and constriction factors in particle swarm optimization", *Proc. Conference Evolutionary Computation 2000*, San Diego, CA: 84-88, 2000.
13. [13] J. Kennedy, "Stereotyping: Improving particle swarm performance with cluster analysis," *Proc. 2000 Conference Evolutionary Computing*: 1507–1512, 2000.
14. J. Kennedy, R. Mendes, "Population structure and particle swarm performance", *Proc. 2002 World Conference Computational Intelligence*, Honolulu, HI: 1671–1676, 2002.
15. Jing Ke, Yizheng Qiao, Jixin Qian, " Identification of Time-varying Delay Systems Using Particle Swarm Optimization", *Proc. the 5th World Congress on Intelligent Control and Automation*, Hangzhou, P.R. China: 330-334, 2004.

39 REMOTE MONITORING AND CONTROL OF MANUFACTURING SYSTEM

E. Villani[+], R.A. Castro*, F.M. Marques*, P.E. Miyagi*
*Escola Politécnica, University of São Paulo, Brazil
[+]Instituto Tecnológico de Aeronáutica, Brazil
e-mails: evillani@ita.br, pemiyagi@ita.br

This paper presents the development of a remote monitoring and control system for a CIM plant. It discusses the main steps for the system specification. The purpose of the remote access system is to provide the user with facilities necessary to understand the system behavior, propose supervisory control strategies and test them.

1. INTRODUCTION

Generally, e-manufacturing is described as the use of Internet to exchange information and achieve rapid response in a distributed and disperse manufacturing environment. The concept of e-manufacturing has emerged during the last years as a consequence of the popularization of Internet applications in business processes. The so-called e-business has added speed to many activities related to the manufacturing industry, particularly to those related to the interaction with suppliers, partners and customers. The high-level of competition among industries, the need of minimizing lead-times, and the demand for optimized scheduling procedures are among the reasons that have been pushing the development of Internet-based applications to exchange information in all levels of an organization.

According to (Lee, 2003), e-manufacturing is a recent concept developed to answer the needs of e-business strategies and meet the requirements for the complete integration of all business elements including all suppliers, customer service networks and manufacturing units through the effective use of web-enabled computational tools and tether-free technologies. E-manufacturing includes remote facilities with the ability to monitor the plant floor assets, predict variation in performance, dynamically reschedule production and maintenance operations, and synchronize related and consequent actions in order to achieve a complete integration between manufacturing systems and upper-level enterprise applications.

E-manufacturing is a broad field of research. A number of examples of successful applications and researches can be found in literature and in industry.

In (Hao et al, 2005), the problem of factory integration is approached by proposing a framework based on Web Services and agents that actuates in all levels of the organization – from the virtual enterprise (inter-enterprise), to the enterprise (intra-enterprise) and shop floor levels. (Wang et al, 2001) explores the Internet for product data management. It proposes an integrated data model to the information integration for remote robot manufacturing. The data model is exchanged among geographically spread customers, suppliers, design and manufacturing companies,

Please use the following format when citing this chapter:

Villani, E., Castro, R. A., Marques, F. M., Miyagi, P. E., 2006, in IFIP International Federation for Information Processing, Volume 220, Information Technology for Balanced Manufacturing Systems, ed. Shen, W., (Boston: Springer), pp. 369–376.

through the lifecycle stages of the product, from requirement specification and conception design, detailed design, fabrication and assembly, installation and operation.

Other web-based solutions focus on Supply-Chain Management (SCM) and Enterprise Resource Planning (ERP) systems. In (Frohlich, Westbrook, 2002) different strategies for demand and supply integration are discussed. Customers and suppliers are linked together into tightly integrated networks where real-time information travels immediately backwards through demand-driven supply chains while inventory flows swiftly forwards.

Another area of research is the development of a collaborative design environment by interconnecting CAD/CAE/CAM systems. Some examples are (Li et al, 2004; Zhan et al, 2003). These works provide an Internet based computational architecture that supports the sharing and transferring of knowledge and information amongst geographically distributed and disperse centers.

Finally, some works focus on the development of remote monitoring and control of manufacturing process. Among them, (Muto, 2003) presents a XML system called @factory that provides remote surveillance, video camera monitoring, schedule management and data analysis for CNC machines.

In the context of this last area of research, the purpose of the paper is to systematize the specification phase of remote monitoring and control systems. It proposes the division of the specification phase into a set of steps. Each step approaches a different problem of the specification of remote monitoring and control systems. The paper presents as an example the application of the e-manufacturing paradigm to a CIM (Computer Integrated Manufacturing) plant of a research laboratory. The focus is on the development of a computational tool for remote monitoring and control of the CIM. This work is been developed as part of the Brazilian Government Program TIDIA/KyaTera (TIDIA-KyaTera, 2006), which connects a number of research laboratories through an advanced high-speed optical network. The TIDIA/Kyatera network is been used as a testbed for research in different areas including the remote control of manufacturing systems.

The paper is organized as follows. Section 2 approaches the problem of specifying remote monitoring and control systems. Section 3 presents the remote control system developed for the CIM plant and Section 4 presents some conclusions.

2. SPECIFICATION OF REMOTE MONITORING AND CONTROL SYSTEMS

The specification of remote monitoring and control system can be organized into a sequence of steps. Each step approaches the system under a different perspective and increases the level of detail of the system specification. In each step, a set of questions must be answered, analyzing a different view of the remote system to be developed. The steps are organized in questions in order to guide and facilitate its applicability.

2.2 Step 1 – Monitoring or control?

The first step consists of defining the purposes of the remote access system. The first point to be defined is if the system will have only monitoring functions or also control facilities. The answers given at this point may change due to limitations imposed by the characteristics of the manufacturing system and the available resources.

In a remote access system, the information collected from the local manufacturing system is available in real-time in the remote destination. So, a remote access system is justifiable only when the data collected is also *processed and used in real-time* in the destination. When the real-time availability is not a key issue, a simple solution is to store the data locally and use conventional tools for sharing databases through the Internet. The first questions to be answered are:

***Question 1.1** – What are the advantages of making information about the remote system available in real-time?*

***Question 1.2** – What kind of decision can be taken based on the available data in the remote destination?*

The motivations for providing control facilities are investigated by establishing what decisions are taken based on the data provided by the remote access system. Basically, the first point is to determine what system may be affected by the decisions of the remote system. Figure 1 illustrates some of the possible configurations for a remote access system. In Figure 1a) the decisions interfere in the evolution of a remote system. An example is when the parameters of the local production are used to control the evolution of a remote system. In Figure 1b) the data is processed by a remote computer or user and affects the evolution of the local system – this is the case of remote control. An example is when a remote operator does the maintenance of a local machine. A third kind of configuration is when the remote decisions are taken based on the combined information from several systems, which are also affected by the decisions. An example is when the scheduling and resource management of geographically spread plants are adjusted in real-time based on their current performance.

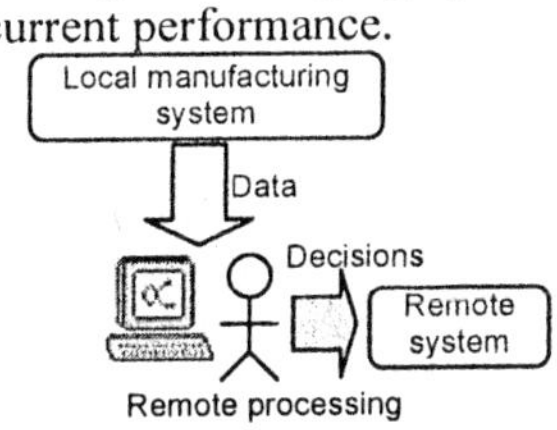

a) Candidates to remote monitoring

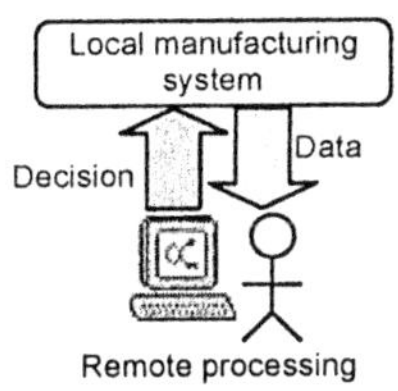

b) Candidates to remote monitoring and control

Figure 1. Configurations for remote monitoring and control.

2.3 Step 2 – Specification of use cases and exchanged information

In order to determine the data to be transmitted between the remote and local system the first question of Step 2 is:

***Question 2.1** – What are the system use-cases and who are their actors?*

This first question can be documented using the Use-Case Diagram of UML (Unified Modeling Language) (Rumbaugh et al, 2004). Among the points to be analyzed when specifying the use-cases is the maintenance of history records and databases, which can be in the local system or in a remote system. At this point, the functional requirements of the system are considered at a high-level of abstraction. Other requirements, such as real-time restrictions are treated in the next steps.

Once the use-cases have been specified the next step is to make a list of the data needed on the remote processing point to take the decisions specified in Step 1. The second question of Step 2 is:
***Question 2.2** – What information is exchanged between local and remote systems?*

2.4 Step 3 – Hardware analysis

Step 3 analyzes the viability of the remote access from the hardware point of view:
***Question 3.1** – What are the system nodes and how they communicate among them?*

In the case that there is more than one node in the local system, the local nodes may communicate using local networks. Each local node can communicate directly to the remote system or they may be connected to a local server that centralizes and manages the communication via Internet with the remote system. Another point is how the local nodes are connected with databases and history recorders.

The next questions are:
***Question 3.2** – What is the hardware for communicating with the local manufacturing system by Internet?*
***Question 3.3** – What is the kind of programming language used for developing the remote access system and what are the technologies available for implementing the communication via Internet?*
This last question must specify if the remote access system will be a web-page, software developed using general purposes programming languages.

2.5 Step 4 – Software refinement

In this step each use-case is detailed. This can be done by representing the sequences of activities in UML Activity Diagrams. The question of Step 4 is:
***Question 4.1** – What are the sequence of activities for each use-case?*
***Question 4.2** – Can the local manufacturing system be accessed by multiples remote system (simultaneously or not)?*
***Question 4.3** – In the case of remote control system, how conflicts are managed?*

2.6 Step 5 – Requirements for remote access

The last step of the remote system specification regards the requirements related to the remote nature of the monitoring and control system. The questions to be answered are:
***Question 5.1** – What happens in the case of communication failure: is the local system able to detect communication failure and put the system in a safe state?*

***Question 5.2** – Are the communication delay critical to the system operation and does the local system check the communication delay, taking the appropriate actions?*

***Question 5.3** – What are the facilities available for the remote system reacting to failures in the local manufacturing system?*

***Question 5.4** – Does the remote nature compromise the system safeness?*

3. CIM SYSTEM AT LSA/EPUSP

3.1 Description of the System

This system to be remote accessed is a computer integrated manufacturing plant installed for research and didactic purposed at Escola Politécnica of University of São Paulo. The CIM system is composed of four stations (Figure 2): a storage station, an inspection station, a transportation station and an assembly station.

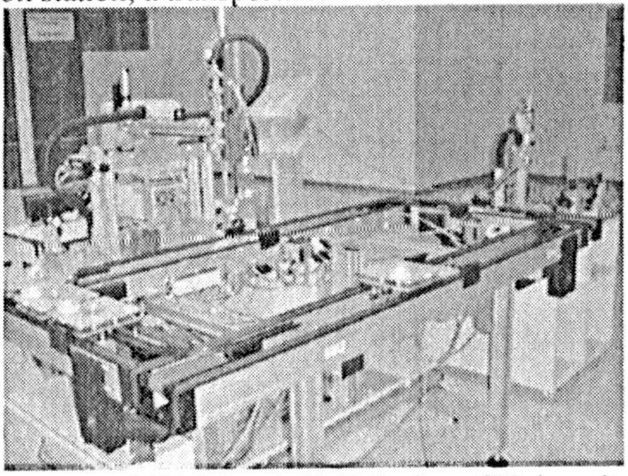

Figure 2. CIM system at LSA/ Escola Politécnica of University of São Paulo.

The purpose of the system is to assembly four components into a single product: a cylinder, an internal pine, a spring and a cover. The components are provided in different colors and must be combined according to pre-defined rules. The system is currently been used for evaluating new techniques proposed for the design of supervisory control strategies. The CIM system can operate in two configurations: a partial configuration where only the storage and inspection stations are active and a full configuration where the four stations are active.

Due to the limited space, this paper will focus on operation under the partial configuration. Basically, the storage station has a buffer of cylinders operated by a pneumatic piston. The cylinders are removed from the buffer and transported to the inspection station by a rotating arm with a vacuum vent. In the inspection station, the color of the cylinder is determined by a set of sensors, as well as its height. If the cylinder has the appropriate height it is sent to the transportation station by a pneumatic actuator, otherwise it discarded in a refuge area.

3.2 Specification of the CIM remote access system

Step 1 – Monitor or Control?
The purpose of the remote access system is to provide the user with facilities necessary to understand the system behavior, propose supervisory control strategies and test them. The user must be able to monitor the system behavior and control the execution of events in each module. This is clearly the case of developing a remote monitoring *and* control system, as the remote user must interfere in the local manufacturing system.

Step 2 – Specification of use cases and exchanged information
There are two actors: local user and remote user. The system can operate in one of the following modes: automatic, step-by-step and testing. Figure 3 details the use

cases for the partial configuration. Basically, in the automatic mode the system performs a pre-defined sequence of operations that sends a cylinder to the transportation module. This is the mode used for understanding the system behavior. In the step-by-step mode the same sequence is performed but the user fires each step, this is the mode used for collecting data about the system, such as the time for performing each operation. In the testing mode the user select the operation to be performed. This is the mode used for testing supervisory control strategies.

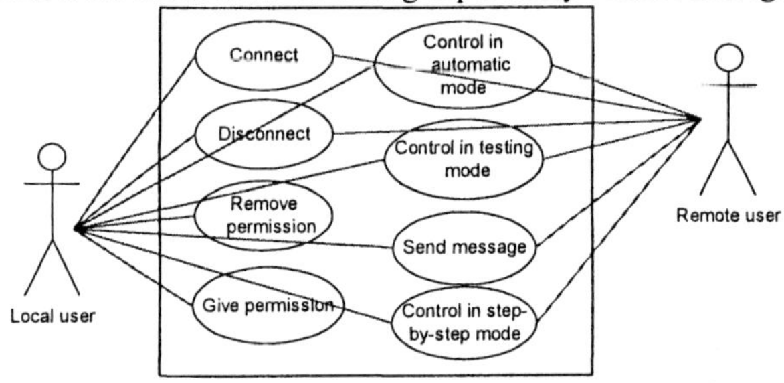

Figure 3 – Use-Case Diagram for partial configuration.

The local user has priority over the remote user. Although a remote user can always connect to the system and monitor it, he can only control the system when the local user gives permission. The local user can also remove the control of the system from the remote user. All the local and remote users connected to the system can exchange messages through a chat area – even when they do not have the control of the system.

The information to be transmitted from the local manufacturing system to the remote user is:

- if the remote user is connected or not; if he is controlling the CIM or not;
- the current state of the CIM system;
- a real-time video to understand the CIM operation and detect problems;
- text messages elaborated by the local user.

The information to be transmitted from the remote user to the local manufacturing system is:

- a request to connect or disconnect to the CIM system;
- the current mode of control chosen by the remote user;
- commands to start and interrupt the sequence of operation in the automatic mode, perform another step in the step by step mode, and perform each operation in the testing mode.
- text messages elaborated by the remote user.

Step 3 – Hardware analysis

The system architecture is illustrated in Figure 4. Each station has a PLC that controls its operation. A local PC centralizes the communication between the four stations and the Internet (remote PC). In the partial configuration the two stations are connected with the local PC by serial communication. In the case of full configuration, a Profibus network connects the four stations. The bus master is station 4. The remote access system is implemented in using generic purposes programming language (Borland C++), while communication via Internet is performed using sockets.

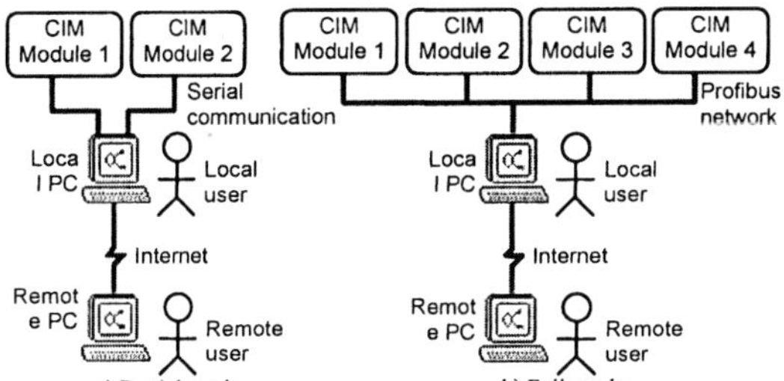

Figure 4. Configurations for remote monitoring and control.

Step 4 – Software specification

Each use-case is detailed in an UML Activity Diagram. As an example, Figure 5 presents the UML Activity Diagram for the 'control in testing mode' use-case.

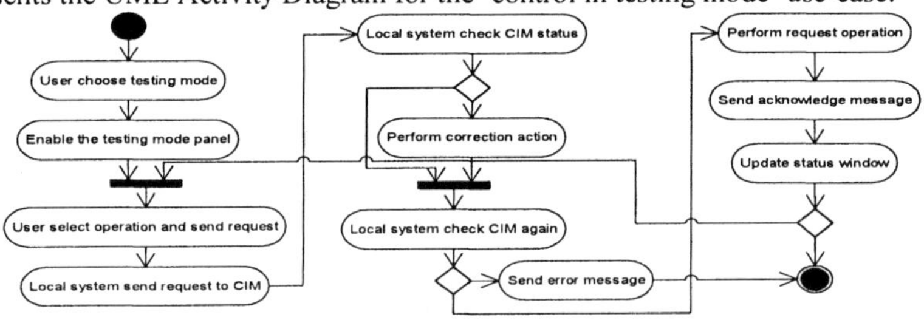

Figure 5. Activity Diagram for the use-case 'control in testing mode'.

Although multiples remote users can be connect to the CIM system at the same time, visualize the CIM evolution and exchange message; only one user at a time has the control of the CIM system. The local user chooses the remote user that has the control.

Step 5 – Requirements for remote access

The last step regards the requirements related to the remote nature of the system. The communication delay is not critical for its operation because its evolution is driven by discrete events. The following requirements are specified, among others:

- The local system must send acknowledge messages with a certain frequency in order to detect when the remote system lose the connection the local system. In this case, the local PC must warn the local user in order to give the control of the CIM system to another remote user.
- The local PC must check the current state of the CIM before requiring an operation. When the CIM is not on a safe state, an adequate correcting operation must be performed. An example is when a request to send a cylinder to the inspection module is emitted but it is already occupied by another cylinder – in this case the cylinder in the inspection module must be sent to the refuge area.
- When an operation is required but is not performed in a certain time interval, the local and remote users must be warned of the occurrence of a fault.

The requirements of Step 5 must be verified in order to validate the specification. For this purpose, a model of the remote access system and local manufacturing systems is build and analyzed using formal techniques and simulation tools.

The remote control system developed based on the specification describe in this section is current under operation. An example of the remote control system interface is presented in Figure 6.

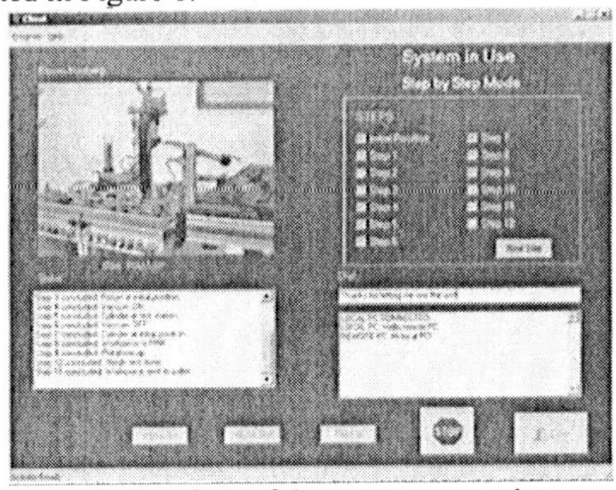

Figure 6. Interface of the remote control system.

4. CONCLUSION

This paper introduces a systematic approach for the specification of remote monitoring and control systems. In order to illustrate the approach the development of a remote monitoring and control system for a didactic assembly plant is presented. Currently, the specification method is been applied to a number of case studies with different characteristics in order to validate it.

5. ACKNOWLEDGEMENTS

The authors would like to thank the partial financial support of the governmental agencies FAPESP, CNPq and CAPES.

6. REFERENCES

1. Frohlich MT, Westbrook R. Demand chain management in manufacturing and services: web-based integration, drivers and performance. Journal of Operations Management 20 (2002) 729–745.
2. Hao, Q, et al. Towards a cooperative distributed manufacturing management framework. Computers in Industry 56 (2005) 71–84.
3. Lee J. E-manufacturing—fundamental, tools, and transformation. Robotics and Computer Integrated Manufacturing 19 (2003) 501–507.
4. Li WD, et al. An Internet-enabled integrated system for co-design and concurrent engineering. Computers in Industry 55 (2004) 87–103.
5. Murata, T. Petri Nets: properties, analysis and applications. Proc. of the IEEE, 77 (1989) 4: 541-580.
6. Muto K. Advanced technology for manufacturing engineering development. JSAE Review 24 (2003) 303–312.
7. Rumbaugh, J. et al Unified Modeling Language Reference Manual (2nd edition) Addison-Wesley Longman, Inc. Harlow, 2004
8. TIDIA-KyaTera – Advanced Internet Program, FAPESP, Brazil. http://www.kyatera.fapesp.br (access in 01/2006)
9. Zhan HF, et al. A web-based collaborative product design platform for dispersed network manufacturing. Journal of Materials Processing Technology 138 (2003) 600–604.
10. Wang, C, et al. Implementation of remote robot manufacturing over Internet. Computers in Industry 45 (2001) 215–229.

40

ARCHITECTURE OF A WEB-BASED POWER SCADA SYSTEM USING J2EE TECHNOLOGY

Qizhi Chen[1,2], Hamada Ghenniwa[2], Weiming Shen[2,3]
[1]*Southwest Jiaotong University, Chengdu, China*
[2]*University of Western Ontario, London, Ontario, Canada*
[3]*National Research Council Canada, London, Ontario, Canada*
qzchen1625@gmail.com; hghenniwa@eng.uwo.ca; weiming.shen@nrc.gc.ca

With the deregulation of the electrical power industry, there is a strong need for power SCADA system to interoperate with other information systems. The interoperation between SCADA and all other systems not only needs the information sharing, but also the business - or transaction-based information integration. This paper presents a Web-based architecture for power SCADA system, which is a multi-tier distributed framework using J2EE technology. This architecture combines pervasive browser with flexible user-specified application client, supports Web-service integration and transaction control, and provides a unified security policy. With the advantages of J2EE platform, this new Web-based SCADA system seamlessly integrates traditional SCADA with the enterprise information system, and exhibits more flexibility, scalability, and security than precious Web-browsing-only SCADA system.

1. INTRODUCTION

Supervisory control and data acquisition (SCADA) system plays an important role in a power system. It supports control of remote equipments, gathers data from substations and/or generation stations, and shares all the gathered information with other systems. From its appearance in 1960s to now, SCADA system has been developed from the first generation, mainframe-dominated, centralized computing system to the third generation, the distributed network computing system [1].

Because of the deregulation of electrical power industry, the combination of more dynamic energy markets and the growing importance of E-commerce is driving the greater needs of effective cooperation and information exchange or sharing among different systems or different electrical entities [2]. Because power SCADA system is the most important information resource for power system operation, maintenance as well as marketing, it faces the great challenge to provide information in a flexible, expandable, and standardized way to many other subsystems among the power information system, such as Electricity Trading System, Planning System, and Enterprise Information System (EIS). The rapid

Please use the following format when citing this chapter:

Chen, Q., Ghenniwa, H., Shen, W., 2006, in IFIP International Federation for Information Processing, Volume 220, Information Technology for Balanced Manufacturing Systems, ed. Shen, W., (Boston: Springer), pp. 377–384.

development of information technologies brings the great opportunity to the power SCADA system to deal with this challenge.

The widespread and successful Internet/Intranet and World Wide Web (WWW) technologies makes the new generation of Web-based power SCADA system possible. Various SCADA applications or sub-systems based on Web technologies have been addressed in the past [3-9]. But all these Web-based application simply aims on supplying the easy information access and lower client hardware investment.

This paper proposes a multi-tier distributed Web-based architecture for power SCADA system using Java 2 Platform, Enterprise Edition (J2EE). It combines pervasive browser client with flexible user-specified application client, supports Web-service integration and transaction control, and provides a unified security policy. With the advantages of J2EE platform, this new Web-based SCADA system seamlessly integrates traditional control- and monitoring-oriented SCADA with EIS, and exhibits higher flexibility, scalability, and security than precious Web-browsing-only SCADA system.

The remainder of this paper is organized as follows. The J2EE platform and related technologies are briefly introduced in Section 2. In Section 3, a multi-tier Web-based architecture for power SCADA system using J2EE technology is proposed. The software architecture of this Web-based SCADA system and its characteristics are discussed in Section 4. Section 5 gives a brief review of the related work. Conclusion and future work are included in Section 6.

2. J2EE AND RELATED TECHNOLOGIES

J2EE is an architecture for distributed enterprise applications. It offers a multi-tiered distributed application model, the ability to reuse components, integrated Extensible Markup Language (XML)-based data interchange, a unified security, and flexible transaction control.

Application logic in J2EE is divided into components according to their function, and the various application components that make up a J2EE application are installed on different machines. The location of components depends on the tier to which the application component belongs. Four tiers are defined in J2EE platform, which are client tier, web tier, business tier, and enterprise information system (EIS) tier. The J2EE application architecture is illustrated in Figure 1.

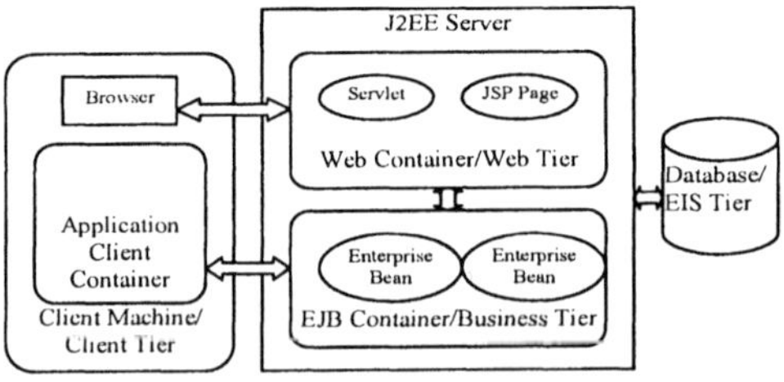

Figure 1 J2EE application architecture

The J2EE specification defines client components, Web components, and business components as illustrated in Figure 1.

Client components: A J2EE client can be a Web client or Application client running on the client machine. A Web client contains dynamic Web pages generated by Web components running in the Web tier and a Web browser. An application client provides a way for users to handle tasks that require a richer user interface, typically a graphical user interface (GUI).

Web components: J2EE Web components are either Servlets or pages created using JavaServer Page technology (JSP pages). Servlets are Java programming language classes that dynamically process http requests and construct responses. JSP pages are text-based documents that execute as servlets but allow a more natural approach to create static content.

Business components: Business logic is handled by enterprise beans running in the business tier. An enterprise Java bean (EJB) receives data, process it, and send it back to client programs. An Enterprise Java Bean also retrieves data from or store data into the EIS tier. There are three kinds of enterprise beans in J2EE: session beans, entity beans, and message-driven beans.

In addition to all above characteristics, the most important and exiting improvement of the latest J2EE platform is substantial support for Web services, and WS-I Basic Profile 1.0 for Web services interoperability. This makes it possible to develop a Web-based application not only supporting Web pages sharing but also transaction-based information integration.

3. SYSTEM FRAMEWORK OF WEB-BASED POWER SCADA SYSTEM

The four-tier system framework of a Web-based power SCADA system is shown in Figure 2.

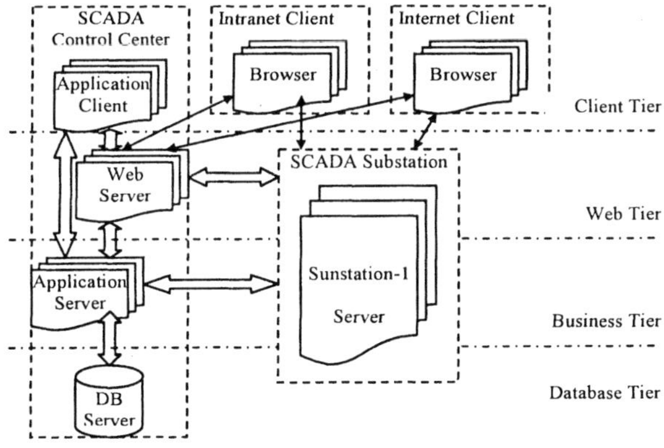

Figure 2 Four-tier framework of the Web-based power SCADA system using J2EE technology

The system framework of this Web-based power SCADA system employs all four tiers defined in J2EE specification, which are client layer, Web layer, business layer, and database layer. In client layer, there are two different client types, one is standalone application client operated by SCADA operator in control center and another is standard thin browser without any plug-in used by Internet or Intranet users.

Each tier in the control center can be deployed into one or more computer nodes, which depends on the system scale, number of operator, and the reliability requirement. In order to keep the high reliability, the business tier and database tier are redundantly configured. The redundant application servers operate in parallel mode or conventional standby mode. The database servers are always configured with storage cluster. Because of the scalability of J2EE, the business layer and the database layer can be deployed into a same server in the case of a small-scale SCADA control center.

Instead of the direct connection to Intranet/Internet for each intelligent electrical device (IED) in a substation [7][9], the substation in Figure 2 is considered as an integrated web server or application server to connect to this Web-based SCADA system. The benefit of this is that no prerequisite to each IED that it must be web-based. Because of its size, the web layer, business layer, and database layer in a substation all be deployed into one physical machine.

The communication infrastructure of a Web-based power SCADA system is illustrated in Figure 3.

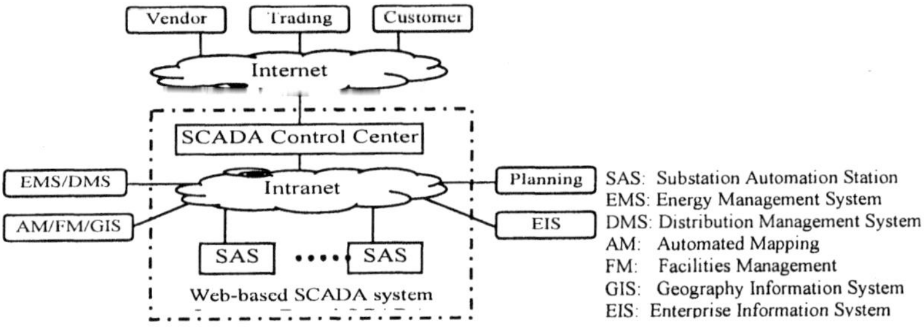

Figure 3 Communication infrastructure of the Web-based power SCADA system

Figure 3 shows the communication relationship between the Web-based SCADA system and other systems in a utility information system. It shows that the intra-enterprise sub-systems, such as EMS, EIS, AM/FM/GIS, and Planning system, all connect with the Web-based SCADA through Intranet. The inter-enterprise sub-system, Trading system, as well as all extra-enterprise users, such as customers and vendors, access this Web-based SCADA system through Internet.

Since J2EE provide the Web services support through integrated data interchange on XML-based standard protocols, the access to SCADA system can be implemented through traditional Web page browsing or novel Web-services integration.

4. SOFTWARE ARCHITECTURE OF WEB-BASED POWER SCADA SYSTEM

4.1 Software Architecture

This control center software architecture of the Web-based SCADA system using J2EE technology is illustrated in Figure 4. The software architecture for the substation server is similar with this except that all J2EE components all resides on the same server.

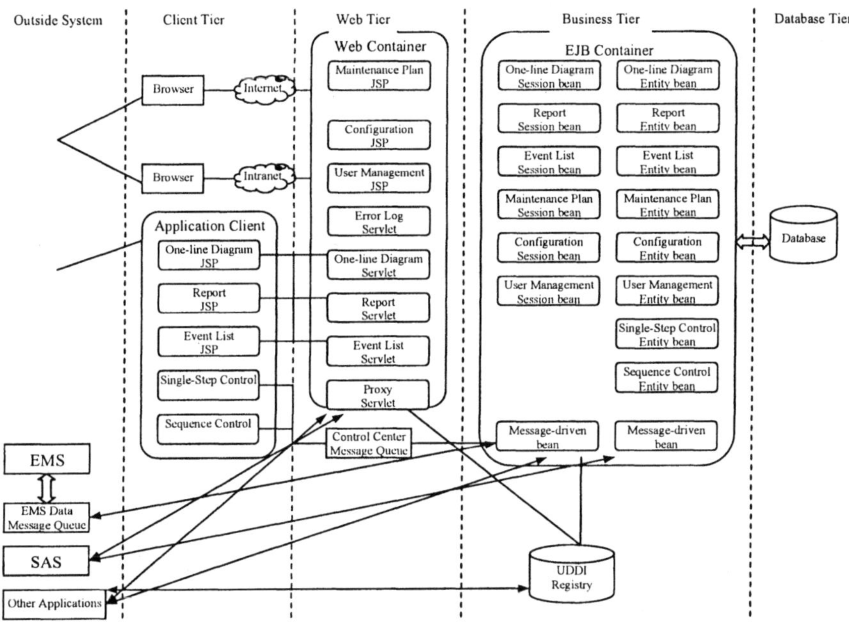

Figure 4 Control center software architecture of the Web-based SCADA system using J2EE technology

The key characteristics of this software architecture described in Figure 4 are:

The software architecture can be described as Client/Server (C/S) structure, but under the C/S structure there are two different client components in this Web-based SCADA system. One is the prevalent thin Web browser, another is the independent user-specified application client. Internet or Intranet users use browser to get the real-time or history data, make maintenance plan, browse or modify system configuration, or trace system error logs. This allows a lot of widely distributed users can access the SCADA with no special application installation. The independent application client provides flexible and friendly man machine interface (MMI) to the SCADA operator.

The main functions of application client are electrical equipment control, real-time or history information display. Except directly accessing the message-driven bean running in the business tier, the application client shares the several servlets running in the Web tier for information display with the Web client.

The Web tier includes both JSP pages and servlet components. JSP pages are used for its natural approach to creating static content, and servlets are used for the purpose that it can be accessed not only by browsers but also by application clients.

A proxy is implemented in the Web tier for two advantages. The first is that it insulates the SAS from the direct connection to the Internet. Proxy redirects all Internet accesses to SAS to their destination substation so as to reduce the access points to this Web-based SCADA system. Another advantage is that the Web tier in the control center does not need implement the same JSP if it has been deployed in a substation server.

The application client and EMS interact with message-driven bean running in the business tier through the JMS queues.

The control operation in SCADA is a three-step operation, which follows the order of selecting first, executing later, and certifying last. In order to ensure the data and control integrity, transaction control is employed for the SCADA equipment control operation.

SAS has the similar software architecture as the control center except that all J2EE components are deployed into one machine. SAS communicates with message-driven bean running on the J2EE server in the control center through the TCP-IP based standard protocols, such as IEC60870-5-104. In the mean time, it provides the dynamic Web contents to Intranet through JSP pages or sevlets.

To those interfaces which are need to be open among the SCADA or to other systems, they will be encapsulated as Web services and their corresponding Web Service Description Language (WSDL) file will be published to the Universal Description Discovery & Integration (UDDI) registry. Thus every application that has the right to access the SCADA system can find the Web services and using it. The endpoint of Web services in this system can be a JAX_RPC Web tier component or session bean. Service endpoint.

4.2 System Security

The security implementation in this Web-based distributed SCADA system only adopts the declarative security in J2EE specification. Different security groups such as operator, maintenance engineer, manager, custom, and vendor, are defined. Each group owns different permission to access the SCADA resources and includes a set of authenticated users. For example, the operator group can access the information display Web components, and control EJB, but it does not have the right to access system configuration Web components.

The other security policy adopted by this Web-based SCADA system is that SAS does not directly connect to Internet but all Internet accesses to it are forwarded by the proxy in the control center. This greatly reduces the Internet access point to the SCADA so as to reduce the possibility of being attacked by malicious users.

5. RELATED WORK

There are significant research efforts focused on proposing Web-based SCADA for power system, which covers the sub-system in SCADA, intelligent electric device (IED), and dependent applications in SCADA system. The IT technologies employed by these works vary from Java, ASP, Applet to J2EE. The followings are brief reviews of some related research efforts.

Two Web-based sub-systems in power SCADA system were discussed in [3]-[4]. The work of Leou *et al.* [3] described a Web-base 3-tier power quality monitoring system, which used ASP technology to connect to and read from database server, created dynamic html pages. Compared with our proposal, it only implemented the Web client and did not support Web-service interface. The work of Qiu *et al.* [4] introduced an Internet based frequency monitoring network. It emphasized the Internet as the physical communication infrastructure more than the Internet-based software implementation.

The work of Qiu *et al.* [5] implemented a Web-based SCADA display application via Internet access, which was a typical three-tier client/server applications containing a service tier, a data store tier, and a Man Machine Interface (MMI) tier. The MMI was implemented by Java language without using any advanced Web technologies. Compared with our full implementation of SCADA system, the WWW displays in this system only limited to the tabular displays and one-line diagrams.

Some issues about Web-based power SCADA system were discussed in [7]-[9]. Ebata *et al. in* [7] used a trial system to discuss how to improve the real-time performance and reliability of Internet-based SCADA, but it did not discuss the technology implementation, which is the emphasis of our paper. The work of Medida *et al.* [8] just employed a gateway as Web server in the conventional SCADA center in order to support information browsing. It just was a temporary solution for the SCADA information sharing and not an ultimate solution as ours. The work of Li *et al.* [9] regards the SCADA control center system and individual intelligent devices (IED) as separate Web-based application. Compared with our proposed architecture, it treated individual IEDs as separate Web components. The shortcoming is that this imposes a prerequisite to each IED, and SCADA system will loose the control if it is not a Web-based device.

Although two systems based on J2EE platform in power system were discussed in [11][12], [11] aimed on a simple power quality monitoring system and [12] was about the electricity market operation system, which were all different from the SCADA system.

6. CONCLUSION AND FUTURE WORK

The proposed Web-based architecture for power SCADA system is a multi-tier distributed framework using J2EE technology. This architecture combines pervasive browser client with flexible user-specified application client. Since the browser implemented by Java Servlet or JSP has not any needs for Java Plug-in or any possible security policy in the client site, it makes the browser thinner than any of previous implements using Applets for dynamic Web content. On the other hand, the standalone application client provides more flexible user-specified GUIs to the operator than just using marked language-based pages.

With the advantages of J2EE technology, this new Web-based power SCADA architecture facilitates transaction control, Web-service support, a unified security policy, and exhibits higher flexibility, scalability, and security. It makes power SCADA system easily achieve the information sharing as well as the business logic- and transaction-based information integration with other systems.

Future works will include implementing a prototype system, design the policy that which kind of services in SCADA should be published as Web services, and search the possible solution to improve the real-time response capability caused by the lower efficiency of Java.

7. REFERENCE

1. McClannhan, RH. "SCADA and IP: Is Network Convergence Really Here?", *IEEE Industry Applications Magazine,*2003; 9(2): 29-36.
2. Becker, D, Falk, H, Gillerman, J, Mauser, S, Podmore, R, Schneberger, L. "Standards-Based Approach Integrates Utility Applications", *IEEE Computer Applications in Power,* 2000; 13(4): 13-20.
3. Leou, RC, Chang, YC, Teng, JH. "A Web-based power quality monitoring system", Power Engineering Society Summer Meeting, 2001, IEEE, Vol. 3, pp. 1504-1508.
4. Qiu, B, Chen, L, Centeno, V, Dong, X, Liu, Y. "Internet based frequency monitoring network (FNET)", Power Engineering Society Winter Meeting, 2001. IEEE, Vol. 3, pp.1166-1171.
5. Qiu, B, Gooi, HB. "Web-Based SCADA display systems (WSDS) for access via Internet", *Power Systems, IEEE Transactions on,* , 2000, 15(2): 681-686.
6. Ishihara, Y, Shirota, Y, Sekiguchi, K, Sato, S, Sawai, K. "Recent trends in the implementation of Intranet based measurement and monitoring", *Transmission and Distribution Conference and Exhibition 2002: Asia Pacific. IEEE/PES,* Vol. 3, pp. 2261-2266, 2002.
7. Ebata, Y, Hayashi, H, Hasegawa, Y, Komatsu, S, Suzuki, K. "Development of the Intranet-based SCADA (supervisory control and data acquisition system) for power system", *IEEE PES Winter meeting 2000,* Vol. 3, pp. 1656-1661, 2000.
8. Medida, S, Sreekumar, N, Prasad, KV. "SCADA-EMS on the Internet", *Energy Management and Power Delivery, 1998. Proceedings of EMPD '98. 1998 International Conference on,* Vol. 2, pp. 656-660, 1998.
9. Li, D, Serizawa, Y, Kiuchi, M. "Concept Design for a Web-based Supervisory Control and Data-Acquisition (SCADA) System", *Transmission and Distribution Conference and Exhibition 2002: Asia Pacific, IEEE/PES,* Vol. 1, pp. 32-36, 2002.
10. Bodoff, S, Armstrong, E, Ball, J, Carson, DB, Evans, I, Green, D, Haase, K, Jendrock, E. "The J2EE Tutotial: Second Edition", Sun Microsoft, 2004.
11. Leou, RC, Tsai, WR, Chang, YN. "A Power Quality Monitoring System Based on J2EE Architecture", *TENCON 2004, 2004 IEEE Region 10 Conference,* Vol. C, pp. 291-294, 2004.
12. Song, Y, Tang, G, Yang, Z, Hu, J. "The Technical Implementation of the Electricity Market Operation System", *Transmission and Distribution Conference and Exhibition: Asia and Pacific, 2005 IEEE/PES,* pages:6, 2005.

41

A WEB-BASED CSCW SYSTEM FOR REMOTE SUBSTATION FAULT DIAGNOSIS

Hongsheng Su, Jianwu Dang, Feng Zhao

School of Information and Electrical Engineering, Lanzhou Jiaotong University, Lanzhou 730070, P.R. China, shsen@163.com

Relying on Computer Supported Collaborative Work (CSCW) technology, expert system (ES) can extend its capabilities from only one working to set up a cooperative labor environment for a group of expert systems working together. In order to more effectively support those remote substations fault diagnosis as well as related departments involvements, Web-based CSCW systems are intensively recommended. The paper for that proposes a framework of CSCW system with Web-based for remote substation fault diagnosis. The system applies Multi-Agent (MA) technology to construct distributed expert systems platform, and presents the collaboration and communication framework of the platform using J2EE. The system is applied in remote substation fault diagnosis, and is proved very effective

1. INTRODUCTION

In recent years, with the fast advancement of Computer Supported Collaborative Work (CSCW) technology and the exponential growth of Internet as well as increasingly broad application of Multi-Agent (MA) technology, an effective support has already been supplied to those far-reaching large-scale distributed expert systems. With respect to those remote or unattended substations fault diagnosis, a general requirement is to yield a cooperative working environment that allows related groups such as fault diagnosis experts in diverse fields, facility manufacturers, management departments, maintenance departments, employment departments, relevant government officers and some persons related to it to work together to cooperate with one another to dispose the accidents while substation faults happen. Supported by Internet and CSCW and MA technologies, expert systems (ES) can extend its abilities from only one working to setup a cooperative working environment, and make diagnosis process possess initiative characteristics. CSCW supplies knowledge share and resource complementation, and also supports cooperative working environment. A Web-based CSCW system can provide user with a friendly human-computer interface and well-regulated cooperative working environment. In addition, it still can integrate each independent small-scale network in Internet such as LAN, WAN, ISP servicing networks to realize joint aims. MA is

Please use the following format when citing this chapter:

Su, H., Dang, J., Zhao, F., 2006, in IFIP International Federation for Information Processing, Volume 220, Information Technology for Balanced Manufacturing Systems, ed. Shen, W., (Boston: Springer), pp. 385–394.

applied to simulate human expert intelligence and behavior to resolve more complicated problem. MAS-based substation fault diagnosis methods have already been reported (Liu *et al.*, 2003; Dong, *et al.*, 2003; Zhao *et al.*,2004; Mao *et al.*,2005; Wang *et al.*,2005), but they expose some problems in process of applications more or less(Su, *et al.*, 2005). Hence, based on Internet, MAS and CSCW technologies, in the paper we propose a Web-based CSCW system for remote substation fault diagnosis. It can effectively apply resources in Internet to improve the serviceability. Meanwhile, the efficiency and speed of substation fault diagnosis are also improved, dramatically.

The paper describes our original work on the design of system architecture, system function, Web-based CSCW and MA technique ensemble as well as some key technique issues to realize it.

2. MAS AND CSCW WITH WEB-BASED

With the rapid development of distributed artificial intelligence(DAI) technology, agent, as an independent calculation entity with cognizing attributes such as faith, aim, intention and behavior properties such as programming, negotiation and interaction, has been broadly considered a very ubiquitous guidance idea for the complicated advanced problem solving(Jennings, *et al.*, 1998). Particularly, the relationships between agent as well as multi-agent and special field knowledge are desired to be able to yield higher intelligent agent (Witting, *et al.*, 1994; Hyacinth, 1996), which is called a super agent. With the farther development of Internet technology, Web-based Multi-Agent Systems (MAS) have been reported in recent years(Wooldridge, 2003), MAS stresses that a large-scale complicated problem may be resolved using cooperative distributed problem solving methods, hence, relevant platform and system architecture as well as communication and cooperation modes have been described(Liu *et al.*, 2003; Wooldridge, 2003).

In view of expansion of system scale, many diverse computers and working stations as well as servers as well as broadly heterogeneous systems in Internet need to be connected for collaboration work, the overall system therefore becomes extremely complicated and considered heterogeneous. To meet the requirement of integrating numerous computers with diverse operating systems in networks for work, CSCW technology is needed to construct enterprise level systems, whose core content is J2EE(Liu, *et al.*, 2003). Certainly, in order to connect more wide-area heterogeneous systems and computers, Autonomous Decentralized Systems (ADS) has already been reported(Mori, 1984), it has no centre but its architecture is more flexible, whether existence of each autonomous agent or not has no important influence on the overall systems, more robust characteristic makes it very adapt to dynamic indeterminate environment. Hence, ADS can be applied to link the heterogeneities.

CSCW provides a means for people to collaborate, which makes many people in different places share the environment and information, engage common task, e.g., decision support, etc. With the fast development of Internet, networks and multimedia technologies, that public takes part in group decision over networks is made possible. With the advancement of MA technology, due to its prominent characteristics such as initiative, autonomy, sociality, intelligence, coordination, cooperation and negotiation etc, which makes the efficiency and speed of MAS-based diagnosis systems are greatly improved. Hence, Web-based CSCW systems supported by MAS benefit form Web and CSCW and MA for public participating in collaborative work for task solution.

3. CSCW FRAMEWORK WITH WEB-BASED

The aim of the Web-based CSCW systems for remote substation fault diagnosis is to establish a visual cooperative work environment for diverse departments, experts, public to collaborate in the activities about remote substation fault diagnosis, whose architecture, function and security are investigated in the section

3.1 Design of System Architecture

The Web-based CSCW system enable more departments, experts, groups and public related to it involved, for example, management, employing and servicing departments, diverse field experts or expert systems, and all concerned persons. The Web-based CSCW system can supply the assistance for communication, interactive operation and data disposal etc. In addition, it also support those distributed heterogeneous systems in communication, collaboration and data disposal. The architecture of the overall systems is shown in Figure 1.

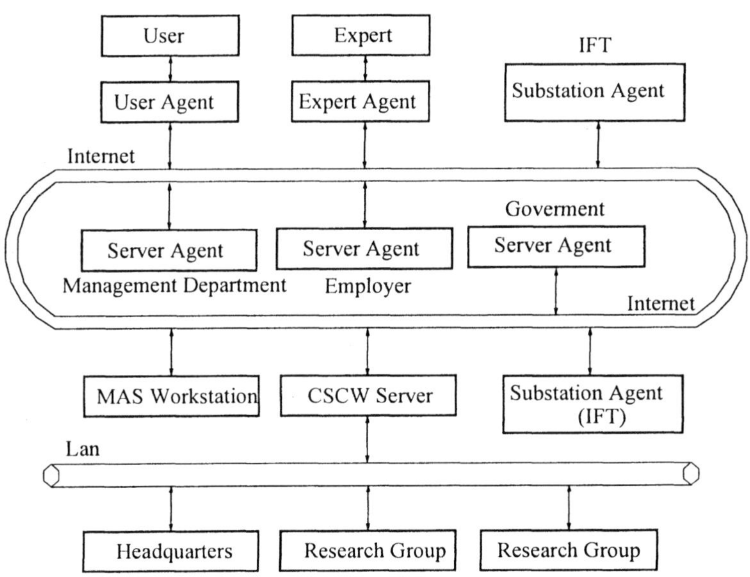

Figure 1 – Architecture of CSCW systems with web-based

In Figure 1, IFT means an Infinite Field Terminal, which usually is used to monitor and diagnosis home substation faults, called a super agent. MAS workstation is a core unit of the overall agent system. It is the centre of agent management and the centre of collaboration information exchange. It supports naming, log-on, log-out service for the agent in the platform, and also supplies service management and message transmission services for the agent in platform. CSCW server is a node, through it a visual collaborating work environment may be set up for different departments, public, research groups, headquarters and government departments involved to collaborate in the activities about substation fault diagnosis.

The system software architecture is composed of three levels as shown in Figure 2. The 1st layer is database layer, including related diagnosis knowledge base, CSCW data, etc. The 2nd layer is transaction layer, including diverse systems function module such as user management, transmission information collecting, collaborative system functions, etc. The 3rd layer is user interface, including CSCW workstations, user interface agents, etc. Web-based user agents can get and display related diagnosis information in Web-site.

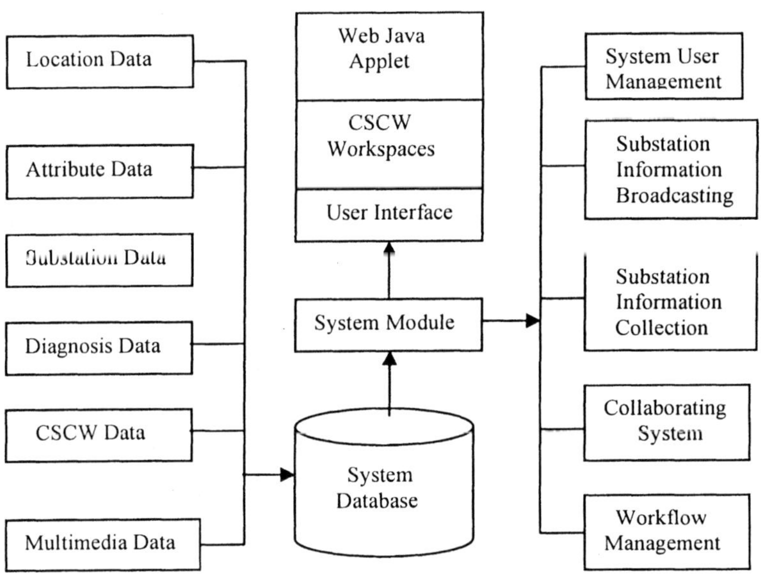

Figure 2 – Software architecture of CSCW systems with web-based

3.2 Design of System Function

In a Web-based CSCW system, the basic principles of system are described below.

(1) Information Sharing: The data base of expert systems, substation information, and CSCW knowledge base are the centre of the systems. All the cooperation and information sharing rely on the database agents.

(2) Collaboration: All departments related to fault diagnosis must be integrated through the centre databases. Substation fault diagnosis information, CSCW information can be shared and exchanged through Web. All collaborations are realized by Multi-agent-system (MAS).

(3) Transactions Broadcasting: In this system, every one can participate in the activities to cooperate, broadcast the data of substation fault diagnosis through Web-agent system.

(4) User Agent: In this system, everyone participates in cooperation and broads the data of diagnosis through user agent. It supplies a basic function in a flexible way in cooperative environment, e.g., friendly human-computer interface, etc.

3.3 J2EE Framework

In a Web-based CSCW system, for there are many heterogeneous computers, systems, workstations and servers that operate in diverse system platforms, the decentralized distributed platforms independent of any system architectures need to be considered to support them for collaborative working and future application and rectification. To satisfy and adapt the situation of integrating heterogeneous systems in Web to work, the J2EE and EJB and Java Servlets technologies have been reported (Xie, *et al.*, 2005). In addition, ADS has already been reported for many years for integrating heterogeneous systems in networks while every system's characteristic is preserved. Because Internet is a complicated large-scale heterogeneous system, the Web-based CSCW systems should ensure the collaboration among all independent systems without knowing the whole system architecture to realize the synergistic effect.

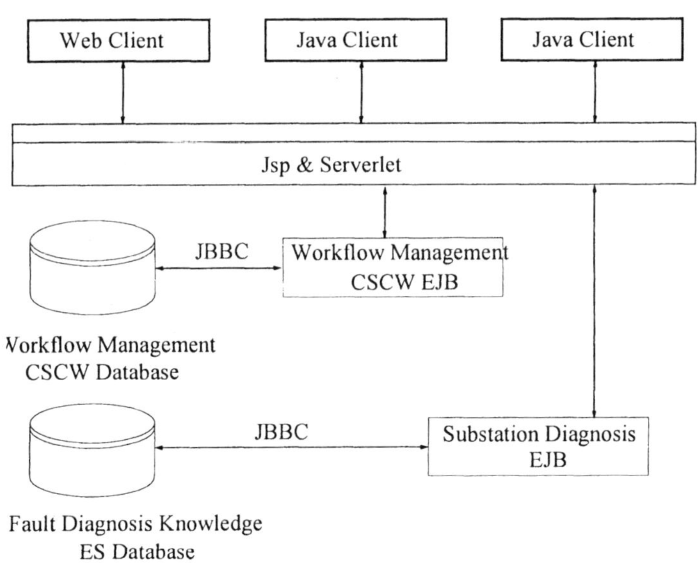

Figure 3 – System framework with J2EE-based

Figure 3 shows the infrastructure of the systems. The middleware of each subsystem such as CSCW, expert systems and other some subsystems, offers a distributed transaction processing. Many host computers can be added to supply many services. Compared with other distributed system, EJB has hidden lower details, for instance, the target management, multi-thread management and some events management etc. Moreover, J2EE offers many sorts of diverse middleware to be applied to enterprise logic. The data are managed and stored with EJB. Distributed computing enables users to operate in remote virtual servers to achieve business logic and data disposal anytime, anywhere. Distributed systems enable the databases and services in same and different computers. The database employs JBBC to communicate with EJB (back-end server). The system front-end adopts JSP and Servlet that employ back-end job to supply service. In the front-end java clients apply to communicate with JSP/Servlet and back-end EJB in Browser.
The system infrastructure has the following advantages:

(1) Platform Independence and good human computer interaction: Codes written in Java could run on any Java Virtual Machine (JVM) and any platform supported by JVM, This happens to be a boon for all, for it allows virtually any available computer systems to be home for the agent in distributed system. Once the parts of the system have been specified using by Java classes and compiled into Java codes, they can migrate to any of the hosts without recompilation. This makes it easy that data and loads balance across the networks. To improve interaction between human and computer, the front-end embedded techniques such as Active X and Java Applet are employed. Active X is constructed on COM, and not only applied in application program, but also could embed inside webpage to realize very complicated functions. Java Applet is a small application program in Java, it could run in any browsers supported by Java, this is equal to a self-inclusion web client program. Active X and Java Applet own the good connected and mobile characteristics. They run in client end and accomplish interaction and simulation.

(2) Network Support and Security: Java programming language API includes multi-level support for network communications. Lower-level sockets can be established between agents and data communication protocol can be layered on top of the socket connection. The Java.io package contains several stream classes intended for filtering and disposing various input and output data streams. APIs built on top of the basic networking support in Java provide higher-level networking capabilities, such as distributed objects, remote connections to database servers, directory services, etc.

(3) Runtime Environment and Remote Transactions: In addition to that Java facilitates the distribution of system elements across the networks. It makes it easy for the recipient of these systems to verify that they can't compromise the security of the local environment. If Java code runs in the context of an applet, then the Java VM places rather strict restrictions on its operation and capabilities. Also, any class denotations loaded on the network, whether from a Java Applet or application, have to go through a stringent byte code verification process.

With the advantages, J2EE can help us to design enterprise-level, reusable and distributed applications compare to other language.

System architecture is platform independence and multi-layer distributed structure. It can combine expert database, CSCW database and lots of one of the host computers together and share the system resources and data.

3.4 System Security Consideration

The security problem is very important because the system operates on the Internet. To ensure the system, we applied following three measures. First, user's identification authentication and authorizing process were served to control the safety and users identity authentication. Second, the firewall is used to set up the high-efficient and practical safe protection for the attack from the outside network and isolate system from the external environment, prevent the external threat and realize the strict system safe border. Finally, the critical information was encrypted. The encryption enabled the sensitive information stored and transmitted safely in communication in Internet safely. In this way, the system's security has been increased, greatly.

4. SYSTEM IMPLEMENTATION

Based on the infrastructure for Web-based remote substation fault diagnosis mentioned in the text, we are studying and developing a prototype design for remote substation fault monitoring and diagnosis. The system adopts the three layers structure of J2EE, that is, relevant diagnosis knowledge and CSCW data are placed in the related databases, User interface is shown with Java applet webpage, Active X embedding into the webpage adopts Delphi and Visual C++ program language. Microsoft Windows 2000 Advance Server acts as operation system of the web server, and SQL server 2000 for back-end databases. Initial experiment proves the availability of the method. Later, diverse system platforms are located at the heterogeneous servers, the experiment results display excellent.

In process of experiment, we still design IFT shown in Figure 1 for each substation, which mostly presides over on-line monitoring substation fault in advance and diagnosis fault, more detail description on which can be seen in our early time work (Su, *et al.*, 2005).

An example below is presented to understand the point.

The tested object in the paper is local main connection diagram of a substation is shown in Figure 3(Su, *et al.*, 2005), five components in Figure 4 respectively expresses bus M and N, transformer T as well as circuits L1 and L2. CB1 to CB4 are breakers. Another 10 protection relays are Mb, Nb, TAm, TBm, TAp, TBp, L1m, L2m, L1p,ml2p, where Mb, Nb are bus protections, TAm, TBm are transformer main protections and TAp, TBp is spare protections, L1m, L2m are circuit main protections and L1p,ml2p are spare protections.

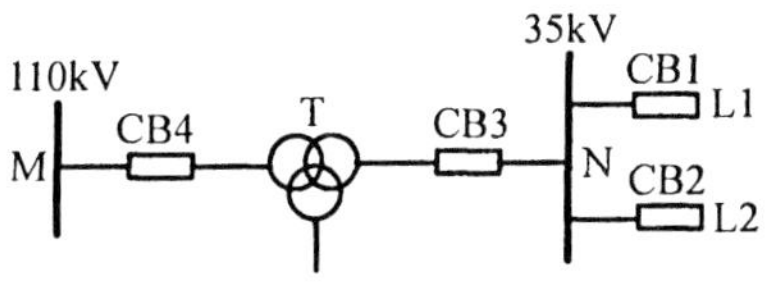

Figure 4 – Local wiring diagram of substation

According to substation fault information collected by home IFT, by data mining, we gain decision rules below (Su, *et al.*, 2005).

Table 1 – Diagnosis decision table

Condition Attribute			Decision Attribute
CB1	*CB2*	*CB3*	Fault Component
0	0	0	*M*
-	0	1	*T*
-	1	1	*N*
1	0	0	*L1*
-	1	0	*L2*

Assume that at one time following fault information is sensed, diagnosis process then is described as follows.

(1) Fault change-site information: *CB3*=1, *TA2*=1, *TB1*=1, *CB4*=1, the rest is no conversion information. According to Table 3, after pattern match, home IFT can judge transformer *T* is fault component. Fielded inspection proves the diagnosis result is right.

(2)Set change-site information is invariable, but information of *CB4* and *TB1* can't be sent over, but at the moment home IFT still can judge that fault component is *T*. This shows noisy tolerance of IFT

(3)Set change-site information is still invariable, but information of *CB1* and *TB1* can't be sent over, thus home IFT has no way to identify, home IFT may seek the solution for that from foreign agents through Web-based CSCW systems. Then foreign agents send back identification information by cooperative working, thus problems are resolved and native knowledge base is also updated.

Compared with the previous method (Dong, *et al.*, 2003), the proposed method in this paper can identify the fault class (3) by CSCW systems, while the previous method only can diagnose fault classes (1) and (2). Hence, with the aid of CSCW systems, ES expands its application range. Meanwhile, the efficiency and correctness of diagnosis are also improved, dramatically.

The system proposed in this text provides an efficient means for remote substation fault diagnosis. First, the proposed system provide a virtual CSCW environment to meet the practical demands in process of substation fault diagnosis, Java-based technique enables system to run in different platforms ubiquitously. Second, Due to operating in Internet, the system can be accessible from anywhere in the world. Third, since substation fault diagnosis activities involve many different departments and diverse public, the collaborative work environment makes it easy to speed up fault diagnosis process. All the jobs in process of diagnosis could be done in real time. Experts and public could take part in activities and report information in different geographical position at the same time. Data collection, analysis, and issue of the fault diagnosis and relevant information will be coordinated operation in diverse departments, and the repeated labor therefore may be greatly reduced. It can ensure and facilitate the advancement of substation fault diagnosis job.

To more effective support and supply serviceability for the remote substation fault monitoring and diagnosis in heterogeneous environment, ADS is expected to be adopted. In ADS environment, the system may be modified and changed on-line dynamically without interrupting the overall systems. The integration of the heterogeneous systems can yield the synergetic effect, each subsystem characteristics may be preserved and subsystem can make its decision autonomously. Therefore, the systems realized by ADS own more flexible and robust as well as fault-tolerance ability, without question, it may supply more effective supports to group decision. ADS still can supply effective help to those heterogeneous databases, data-structure and different application models, and it therefore is intensively recommended for future application.

Although the proposed system would be beneficial to remote substation fault diagnosis, this paper only presents its system infrastructure and prototype of Web-based CSCW. Further works should be done and investigated in several key technologies such as the Java applet, cooperative diagnosis model and data type and so forth.

5. CONCLUSION

A Web-based CSCW system for remote substation fault diagnosis provides an easy route for equipment fault diagnosis. It implements a distributed far-reaching multi-expert collaborative diagnosis method, and realizes resource share and information complementation, the efficiency and speed of facility fault diagnosis are greatly improved. In this system, MAS are also applied to realize the initiative and independence of fault diagnosis. CSCW system with Web-based provides user with friendly interfaces and well-regulated cooperative working environment with many advantages such as open communication platform, substation information query, diagnosis task management, resource management and general project management, etc. In addition, it still realizes the departure between exploitation and application environment. Hence, it is very convenient for servicing, expansion and management of the equipments. Combining with exploitation and application, eventually, the paper validates that the method is available and feasible.

6. ACKNOWLEDGMENTS

The authors wish to acknowledge the 'Qing Lan' Talent Engineering Funds supported by Lanzhou Jiaotong University.

7. REFERENCES

1. H Dong, JS Bai, JY XUE. An Approach for substation fault diagnosis based on multi-agent. Advanced Technology of Electrical Engineering and Energy 2003; 22:21-24
2. LP Zhao, SF Xie QZ Li, XC. Chen. Study on multi-agent based power quality monitoring and control system for electrified railway. Power System Technology 2004; 24:69-73.
3. H Mao, Q Xu, W Liang. Research on remote fault diagnosis technique based on web model. Computer Engineering and Design 2005; 26: 686-687.

4. H Wang, Y Wang. "MAS-based Distributed Expert System Platform Research". In Proceedings of the 8th International Conference for Yong Computer Scientists, JS He, ed. , Beijing: IAP/WPC, 2005.
5. L Liu, K Liu, Y Hu. Architecture of multi-agent based distributed monitoring and control system for substations. Power System Technology 2003 ; 6: 26-28.
6. NR Jennings, K Sycar , M Wooldridge. A roadmap of agent research and development. Auto Agents & Multi-agent Syst 1998; 1:7-38.
7. T Wittig, NR Jennings, EH Mamdani. Framework for intelligent cooperation. Intelligent Systems Engineering 1994; 3:168-179.
8. SH Hyacinth. Software Agents: An Overview. The Knowledge Engineering Review 1996; 3: 205-244.
9. M Wooldridge, Writing; CY Shi, W. Zhang, J Xu Translation. An Introduction to Multi-Agent-System. Beijing: Publishing House of Electronics Industry, 2003.
10. X Liu. Infrastructure of United Network Management System Driven by Web Technology. In Proceedings of the Fourth International Conference on Parallel and Distributed Computing, Chengdu: IEEE, 2003
11. L Xie, H Qiu, L Qin and L Li. "The Development of XMSF Based on J2EE". In System Simulation Technology & Application, ZH Chen, Hefei: USTC, 2005.
12. K. Mori. A Pproposal of the Autonomous Decentralized System Concept. IEICE-C 1984, 12: 303-310.
13. Hongsheng Su, Qunzhann Li. "Study on Substation Fault Method based on Multi-Immune-Agent." In Proceedings of the XIVth International Symposium on High Voltage Engineering, ZC Guan, ed., Beijing: Tsinghua University Press, 2005.

42 A SERVICE-ORIENTED FRAMEWORK FOR INTEGRATION OF SHOP FLOOR SCHEDULING AND CONTROL

Kewei Li[1], Chun Wang[1],
Hamada Ghenniwa[1], Weiming Shen[1,2]
[1]*University of Western Ontario*
[1,2]*National Research Council Canada*
kli64@engga.uwo.ca; cwang28@engga.uwo.ca;
hghenniwa@eng.uwo.ca; weiming.shen@nrc.gc.ca

This paper presents an agent-based service-oriented framework to address the challenges for integrating real time shop floor scheduling and control in a multi-workcell environment. The proposed framework enables shop floor to adapt to the changes in the environment dynamically. Shop floor scheduling and control can address dynamic changes either within a workcell, such as machine breakdowns, or cooperatively amongst multiple workcells, for example to delegate a task that cannot be performed locally or to improve the performance of the workcell. A software simulation environment has been implemented to validate the proposed approach.

1. INTRODUCTION

Globalization has driven manufacturing enterprises to shed the security of mass production and shift to a new paradigm which is referred to by different names such as lean production and agile manufacturing. An agile manufacturing system needs to respond to dynamic changes in a timely and cost effective manner.

Dynamic changes can derive from either outside parties in the market, such as the supply side (representing suppliers), demand side (representing customers) or within the enterprise, such as real-time events from the shop floor. To deal with the dynamic changes within the enterprise, the monitoring and control of shop floor has to be fully integrated with manufacturing scheduling system to provide the manufacturing enterprises with the capability to survive in today's dynamic market environments. This research is concerned with integrating Real time shop floor monitoring and control system with Scheduling system. In order to achieve effective real time Shop floor monitoring and control in dynamic manufacturing environments, we propose a service-oriented integration framework including interaction mechanisms which coordinate the behaviors of scheduling system and shop floor control system in a flexible manufacturing multi-workcell environment.

Please use the following format when citing this chapter:

Li, K., Wang, C., Ghenniwa, H., Shen, W., 2006, in IFIP International Federation for Information Processing, Volume 220, Information Technology for Balanced Manufacturing Systems, ed. Shen, W., (Boston: Springer), pp. 395–404.

The manufacturing environment is typically seen as a large, complex man-made system of heterogeneous, interrelated activities (STANESCU, 2002). The close monitoring of the operational performance of various plants, manufacturing shop-floors / cells / machines, as well as their associated instrumentation and control is of increasing strategic importance. Industrial plants are complex open system, the integration of scheduling and control are based on wide-are network of heterogeneous platforms. The integration needs to address the issues such as distribution, heterogeneous, interoperability and open system environment.

This paper presents our research work on the integration for real time shop floor monitoring and control. We apply agent technology to build up a multi-agent system for a single workcell, and Web services technology for the integration of distributed multiple workcells (or plants). The rest of the paper is organized as follows. Section 2 provides a literature review. Section 3 proposes a agent-based service-oriented framework for integration of shop-floor scheduling and control. Section 4 presents our prototype implementation. Section 5 concludes the paper.

2. A REVIEW OF INTEGRATION APPROACHES

2.1 MIDDLEWARE BASED APPROACHES

The integration of the scheduling and control system is a form of distributed information system. There are kinds of ways trying to address the problems and constraints existing in the design and implementing of distributed information systems. The system architecture may vary from 1-tier to N-tier designed in either a bottom-up or top-down manner. Their communication patterns are between synchronous and asynchronous interaction.

Distributed information systems have evolved in response to improvements in computer hardware and networks. Middleware constitutes the basic infrastructure behind any distributed information system (Alonso, 2004). Middleware facilitates and manages the interaction between applications across heterogeneous computing platforms. Conventional middleware platforms can be used for the integration for manufacturing system, such as RPC and related middleware, object brokers, object monitors and message-oriented middleware (MOM).

EAI can be seen as a step forward in the evolution of middleware, extending its capabilities to cope with application integration. When the systems involved were compatible and comparable in their functionality and did not involve many platforms, middleware could be used with out further ado to integrate the servers. Unfortunately, for more ambitious projects, plain middleware platform was not enough. The main limitation was that any concrete middleware platform makes implicit assumptions about the nature of the underlying systems. When these systems are very different in nature and functionality, using conventional middleware to integrate them becomes rather cumbersome, and in some cases simply infeasible. Message brokers may be the versatile platform for Enterprise Application Integration. Message brokers are direct descendants of the platforms for message oriented middleware. They are derived from the new requirements posed by EAI, in terms of supporting the integration of heterogeneous, coarse-grained enterprise applications. Message brokers can address one of the enterprise process automation problems: that of hiding the heterogeneity and the distribution of enterprise systems to provide a uniform view to the applications that integrate those systems. Examples of leading commercial implementations of EAI platforms today are Tibco ActiveEnterprise, BEA Weblogic Integration, WebMethods Enterprise, and WebSphere MQ. Integration through a message broker entails a number of benefits which are typically characterized as lower development cost, lower opportunity costs and lower maintenance effort. Despite these advantages, message brokers are not a panacea for all application problems. Indeed, there are many drawbacks to implementing EAI solutions using message brokers. The main issue is that software licenses are extremely expensive. Besides licensing costs, companies need to invest in the training of IT personnel and acquire the resources necessary for the installation and operation of the tool. These reasons often discourage small and medium enterprises from adopting EAI platforms, and even from performing the integration at all.

Workflow management systems are the tools used to make the integration logic explicit and more manageable. Workflow management systems (WfMSs) tackle the other side of the application integration problem: that of facilitating the definition and maintenance of integration logic. Examples of leading commercial workflow systems include WebSphere MQ Workflow by IBM, Vitria BusinessWare, Tibco BPM, BEA WebLogic Integration, and Microsoft BizTalk Orchestration.

On the one hand, the type of integration discussed so far has been implicitly limited to LANs. As it is easy to imagine, the problems of application integration are significantly amplified when we consider integration across enterprises and over the Internet. On the other hand, although current technology provides tremendous benefits, existing platforms are still very expensive and quite complex to use and maintain.

2.2 Web-Based Approaches

The increasing use of the Web as a channel to access information systems forced middleware platforms to provide support for Web access. This support is typically provided in the form of *application servers* (SUN's J2EE or Microsoft's .NET based). Application servers are equivalent to the middleware platforms discussed in earlier section. The main difference is the incorporation of the Web as a key access channel to the services implemented using the middleware.

At the presentation layer, application servers conceptually resemble conventional middleware. The functionality provided is similar to that of CORBA, TP monitors, and message brokers. The support for the presentation layer and for the document as the basic unit of transfer is what differentiates application servers from conventional middleware.

Firewalls present a special challenge to integrating inter-enterprise systems across the Internet. Almost all forms of communication usually employed by conventional EAI products cannot traverse a well-configured firewall. The widely accepted solution to get through the firewall is done by a technique known as tunneling, whereby protocols which would be blocked by the firewall are hidden under protocols that accepted by the firewall. Tunneling is used by conventional middleware systems (e.g., GIOP/IIOP over HTTP) to bypass firewalls.

The widespread use of the Web made it very difficult to standardize all forms of Internet exchanges. In the context of the Web, the answer to this problem is the eXtensible Markup Language (XML). XML addresses the data representation problem by focusing on the syntax, rather than the semantics, of the documents exchanged. However, by providing syntax along with parsing and validation tools, XML lays the foundation on which semantics can be defined.

2.3 Agent-Based Approaches

Multi-agent systems have been the subject of massive amounts of research in recent years and are beginning to find their way into commercial applications (Shen, 2001). An exhaustive overview is beyond the scope of this paper, but essential pointers include (Wooldridge, 2002) and [Luck, 2003].

Early research work on agent-based intelligent manufacturing scheduling and factory control was reported almost 20 years ago (Shen, 1999). Agent Based Manufacturing has become a new paradigm for next generation manufacturing systems (Shen, 2001). Researchers have been applying agent technology to manufacturing enterprise integration and supply chain management, manufacturing planning, scheduling and execution control, materials handling and inventory management. Researchers have conducted an extensive literature review and published the review results already (Shen, 1999; Shen, 2001.).

Many researchers are probing into solutions for enterprise integration and some reached the conclusion that the agent technology provides a nature way to realize enterprise integration effectively. Agent-based enterprise integration has been a very active research area recently. In addition to significant academic researches, some projects have attracted active industrial participation and developed industrial applications. In most projects, software agents are used to encapsulate existing legacy software systems using various middleware approaches.

Security and privacy are critical issues in implementing Internet-enabled agent-based manufacturing systems. Socket communication is actually used for inter-agent message exchange. Although secure sockets could be used to enhance the communication security, an implementation in a real shop floor may have some difficulties because of firewalls;

The agent technology is still considered to be "nice-to-have" in industrial applications by industrial people. It seems to be still a long way to go before it is considered to be "must-have". We believe that the agent technology must be integrated with other technologies such as Web based technologies (including Web Services and Semantic Web) and Grid computing for its wide and successful applications in industry in a near future.

Web services are emerging as a major technology for achieving automated interactions between distributed and heterogamous applications (Tsalgatidou, 2002). Web Services technology is part of the SOC (Service Oriented Computing) paradigm and can be considered as an implementation of the SOC model providing Web applications with a loosely coupled integration approach (Shen, 2005). The W3C (The World Wide Web Consortium) defines a Web service as "... a software system designed to support interoperable machine-to-machine interaction over a network. It has an interface described in a machine-processable format (specifically WSDL – Web Service Definition Language). Other systems interact with the Web service in a manner prescribed by its description using SOAP messages, typically conveyed using HTTP with an XML serialization in conjunction with other Web-related standards." (W3C, 2004).The advantage of Web services have already been demonstrated and highlight their capacity to be composed into high-level business processes (Benatallah, 2002). It is argued that composition via service interconnection allows more sophisticated services and applications to hierarchically constructed from primitive ones (Maamar, 2003). Web service can be independent as much as possible from specific platforms and computing paradigms, in addition, it can be easily compassable.

The integration using software agents and Web Services technologies together is able to avoid the weakness of each individual technology, while reinforcing their strengths. It can be a good approach to address the issues which remains in the real time shop floor monitoring and control.

3. SERVICE-ORIENTED FRAMEWORK FOR INTEGRATION OF SCHEDULING AND CONTROL

For the scheduling and control aspect, developing multi-agent systems is carried out in cooperative distributed multi-workcell manufacturing environments. In this context, each work cell (or plant) have a designated autonomous scheduling system based on the local constraints and objectives of individual manufacturing resources (represented by software agents), and is directly connected to the shop-floor's control level.

Based on the analysis in the previous section, we propose this approach, and its name is Service-Oriented Framework for Integration of Scheduling and Control (FISC). It has two parts: a Platform for Integration of Scheduling and Control (PISC) and a Mechanism for Integration of Scheduling and Control (MISC). Technically, the FISC is an abstract extensible skeleton with a set of meta-components and a higher-level management component. According to the particular Multi-workcell scenarios, the FISC can dynamically extended and deployed to multiple work cells, adjust or re-configure itself to initialize the schedule and control parameter, to accept new order or outsource corresponding job to remote work cell in response to the dynamic event happened in the shop floor level (such as machine break down). The PISC is designed to improve the adaptability of integrating platform in the heterogonous distributed multi-workcell environment. The purpose of the MISC is to provide the adaptability from the interaction protocol point of view. The relations among the FISC, MISC and PISC are demonstrated in Figure 1. The PISC includes five parts: the agent factory, general utility, schedule and control platform, web service run time environment, web service interface. The MISC has one major component: the knowledge base. Further detailed description will be given in the following subsections.

We adopt the scheduling model, shop floor control model, Coordinated Intelligent Rational Agent (CIR-Agent) model (Hamada, 2000), and the web service integration approach as the base to construct the FISC. The CIR-Agent model is adopted to design the work cell resource level agent (such as machine agent and operator agent). The web service integration approach is used to construct the multi-workcell environment, and the PISC is considered to be settled on it. The scheduling and control model are utilized to design the mechanism related to the protocols between agents (e.g., protocol between scheduler agent and controller agent, protocol between control agent and resource agent) and protocols between multiple workcells (e.g., Protocol used for one work cell out source job to a remote work cell). The web service coordination model and service composition model are used as the foundation when designing the algorithms for the multi-workcell integration mechanism determination and composing mechanism in both the PISC and MISC.

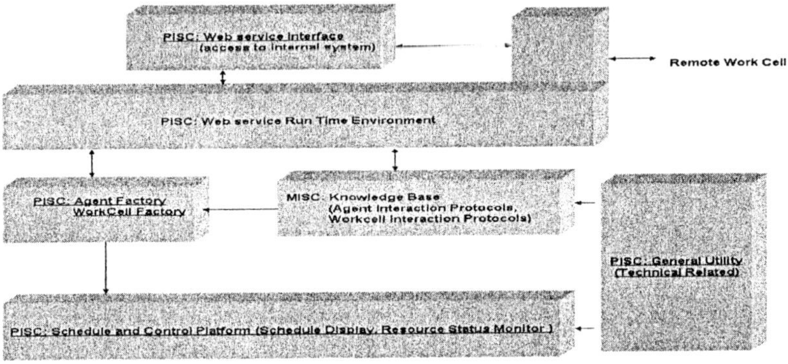

Figure 1 - Service-Oriented Framework for integration of scheduling and control

3.1 Framework for Integration of Scheduling and Control (FISC)

A framework is an extensible high-level structure including a set of reusable components, which can be dynamically reused and instantiated to build a concrete application. The main advantage of a framework is that it can successfully decrease the systematic complexity without sacrificing the flexibility and adaptability of the generated application.

The main goal of the FISC is to provide an open, flexible, Service-oriented and FIPA-compliant (FIPA, 2005) framework to improve the adaptability of the integration platform in the complex flexible manufacturing system (such as multiple workcells in wide area network). In order to achieve the goal, the FISC is structured as two parts: the Platform for Integration of Scheduling and Control (PISC) and Mechanism for Integration of Scheduling and Control (MISC), by which it can improve the adaptability at two levels: the integration platform and participated workcells. In this paper, the adaptability of an integration platform indicates that it can support integration in a heterogeneous, distributed and Wide area network environment For a participated work cell (including the work cell controller and shop floor level resources), the adaptability means that a new instance of controller or resource agent can be instantiated and configured to join the flexible manufacturing system. The PISC provides a scalable and flexible integration platform with the capabilities to schedule, monitor and control in a multi-workcell manufacturing environment. The MISC will include mechanism for protocol selection, service coordination and service composition.

3.2 Platform for Integration of Scheduling and Control (PISC)

The overall usage of the PISC is to provide a flexible platform for scheduling and control integration. In terms of multi-workcell environment, when a new work cell need to be integrated, the platform can be easily deployed on the new work cell, what need be done by the user is to configure the capability of the resource agent through GUI, then register with the work cell controller agent, the new work cell then is integrated into the framework. For the interaction between work cells which happened at the web service level or the interaction between scheduler, controller and resources agent in a single work cell which happened at the agent communication level, MISC is used to deal with the coordination and composition function.

3.3 Mechanism for Integration of Scheduling and Control (MISC)

As briefly mentioned previously, the Mechanism for Integration of Scheduling and Control (MISC) is designed to realize all key functions, such as the function for service coordination, function for service composition and function for agent negotiation protocol selection.

4. PROTOTYPE IMPLEMENTATION

Figure 2 demonstrates a sample deployment of the prototype in a two work cell environment. Currently the cells are simulated in a computer environments, not connected to the real machines directly. The used key implementation technologies are listed below:

- System platform: Linux / Windows
- Programming language: Java 2 Platform Standard Edition 1.4.2 (J2SE)
- Web Service middleware tool: Java WSDP 1.6
- Web Service container: Tomcat 5.0 for Java WSDP 1.6
- Multi-agent Platform: JADE

Within the scope of integration, the dynamic state change of the multi-workcell control system is modeled as a Finite State Machine as shown in Figure 3.

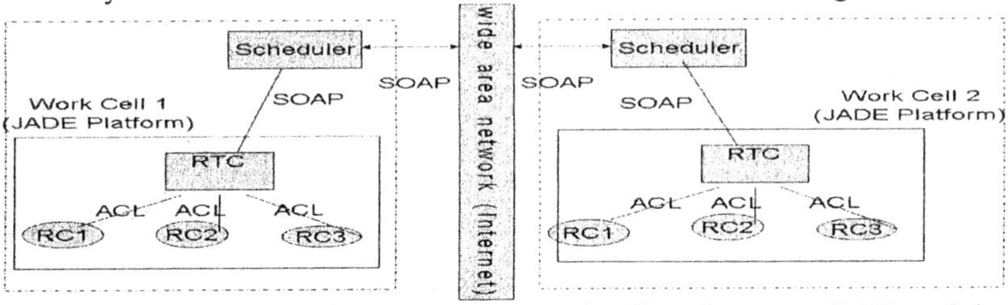

Figure 2 - A sample deployment on two work cell environment (RTC: real time controller; RC1: resource controller 1; RC2: resource controller 2; RC3: resource controller 3; they exist as agents. SOAP: Simple Object Access Protocol; ACL: Agent Communication Language)

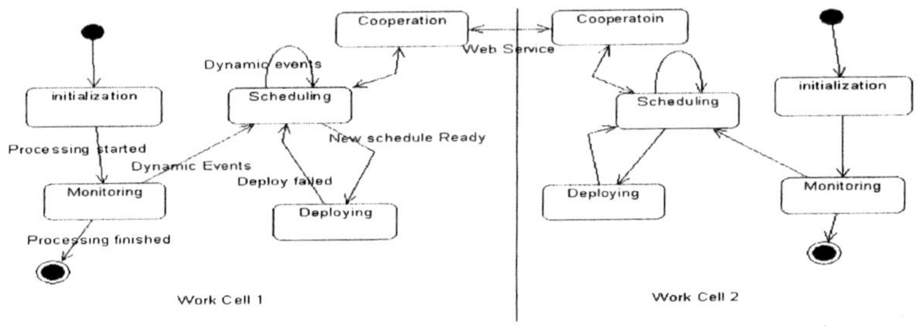

Figure 3 - Finite State Machine model of two work cell control

Figure 4 depicts the multi-workcell service protocol. In this figure, we show two work cells. Real Time Controller of work cell 1 provides Report machine capability service and Report dynamic event service to the Scheduler of Work Cell 1. Scheduler provides Deploy Schedule Service to Real Time Controller. All of these happen within one work cell.

Under certain circumstances, scheduler of Work Cell 1 receives some Dynamic event such as machine down, then scheduler found it is not able to redeploy schedule within this work cell because the resources is not available anymore. Then Scheduler 1 invokes Distribute Job Service of Scheduler 2 in Work Cell 2, assign the needed job to Scheduler 2, Scheduler 2 then deploy schedule within Work Cell 2. This will achieve the goal to distribute work load between multiple workcells.

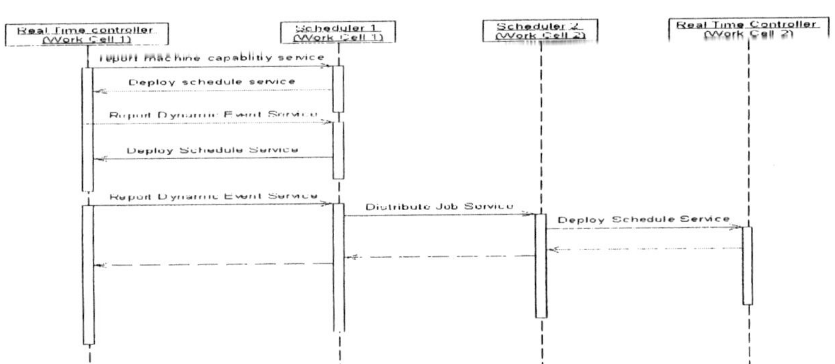

Figure 4 - Two Work Cell's Service protocol in the sample prototype

An implemented working scenario can be represented as follows:

- There is a Work Cell including a RTC Agent, three Machine Agents and one Operator Agent, these agents sit on an Agent platform (JADE Platform 1), a Scheduler Agent which is designated for this Work Cell is running on a remote Agent Platform.
- RTC Agent look up the service in UDDI, and find this designated Scheduler, then bind with the Scheduler.

- RTC Agent report the Work Cell's capabilities to the Scheduler through Web service.
- Scheduler deploys the Work Cell's schedules to RTC through Web Service.
- Within the Work Cell, RTC Agent deploys schedule to different resource Agent through ACL Messages. (Agent communication)
- RTC Reports Work Cell disturbance (such as machine down) in real time to Scheduler through Web service.
- Scheduler receives the disturbance, generates new schedule, and deploys new Schedules to RTC Agent through Web service.
- RTC agent deploys new schedules to different Resource Agent through ACL Messages in real time.
- Work cell 1 resources down, not available anymore; Scheduler 1 assign Job to scheduler 2 in work cell 2 through Web Service. Work Cell 2 accepts the job. This is an example of the multi-workcell scenario.

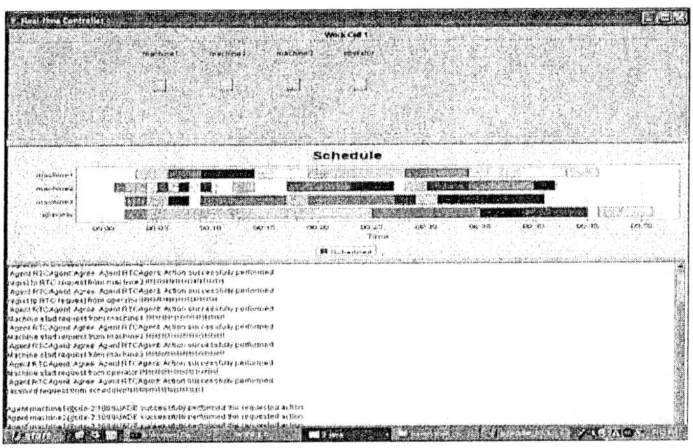

Figure 5 - RTC screen shot (all resources are working properly)

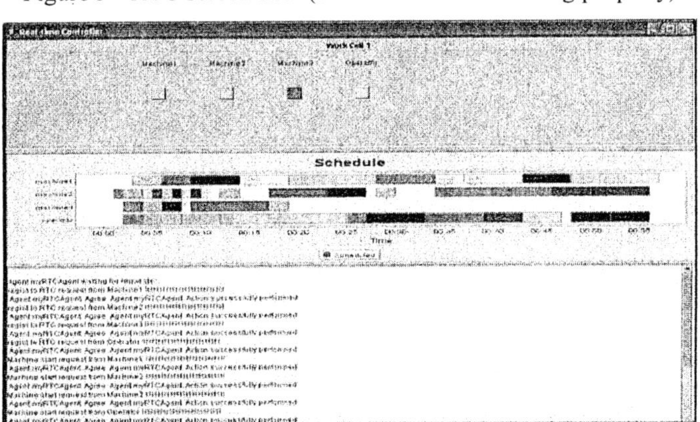

Figure 6 - RTC screen shot (Machine 3 is down)

The above two figures are the screen shot of the RTC (Real Time Controller). The RTC monitors and controls 3 machines and 1 operator. Figure 5 shows that every resource' status is well and their related schedule; Figure 6 shows machine 3 is down at certain moment and the re-deployed schedules.

5. CONCLUSIONS AND PERSPECTIVES

This paper is concerned with developing a framework using software agent and web service technologies for the integration of real-time shop-floor monitoring and control in multi-workcell manufacturing environment.

Web Services offer fundamentally new ways of doing business through a set of standardized tools, and support a service oriented view of distinct and independent software components interacting to provide valuable functionality. The supporting of interoperation makes Web Service a good approach to integrate heterogeneous systems. A Web Service infrastructure solution provides a rapidly deployable and re-configurable cooperative distributed manufacturing environment.

The experimental deployment prototype of the service-oriented integration framework for scheduling and control has shown that it is a promising approach for building up information integration system in flexible manufacturing environments.

6. REFERENCES

1. Alonso, G, Casati, F, Kuno, H, Machiraju, V. Web Services: Concepts, Architectures and Applications, Berlin : Springer, c2004. pp. 67.
2. Benatallah, B, Casati, F. (Guest Editors). "Special Issue on Web Services". Distributed and Parallel Databases, Kluwer Academic Publishers, 2002; 12(2-3).
3. FIPA. The Foundation for Intelligent Physical Agents. http://www.fipa.org/
4. Ghenniwa, H, Kamel, M. "Interaction Devices for Coordinating Cooperative Distributed System". Journal of Intelligent Automation and Soft Computing, 2000; 6(2): 173-184.
5. Luck, M, McBumey, P, Preist, C. Agent technology: Enabling Next Generation Computing. http://www.agentlink.org/roadmap, 2003
6. Maamar, Z, Sheng, QZ, Benatallah, B. "Interleaving web services composition and execution using software agents and delegation". In AAMAS'2003 Workshop on Web Services and Agent-based Engineering, 2003.
7. Shen, W, Li, Y, Hao, Q, Wang, S, Ghenniwa, H. "Implementing Collaborative Manufacturing with Intelligent Web Services". Proceedings of CIT 2005, 2005; pp. 1063 – 1069.
8. Shen, W, Norrie, DH. "Agent-based systems for intelligent manufacturing: A state-of-the-art survey". Knowledge and Information Systems 1999; 1(2): 129-56.
9. Shen, W, Norrie, DH, Barthes, JP. Multi-Agent Systems for Concurrent Intelligent Design and Manufacturing. London: Taylor and Francis; 2001.
10. Stanescu, AM, Dumitrache, I, Curaj, A, Caramihai, SI, Chircor, M. "Supervisory control and data acquisition for virtual enterprise". International Journal of Production Research, 2002; 40(15): 3545-3559.
11. Tsalgatidou, A, Pilioura, T. "An Overview of Standards and Related Technology in Web services". Distributed and Parallel Databases, 2002; 12(2-3): 135-162.
12. W3C Web Service Architecture Working Group. Web Service Architecture Recommendation. http://www.w3.org/TR/2004/NOTE-ws-arch-2040211/. 2004.
13. Wooldridge, M. An Introduction to Multi-Agent Systems. Wiley and Sons, 2002.

43 COMPUTATIONAL REPRESENTATION, HETEROGENEITY AND INTEGRATION OF PRODUCTION SYSTEM MODELS

Wilson M. Arata and Paulo E. Miyagi
Escola Politécnica da Universidade de São Paulo
wimarata@usp.br, pemiyagi@usp.br

This work discusses important aspects in the computational representation of models, focusing on the treatment of the heterogeneity and the integration of models. The relevance of this topic lies in the necessity of achieving an efficient workflow when handling heterogeneous information structures and of giving proper answer to the involvement of new types of models driven by the increasing demand for enhanced information handling capabilities in the planning and the control of production systems.

1. INTRODUCTION

This work presents some points to consider when planning and designing computational support systems that make intensive use of models, taking into account the heterogeneity and integration models. In the case of this text, these issues are discussed in the context of the so called Discrete Event Dynamic Systems (DEDS), from which perspective important results for production systems have been obtained. DEDS are systems whose dynamics presents discrete states and the transitions between the states are associated to the occurrence of discrete events (Cassandras, 1993). In the case of production system, the coordination of operations and the resource utilization patterns can be approached from the perspective of DEDS.

Models are an important means to approach systems, providing an effective way to deal with their complexity. They are abstract descriptions of systems, that consider only the relevant aspects (for a certain purpose), ignoring those that are irrelevant.

Models are inherently specific in the sense that their structure and content strongly depend on the aspects that are considered, the mathematical tools used to deal with them and the purposes of their utilization. This specificity is a major factor in the heterogeneity, that is, existence of a wide variety of models, expressed in terms of different modeling formalisms. In the case of DEDS, this heterogeneity is remarkable, with some examples presented in Table 1.

Please use the following format when citing this chapter:

Arata, W. M., Miyagi, P. E., 2006, in IFIP International Federation for Information Processing, Volume 220, Information Technology for Balanced Manufacturing Systems, ed. Shen, W., (Boston: Springer), pp. 405–414.

Table 1 – Brief descriptions of some DEDS types of models

Petri Nets (Murata, 1989)	Modeling of causal relationships between different state variables, providing information on qualitative features.
Stochastic Petri Nets (Molloy, 1980)	Extension of Petri Nets with stochastic timings associated to state transitions, to approach performance features of systems.
Markov Chains (Kulkarni, 1995)	Description of state transitions and the timings associated to them, for the study of state probabilities.
Queueing Networks (Bolch, 1998)	Analysis of congestion phenomena as entities seeks for service when accessing a network of limited resources.
Mark Flow Graph (Hasegawa, 1988)	Extension of Petri Nets considering discrete production system control elements.

The idea behind computationally supporting model heterogeneity is to take advantage of the best that each type of model ca provide and, thus, enhance the planning and the operation of production systems. It is also important to remember that the diversity of modeling techniques is continuously growing. Therefore, it is important to develop an adequate computational framework that does not amplify the complexity inherent to the model heterogeneity, providing a favorable cost-benefit ratio.

Besides the existence of heterogeneous models, there is the issue of integration of models, that is based on the consideration of the relationships between different models. A well-known example is the case of the isomorphism between a Stochastic Petri net and a Markov chain (Molloy, 1980): if both models refers to a same system dynamics, it is very likely that the former constitutes a much more concise description, while, from the latter, that can be computationally obtained from the former, one can calculate quantitative information on the system dynamics. So, another aspect to consider is the adequate treatment of the relationships between models, in order to maximize the advantages that those relationships can provide.

The focus of this work is on the issue of the computational representation of the information conveyed by the models and how it can put into effect the benefits of an adequate treatment of heterogeneity and integration. So, at first, a structured discussion about the characteristics a modeling language should have to deal with model heterogeneity is carried out; for that propose, set-based diagrams are presented as a means to visualize relevant aspects in modeling and analysis and the relationships between them. In the second part, the representation of semantic information associated to models is presented as a means of dealing with the integration of models, showing that, if adequately elaborated, it can provide a unified view of different models referring to a same system dynamics, where the diverse relationships between models are shown without ambiguity, redundancy and concealment of relevant information. Besides analytical models, these ideas can be extended to other information structures, such as those containing sensorial data, so that it can be treated together with other models within a computational framework.

2. HETEROGENEITY AND WORKFLOW EFFICIENCY

This section introduces an abstract model of modeling and analysis tools to serve as an instrument to achieve a structured discussion about the characteristics of computational representations of models and of their handling. It uses basic elements of Set Theory and provides a useful way to visualize important relationships, being general enough to approach a wide range of configurations of computational modeling and analysis frameworks or schemes.

So, consider that all computational modeling and analysis related activities take place within Modeling and Analysis Environments (MAEs), whose constitution is described in detail in the next subsection. Also, models are computationally handled in the form of numerical-symbolic constructs. Ultimately, the role of a MAE is to process numerical-symbolic constructs and generate numerical-symbolic constructs. The numerical-symbolic constructs are given the generic denomination of *structures* in the text. In particular, the structures containing all the information provided by a model are called *representations*.

2.1 Mathematical description of MAEs and mapping models into them

Mathematically, a MAE can be described by a triple $\langle L_C, L_B, C_A \rangle$, whose elements, explained in the following paragraphs, can be represented by diagrams, like those present in Figure 1.

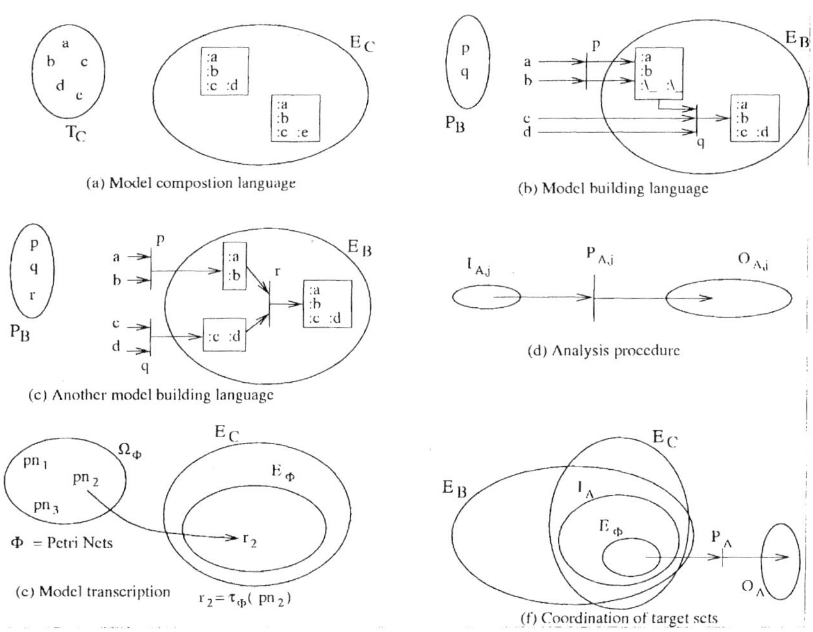

Figure 1. Diagrams involving target sets

$L_C=\langle T_C,E_C\rangle$ is a *model composition language* that provides the descriptive elements that comprise model representations, where T_C is the set of terms from which structures are made and E_C is the set of all valid structures (according to a certain criterion) — Figure 1(a) shows examples of T_C and E_C, the latter being a set of model representations expressed, in this example, by forms whose fields are correctly filled with the elements from the former.

$L_B=\langle P_B,E_B\rangle$ is a *model building language*, such that P_B is the set of operations on T_C (in L_C) provided by this language and E_B is the set of all structures that can be built using operations in P_B and terms in T_C — Figure 1(b) shows examples of P_B, E_B and the application of some operations in P_B in the construction of filled and semi-filled forms that belong to E_B; in Figure 1(c), another example shows a building language that provides operations to construct fragments of forms (operations *p* and *q*) and an operation *r* to compose complete forms from fragments.

$C_A = \{C_{A.1}, ..., C_{A.n}\}$is a set representing the *analytical capacity* of the MAE, where $C_{A.j}$ is a triple $\langle P_{A.j}, I_{A.j}, O_{A.j}\rangle$, where $P_{A.j}$ is a *procedure* implementing an analysis, $I_{A.j}$ and $O_{A.j}$ are, respectively, the sets from which inputs are taken and to which the results from the procedure belong — Figure 1(d) schematically illustrates $C_{A.j}$.

In this text, mentions to modeling language usually refer to the pair $\langle L_C,L_B\rangle$.

Along with mathematical models of MAEs, types of models are introduced by means of a *transcription function* $\tau_{\Phi}: \Omega_{\Phi} \rightarrow E_C$, where Φ stands for a model type (like Petri Nets or Markov Chains), Ω_{Φ} for the set of all models of type Φ and E_C for the set of all valid models in L_C. E_{Φ} is the image of τ_{Φ} , i.e. the set of all representations of models of type Φ in the language L_C.

Some of the sets just introduced are called *target sets* and they are specially interesting for the purposes of this work, for representing four essential dimensions in the relationships between MAEs and types of models:

- E_C is the target set of the composition language, representing the model expressivity of the MAE;
- E_B is the target set of the model building language, representing the model building features of the MAE;
- $I_{A.j}$ is the target set of procedure $P_{A.j}$, referring to the analyzability of structures as provided by the implementation of analysis methods;
- E_{Φ} is the target set of the transcription of models of type Φ, that is, indicating how models of this type are represented within a MAE.

An effective MAE must have its target sets duely coordinated, so that, from the point of view of the representations of the models of a type Φ to be considered, they can be expressed, built and analysed. The configuration illustrated in Figure 1(f) meets these conditions: $E_{\Phi} \subset$ in E_C means that the MAE can effectively express such models of type Φ; $E_{\Phi} \subset E_B$, that it can build representations of these models; $E_{\Phi} \subset I_A$, they can computationally processed by the analysis implemented by the MAE.

An interesting feature that can be observed in certain model building languages (L_B) is that, if appropriate operations are provided in P_B, the construction of model representations can be performed with the use of several operations. In this case, besides the aimed model representation, the process generates intermediary structures, as illustrated in Figures 1(b) and 1(c). The availability of such structures grants flexibility to the model building process, since more building paths can be followed and diverse structure reuse pattern can be employed.

When dealing with heterogeneous models, Figure 2 shows two ways of handling them. In Figure 2(a), each type of model is handled by a specific MAE, as indicated by the target sets. In Figure 2(b), the configuration that is the object of interest of this work is shown, where all types of models are represented using a comprehensive modeling language. A modeling language is said to be *comprehensive* if it can express and build models of multiple types (it is interesting to notice that Figure 2(b) also includes the results of the procedures). The problem with the specific MAEs configuration is that, in the general case, implementation and learning costs with respect to model composition and building languages can be significant and the transition from working with a specific MAE to another can be cumbersome. The configuration described in the next paragraph can minimize these problems.

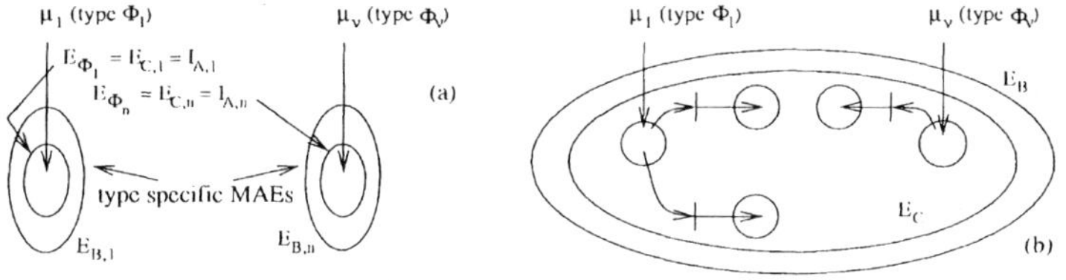

Figure 2. Configurations for dealing with heterogeneous models: (a) specific MAEs and (b) comprehensive MAE.

Thus, going further, a modeling language is ideally *comprehensive and uniform* if, with the same set of elements and composition rules, is capable of building representations of models of an indefinite number of types. This is particularly important when a new type of model is to be considered: a uniform MAE (Figure 3(a)) incorporates new model types without changes; a nonuniform MAE (Figure 3(b)) probably will need extensions in the expressive part, which will likely induce a need for extensions in the building language as well.

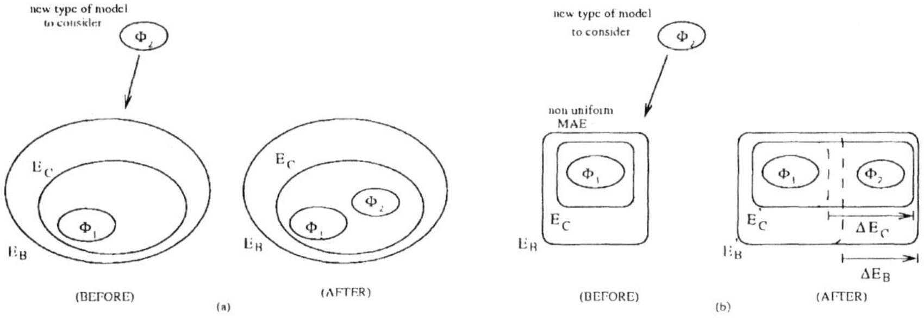

Figure 3. Comprehensive MAEs: (a) uniform and (b) nonuniform.

Then, at this point, it is possible to enumerate important features that an extensible MAE for heterogeneous models should present: coordination between the elements of MAE and the transcriptions of different types, comprehensive and uniform model composition language, flexible model building language and analytical extensibility (to incorporate new analysis procedures). Also, there are issues like representation ambiguity and redundancy, that are treated in (Arata and Miyagi, 2003).

3. INTEGRATION OF MODELS AND THEIR SEMANTICS

Comprehensive and uniform modeling languages are important to avoid the amplification, by the computational modeling and analysis frameworks, of the complexity due to model heterogeneity. However, even languages presenting such features cannot prevent the inherent formal heterogeneity of models. This heterogeneity means that, if no support is available, dealing with heterogeneous models means having to deal with information organized in different structures, what can lead to many very different procedures to access and manipulate the contents of each of them.

In the direction of such support, it is interesting to consider the fact that the integration of models (specially those of different types) is based on the semantic bindings between them (Arata and Miyagi, 2005). In this context, semantics refers to the meaning, that is, to what is being represented by the models; therefore, the semantics of models corresponds to what is observed in the system and in the dynamics being treated. Indeed, the simple fact that different models refer to one same dynamics suggests the existence of such connections.

A notorious case of model integration involves the isomorphism between Generalized Stochastic Petri Nets (GSPNs) and Continuous-time Markov Chains (CTMCs) (Marsan, 1984): if a GSPN is isomorphic to a certain CTMC, there is an one-to-one mapping *f* between GSPN markings and CTMC states so that transitions from markings *M* to *M'* mean that there are transitions from CTMC states *S=f(M)* and *S'=f(M')*. In formal terms, GSPN marking is a concept strange to CTMCs in the same way a CTMC state is with respect to GSPNs; so, the nature of this binding is essentially semantic, that is, the only fact that links related GSPN markings and CTMC states is that they refer to the same entities, i.e., the same dynamic states.

Based on the fact that semantic bindings are a major component behind model integration, it is natural to expect that a proper representation of the semantic information associated to the integrating models plays a role in the development of a description that reflects and makes explicit the integration of models. In the next section, this representation is elaborated in the form of a set of predicates and it is shown that, while formal comprehensiveness and uniformity of modeling languages provides an effective coexistence of heterogeneous models, a coordinated representation of semantic information provides a straightforward way to present the relationships between different models, overcoming a barrier due to heterogeneity.

A major issue in building such descriptions is dealing with semantic conflicts. In particular, two kinds of name conflicts can make the descriptions ineffective: synonyms and homonyms.

In the case of synonyms (multiple names, one entity), the description may not completely reflect the integration, since certain relationships can be interpreted as involving different entities, while, instead, they involve one same entity (for example, if an entity A is denoted by both X and Y, the expressions "X is being processed" and "Y is being inspected" give no clue about the fact that both situations refers to entity A; allowing synonyms requires additional elements to handle their occurrences, increasing the complexity of the descriptions).

If homonyms (one name, multiple entities) are allowed, semantically ambiguous or senseless statements can be made (if X refers to a machine and also to a part, an order to "mill X" can lead to undesired situations).

The consideration of semantic bindings is an important element in the representation of integration of models. Although not explicitly mentioned, this is also the mechanism used by the hierarchical model composition in the SHARPE system (Trivedi, 2002), where results of model analysis are employed as elements of other models, and in the Möbius framework (Sanders et al., 2003), where formalisms are described in terms of the components provided by the framework.

4. APPLICATION OF THE CONCEPTS

In this section, the concepts thus far develop are illustrated by means of the example of the analysis of GSPNs via isomorphism with CTMCs (Marsan, 1984), similar to the case described in the previous section. Briefly describing, in this analysis, a GSPN model is used to enumerate the relevant conditions to the dynamics being modeled, how these conditions enable state transitions and what conditions these transitions maintain, activate or deactivate; it also specifies the duration of the activation of the conditions. Then, a timed reachability graph is built, describing what conditions are active in each reachable state, indicating the elements associate to each state transition. From this graph, a CTMC model is generated and, then, its steady-state probability distribution is calculated (providing the steady-state probability of each reachable state).

A comprehensive and uniform modeling language is presented in (Arata and Miyagi, 2003). The language provides three kinds of constructs that are used to construct all the data types that can be expressed by the language: atoms, homogeneous sets (whose elements are all of the same type) and tuples (ordered aggregates of objects), where the latter two can be nested one inside another indefinitely. In Figure 4, using these constructs, metamodels for GSPN, timed reachability graphs, CTMC and steady-state probability distributions are defined.

```
P is atomic; TT is atomic; IT is atomic;
MarkingType is setof (place is P,
                      tokens is Integer);
MarkedGSPNType is (
   Places is setof P,
   TimedTransitions is setof TT,
   ImediateTransitions is setof IT,
   ITT is setof (arc is (place is P,
                         transition is T),
                 weight is Integer),
   IIT is setof(arc is (place is P,
                        transition is T),
                weight is Integer),
   OTT is setof(arc is (transition is T,
                        place is P),
                weight is Integer),
   OIT is setof(arc is (transition is T,
                        place is P),
                weight is Integer),
   Marking is MarkingType
   );                                    (a)
```

```
TimedReachabilityGraphType is
   setof (origin is MarkingType,
          end is MarkingType,
          timedTranstion is TT,
          imediateTranstitions is
             setof (order is Integer,
                    transition is IT)
         );                              (b)
```

```
S is atomic;
CTMCType is
   (states is setof S,
    transitions is
    setof (origin is S, end is S, rate is Real)
   );                                    (c)
```

```
SteadyStateDistribution is
   setof (state is S, probability is Real);
                                         (d)
```

Figure 4. Metamodels describing representations of models

Examining the first line of the metamodel in Figure 4(a), three atomic types (referring to atomic objects or atoms) are defined: P, TT and IT; the idea is that other data types are built using atoms of these three type. The type MarkingType is defined to be a set (using the keyword **setof**) of elements, whose type follows the keyword **setof**: in this case, the elements are tuples (that are specified by a list of type specification enclosed by parentheses); these tuples are of a data type with two elements: the first one, that can be referred by the name place, is of type P, and the second one, that can be referred by the name tokens, is of type Integer (a number, that the language treats as a special kind of atom). The same can be analogously said about the other metamodels. So, this is a modeling language that can express a wide variety of types of models (comprehensiveness), using the same simple elements and constructs (uniformity).

In Figure 5, instances of these types of models implementing such analysis are shown. Figure 6 presents a set of predicates, similar to those in the Prolog language (Deransart, 1996), that is a representation of the semantic information associated to the models in Figure 5. The "rate" predicates specify the rate of completion of an activity, described by a timed transition in the GSPN. The "in" predicates state that a certain dynamic state belongs to the state space of the system (denoted by the term "StateSpace"), what can be observed from both GSPN and CTMC. The "at" predicates indicates that a certain activity occur in a certain state, what can be concluded from the places of the GSPN and the marking in the timed reachability graph. The "prob" predicates indicates the probability of occurrence of a certain state, obtained from the steady-state probability distribution of the CTMC. Consistency in this representation is achieved by using the same terms for states and the same expressions for activities in the different predicates obtained from different models

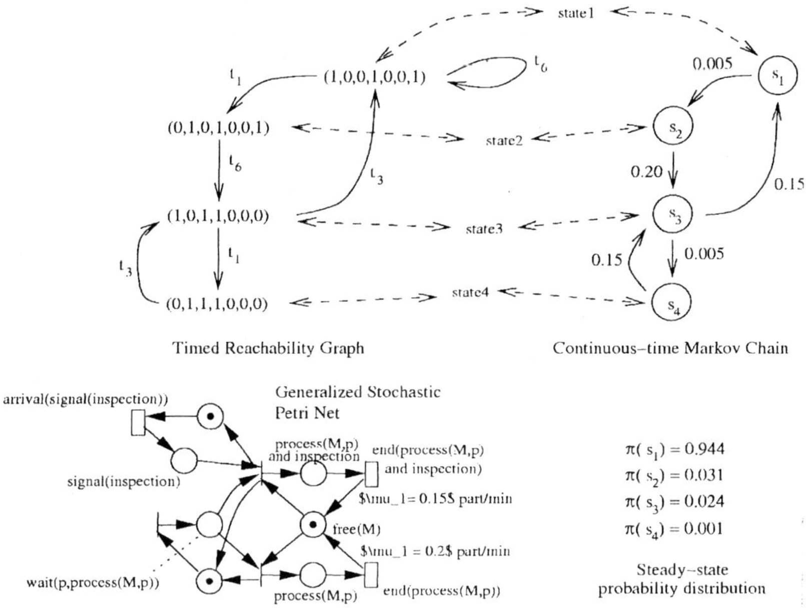

Figure 5. Models involved in the analysis of GSPN via isomorphism

```
rate(process(M,p),.2).
rate(process(M,p) and nspecting,.15).
rate(arrival(signal(inspection)),.005).

in(state1,StateSpace).
in(state2,StateSpace).
in(state3,StateSpace).
in(state4,StateSpace).

at(process(M,p),state1).
at(process(M,p) and inspecting,state2).
at(process(M,p) and inspecting,state3).
at(process(M,p),state4).

prob(state1,0.944).
prob(state2,0.031).
prob(state3,0.024).
prob(state4,0.001).
```

Figure 6. Representation of semantic information via predicates.

The use of predicates in the representation of semantic information shows that information carried from different models can be put in one same description, in a way that the heterogeneity of models is no longer an obstacle to visualize the relationships between the entities that participate in the system dynamics, as long as consistency in the representation is observed.

5. CONCLUSIONS

This work presents some guidelines to be observed in the design of computational systems dealing with heterogeneous models. Adequate treatment of heterogeneous models is relevant as it enables an efficient usage of computational and human resources by providing an infrastructure for a smoother workflow when dealing with model heterogeneity. It also covers the case where new models need to be considered (for instance, because new features are to be included in a system); thus, supporting heterogeneity increases the likelyhood of a computational tool to evolve following a smooth path. Diagrams involving target sets have shown to be a useful means to visualize features such as comprehensiveness and uniformity. Also it has been shown that an adequate representation of semantic information associated to model integration leverages the contribution that multiple models working together can give, providing more resources to query and manipulate those information.

6. ACKNOWLEDGEMENTS

The authors gratefully acknowledge the financial support to the present project provided by the Brazilian Governmental Agencies CNPq, CAPES and FAPESP. Particularly, the authors would like to thank TIDIA/KyaTera program under which this work is developed.

7. REFERENCES

1. Arata WM, Miyagi PE. Uniform computational treatment of heterogeneous discrete-event dynamic system models. Proceedings of 9th IEEE International Conference on Emerging Technologies and Factory Automation, Lisbon, Portugal, 2003; p.47-53.
2. Arata WM, Miyagi PE. Formal comprehensiveness and uniformity and semantic intra and intermodel consistency in the representation of Discrete Event Dynamic System models. Proceedings of the 18th International Congress of Mechanical Engineering, COBEM 2005, Ouro Preto, Brazil, 2005.
3. Bolch G, Greiner S, Meer H, Trivedi KS. Queuing Networks and Markov Chains: modeling and performance evaluation with computer science applications. New York: Wiley-Interscience, 1998.
4. Cassandras CG. Discrete Event Systems: Modeling and Performance Analysis. Burr Ridge: Richard D. Irwin Inc,1993
5. Deransart P, Cervoni L, Ed-Dbali A. Prolog: the standard: reference manual. London: Springer-Verlag, 1996.
6. Hasegawa K, Takahashi K, Miyagi PE. Application of the Mark Flow Graph to represent discrete event production systems and system control. Transactions of the Society of Instrument and Control Engineers 1988: 1, 69-75.
7. Kulkarni VG. Modeling and Analysis of Stochastic Systems. London: Chapman & Hall, 1995.
8. Marsan MA, Conte G, Balbo G. A class of generalized Stochastic Petri Nets for the performance evaluation of multiprocessor systems. ACM Transactions on Computer Systems 1984: 2, 93-122.
9. Molloy MK. Performance analysis using Stochastic Petri Nets. IEEE Transactions on Computers 1980: 9, 913-917.
10. Murata T. Petri Nets: Properties, Analysis and Applications. Proceedings of IEEE 1989: 4, 541-580.
11. Sanders WH, Courtney T, Deavours D, Daly D, Derisavi S, Lam. Multi-formalism and Multi-solution-method Modeling Frameworks: The Möbius Approach. Proceedings of the Symposium on Performance Evaluation - Stories and Perspectives, Vienna, Austria, 2003: 241-256.
12. Trivedi KS. SHARPE 2002: Symbolic Hierarchical Automated Reliability and Performance Evaluator, Proceedings of 2002 International Conference on Dependable Systems and Networks (DSN 2002), http://csdl.computer.org/comp/proceedings/ dsn/2002/1597/00/15970544.pdf.

44

TIMED-EVENT-STATE-BASED DIAGNOSER FOR MANUFACTURING SYSTEMS

M. Sayed-Mouchaweh, A. Philippot, V. Carré-Ménétrier, B. Riera
Université de Reims, CReSTIC - LAM
Moulin de la Housse B.P. 1039
51687 REIMS Cedex 2, FRANCE
Moamar.sayed-mouchaweh@univ-reims.fr

This paper proposes an adapted diagnoser for manufacturing systems. This diagnoser combines event and state based models to infer the fault's occurrence using event sequences and state conditions characterized by sensor's readings and commands issued by the controller. Furthermore, this diagnoser uses expectation functions to capture the inherent temporal dynamics of the system represented by time delays between correlated events.

1. INTRODUCTION

Fault diagnosis is defined as the operation of detecting and isolating faults. Manufacturing systems are an example of Discrete Event Systems (DES). Their behavior is generally based on two characteristics: all information regarding their operation is determined from the event sequencing and timing of events and their initial state may not be known.

The majority of DES fault diagnosis methods are based on a finite-state automaton. Some examples of the use of automata can be found in (Philippot, 2005), (Sampath, 1994), (Tripakis, 2001) and the references therein. This model accounts for the normal and failed behaviors of the process.

Since not every process is diagnosable, a notion of diagnosability must be defined (Lin, 1994), (Sampath, 1994), (Zad, 2003), (Mouchaweh, 2005) to determine if a process is diagnosable according to a certain set of observable events and pre-defined partitions of faults. It defines a diagnoser that must be able to infer the fault's occurrence after both the occurrence of the fault and the initialisation of the diagnoser and within a finite delay. This notion is formalized differently according to whether the fault model is event-based or state-based ones.

In (Sampath, 1994) a diagnosability notion using an event-based model is defined. All information relevant to the diagnosis is captured in the event set of the model. A process model is diagnosable if and only if any pair of faulty/non-faulty behaviors can be distinguished by the diagnoser using their projections to observable behaviors. The diagnoser, defined by this notion, can handle the actuator and sensor

Please use the following format when citing this chapter:

Sayed-Mouchaweh, M., Philippot, A., Carré-Ménétrier, V., Riera, B., 2006, in IFIP International Federation for Information Processing, Volume 220, Information Technology for Balanced Manufacturing Systems, ed. Shen, W., (Boston: Springer), pp. 415–424.

faults. However, the diagnoser and the process model must be initiated at the same time to allow the process model and diagnoser to response simultaneously to events. This initialization is hard to obtain in manufacturing systems since their initial state may not be known. To enhance the diagnosability, the above framework is extended to dense-time automata (Tripakis, 2001). This extension is useful since it allows diagnosers to base their decisions not only on the sequences of observed events, but also on the time occurrence of these events.

To find a remedy to the initialization problem, a diagnosability notion using a state-based model is proposed in (Lin, 1994). In this notion, since the process states describe the conditions of its components, diagnosing a fault can be seen as the identification in which state or set of states the process belongs to. In (Zad, 2003), a diagnoser using a state-based model is proposed. However, the diagnosis is limited to the case of actuator faults. While manufacturing systems use many sensors entailing the necessity of diagnosing their faults.

In (Pandalai, 2000) Pandalai *et al.* use template models for monitoring DES, specifically manufacturing systems. Template models are based on the notion of expected event sequencing and timing relationships. However, these models do not allow the analysis of diagnosability properties based on a diagnosability notion, as the case for event or state based models.

This paper presents a diagnosability notion that defines a diagnoser combining event and state based models. Furthermore this diagnoser uses expectation functions in order to take into account the inherent temporal dynamics of processes. Consequently this diagnoser is adapted to diagnose manufacturing systems.

The paper is structured as following: firstly, the proposed diagnosability notion is introduced. Secondly, the construction of the diagnoser, defined by this notion, is explained using an illustration example of manufacturing systems. Finally, a conclusion and the perspectives of our future work end this paper.

2. TIMED-EVENT-STATE-BASED DIAGNOSABILITY NOTION

In this paper, we consider the problem of diagnosing DES, specifically manufacturing systems with discrete sensors and actuators.

2.1 Notations and definitions

2.1.1 Process components model

Let G and its corresponding prefixed closed language, $L = L(G)$, be the process model. This model accounts for normal and failed behaviors of the process. $G = (\Sigma, Q, Y, \delta, h, q_0)$ is a Moore automaton. Σ is the set of finite events, Q is the set of states, Y is the output space, $\delta : \Sigma \times Q \rightarrow Q$ is the state transition function. $\delta(\sigma, q)$ gives the set of possible next states if σ occurs at q. h: $Q \rightarrow Y$ is the output function. $h(q)$ is the observed output at q. q_0 is the initial state.

Balemi *et al.* (Balemi, 1993) define controllable events $\Sigma_c \subseteq \Sigma$ as the controller's outputs sent to actuators, and uncontrollable events $\Sigma_u \subseteq \Sigma$ as the controller's inputs coming from sensors. ($\Sigma_o = \Sigma_c \cup \Sigma_u$) $\subset \Sigma$ is the set of observable events.

We use the Boolean DES (BDES) model, introduced in (Wang, 2000), to construct G. This modelling was initially used for the supervisory control of DES. We develop it to realize the diagnosis of DES. Each state of G is represented by an output vector, h_j, considered as a Boolean vector whose components are Boolean variables characterizing the state variables. Let n denote the number of state variables of G, the output vector, h_j, of each state, q_j, can be defined as :

$$\forall q_j \in Q,\ h(q_j) = h_j = (h_{j1},\dots,h_{jp},\dots,h_{jn}),\ h_{jp} \in \{0,1\}, 1 \le j \le 2^n,\ h_j \in Y \subseteq \mathrm{B}^n$$

A transition from one state to another is defined as a change of the variable state value from 0 to 1, or from 1 to 0. Thus each transition produces events characterized by either rising, $\uparrow\alpha$, or falling edges, $\downarrow\alpha$.

To describe the effect of the occurrence of an event, $\alpha \in \Sigma_0$, a displacement vector, E_α, is used. It is defined as a Boolean vector, $E_\alpha = (e_{\alpha 1},\dots, e_{\alpha p},\dots, e_{\alpha n})$, in B^n. If $e_{\alpha p} = 1$, then the value of pth state variable, h_{jp}, will be set or reset when α occurs. While if $e_{\alpha p} = 0$, the value of pth state variable, h_{jp}, will remain unchanged when α occurs. Consequently we can write the transition function as:

$$\forall q_i, q_j \in Q : q_j = \delta(\alpha, q_i) \Rightarrow h_j = h_i \oplus E_\alpha \qquad (1)$$

Similarly, we can define the displacement vectors for the other events. The set of all the displacement vectors of all the events provides the displacement matrix E.

For each event, $\alpha \in \Sigma_0$, an enable condition, $en_\alpha(q_i) \in \{0,1\}$, is defined in order to indicate if this event can occur at the state q_i, $en_\alpha(q_i) = 1$, or not, $en_\alpha(q_i) = 0$. Consequently, (1) can be re-written as :

$$\forall q_i, q_j \in Q : q_j = \delta(\alpha, q_i) \Rightarrow h_j = h_i \oplus (E_\alpha . en_\alpha(q_i)) \qquad (2)$$

2.1.2 Constrained-process model

Let $S = (\Sigma, Q_S, Y, \delta_S, h, q_0)$ denote the constrained-process model, characterized as a Moore automaton. It defines the desired behavior of the process which is represented by the prefixed closed specification language, $K = L(S) \subseteq L(G)$. S and its language, K, are constructed by experts. The set of states, Q_S , is included in Q.

The user provides the automaton S representing the desired behavior that the process should follow. We represent S as a BDES model. To obtain the transition function, δ_S, the enablement conditions for all the process events, $\forall \alpha \in \Sigma_0$, must satisfy all the specifications, K, representing the desired behaviour:

$$\forall \alpha \in \Sigma_0, \forall q_i, q_j \in Q_S : q_j = \delta_S(\alpha, q_i) \Rightarrow en_\alpha(q_i) = 1, h_j = h_i \oplus (E_\alpha . en_\alpha(q_i)).$$

2.1.3 Definitions and conditions of Diagnosability

Let $F = \{F_1, F_2,\dots, F_r\}$ define the set of fault modes to be diagnosed. Each fault mode corresponds to some kind of faults in a component (sensor, actuator) or a set of such faults. We will take the case of simple fault for which a fault mode corresponds to a simple fault, $F_i = \{f_i\}$. Additionally, we assume at most one of the fault modes may occur at a time. Let Ψ_{f_i} define the set of all the event sequences ending by the fault f_i. Thus $\Psi_f = \bigcup_{i=1}^{r}(\Psi_{f_i})$ denotes the set of all the event sequences ending by a fault from F. Consequently $\Psi_f \subseteq (L - K)$, i.e., all the faulty sequences ending by a fault of F are considered as violation of the specification language K. An observation mask is defined as : $P: \Sigma^* \cup \{\varepsilon\} \rightarrow \Sigma_o^*$, where Σ^* is the set of all event sequences of the language $L(G)$ and Σ_o^* is the set of all observable sequences. The inverse mask is defined as: $P_L^{-1}(u) = \{s \in L : P(s) = u\}$. The set of faulty states is defined as $S_f : \bigcup_{i=1}^{r} (S_{f_i})$ where S_{f_i} is the set of states reached by the occurrence of the fault f_i. Let H_{f_i} denote the set of all state output vectors of the faulty state partition S_{f_i}, then the output partition H_{f_i} can be defined as: $\forall q' \in S_{f_i}, h' = h(q') \Rightarrow h' \in H_{f_i}$. Since the state output vectors are used to diagnose fault's occurrence in state-based model, then two states belonging to different faulty state partitions must have different output vectors.

We define the set of fault labels $\Lambda_f = \{f_1,...,f_r\}$ in order to indicate the occurrence of a fault belonging to one of fault partitions $F = \{\{f_1\}, \{f_2\},...,\{f_r\}\}$. In adding the normal label N, which indicate the non presence of any fault, we can obtain the set of all the labels used by the diagnoser : $\Lambda = \{N\} \cup \Lambda_f$.

2.1.4 Events timing delays modelling

The majority of sensors and actuators in manufacturing systems produces constrained events since state changes are usually effected by a predictable flow of materials (Pandalai, 2000). Therefore, a temporal model centered on the notion of expected event sequencing and timing relationships can be used. This expectation function is constructed for observable events and it describes the next events that should occur and the relative time periods in which they are expected.

In this paper, we define an expectation function for each controllable event, $\beta \in \Sigma_c$, in order to predict uncontrollable but observable consequent events within a pre-defined time periods. These pre-defined time periods are determined by experts according to the process dynamic and to the desired behaviour S. If $u = \beta u_l \alpha$ is an observable sequence starting by a controllable event β, and ending by the consequent event α and the observable events $u_l \alpha$ occur at the states $q_i, \ldots, q_j$, then each expectation function $EF(u)$ is created when the event β occurs. This function has the following form: $EF(u) = (\{\{q_i\},..., \{q_j\}\}, \alpha, [t1, t2], \{\{f_i\},..., \{f_j\}\}), \forall f_i \in \Lambda_f$. This expectation means that when β has occurred, the event α should occur at any instant between t_1 and t_2. If it is the case then the expectation function is satisfied and it will provide the value 0. If the event α has occurred before t_1 or after t_2 then the expectation function is not satisfied and it will provide the value 1. If this non satisfaction occurs at the state q_i then this expectation function provides the fault label, $l = \{f_i\}$, to indicate that the cause of this non satisfaction is the occurrence of the fault f_i. Similarly the non satisfaction of the expected function at the state q_j indicates the occurrence of the fault f_j and it provides the fault label $l = \{f_j\}$. The expectation function is deleted when it is satisfied.

2.2 Timed-Event-state-based diagnosability notion formulation

The fault diagnosis problem is to diagnose unambiguously the occurrence of the fault f_i entailing a faulty behavior belonging to $(L - K) \cap \Psi_{f_i}$, by the diagnoser within a bounded delay. Consequently the timed-event-state-based diagnosability notion is defined as follows: a process model G with its language L, and a specification language K is diagnosable according to the observation mask P and the fault partitions F iff:

$$\left.\begin{array}{l}\exists k \in \mathrm{N}, \forall i \in \{1,2,\ldots,r\}, \forall st \in (L-K) \cap \Psi_{f_i}, \\ |t| \geq k, \forall u \in P^{-1}P(st) \cap (L-K)\end{array}\right] \Rightarrow u \in (L-K) \cap \Psi_{f_i}$$

$$\forall q \in Q, q' = \delta(st, q), q'' = \delta(u, q), h' = h(q'), h'' = h(q'') \Rightarrow h', h'' \in H_{f_i} \qquad (3)$$

$$\exists z \in \{1,2,\ldots,m\} \Rightarrow EF_z(P(st)) = 1 \wedge l = \{f_i\}$$

The satisfaction of (3) means that the occurrence of the fault f_i is diagnosable by one or more of the following three ways :

- If the faulty event sequence, s, ending by the fault f_i is distinguishable by the diagnoser after the execution of $k = |t|$ transitions where t is a continuation of s. Then, any other event sequence, u, belonging to $(L - K)$ and producing the same observable event sequence as st, $P(u) = P(st)$, should contain in it the fault f_i.
- If the state q' is reached by an event sequence containing the fault f_i and possessing an output vector $h' = h(q')$ belonging to the output partition H_{f_i}, then any other state reached by any other event sequence containing the same fault, f_i, should possess an output vector $h'' = h(q'')$ belonging to the same output partition H_{f_i}.
- There is at least one expectation function, defining a temporal constraint between the occurrence of the observable events $P(st)$ not satisfied due to the occurrence of the fault f_i. This expectation function should provide the fault label $l = \{f_i\}$ as the cause of this non satisfaction.

2.3 Timed-Event-state-based diagnoser definition

The diagnoser D defined by (3) is considered as a Moore automaton : $D = (\Sigma_o \cup \{e_{EF1},\ldots, e_{EFm}\}, Q_D, Y, \delta_D, h, q_{0D})$. The set of events, $(e_{EF1},\ldots, e_{EFm})$, corresponds to the non satisfaction of expectation functions indicating a fault. Each diagnoser state, $q_{iD} \in Q_D$, is of the form: $q_{iD} = (h_i, L_i)$, where h_i is the output vector of the diagnoser states q_{iD}, characterized by sensor's readings and commands issued by the controller. L_i is a subset of the set, Λ, of faults labels with the normal label. We can obtain $(2^{|\Lambda|}-1)$ possible subsets of labels for L_i. If $L_i = \{N\}$ or $\{f_j\}$ then the diagnoser, when it reaches the state q_{iD}, can decide with certainty the non presence of fault or the occurrence of the fault f_j. If L_i contains the label N and any other fault label then the diagnoser, at the state q_{iD}, cannot decide whether a fault has occurred or the system is in normal function, i.e., ambiguity case. The transition function δ_D is based on (1) and is defined as, $\delta_D : \Sigma_o \cup \{e_{EF1},\ldots, e_{EFm}\} x Q_D \rightarrow Q_D$. (1) calculates the output vectors without verifying the enablement conditions. Thus all the output vectors of faulty states reached by a fault, characterized by the occurrence of an event not authorized at this state, can be calculated.

To construct the diagnoser, each diagnoser state must be defined. This state must be reached by an observable event belonging to $\Sigma_o \cup \{e_{EF1},..., e_{EFm}\}$. The construction of this diagnoser will be explained in detail using the illustration example of the next subsection.

3. ILLUSTRATION EXAMPLE

To explain the construction of the timed-event-state-based diagnoser, let take the example of a wagon moving from an initial position, state *A*, measured by a sensor *a*, towards a terminal position, state *B*, measured by another sensor *b*, passing by the state *A-B* indicating that the wagon is located somewhere between *A* and *B*. This movement of two directions (right and left) is realized by two commands: left, *L*, and right, *R* as it is depicted in Figure 1.

The following hypotheses are verified for this example :

- There is one part (i.e. one wagon),
- Accepted response time is defined for the actuator by the designer,
- The wagon inertia is null,
- The actuator does not fail during operation, i.e., if it does fail, the fault occurs at the start of operation,
- The operating conditions of the process initial state are normal,
- The occurrence of any fault can be expected only after the activation of a command by the controller.

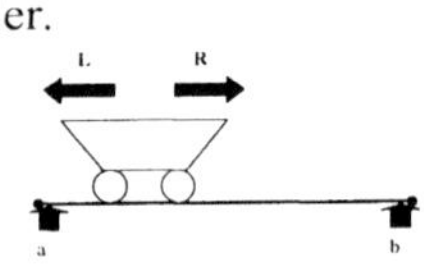

Figure 1 – Example of a wagon with two directions of movement

3.1 Component Boolean DES models

The wagon example consists of two components: the change of the wagon location measured by the sensors *a* and *b* and the wagon motor behavior. Since each BDES model must account for the normal and failed behaviors, the fault partitions to be diagnosed must be defined. Let $F = \{\{f_1\}, \{f_2\}, \{f_3\}\}$ be the set of faults to be diagnosed. f_1, f_2, and f_3 indicate, respectively, fault in sensor *a*, fault in sensor *b* and the wagon motor is stuck-off in one of the two senses right or left. The Figure 2 shows the BDES models for the change of process location and the behaviour of the wagon motor due to the commands *R* and *L*. Four Boolean state variables *a*, *b*, *R* and *L* are used to describe the overall wagon behaviour. State variables *a* and *b* are true when the wagon is located, respectively, in the position *A* or *B*. The value of *R* or *L* is 1 when they are enabled and 0 when they are disabled. If the fault f_1 occurs at the state *A*, then the model will transit to the state *a_fault* with the state variables $(a\ b) = (1\ 0)$. When the wagon arrives at the position *B*, the state *a_fault* will be characterized by the state variables $(a\ b) = (1\ 1)$. Similarly if *R* is enabled then the

BDES motor behaviour model will transit to the state characterized by $(R\ L) = (1\ 0)$. If the motor is stuck-off at this instant, then the model will transit to the state *stuck_off* with the same output vector $(R\ L) = (1\ 0)$. The same reasoning can be applied for the occurrence of the other faults. The sets of observable and controllable events are: $\Sigma_o = \{\uparrow R, \downarrow R, \uparrow L, \downarrow L, \uparrow a, \downarrow a, \uparrow b, \downarrow b\}$ and $\Sigma_c = \{\uparrow R, \downarrow R, \uparrow L, \downarrow L\}$.

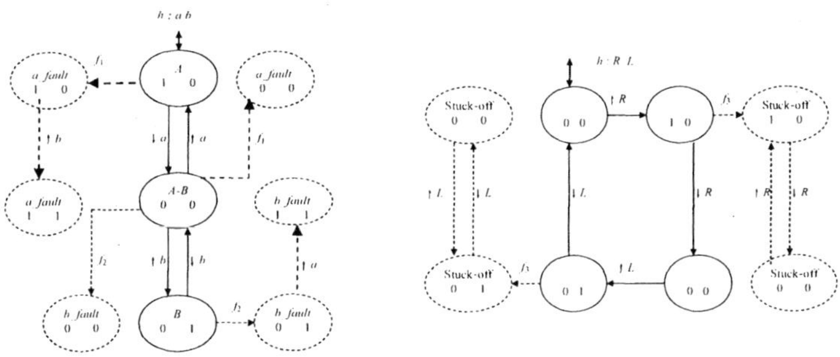

a) b)

Figure 2 – Boolean DES models for the change of wagon location, a), and the wagon motor behavior, b)

3.2 Constrained-process model construction

The constrained-process model, S, for the wagon example is depicted in the Figure 3. It corresponds to the desired behavior provided by an expert. In BDES modeling, this desired behavior can be described using two tables; the first explains the enablement conditions for the occurrence of each event, Table 1, and the second is the displacement matrix E, Table 2. We can notice that the only event allowed to take place in the initial state, characterized by the output vector $h_1 = (abRL)$=(1000), is the enablement of the controllable event $\uparrow R$ since its enablement condition, $en_{\uparrow R}(q_1) = 1$, is satisfied at this state and the enablement conditions for all the other events are false. The output vector of the next state can be calculated using (2): $h_2 = h_1 \oplus (E_{\uparrow R}.en_{\uparrow R}(q_1)) = (1000) \oplus ((0010).1) = (1010)$. Similarly we can calculate output vectors for next states after the occurrence of each authorized event.

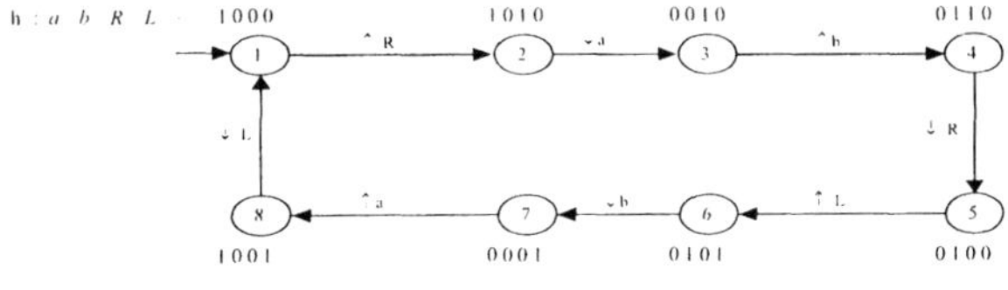

Figure 3 – Constrained-process model, S, for the wagon example

Table 1 – Enablement conditions for the events of the wagon example

Event σ	Enable condition en_σ	Event σ	Enable condition en_σ	Event σ	Enable condition en_σ	Event σ	Enable condition en_σ
$\uparrow a$	$\overline{a}.\overline{b}.\overline{R}.L$	$\uparrow b$	$\overline{a}.\overline{b}.R.\overline{L}$	$\uparrow R$	$a.\overline{b}.\overline{R}.\overline{L}$	$\uparrow L$	$\overline{a}.b.\overline{R}.\overline{L}$
$\downarrow a$	$a.\overline{b}.R.\overline{L}$	$\downarrow b$	$\overline{a}.b.\overline{R}.L$	$\downarrow R$	$\overline{a}.b.R.\overline{L}$	$\downarrow L$	$a.\overline{b}.\overline{R}.L$

Table 2 – Displacement matrix E of the wagon example

State variable	$\uparrow a$	$\downarrow a$	$\uparrow b$	$\downarrow b$	$\uparrow R$	$\downarrow R$	$\uparrow L$	$\downarrow L$
a	1	1	0	0	0	0	0	0
b	0	0	1	1	0	0	0	0
R	0	0	0	0	1	1	0	0
L	0	0	0	0	0	0	1	1

3.3 Expectation functions construction

The faults entailing the actuator to be stuck-off can be diagnosed using the principal of watchdogs to watch the actuator response times. We use expectation functions to model the actuator response times which can be obtained by learning, and/or by technical documentation. For the wagon example, we define 2 expectation functions, one for each command enablement: $EF(\uparrow R \downarrow a \uparrow b)$, $EF(\uparrow L \downarrow b \uparrow a)$.

To construct these expectation functions, we use the desired behavior, S, provided by the user, see Figure 3. The enablement of R, entails the event $\downarrow a$ at the state q_2 and the event $\uparrow b$ at the state q_3. We expect $\uparrow b$ to occur within the time period [3, 5] according to the process dynamic. If this event does not occur at q_2 then the wagon motor has not responded since $\downarrow a$ did not occur. Thus the non satisfaction of the expectation function at this state indicates the occurrence of the fault f_3, i.e., $l = \{f_3\}$. If $\downarrow a$ has occurred, then S will transit to the state q_3. If $\uparrow b$ has not occurred then the non satisfaction of the expectation function provides the label $l = \{f_2\}$ to indicate that the sensor b is faulty since the wagon has responded, because the event $\downarrow a$ has occurred. While if $\uparrow b$ has occurred too early, then the model will transit to the state q_4 and the expectation function will not be satisfied at this state and will provide the fault label $l = \{f_2\}$. Consequently the expectation function can be written as following :

$EF(\uparrow R \downarrow a \uparrow b) = (\{\{q_2\},\{q_3\},\{q_4\}\}, \uparrow b, [3, 5], \{\{f_3\},\{f_2\},\{f_2\}\})$. Similarly the expectation function for the enablement of the command L can be written as following : $EF(\uparrow L \downarrow b \uparrow a) = (\{\{q_6\},\{q_7\},\{q_8\}\}, \uparrow a, [3, 5], \{\{f_3\},\{f_1\},\{f_1\}\})$.

3.4 Timed-Event-state-based diagnoser construction

The diagnoser D for the wagon example is depicted in Figure 4. It contains, besides the states of the desired behavior model S, all the faulty states reached by the occurrence of a fault belonging to F. Each one of these faulty states is reached due to the non satisfaction of the enablement condition of an event or of an expectation function. This diagnoser is constructed using the following steps :

1. We start from the first diagnoser state, q_{1D}, corresponding to the process initial state, which is characterized by the state vector $h_1 = 1000$. At this state we suppose that the diagnoser is in normal state $L_1 = \{N\}$.
2. All the enablement conditions for all the wagon controllable events will be tested in order to find the authorized event to occur at this state. We can find that the activation of R is the only event authorized to take place since its enablement condition, $en_{\uparrow R}(q_{1D}) = a.\bar{b}.\bar{R}.\bar{L}$, is equal to 1, (see Table 1).
3. The occurrence of $\uparrow R$ transits the diagnoser to the second diagnoser state,

q_{2D}. The output vector for this state can be calculated using (1): $h_2 = (h_1 = 1000) \oplus (E_{\uparrow R} = 0010) = (1010)$. Since $en_{\uparrow R}(q_{1D}) = 1$, then this state corresponds to a state of the desired behaviour S.

4. After each command enablement, all the fault labels and the label N will be included in the label of this state : $L_2 = \{f_1, f_2, f_3, N\}$ since after each command enablement, any fault belonging to F can occur. Then, all the enablement conditions of the possible observable events will be tested. Here there are three possibilities: either the event $\downarrow a$, or $\uparrow b$ or the event $e_{EF(\uparrow R \downarrow a \uparrow b)}$ corresponding to the non satisfaction of the expectation function $EF(\uparrow R \downarrow a \uparrow b)$. The event $\downarrow a$ is authorized, $en_{\downarrow a}(q_{2D}) = a.\overline{b}.R.\overline{L} = 1$ and it transits the diagnoser to the state q_{3D}. The output vector for this state can be calculated using (1) :
$h_3 = (h_2 = 1010) \oplus (E_{\downarrow a} = 1000) = (0010)$. The occurrence of $\uparrow b$ conducts D to the state q_{9D}. This event is not authorized at q_{2D} , $en_{\uparrow b}(q_{2D}) = \overline{a}.\overline{b}.R.\overline{L} = \overline{1}.\overline{0}.1.\overline{0} = 0$ and the only reason of this non enablement is the variable state of the sensor a. Thus the label of this state contains only the label f_1, i.e., $L_9 = \{f_1\}$. The output vector for this state is: $h_9 = (h_2 = 1010) \oplus (E_{\uparrow b} = 0100) = (1110)$. If $EF(\uparrow R \downarrow a \uparrow b)$ is not satisfied at q_{2D}, then the event $e_{EF(\uparrow R \downarrow a \uparrow b)}$ will occur to indicate the occurrence of the fault f_3. Thus this event will transit the diagnoser to the state, q_{11D}, with $h_{11} = h_2$ and $L_{11} = \{f_3\}$. Consequently, the state corresponding to the one with the authorized event, q_{3D}, will possess the label $L_3 = L_2 - L_{11} - L_9 = \{N, f_2\}$.
5. Similarly, the other diagnoser states can be calculated to obtain, at the end, the diagnoser of the Figure 4.

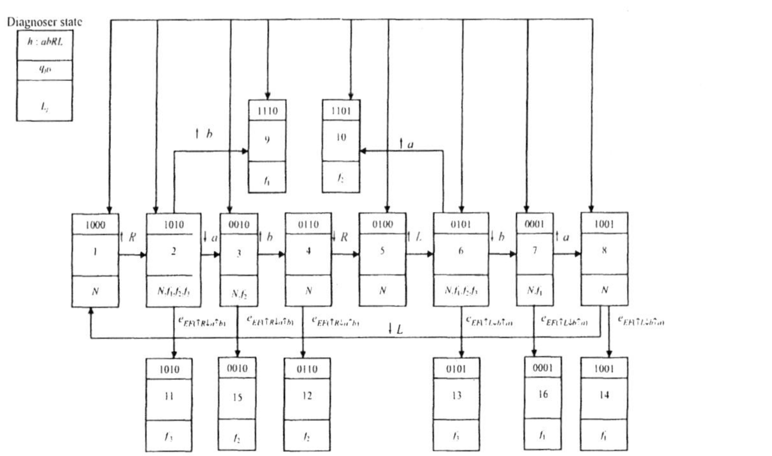

Figure 4 – Timed-event-state-based diagnoser for the wagon example

The diagnoser can be initiated at any state distinguished by its output. While if the diagnoser was initiated at any state distinguished by an event, the diagnoser cannot diagnose a past occurrence of a fault. As an example, if the fault f_3 has occurred before the diagnoser initialisation, then the diagnoser cannot distinguish between q_{2D} and q_{11D} since they have the same output vectors : $h_2 = h_{11}$. The state 2 has the labels $L_2 = \{N, f_1, f_2, f_3\}$ which means an ambiguity and thus the diagnoser cannot decide if a fault has occurred or not.

4. CONCLUSION

In this paper, a timed-event-state-based diagnoser is proposed for manufacturing systems. The goal of this combination is to find a remedy to the problem of synchronisation between the process and its diagnoser. This problem can be often seen in manufacturing systems. To enhance the diagnosability and to capture the inherent temporal dynamics of the process, expectation functions are used. They model the temporal information about actuator's minimal and maximal response times.

Since manufacturing systems are modular, we are developing a distributed diagnosis module to perform the diagnosis. This module uses the timed-event-state-based diagnoser, proposed in this paper, as a local diagnoser in a distributed structure. In addition, we are developing the diagnosis module to relax the hypothesis of unique source or part in order to be adapted to manufacturing system.

5. REFERENCES

1. Balemi S, Hoffmann GJ, Gyugyi P, Wong-Toi H, Franklin GF. Supervisory control of a rapid thermal multiprocessor, IEEE Transactions on Automatic Control, vol. 38, n°7, pp. 1040-105, 1993.
2. Lin F. Diagnosability of Discrete Event Systems and its Applications, In *Discrete Event Dynamic Systems4*, Kluwer Academic Publishers, Boston, USA. 1994.
3. Pandalai D, Holloway LEN. Template Languages for Fault Monitoring of Timed Discrete Event Processes, In *IEEE Transactions On Automatic Control 45(5)*, 2000.
4. Philippot A, Sayed Mouchaweh M, Carré-Ménétrier V. Multi-models approach for the diagnosis of Discrete Events Systems, In *IMACS'05, International conference on Modelling, Analyse and Control of Dynamic Systems*, Paris-France, 2005.
5. Sampath M, Sungupta R, Lafortune S, Sinnamohideen K, Teneketzis D. Diagnosability of discrete event systems, In *11th Int. Conf. Analysis Optimization of Systems: Discrete Event Systems*, Sophia-Antipolis, France, 1994.
6. Sayed Mouchaweh M, Philippot A, Carré-Ménétrier V. Detectability and Diagnosability of Discrete Event Systems: Application on manufacturing systems, 2nd ICINCO/IFAC International Conference Informatics in Control, Automation and Robotic, 14-17 September 2005, Barcelona, Spain.
7. Tripakis S. Fault Diagnosis for Timed Automata, VERIMAG (www-verimag.imag.fr). 2001.
8. Wang Y. *Supervisory Control of Boolean Discrete-Event Systems*, Thesis of Master of Applied Sciences, University of Toronto, Canada, 2000.
9. Zad SH, Kwong RH, Wonham WM. Fault Diagnosis in Discrete Event Systems: Framework and model reduction, *IEEE Transactions On Automatic Control 48(7)*. 2003.

45

NFC TECHNOLOGIES IN MOBILE PHONES AND EMERGING APPLICATIONS

Diogo Remédios, Luís Sousa, Manuel Barata and Luís Osório
GIATSI research group at ISEL,
Instituto Superior de Engenharia de Lisboa, PORTUGAL
dremedios@deetc.isel.ipl.pt, lsousa@isel.ipl.pt,
mmb@isel.ipl.pt, aosorio@deetc.isel.ipl.pt

The main purpose of this paper is to introduce some new non-conventional applications to allow the exploitation of today's mobile phone resources, namely their connectivity capabilities to other devices. There is a great potential for applications to take advantage of these available resources, to perceive and interact with the world around us, namely to serve as a user interface to interact with controlled devices. Recently beyond the necessary speaker and microphone, cell phones incorporate more sensing, processing and storage capabilities as well as alternative communication systems, e.g. digital camera, Bluetooth. This paper addresses, in particular, the Near Field Communication (NFC) incorporated in cell phones and enumerates some of its main promising applications like in device monitoring and control.

1. INTRODUCTION

In today's world, where everyone walks around with a mobile phone in their pockets, one question arises: Are we taking advantage of their full potential?

Mobile phones are becoming ubiquitous, in the sense that they are everywhere, making it possible and easier to develop non-conventional ubiquitous applications. Traditionally Ubiquitous Computing (UbiComp) is divided into several areas of expertise (Weiser, 1993) trying to solve different problems like: Context Awareness (Yau, 2002); Location Awareness (Hightower, 2001); Data availability and Storage (Bindel, 2000).

On the other hand, Mobile computing (Forman, 1994) – the use of a portable computer capable of wireless communication – consists on applications for devices that are mobile having wireless connectivity, or at least having to deal with some of the following problems: Disconnection without warning (Frequent loss of connectivity); Low Bandwidth; High Bandwidth variability; Heterogeneous network; Security risks; Mobility (address and location migration); Low Power restriction; Risk to data Small user interface; Small storage capacity.

A mobile phone, a PDA and a LAPTOP are some examples of this kind of devices. In this category of devices one can also include sensors and actuators, for

Please use the following format when citing this chapter:

Remédios, D., Sousa, L., Barata, M., Osório, L., 2006, in IFIP International Federation for Information Processing, Volume 220, Information Technology for Balanced Manufacturing Systems, ed. Shen, W., (Boston: Springer), pp. 425–434.

example, in a smart building there can be sensors spread across the entire area keeping measure of the temperature, lighting, personnel presence, surveillance cameras, access control mechanisms, etc.

Mobile computing experienced a great evolution since its birth, and still continues to, due to the fact of the ever-growing resources available in mobile devices. These include: Photographic camera; InfraRed (IrDA); Bluetooth; NFC (Near Field Communication); High resolution Color Displays; Generating computer Graphics; Speaker and Microphone; Growing storage capacity; Growing computing capacity; Human Machine Interaction – HMI.

Some of these resources confirm Moore's law (Moore, 1965) (which says the size and cost of electronics shrinks by a factor of two every eighteen months) will provide for more processing power, memory, and storage space. Though, the progress in batteries technology does not follow Moore's law (it improves only perhaps 10% each year), Gene Frantz's law (Pulli, 2003) states that the power consumption of integrated circuits decreases exponentially; hence the batteries will last longer.

However, mobile terminals are already in many ways equivalent or even superior to home computers of ten or fifteen years ago.

These resources extend the ability of the devices to interact and perceive the surrounding environment. As they evolve, so will the applications capacity to help people in their every day tasks. Near Field Communication (NFC) in concrete allows, just by touching or holding a mobile phone within a range of few centimeters of a tag or reader, to be turned into a cashless wallet, an information source, a privileged human machine interface to monitor and control embedded devices or a tool to help make business more effective.

The range of applications that can be developed taking advantage of this privileged interaction is enormous. Keeping those as goal objectives in the horizon let's not forget today's limitations and walk one step at a time in the long walk ahead of us.

A cell phone can be used not only to make phone calls, but also to interact with the environment controlling and monitoring surrounding entities providing services of authentication and control. For example, the cell phone can provide a universal interface to open garage doors, yard gates, front door and turning on/off alarm systems. Moreover the integrated functionalities for the support of human to machine interface can be exploited for implementation of monitoring and controlling functions in manufacturing environments, like the use of the screen and keyboard of the cell phone as small portable contact less terminal.

These kinds of applications have to deal with problems like integrating different systems and security issues, namely, guarantying that the communication is secure and not replicated by an intruder.

In this paper we will address, in particular, how the use of NFC can enhance and facilitate every day tasks, presenting several different usage scenarios and describing in detail an implementation of one case study where a cell phone using its NFC capabilities is used to pay a parking meter (P&D machines).

NFC technology is not at a point where it is feasible to implement all these usage scenarios because it is not available to the major public, yet. Hopefully it will get as widespread as Bluetooth is today, and much quicker. Nokia is being very aggressive in the promotion of NFC technology and is engaging in several field tests like in the

Atlanta stadium and in some commercial environments like fast food restaurants and pharmacies (Nokia, 2005).

The next chapter explains the NFC technology and why we believe that in the near future a major part of all mobile phones will be equipped with it.

Chapter 3 presents several examples of applications where NFC can be a facilitator of everyday tasks.

Our case study is an evolution from a system that enabled the payment of services using the mobile phone Bluetooth connectivity (Remédios, 2005). In this first approach, several problems were identified due to the distance of the communication that can range from 10 to 100 meters. The main problem is the association of the cell phone and the service being paid; if there is more than one service and cell phone, in a limited area, an ambiguity problem arises due to the several possibilities of combinations. An example of this problem is a highway toll where the user through the mobile phone using Bluetooth makes the toll payment, but the system doesn't know in which lane he is to clear his passage. Another identified problem was the initial time needed for the discovery of the surrounding devices. With NFC the proximity requirement solves all of the identified problems.

2. NFC (NEAR FIELD COMMUNICATION)

NFC is the worldwide standard (ECMA, 2003; ECMA, 2004) for mobile phone short-range communication, with the leading mobile phone manufacturers as members of the industry committee (panel includes Nokia, Philips, Samsung, and Sony) (Forum, 2004).

Near Field Communication (NFC) is a wireless technology for short-range communications designed to enable interaction between electrical devices, allowing the exchange of data between NFC enabled devices, assuring interoperability between vendors.

Although NFC has many advantages and solves problems exposed in other wireless communication technologies, its main goal is to simplify communications.

NFC devices operate under the 13.56 MHz frequency, thus allowing interaction between NFC devices and existing contactless cards, since the underlining protocols are those used by FeliCa and Mifare (ISO, 2000) Cards. Expected data rates range from 106 kbps, up to 424 kbps.

This standard (ECMA, 2004) started as a joint effort from Phillips and Sony, to provide a unified interface between electrical devices and also Smart Cards. Nokia joined one year later. Together with Phillips and Sony founded the NFC–Forum. Nokia quickly introduced the technology in Mobile Phones.

NFC is now starting to be deployed in commercial devices. Nokia, for instance, introduced NFC as an add-on shell for Nokia 3220 phones. Philips is producing microcontrollers with NFC capabilities built in that, although not yet stable, it is already being integrated in embedded devices.

3. APPLICATIONS

In this chapter, we present a survey of several emergent applications made possible by the NFC connectivity. This chapter is divided into four subsections, each representing a different type of usage scenario.

3.1. Authentication

Authenticity is very important whether we are authenticating a person or a product. The authentication of a person has great potential, imagine if you could approach your cell phone to a NFC reader and instantaneously the system knows who is trying to use a specific resource allowing or denying access depending on the predefined permissions.

This could be used for access control to the workplace, home, building, garage, etc. This application could replace the use of keys, cards and passwords.

Another example is the authentication of products with an NFC tag. This could lead to better tracking of goods and possibly identify vulnerable points in the supply chain, either detecting damaging or theft hazards. Additionally there is the possibility of the final consumer being able to read from the product relevant information, such as, its authenticity and validity dates. As the tags get cheaper the number of products labeled with them will increase. As an example of this tendency, in Florida, it has been approved a law requiring tagging on all pharmaceutical products starting in 2007. (Kanellos, 2006)

3.2. Interface with the environment

A mobile application running on the mobile phone has the potential to be a privileged access point to interact with systems in the surrounding environment. This could provide monitoring and control to every kind of systems from a device the user is already familiar with. There are several usage scenarios that fit in this category:

The first example is home automation where we don't have to use several different remote controls for different systems like stereo, television, air-conditioned, illumination controls, etc. The only device needed is our mobile phone that we already carry voluntarily everywhere.

Another kind of application is for enabling the collecting of particular data/information relevant for a person like, for instance, description about a monument, a tourist path, museum's visit, advertisement poster, etc. In these situations, the NFC technology will make easy the transfer of the detailed information wanted, and display it on the mobile phone screen, by merely approaching the device near the information supply device installed in these places.

3.3. Token Storage

This usage scenario implies the transfer of a token that by itself represents a digital ticket. The idea is when you buy a plane or a concert ticket instead of receiving a paper ticket, receive a token representing the digital ticket stored in the mobile phone. This allows the token to include some extra information that can be viewed at any time, like detailed information about the event, etc. NFC should act as the communication medium between the mobile phone and the tickets machine.

After use the ticket expires, it can be deleted or maintained as a customer relationship card where information about the client can be stored to offer future discounts and promotional offers.

3.4. Payment Service

If the mobile application has the ability to guarantee the authenticity of a person then why not use that information in a monetary transaction for a service payment to a given clearing company like *Via Verde (www.viaverde.pt)*, *VISA* or *MASTERCARD*. Security concerns are of the utmost importance and must be guaranteed, just like in the first usage scenario for authentication. The mobile application must have mechanisms to become disabled when copied to other devices. The communication between the mobile phone and the point of sale must be secure.

We choose this usage scenario for our case study to evaluate the technologies and issues involved. The next chapter will describe in detail the architecture of our proposed solution and the developed prototype to provide the users of *Pay & Display* (P&D) machines the possibility to pay with the mobile phone, dispensing the always problematic coins.

4. PROTOTYPE

4.1. Overview

The proposed application displayed in **Error! Reference source not found.** consists in a new method for paying street parks, using a NFC enabled mobile phone to interact with the P&D machines. This gives users a more convenient and comfortable payment procedure. Current P&D machines accept only coins or credit cards as the payment method.

This new approach to street park payments allows the user to specify the exact amount they want to pay and, gives the park operators a great advantage by reducing the amount of coins in P&D machines safe. This leads to less vandalism but also to a cost reduction by decreasing the cash collection rate, from the P&D machines.

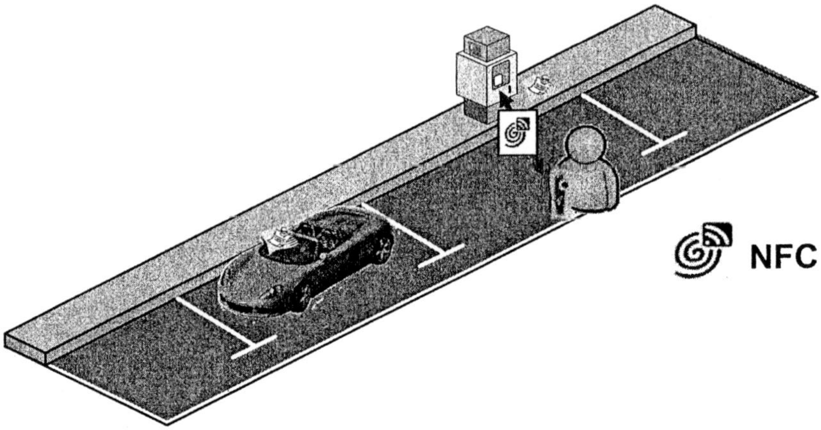

Figure 1 – NFC enabled phone performing a P&D parking payment.

4.2. Architecture

The proposed architecture will use a third party entity as money broker. This model is adapted to the Portuguese scenario where such an entity already exists and is present in every national toll, providing electronic payment alternatives (www.viaverde.pt) based in DSRC technology (CEN, 1999).

The payment scenario presented uses a mobile agent that will be installed in mobile phones, trough the broker's web site. This agent will handle the communication with the P&D machine.

RF communication in the mobile phone side is handled by the phone shell; this shell is connected to the mobile phone by its expansion bus. On the park equipments, an embedded device with NFC capabilities has been added.

The mobile agent has to be installed in the mobile phone. Since this is a piece of software that contains personal user data, and authenticates the user before the system, there are security issues that should be taken into consideration. The most important are: mobile phone theft; mobile agent theft; user ID replacement; mobile ID replacement; and all of the previous combinations.

The mobile agent has to be personalized in a per user basis, to insure that only the subscribed user can use the mobile agent with its user ID. Each agent must contain the unique user ID, assign by the broker, the mobile phone unique identification and the SIM card unique identification.

All this information has to be bound in a single file, and signed using Public Key Cryptography. The signed file will be stored inside the mobile application package. This way the combined user information can be validated by the mobile phone and the P&D machine.

With the schema presented it is ensured that, if the agent is cloned to another mobile phone, it won't work because the mobile phone ID won't match the one in the file. Even if the attacker manages to replace its mobile phone ID to match the one in the file, the SIM card ID will fail to match. The user ID has to be stored in a signed location to assure that a misbehaved user does not change its user ID, trying to impersonate some other user. All this information must also be sent to the P&D Machine allowing for it to validate all the information and its signature.

4.3. Deployment

With the schema presented above, the application must be packaged on a per user basis. Nevertheless the application installation should be done with the minimum user intervention. Ideally the user should be prompted for his login in the broker's site and based on this information the security file should be generated and signed; the mobile agent should be packaged and then deployed to the user Over the Air (OTA) like demonstrated in Figure 2 and Figure 3.

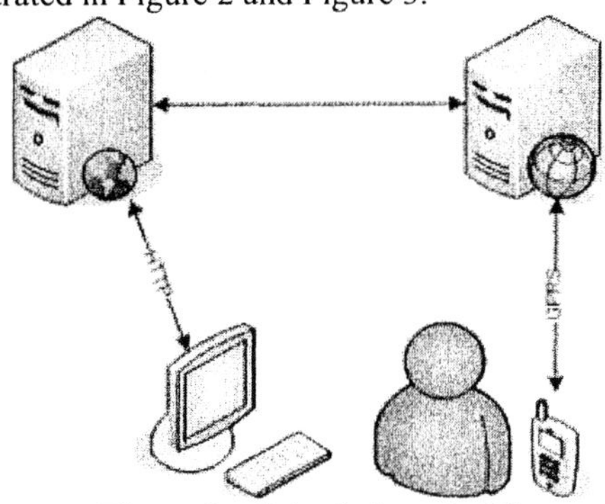

Figure 2 – Submission via PC

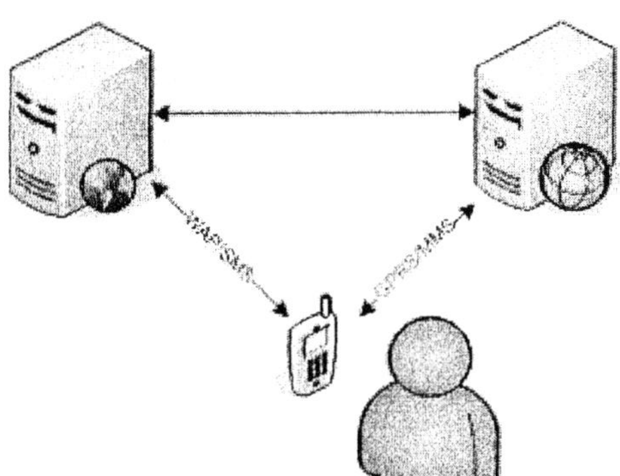

Figure 3 – Submission via Mobile Phone

Over-The-Air deployment, allows the application deployment without user intervention on the mobile phone. Since the user is required to have the mobile agent to interact with the P&D machines, it should first visit the brokers website, login, optionally enter his mobile phone information (all other relevant user information should already be stored in the broker's database) and then receive on the cell phone the mobile agent. Ideally the user should be able to visit the broker's web site in his

WAP enabled phone and start the submission procedure without the need of any other device.

4.4. Test Results

Our current implementation demonstrates the feasibility of the concept and allows users, equipped with NFC enabled phones, to pay their parking. NFC proved to be a good alternative to the Bluetooth implementation solving the setup time problem (Remédios, 2005), reducing the user interaction time with the P&D machines radically.

Using a Nokia 3220 with a NFC shell the total setup and communication time is an average of 600 ms for each transaction. This translates to data transfer rates of around 1 kbps. For the prototype application the observed transfer rates were sufficient despite of according to the NFC specification it can go up to 106kbps.

In a monitoring and control context where we might use the NFC enabled cell phone as an interaction terminal to other devices this transfer rates also seem adequate.

This observation clarifies that NFC has a fast setup time, but a slow data transfer rate. On the other hand Bluetooth has a slow setup time, mainly due to the service discovery protocol taking up to 14 seconds to discover surrounding devices. With the connection established the data transfer rate is much higher than NFC, and with every new version of the BT specification the maximum data transfer rate increases.

Communication with passive equipments showed even slower data transfer rates but its accuracy is not good enough to be quantified.

5. RELATED WORK

The Atlanta pilot launched by *VIVOtech*, *Nokia*, *Phillips*, *Visa* and others demonstrated, in December 2005, how NFC could ease and speedup monetary transactions in a full Hockey Stadium where a test case involving 250 consumers, 150 points-of-sale contactless readers and about 60 smart posters equipped with NFC.

Philips started in 2005 a project to use NFC in the public transport system of Frankfurt, using the mobile phone to access an exiting contactless smart card infrastructure.

6. FUTURE WORK

The next step will be to combine the NFC technology with other already common technologies like Bluetooth. Alone each has disadvantages, but combined together they solve many of the obstacles existing today preventing the realization of the earlier mentioned emergent applications.

Another interesting project would be to implement a mobile application with a consistent look and feel and user interface to use in a manufacturing environment allowing authentication, monitoring and control over different kind of machinery.

To simplify the deployment process on the mobile phone, NFC deployment of the application should be investigated. Although this would require changing the submission process, allowing for the P&D Machines to generate and package the mobile agent application, user experience would be greatly improved.

7. CONCLUSIONS

NFC (Near Field Communication) proved adequate to exchange information at very small distances guaranteeing that the communication is being done only by the interested parties. The location problem now solved with NFC was initially detected when implementing a mobile application to pay services via Bluetooth (Remédios, 2005). Another problem was the time it took for the discovery of neighboring devices. To solve this problem and the NFC limited bandwidth, it should be considered a combined use of NFC with other communication protocols like Bluetooth. Exchanging through NFC only the necessary setup information to establish a broader band communication channel. This setup information for Bluetooth could be the Bluetooth address and encryption keys, etc.

Our experience tells us that in the future, when mobile devices permit it, we will see a symbiotic combined use of several different technologies like NFC, Bluetooth and Wi-Fi to provide the described mobile applications. These usage scenarios refer to authenticity of people and products, interaction with the surrounding environment, token transfer and storage allowing the procurement of digital tickets, and finally the possibility to make service payments.

A big driver leading people to use these applications will be the ease of use of the touch and go philosophy. The added value that we see for the future is the combined use of different technologies to provide in a single mobile application all of the services described before.

8. ACKNOWLEDGMENTS

We acknowledge the valuable contributes of Rogério Rocha, with the base code of the first prototype.

9. REFERENCES

1. Bindel D, Chen Y, Eaton P, Geels D, Gummadi R, Rhea S, Weatherspoon H, Weimer W, Wells C, Zhao B, and Kubiatowicz J. "Oceanstore: An Extremely Wide-Area Storage System". Proceedings of the Nine International Symposium on Architectural Support for Programming Languages and Operating Systems (ASPLOS IX), November 2000.
2. CEN. "Road Transport and Traffic Telematics". CEN/TC 278, 1999.
3. Ecma International. "ECMA-352: NFCIP-2 Interface and protocol". Geneva, 1st Ed. December 2003. http://www.ecma-nternational.org.
4. Ecma International. "ECMA-340: NFCIP-1 Interface and protocol". Geneva, 2nd Ed. December 2004. http://www.ecma-international.org.
5. Ecma International. "Near field communication (NFC)". ECMA/TC32-TG19/2004/28; ECMA/GA/2004/67, June 2004. http://www.ecma-international.org/activities/Communications/2004ga-067.pdf.

6. Forman G, Zahorjan J. "The Challenges of Mobile Computing". IEEE Computer, vol. 27, no. 4, pp. 38-47, April 1994.
7. Forum NFC. "Near Field Communication (NFC) Forum". 2004. http://www.nfc-forum.org.
8. Hightower J, Borriello G. "A Survey and Taxonomy of Location Systems for Ubiquitous Computing, Extended paper from Computer". 34(8), pp.57-66, August 2001.
9. ISO/IEC. "Identification cards - contactless integrated circuit(s) cards - proximity cards" part 1 to 4, .ISO/IEC 14443, 2000.
10. Kanellos, Michael. "RFID tweaked for item-level tracking". CNET News.com, 2006. http://news.com.com/RFID+tweaked+for+item-level+tracking/2100-1008_3-6041718.html.
11. Moore, G. "Cramming more components onto integrated circuits". Electronics 38, 8, 1965. ftp://download.intel.com/research/silicon/moorespaper.pdf.
12. Nokia. "Nokia Says Look for New Phone, Busy Stores in '06 NFC Payment Pilots". December 2005. http://www.digitaltransactions.net/newsstory.cfm?newsid=809.
13. Nokia Corporation. Nokia Mobile RFID Kit and Nokia NFC (Near Field Communication). http://www.nokia.com/rfid and http://www.nokia.com/nfc.
14. Pulli K. "The Rise of Mobile Graphics". Featured article of Nokia Mobile Phones, May 2003. http://www.nokia.com/nokia/0,,53718,00.html
15. Remédios D, Barata M, Osório A. "EXTENDING MOBILE PHONE USAGE FOR NON-CONVENTIONAL APPLICATIONS - WIRELESS IDENTIFIER". Proc International Conference on Electrical Engineering - CEE'05, Coimbra, Portugal, Vol. I, pp. 39, October, 2005.
16. Weiser, M. "Some computer science issues in ubiquitous computing". Communications of the ACM, vol. 36, issue 7, pp. 75-84, July 1993.
17. Yau SS, Karim F, Wang Y, Wang B, Gupta SKS. "Reconfigurable Context-Sensitive Middleware for Pervasive Computing". IEEE Pervasive Computing, IEEE Computer Society Press, pp. 33-40, July-September 2002.

46

INTEGRATED KNOWLEDGE- AND SIMULATION-BASED FACILITY SUPERVISION AND CONTROL

Gerhard Schreck, Alexei Lisounkin, Jörg Krüger
Fraunhofer-Institute for Production Systems and Design Technology
Pascalstraße 8-9, D-10587 Berlin, Germany
e-mail: { gerhard.schreck, alexei.lisounkin, joerg.krueger }@ipk.fraunhofer.de

This paper presents the concept of integrated knowledge-based and simulation-based supervisory control for process facilities. Linked to a common SCADA system, such knowledge-based decision procedure, which includes process simulation, assists the operator and facility engineer in operating the facility. The functional chain, which begins with operator knowledge acquisition and moves to knowledge and facility modeling and finally to their integration into a real SCADA system, have been implemented and tested for a water treatment and supply plant.

1. INTRODUCTION

The increasing complexity of modern process plants and the demand for energy conservation, product quality, environmental protection, safety, and reliability call for new approaches to process automation. Along with decision-making and control procedures based on mathematical models and optimization solvers, knowledge-based systems involving the experience of human operators are a promising approach.

The aim of the integrated knowledge-based and simulation-based method is to combine available formal information and process data from the facility, e.g., information on plant models and historical operation profiles, with human operation knowledge. This calls for experience-based evaluation and active assessment of operation artifacts through an operator team and further use of this information in a regular manner.

This paper is devoted to advanced facility supervision and control concepts that simultaneously embrace simulation- and knowledge-based components. Focus is given to the decision-making processes of human operators and the respective modeling of knowledge objects for automatic processing. The aspects of system implementation and the example of their application presented are related to a real industrial environment.

Please use the following format when citing this chapter:

Schreck, G., Lisounkin, A., Krüger, J., 2006, in IFIP International Federation for Information Processing, Volume 220, Information Technology for Balanced Manufacturing Systems, ed. Shen, W., (Boston: Springer), pp. 435–442.

2. ADVANCED FACILITY SUPERVISION AND CONTROL

2.1 Control Levels and Degree of Automation

The presented work focuses on the application of simulation- and knowledge-based systems in a process industry. Figure 1 shows the typical control and supervision levels in this field. SCADA systems (Supervisory Control and Data Acquisition) are usually introduced for high-level process control and automation. They play an important integrative function in distributed, multilevel control environments and provide the appropriate operator panels (HMI). Typical functionalities include the monitoring of process states, the recording of alarms and events, the activation of control actions, emergency shutdown, etc.

At control levels below SCADA, the favored system modularity, functional encapsulation, and real time requirements are considered and implemented. Furthermore, the aggregation of data on the way from process-level to high-level control is a main aspect of the multilevel approach (see Figure 1). Therefore, appropriate abstraction levels must be engineered to provide suitable views on plant conditions and process states for supervisory control and decision-making by human operators.

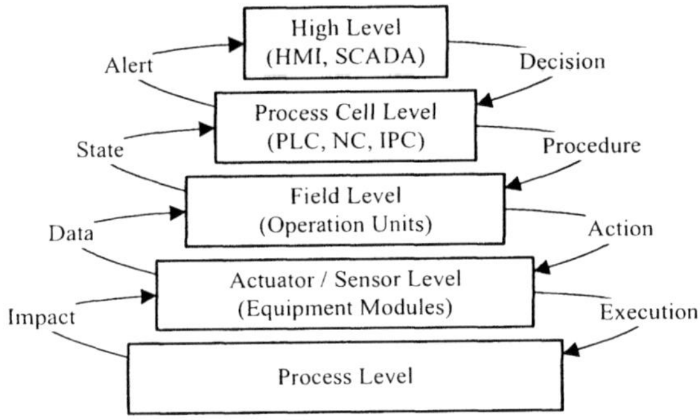

Figure 1 – Typical control and supervision levels in a process industry

2.2 Operator Integration

Even with a high level of automation of process supervision, diagnostics, and control, the facility operator's role is continually expanding. Although local automation tasks are covered by an installed control system, the facility operation staff must take precautions with high-level functional, technological, strategic objectives.

The human operator team is responsible for the high-level process management which includes tasks such as

- Assessment of process situations
- Selection of operating points
- Consideration of different modes of working and use of resources
- Reaction to changing requirements/demands
- Ensuring a continuous and smooth running of the system.

The decisions of operators are based on their knowledge of process situations, process trends, and process control. Here past gained experience, i.e., historical data and knowledge of situations, plays an important role in allowing a high level of performance to be reached (Lisounkin, 2004).

In our concept of an advanced supervisory and control system, the provision of support functionalities to the operator team plays an important role. It includes not only tools for decision support, e.g., through integrated simulation functions, but also methods and tools for arranging the automatic execution of favored process patterns. Here the need for optimization of system operation and the demand for a high level supervisory control should be met without compromising the benefits of the flexibility and knowledge of human operators.

2.3 Requirements on Simulation and Knowledge Processing

Specific technical realization, technological and administrative requirements, operative performance criteria, as well as long-term planning result in unique supervisory and control schemes for each facility to be controlled. Usually, the design, development, and maintenance of such supervisory and control schemes employ a significant amount of human resources for each application. Reasonable methods for modeling a corresponding knowledge base and techniques for its application and maintenance will lead to a high acceptance of such systems and efficiency in use.

Considering the acceptance by operators as a key aspect, an extensive use case analysis must be foreseen even at the very beginning of the development. Therefore, the system must enable different operation modes, such as manual mode, decision support mode, and automatic execution mode. Furthermore, automated functions must provide feedback on decisions in order to establish confidence and also to support the tuning of system behavior.

From the system integrator's point of view, two classes of requirements can be identified. On the one hand, there are clear technical requirements concerning functional integration concepts and interfaces with different SCADA systems. On the other hand, requirements concerning the acquisition and engineering of knowledge, the re-use of models and knowledge components, maintenance, and lifecycle support become of increasing importance.

3. OPERATION KNOWLEDGE MODELLING

3.1 Decision Making

The analysis and modeling of role-dependent staff responsibilities and activities in the context of process facilities and power stations have been studied by Rasmussen (Rasmussen, 1986). His findings continue to influence the development of decision support systems for technical facilities.

Within the facility operation context, the decision-making procedure consists of the following steps. First, the decision maker must detect the need for intervention by observing actual process data. For this, the operator has features at his disposition – data evaluation functions and criteria – which give him an estimate of the optimality and regularity of the system's operation or of a malfunction in the system. In regular cases, this information can give the direction for subsequent activities. Based on this evaluation, a target state into which the system should be put will be chosen, and the task that the decision maker should perform will be selected from a review of available resources. Once such a task is identified, the proper execution can be planned and carried out.

If the connection between the data evaluation criteria and the sequence of actions to reach the given goals is clear, the supervision procedure is considered "skill-based". In cases where decision-making involves the evaluation of alternatives, the supervision procedure is considered a "rule-based" behavior. A decision-making procedure which involves the comparison of models and analysis of goals and depends on the know-how of the operator is considered "knowledge-based". These three levels of the decision-making procedure are depicted in Figure 2.

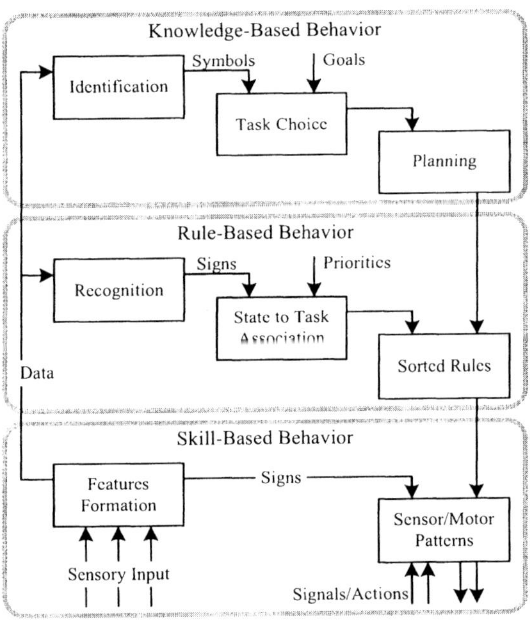

Figure 2 – Decision-making behavior model (according to Rasmussen)

3.1 Knowledge Levels

In integrated knowledge- and simulation-based supervision and control we consider the following three knowledge levels (see Figure 3):

- Knowledge based on process history and process states
- Knowledge based on process formal modeling
- Knowledge based on process operation experience.

Based on historical process data available from a SCADA database, typical process situations and patterns can be elaborated using data mining analysis and classification tools (Lisounkin, 2004).

Mathematical analysis and the modeling of process behavior result in numerical simulation tools which are applied extensively in assistance, support, and training tools (Schreck, 2002).

Acquisition and adequate use of operator experience is a rather difficult topic. It is the focus of much current research as well as the subject of our research activities. A rule-based approach to the maintenance and processing of operator knowledge is a promising method (Krüger, 2005).

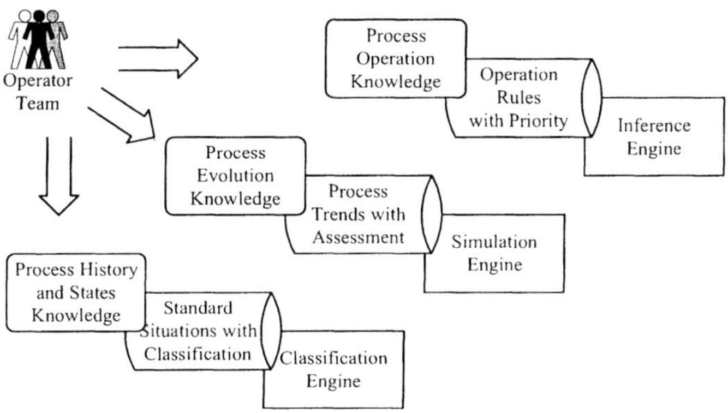

Figure 3 – Structuring of operation knowledge and related tools

3.2 Knowledge Objects

The knowledge objects include the following: resource, state, condition, rule, action, setpoint. "Resource" objects represent atomic processes or groups of processes on the corresponding hierarchical level. The entire facility is also an object of the type "resource". Objects of this type are the source of information and the operation field for the decision-making procedure. For the decision-making procedure, the principle components of the resource objects are "state" objects, which collect measurable signals and indicators for the evaluation of facility states and provide an informational starting point for the decision-making process. For the evaluation of process states, objects of the type "condition" are specified. These objects represent data processing features - operators from a predefined set with "Boolean"-type result values. The arguments for these operators are the state variables of resource objects. The supervised system allows impacts which influence the future system evolution. The set of impacts correspond to the possibilities of the system operator to change the system behavior in order to achieve wished results. These impacts are modeled by objects of the type "action". The action objects include objects of the type "setpoint", where the reference to a corresponding resource object as well as the type of control impact and its parameters are encoded. The objects of the type "rule" link objects of the type "condition" with objects of the type "action". If, in the case of some system states, a condition fails to be met, a corresponding object of the type "rule" provides an automatic reaction. The association of system states to control tasks is defined via the rule objects. In cases where alternative actions can be identified, the rule objects maintain priority over the action objects. A more detailed description of the knowledge objects by means of XML Schema is given in (Krüger, 2005).

4. SYSTEM IMPLEMENTATION

4.1 System Architecture

Several components must be added to the infrastructure of a process control system in order to provide the operator with new knowledge- and simulation-based supervision methods and facilities. This includes components for knowledge modeling, simulation, and visualization. The components can be grouped according to the phase in which they are needed: the development phase or the runtime phase. In this paper we focus on the runtime phase. Its architecture is shown in Figure 4.

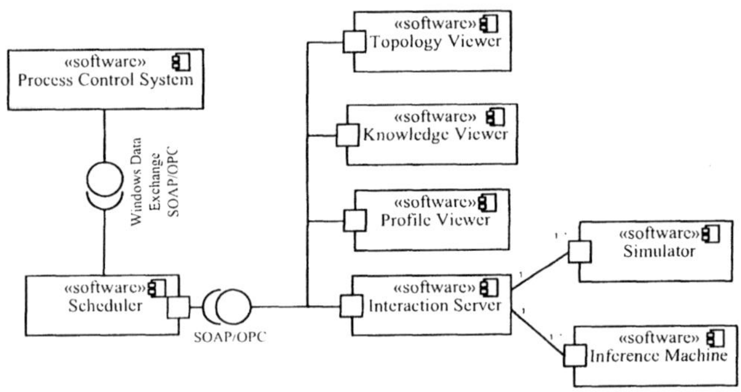

Figure 4 – Runtime components of the knowledge-based supervision system

The "scheduler" is the data and task control unit of the system connecting the existing process control system (SCADA) with the knowledge-based supervision. Typical tasks are the initiation, monitoring, timing and data exchange of components. In addition to process information performed by SCADA, the "topology viewer" provides processing and visualization of process states relevant for knowledge-based supervision, e.g., intermediate results of the calculations. It also reads in the simulation model and visualizes the dynamic behavior of the process. The "knowledge viewer" visualizes the selected and applied rules of the supervision system. This allows the operator to retrace the actions and the reasons they were executed. The "profile viewer" gives the operator an overview of the relevant operation history and actual context. The currently selected as well as all available operation trends are drawn as a diagram on the screen. The "simulator" forecasts the process values for the supervision components. Based on the current process values, the output values for a defined time frame are computed. This allows the "inference machine" to take preventive actions based on the knowledge provided. The current process states are checked constantly for rules violated. The "interaction server" allows the dynamic connection of multiple simulators and inference machines to the system architecture. Simulation and knowledge models can be split into subsystems but still appear as a single system to the scheduler through an encapsulation of the coordination of the subsystems.

The presented architecture has been prototypically implemented and tested within an industrial environment. More detail concerning interface generation and task scheduling is given in (Schmidt, 2005).

4.2 Application Example

A system application and test of the implemented components was performed at a water supply and distribution facility equipped with an advanced SCADA system. The main objective of the facility supervision and control was to ensure a proper operation of the system. For this, the following criteria had to be fulfilled:

- Allocation of the required quantity of pure water
- Consideration of the different modes of working and use of resources
- Compliance with requirements for the quality of water
- Adherence to a minimal consumption of energy
- Maintenance of the purification cycle of individual water filters
- A continuous and smooth running of the system
- Adherence to the prescribed limits of the tanks with a quick reaction to deviations.

Figure 5 shows the user interface for simulation-based operator support and the main topology of the process plant. Real plant operation profiles (historical data) spanning two years as well as knowledge of system operation have been gathered for the implementation of the knowledge-based supervision and control system.

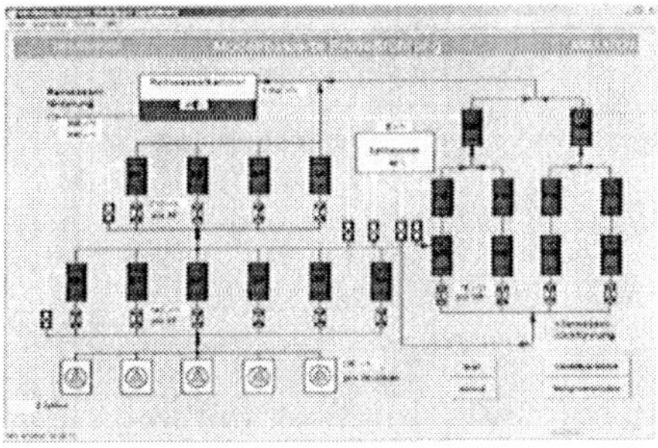

Figure 5 – Example of water facility plant operator interface (SCADA GUI)

Here available simulation models of the facility have been extended, and the required knowledge objects have been implemented. A SCADA module for automatic plant control was used as a reference for specification and testing of operation rules. The challenge was to show the applicability and flexibility of the proposed approach with respect to the integration and adaptation of the operation knowledge available to facility engineering and operation staff.

The set of operation rules includes about 60 items. By means of these rules, the number of possible facility states was reduced from about 10,000 to about 100. During the operation, the inference machine had to take its decision within this pool of 100 allowed states.

The interaction of inference machine and simulator and the functional behaviour were tested, and the tuning to the expected operation profile was able to be shown. Furthermore, the overall functionality with respect to the interaction of all functional components and the interface to the process control system was successfully tested.

5. CONCLUSIONS AND OUTLOOK

The developed concept of integrated knowledge- and simulation-based facility supervision and control was prototypically implemented and tested within a SCADA environment for water supply stations. The expected functionality and interaction of the runtime components in response to the process control system were approved. Further research activity will be devoted to the respective modeling and engineering tools. Additionally, the relevance of this concept to other applications will be investigated: in particular, in regard to gas distribution networks and storage facilities.

6. ACKNOWLEDGMENTS

The study presented was conducted within the scope of the R&D project "Akquisition, Management und Integration von Prozesswissen in eine modellbasierte Prozessführung als Repräsentant einer neuen Generation von wissensbasierten Systemen in der Leittechnik", partially funded by the Federal Ministry of Education and Research, Germany (AMARYL, Reference No. 01INB02B). We thank our industrial partner Elpro Prozessindustrie und Energieanlagen GmbH, Berlin, for numerous contacts with companies in the water industry.

7. REFERENCES

1. Lisounkin A, Schreck G, Schmidt H-W. "Knowledge Acquisition from Historical Data for Case Oriented Supervisory Control.". In Emerging Solutions for Future Manufacturing Systems. Edited by Luis M. Camarinha-Matos, Springer, 2004, pp. 499 – 506.
2. Rasmussen, J. "Information Processing and Human-Machine Interaction". Elsevier Science Publishing Co. Inc., New York, 1986.
3. Schreck G. "Simulation Services for Training of Plant Operators". In Knowledge and Technology Integration in Production and Services: Balancing Knowledge and Technology in Product and Service Life Cycle. Edited by Vladimir Marik, Luis M. Camarinha-Matos and Hamideh Afsarmanesh, Kluwer Academic Publisher, 2002, pp. 79 – 86.
4. Krüger J, Lisounkin A, Sabov A, Schreck G, Pocher M. "Knowledge Modeling and Processing for Supervision of Process Facilities. In Proceedings of the Industrial Simulation Conference 2005 (ISC-2005, Jun.9-11). Berlin, Germany, pp. 435-439.
5. Schmidt H-W, Saffert U, Schreck G, Sabov A, Pleßow M. "Architecture for Simulation & Knowledge Based Supervision for Process Facilities. In Proceedings of the Industrial Simulation Conference 2005 (ISC-2005, Jun.9-11). Berlin, Germany, pp. 353-357.

47

ON STABILITY ANALYSIS OF BEARING-ROTOR SYSTEM

Meng-su Li, Xu-chu Cai
Donghua University, Shanghai 200051, China, Email: lims@opg.cn
Dan Zhang
University of Ontario Institute of Technology, Canada, Email: Dan.Zhang@uoit.ca

The rotor system of the small power unit used in ball pen tip production machine, is discussed in detail. Meanwhile, nonlinear model of the rotor system is established and the related research method is introduced. Furthermore, the stability bifurcation of nonlinear periodic solution is obtained preliminarily. Characteristic comparisons between pseudo-periodic solution and chaotic one are put forward in the paper, which will be of benefit to further engineering applications.

1. INTRODUCTION

The production quality of ball pen tip is directly influenced by the rotating precision of the small power shaft system of its producing equipment. How to improve its rotating precision is therefore a key problem in the pen manufacturing industry. However, due to the fact that the rotating shaft system is unstable during production, the perturbed rotor will deviate from its normal motion orbit more and more as time goes on so that there is no favorable and satisfactory method for the stability analysis of the rotor. According to the linear theory, the amplitude of the perturbed rotor center will increase indefinitely with time, but the actual fact is that the amplitude will be held after reaching a certain limited value, and then the rotor center will more in a closed orbit (as shown in Fig. 1). Obviously, this is a kind of common nonlinear phenomenon in bearing-rotor system.

This paper is attempting to analyze the stability of the restrained rotor by nonlinear method, and provide a better theoretical foundation for the design and manufacture of multi-station machine tool with small power unit for ball pen tip production.

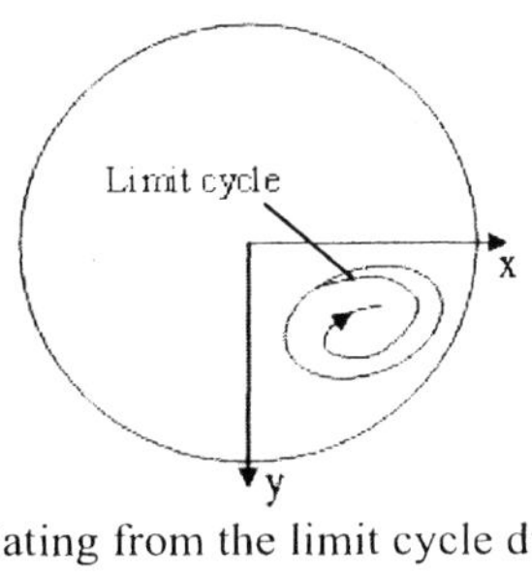

Fig. 1 Equilibrium point deviating from the limit cycle due to destabilization

Please use the following format when citing this chapter:

Li, M., Cai, X., Zhang, D., 2006, in IFIP International Federation for Information Processing, Volume 220, Information Technology for Balanced Manufacturing Systems, ed. Shen, W., (Boston: Springer), pp. 443–452.

2. NONLINEAR FORMULATION OF ROTOR SYSTEM

Bearing-rotor system, a nonlinear dynamic system, can be described by a set of finite-dimensional, second-order and ordinary differential equations with multi-parameters. Since the rotating frequency ω is the main influencing parameter, so the system that can be described only as sets of finite-dimensional, second-order and ordinary differential equations with single-parameter, is studied here. The typical form of its formulations is

$$M\ddot{q} + f_{\text{in}}(\dot{q}, q, t, \omega) = f_{\text{ex}}(t, \omega) \quad (t, q) \in \left(R \times R^{m}\right) \tag{1}$$

where M is the mass matrix; t is the time ($t \geqslant 0$); frequency ω is a parameter on the real number axis in the system; the m-dimensional vector $q(t)$ is unknown; $f_{\text{in}}(\dot{q}, q, t, \omega)$ is the internal force vector which results from rigidity of the uncoupled axis, force of the bearing film, etc. in the system; $f_{\text{ex}}(t, \omega)$ represents the external excitation force vector acting on the system.

Introducing the state variable, $u = \begin{pmatrix} q \\ \dot{q} \end{pmatrix}$, then we can obtain the expression of the system equation in state space,

$$\frac{\mathrm{d}u}{\mathrm{d}t} = F(t, \omega, u), \quad (t, u) \in \left(R \times R^{m}\right), \tag{2}$$

where

$$F(t, \omega, u) = \begin{bmatrix} \dot{q} \\ M^{-1}\left\{-f_{\text{in}}(\dot{q}, q, t, \omega) + f_{\text{ex}}(t, \omega)\right\} \end{bmatrix}.$$

If there is no time variable t in F, and satisfying $F(t, \omega, u) \overset{\text{def}}{=\!=} F(\omega, u) \neq 0$, then Eq. (2) can be simplified as follows,

$$\frac{\mathrm{d}u}{\mathrm{d}t} = F(\omega, u), \quad (t, u) \in (R \times R^{2m}) \tag{3}$$

Now, the nonlinear bearing-rotor dynamic system, corresponding to Eq. (3), will then be autonomous.

If F is a periodic function of time t, namely satisfying $F(t, \omega, u) = F(t + T, \omega, u)$, and $F(t, \omega, 0) \neq 0$, then Eq. (2) can be simplified as follows,

$$\frac{\mathrm{d}u}{\mathrm{d}t} = F(t, \omega, u) = F(t + T, \omega, u), \quad (t, u) \in (R \times R^{2m}) \tag{4}$$

The nonlinear bearing-rotor dynamic system is then nonautonomous.

3. STABILITY ANALYSIS OF THE NONLINEAR SYSTEM

3.1 Mapping

The physical model of bearing-rotor system can be expressed in the forms of Eqs. (1), (2). It is therefore of great significance to study the problems of stability bifurcations of steady-state solutions for this type of system.

If the problems, such as the stability, bifurcations and regions of attraction of the steady-state periodic solutions, are studied directly through the nonlinear dynamic system Eq. (2), it will be quite difficult to obtain the analytic expression of the periodic solution $u(t)$ in phase space, while the using of closed orbit comprised of infinite number of points will also bring comparative difficulties to numerical analysis. Therefore, it is necessary to find a much better way to express $u(t)$. As the nonlinear mapping expression, based on Eq. (1) of the nonlinear dynamic system and its manifold φ_t, can be given as follows,

$$u^{(n+1)} = G(u^{(n)}) \quad \text{or} \quad u \rightarrow G(u) \tag{5}$$

where n=2m, $G=\varphi_t$ is a vector field, and we here are concerned only about the characteristics of the steady-state periodic solutions $u(t) = u(t+T)$, so a point mapping system, converted from the nonlinear dynamic system, will thus be put forward for more convenient studied as follows.

For the autonomous system Eq. (3) deduced from the Eq. (2), i. e.,

$$\varphi_t: \quad \frac{\mathrm{d}u}{\mathrm{d}t} = F(\omega, u), \quad (t, u) \in (R \times R^n),$$

assume that $u(t)$ is a periodic orbit of the manifold field φ_t defined in R^n, namely $u(t)$ is a periodic solution of the above equation. Taking a global cross sectional hypersurface $\sum \subset R^n$ with n-1 dimensions in the n-dimensional state space, this hypersurface $\sum$ doesn't have to be a hyperplane but it should intersect the manifold field φ_t everywhere.

Thus, the number of intersections, formed by the periodic solution $u(t)$ and $\sum$ from the same direction ($u(t)$ goes across $\sum$ on its same side and forms the intersections) is finite. If the number of intersections is K, then $u(t)$ is the KT periodic solution of the system.

Therefore, the mapping P of point $q \subseteq \sum$ can be defined as:

$$P(q) = \varphi_\tau(q) \tag{6}$$

where $\tau = \tau(q)$ is the time used by manifolds $\varphi_{t_0}(q)$ when first return to $\sum$ in the same direction with q as the starting point. Note that some other manifolds $\varphi_{t_0}(A)$, with certain point A on $\sum$ as their corresponding starting point may never return to $\sum$ in the same direction, which represents the corresponding orbits of the divergent transient state. These points and their corresponding orbits in transient state will be disregarded and defined only as puzzled points on $\sum$ without further study anymore.

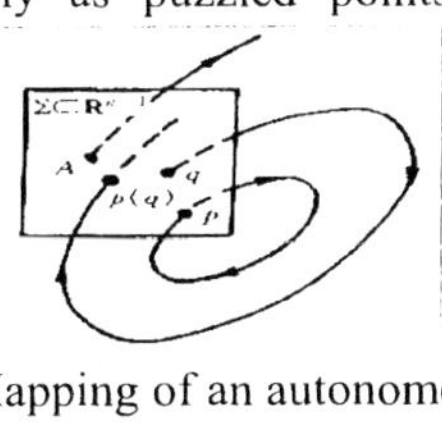

Fig. 2 Mapping of an autonomous system

For the autonomous system Eq. (4) as deduced from Eq. (2),

$$\varphi_t: \quad \frac{\mathrm{d}u}{\mathrm{d}t} = F(t,\omega,u) = F(t+T,\omega,u) \qquad (t,u)\in(R\times R^n),$$

if now taking the time t as a state variable, the above equation can then be converted into the form of autonomous system Eq. (3) by adding one more dimension,

$$\begin{cases} \dfrac{\mathrm{d}u}{\mathrm{d}t} = F(\theta,\omega,u) \\ \dfrac{\mathrm{d}\theta}{\mathrm{d}t} = 1 \qquad (t,\theta,u)\in(R\times S^1\times R^n) \end{cases} \tag{7}$$

$S^1\times R^n$ is the form of manifolds in phase space, and the circle component $S=R(\bmod T)$ shows the periodic dependence of vector field F on the variable θ. Then a global cross section can be defined as follows,

$$\Sigma = \left\{ (\theta,u)\in S^1\times R^n \Big|_{\theta=\theta_0} \right\} \tag{8}$$

Now, all solutions (θ,u) will intersect with Σ. The mapping P: $\Sigma\to\Sigma$ of point $q\in\Sigma$ can therefore be defined as:

$$P(q) = u(\theta_0+T,q) \tag{9}$$

where $u(t,q)$ is a solution of a nonautonomous system, (as shown in Fig. 3) with $u(\theta_0,q)=u_0$ as its starting point. Note that it differs from the autonomous system, and there will be no more puzzled point A on the cross sectional surface Σ now.

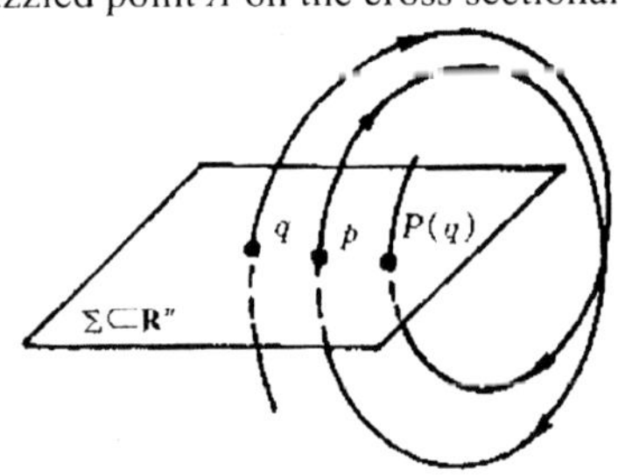

Fig. 3 Mapping of a nonautonomous system

Although the methods adopted to set up a point-mapping system, are somehow different between autonomous system Eq. (3) and nonautonomous system Eq. (4), yet a completely same point-mapping system in form can be obtained finally as follows, without any distinction

$$u^{(k+1)} = P(u^{(k)}) \qquad k\in Z \text{ or } u\to P(u) \tag{10}$$

It is obvious that the intersections p_1, p_2, Λ, p_k, formed by KT periodic solutions $u(t)=u(t+kT)$, $k\in Z$ with the cross section supersurface Σ, are all the P-K periodic points of mapping P, according to the definition mentioned above.

$$P^k(p_i) = p_i, \qquad i=1,2,\Lambda,k \tag{11}$$

The domain of attraction of these P-K periodic points $p_i (i = 1,2,\Lambda ,k)$ on the supersurface $\sum$, is just that part of the attraction domain of the solutions $u(t) = u(t + kT)$ in the state space intersected by the supersurface $\sum$, whose stability with respect to the mappings P reflects the stability of the solutions $u(t) = u(t + kT)$ with respect to the manifold φ_t.

Only the points on the supersurface $\sum$ and their mapping rules are concerned in the mapping system established on the dynamic system, while the characteristics of all other points on the solution orbits in the entire global state space are not concerned. The study of the problem concerned is thus much simplified, which is advantageous for the further application of numerical methods.

3.2 Stability analysis

It can be seen from the above discussion that a discrete dynamic system can be transformed from the continuous dynamic system by means of mapping.

$$u^{(k+1)} = P(u^{(k)}) \quad k \in Z \text{ or } u \to P(u) \tag{12}$$

Consequently, the analysis of continuous dynamic system of Eq. (2) can be obtained through analyzing the discrete dynamic system of Eq. (12).

Assuming there is a stable solution $u^e(t)$ of the dynamic system Eq. (3), then $u^e(t)$ will be unstable if the initial perturbations $u^e(t)$ increase with time; or $u^e(t)$ is stable during perturbations if it is gradually damped. The stability theory related to the infinitely small perturbations is actually a linear theory, since the high-order terms can be ignored in comparison with the linear terms in a perturbation equation. Assuming that $u^e(t,\omega)$ is a stable solution of dynamic system Eq. (2), and δv is a randomly given constant perturbation of u, then we shall obtain:

$$\delta \frac{\mathrm{d}v}{\mathrm{d}t} = F(t,\omega,u+\delta v) - F(t,\omega,u) \tag{13}$$

When $\delta \to 0$, the perturbation tends to be infinitely small, then we can get

$$\frac{\mathrm{d}v}{\mathrm{d}t} = \lim_{\delta\to 0} \frac{F(t,\omega,u^e+\delta v) - F(t,\omega,u^e)}{\delta} \overset{\text{def}}{=\!=\!=} F_u'(t,\omega,u^e|v) = A(t,\omega,u^e)\bullet v \tag{14}$$

where $A(t,\omega,u^e) = F_u'(t,\omega,u^e|\bullet)$, is a linear operator of the variable v after the vertical line. It is an $n\times n$ matrix function for the n-dimensional problems. For nonlinear autonomous system Eq. (3), Eq. (14) can be simplified as,

$$\frac{\mathrm{d}v}{\mathrm{d}t} = F_u'(\omega,u^e|v) = A(\omega,u^e)\bullet v \tag{15}$$

For the fixed point $u*$ of the discrete dynamic system in Eq. (12), we obtain,

$$u*+\delta v_{k+1} = P(u^{(k)}+\delta v_k) \tag{16}$$

where $u* = u^e(t_k,\omega) = u^e(t_k+T,\omega)$

$$v_k = v(t_k);\ v_{k+1} = v(t_k+T)$$

We can therefore obtain the following results with respect to the perturbation when $\delta \to 0$:

$$u*+\delta v_{k+1} = P(u*) + P_u' \bullet \delta v_k$$

so $$v_{k+1} = P'_u \bullet v_k \qquad k \in Z \tag{17}$$

P'_u, the linear form of the fixed point u^* of the mapping, is a constant matrix here.

It can be proven that Eqs. (14), (17), completely equivalent to each other, are perturbation equations in continuous and discrete forms respectively. The stability of the stable solution $u^e(t,\omega)$ in the system can be evaluated through discussing the perturbation equations.

For the autonomous system Eq. (3), when the outer parameter ω is a certain value of ω_m, these following condition can be satisfied,

$$F(\omega_m, u^s) = 0 \qquad u^s \in R^n \tag{18}$$

where u^s is the equilibrium point of the system, so it is considered that the stability of the equilibrium point solutions can be judged by the eigenvalues of the matrix $A(\omega_m, u^s) = F'_u(\omega_m, u^s|\bullet)$ which is obtained when the linear system is at u^s. Here the matrix is an $n\times n$ constant one.

$$\frac{\mathrm{d}v}{\mathrm{d}t} = F'_u(\omega, u^s|v) = A(\omega, u^s) \bullet v \tag{19}$$

The loss of stability of system Eq. (3), as a result from the destabilization of the equilibrium point with the changing of outer parameter ω, will cause bifurcation phenomenon with multifarious forms. Since destabilization of the equilibrium point solutions is one of the main forms of bifurcation in bearing-rotor system, so we study only this form of bifurcation of equilibrium points here. The distinguishing criteria are as follows.

Assuming $\omega = \omega_c$, the equilibrium point solution u^s of the autonomous system Eq. (3) satisfies that:

(1) $F(\omega, u)$ is differentiable in the domains of $(\omega_c, u^s(\omega_c))$;

(2) There is a pair of pure imaginary nonzero eigenvalues $\pm \mathrm{i}\beta_0 (\beta_0 > 0)$ in the matrix $F'_u(\omega_c, u^s(\omega_c)|\bullet)$ obtained when $F(\omega, u)$ is at the stable solution $u^s(\omega_c)$, while the other n-2 eigenvalues all have negative real parts.

(3) $\frac{\mathrm{d}\alpha}{\mathrm{d}\omega}\Big|_{\omega=\omega_c} \neq 0$, where $\alpha(\omega) \pm \mathrm{i}\beta(\omega)$ is the continuous unfolding eigenvalues obtained by $F'_u|_{\omega=\omega_c}$ with respect to $\pm \mathrm{i}\beta_0$. In this case, the equilibrium point solutions are destabilized. A non-constant periodic solution u^H, corresponding to the self-exited limit cycle of the system, will be bifurcated from steady-state equilibrium point solutions of $u^s(\omega_c)$ at $(\omega, u(\omega)) = (\omega_c, u^s(\omega_c))$ in the autonomous system.

According to the different periodic solutions due to bifurcation, it can be further classified into two different cases: supercritical condition and bifurcation, to be described as follows,

(1) If $\omega > \omega_c$, a stable periodic solution u^H is bifurcated from the equilibrium point solution $u^s(\omega_c)$ of the system. The periodic solution $u^H \to u^s(\omega_c)$ is a supercritical bifurcation obtained when $u^H \to u^s(\omega_c)$. Its bifurcation is characterized by the gradually varying periodic solution with the changing of ω, and there is no phenomenon of "jump and decay" in this system, as shown in Fig. 4(a).

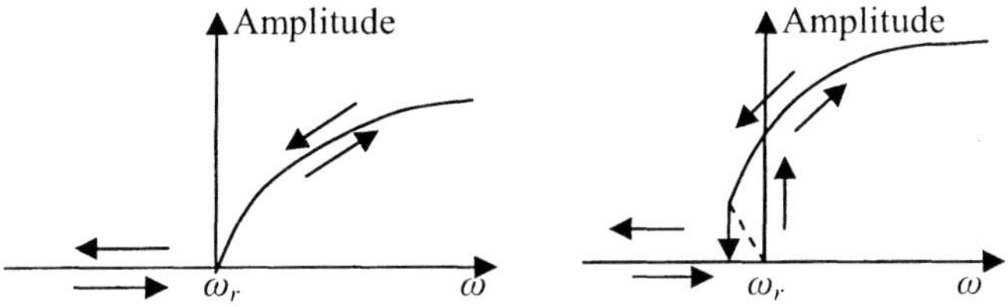

(a) Supercritical bifurcation (b) Subcritical bifurcation
Fig. 4 Bifurcations on equilibrium point solutions

(2) If $\omega < \omega_c$, an unstable periodic solution u^H (with $u^H \to u^s(\omega_c)$ when $\omega \to \omega_c$) is bifurcated from the equilibrium solution $u^s(\omega_c)$ in the system. Its bifurcation is characterized by the abrupt change of periodic solution with the changing of ω, with the phenomenon of "jump and decay" because there often exists a stable periodic solution besides the unstable periodic ones, as shown in Fig. 4(b).

The matrices in the corresponding perturbation Eq. (14) are constant ones, according to the above results. Therefore the perturbation of the stable equilibrium point solutions in the autonomous system Eq. (3) can be directly confirmed by analyzing the Eq. (14).

3.3 Stable perturbation of the nonlinear steady-state periodic solutions

For a given outer parameter ω and its corresponding steady-state periodic solution u_p with period *T*, (i. e., $u_p = u_p(t+T)$), its perturbation equation is as follows,

$$\frac{\mathrm{d}v}{\mathrm{d}t} = F_u'(t,\omega,u_p|v) = A(t)\cdot v, \tag{20}$$

where $A(t) = A(t+T)$ is a matrix function with period *T*. The perturbation equation related both to steady-state periodic solutions in an autonomous system and to a periodic nonautonomous one can be represented by Eq. (20)

If $V(t)$ is a fundamental solution matrix of Eq. (20), there must exist a nonsingular matrix $\Phi(t) = \Phi(t+T)$ with period *T* and a constant matrix *D*, so that

$$V(t) = \Phi(t)\cdot e^{tD} \tag{21}$$

Meanwhile, according to the periodic property of $A(t)$ in Eq. (20), we should have

$$\frac{\mathrm{d}V(t+T)}{\mathrm{d}t} = A(t+T)\cdot V(t+T) = A(t)\cdot V(t+T) \tag{22}$$

consequently, $V(t+T)$ is also the fundamental solution matrix of Eq. (20). From Eq. (20) we can have

$$V(t+T)=\Phi(t+T)\cdot e^{(t+D)D}=\Phi(t)\cdot e^{tD}\cdot e^{TD}=V(t)\cdot e^{TD} \tag{23}$$

The above equation can also be simplified as

$$V(t+T)=V(t)\cdot C \tag{24}$$

where $C=e^{TD}$ is a constant matrix.

Take any two fundamental solution matrices $V_1(t)$ and $V_2(t)$ in Eq. (20), then the constant matrices C_1 and C_2 can be obtained correspondingly, so that

$$\begin{cases} V_1(t+T)=V_1(t)\cdot C_1 \\ V_2(t+T)=V_2(t)\cdot C_2 \end{cases} \tag{25}$$

Meanwhile, there must exist nonsingular constant matrix S, which satisfies:

$$V_2(t)=V_1(t)\cdot S \tag{26}$$

Therefore

$$V_2(t+T)=V_1(t+T)\cdot S=V_1(t)\cdot C_1S=V_2(t)\cdot S^{-1}C_1S \tag{27}$$

That is:

$$C_2=S^{-1}C_1S \tag{28}$$

Consequently, C as well as D, is a group of similar constant matrices. Namely their eigenvalues are independent of the choice of given initial conditions and fundamental solution matrices, but are determined only by $A(t)$ in Eq. (20). C and D, for discrete and for continuous conditions respectively, are the transfer matrices of Eq. (20). Their eigenvalues λ and δ are respectively the multiplier and the exponent index. From Eqs. (21), (22) we can see that they will determine the stability of Eq. (20) with respect to the origin.

From the relations between C and D, we can obtain

$$\lambda=e^{\sigma T}=e^{\mathrm{Re}(\sigma T)}\bullet[\cos(\mathrm{Im}(\sigma T))+\mathrm{i}\bullet\sin(\mathrm{Im}(\sigma T))] \tag{29}$$

and

$$\begin{cases} \mathrm{Re}(\sigma)=\dfrac{1}{T}\bullet\ln|\lambda| \\ \mathrm{Im}(\sigma)=\dfrac{1}{T}\bullet\arctan\left[\dfrac{\mathrm{Im}(\lambda)}{\mathrm{Re}(\lambda)}\right] \end{cases} \tag{30}$$

Note that σ has multiple values, and should be taken in the region $0\leqslant\mathrm{Im}(\sigma)\leqslant\dfrac{2\pi}{T}$ for practical applications.

Furthermore, the stability of any solutions of Eq. (20) with respect to the origin can be judged by λ and σ. The stability criteria and bifurcation conditions of the steady-state periodic solutions can also be obtained then in the dynamic system of Eq. (2).

Assume that the basic solution matrix $V(t)$ is a unit matrix at the initial time t_0, that is

$$V(t_0)=I \tag{31}$$

so $$V(t_0+T)=V(t_0)\bullet C=C \qquad (32)$$

Assume that $v(t)$ is a perturbation at the time t, and $v(t+T)$ is another perturbation after a period of T, then:

$$v(t_0)=V(t_0)\bullet\varphi_i=\varphi_i \qquad \varphi_i\in R^n \qquad (33)$$

$$v(t_0+T)=V(t_0+T)\bullet\varphi_i=V(t_0)\bullet C\bullet\varphi_i=C\bullet\varphi_i \qquad \varphi_i\in R^n \qquad (34)$$

so

$$S_i=\frac{\|v(t_0+T)\|}{\|v(t_0)\|}\leqslant\|C\| \qquad (35)$$

Assume S_{max} is the maximum of S_i, and thus

$$S_{max}=\|C\|=\max_{i=1\Lambda,n}(|\lambda_i|) \qquad (36)$$

Then we can get the following conclusions:

(1) The sufficient condition for the stability of the periodic solution u_p in the dynamic system of Eq. (2), is that its maximum multiplier modulus is less than 1, namely all multipliers are within the unit circle on complex plane.

(2) u_p is unstable when the maximum multiplier modulus is greater than 1.

(3) u_p is critical steady state when the maximum multiplier modulus is equal to 1.

Meanwhile, similar to the equilibrium point solution, there may be destabilization with multifarious bifurcation in the periodic solutions of the dynamic system of Eq. (2). Further study shows that the forms of the destabilization and bifurcation of the periodic solution u_p, according to different positions where the maximum multipliers of the modulus pass through the unit circle, can be classified into three kinds as follows:

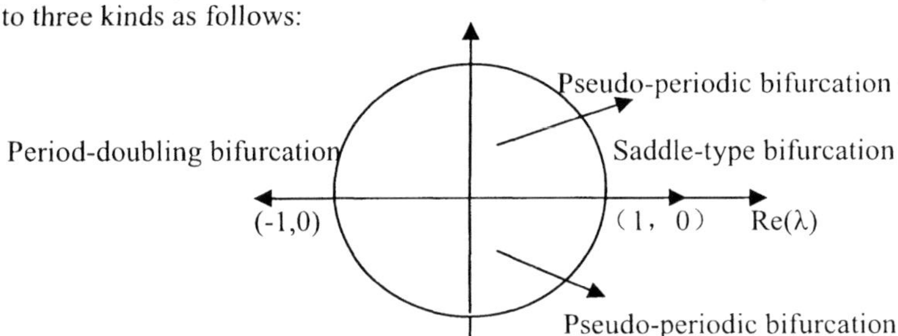

Fig. 5 Three ways of destabilization of periodic solutions with indications of multipliers

(1) When the maximum multiplier of modulus passes through the unit circle on the point of (1,0), the possible ways of destabilization and bifurcation of periodic solutions may involve saddle-type, fork-type, symmetrical breakage- type and so on, depending on different systems.

(2) When the maximum multiplier of the modulus passes through the unit circle on the point of (1, 0), the periodic solution will be destabilized via a period-doubling bifurcation. After passing through the bifurcated point, its period will split

from T into $2T$, and then $4T$ on the orbits. This kind of period-doubling bifurcation will result in the chaotic motion in the system finally.

(3) When the maximum multipliers of a pair of moduli pass through the unit circle in a form of conjugate complex numbers (the imaginary part is nonzero), the pseudo-periodic solutions will be periodically bifurcated and obtained on the basis of period solutions.

3.4 Nonlinear pseudo-periodic and chaotic solutions

Besides the equilibrium points and periodic solutions, the pseudo-periodic and chaotic solutions are two other kinds of steady-state motion forms in the nonlinear dynamic system of Eq. (2), and are comparatively more complex. It should be pointed out that there is a direct relation between the periodic solution and these two kinds of steady-state motion in pseudo-periodic and chaotic solutions. Pseudo-periodic solution in general is periodically bifurcated from the basis of periodic solution, and so also are the three ways of occurrences of chaotic solutions. The period-doubling bifurcation is actually a processing of continuous solution-cracking and period-doubling, and finally becoming chaotic. On the other hand, the pseudo-periodic bifurcated ways can result from such course as: firstly being bifurcated to pseudo-periodic solutions and then generating chaotic ways on the basis of period solutions, while the intermittent form behaves as the alternate appearances of periodic and chaotic solutions. Table 1 is the comparison between the characteristics of periodic, pseudo-periodic and chaotic solutions.

4. CONCLUSION

The bifurcation and regularities of stability variation with the changing of rotating speed ω, are obtained based on the detailed partial analysis on the stability problems of both stable and unstable bearing-rotor system. There will be more than a single periodic solution with the same rotating speed ω in practice. Which solution the system is carrying out will depend on initial conditions. Therefore, it is necessary to study the influence on system solutions due to initial conditions, namely by finding the domains of attraction of every solution in the system. In addition, the study shows that no matter whether the nonlinear bearing-rotor system is, a stable one or not, there are definite relations between destabilized attenuation and the orbits of periodic motion of the system, whenever there exists a certain orbit of the rotor center. Different attenuation rates can be obtained when the system is destabilized at different positions.

5. REFERENCES

1. Wen Xi-sen, et al, Modeling and Dynamic Analysis of Mechanical System, Science Press, Beijing, China, 11, 2004.
2. Iooss G.., Joseph D. D., Elementary Stability and Bifurcation Theory, Springer-Verlag, New York, 1980.
3. Lund J. W., Saibel E., Oil-whip Whirl Orbits of a Rotor in Sleeve Bearings. Journal of Engineer Industry, ASME, 89, 1967.
4. Gunter E. J., Hamphris R. R., Springeer H., Influence of Unbalanced on the Nonlinear Dynamical Response and Stability Flexible Rotor Bearing System. Rotor Dynamical Instability, ASME, 1983.
5. Liu Heng, Chen Shao-ding, PMC Method in the Global Nonlinear Analysis and Its Application in Bearing-rotor System. Journal of Applied Mechanics (in Chinese), 12(5), 1995.

48

BATCH ROUNDNESS CHARACTERIZATION AND EVALUATION

Qimi Jiang, Hsi-Yung Feng
Department of Mechanical and Materials Engineering
The University of Western Ontario
London, Ontario, Canada N6A 5B9
qimi.jiang.1@ulaval.ca, sfeng@eng.uwo.ca

A new concept referred to as the batch roundness is presented in this paper. This concept is derived to illustrate the worst possible form error for a batch of circular features machined under the same conditions. It is to be used as a statistical quality measure for such a batch of circular features and evaluated by analyzing their systematic and random form error components. The definition of batch roundness is introduced first and the associated evaluation algorithm is then presented. The evaluation algorithm starts by characterizing the deterministic profile for the batch of circular features. When the deterministic profile is obtained, the residuals, which are regarded as the random form error component, are available. The batch roundness can then be evaluated and the corresponding confidence level of the batch roundness zone determined. Case studies using both the simulated and experimental data sets have successfully demonstrated that the batch roundness can be reliably estimated from the inspection data of only one circular feature in the batch. This unique feature of the presented algorithm will hold as long as the measurement data size is adequate and the relative magnitude of the random form error component with respect to the batch roundness is not too large.

1. INTRODUCTION

In today's manufacturing industry, discrete parts are often produced in batch under nearly the same conditions. This motivates many research studies that have focused on the related economic or financial issues such as the calculation of production costs (Lee and Leonard, 1992; Primrose, 1993; Shiu, 1991; Koziarski and Królikowski, 1998), the performance analysis of batch production systems (Garavelli, 2001), and batch production scheduling (Heywood et al., 1997). If a batch of mechanical parts with circular features is machined under the same conditions, their form errors (or roundness) should contain very similar systematic and random components. This makes it possible to estimate the worst possible roundness value for parts in the batch based on the circular measurement data from only one of the parts. Since such a roundness value describes the overall quality for a batch of manufactured circular features, it is named here as the batch roundness.

Please use the following format when citing this chapter:

Jiang, Q., Feng, H.-Y., 2006, in IFIP International Federation for Information Processing, Volume 220, Information Technology for Balanced Manufacturing Systems, ed. Shen, W., (Boston: Springer), pp. 453–460.

This batch roundness value is very important to the successful implementation of modern monitoring, diagnostic and control technologies in production. It can be used as a quality measure for conformance checking of a batch of manufactured parts and more importantly, as a feedback input for proper corrective actions to the associated manufacturing processes to improve overall quality of the batch.

According to the ASME Y14.5M-1994 standard, the form error for a circular feature (roundness) is the gap between two concentric circles that completely bound the entire circular feature. However, this standard does not give any guidelines to establish this region. Many algorithms have been developed for roundness evaluation (Dowling et al., 1997). Among these algorithms, minimum-zone fitting and least-squares fitting are two commonly adopted approaches. It has been shown that unlike minimum-zone fitting, least-squares fitting does not yield a minimum roundness value from the inspection data (Lin et al., 1995; Murthy, 1986). To be consistent with the tolerance definition in the ASME standard, most algorithms have focused on obtaining the minimum roundness value from the inspection data.

Many researchers have cast the roundness evaluation as the solution of a general optimization problem (Choi and Kurfess, 1999; Gou et al., 1999; Sharma et al., 2000; Weber et al., 2002; Yau and Menq, 1996). To get a value that is close to the global minimum, some researchers addressed the initial conditions and adopted coordinate transformation techniques (Endrias and Feng, 2003; Lai and Chen, 1996) or suggested approximating orthogonal residuals by functions that are linear in the feature parameters (Shunmugam, 1991).

Other researchers based their algorithms on computational geometric techniques. Some adopted Voronoi diagrams (Huang, 1999; Novaski and Barczak, 1997; Samuel and Shunmugam, 2000; Roy and Zhang, 1992), convex hull or convex polygon techniques (Huang, 2001; Kaiser, 1998), or a data partition approach (Rajagopal and Anand, 1999). Moreover, some artificial intelligence techniques, such as neural networks, were also adopted for roundness evaluation (Suen and Chang, 1997). Chou et al. (2001) considered roundness evaluation in the 3D space as the measured points are impossible to be exactly on the same plane.

All the above algorithms are developed to evaluate the roundness of individual circular feature. So far, there is no concept reported in the literature that describes the overall quality status for a batch of circular features. For quality control, however, it is very important to know the overall quality for a batch of manufactured parts. Hence the concept that describes the worst possible case among a batch of circular features is very useful for the overall quality improvement of the batch.

This paper addresses the characterization and evaluation of batch roundness. The definition of batch roundness and its evaluation algorithm are given in the next section. Sections 3 and 4 present and discuss the implementation results in detail using simulated and experimental data sets, respectively. Section 5 wraps up this paper with concluding statements.

2. BATCH ROUNDNESS

2.1 Definition

The concept of batch roundness presented in this paper is defined as follows:

Batch Roundness is the worst possible roundness value of a circular feature among a batch of the same circular features manufactured under the same conditions.

Such a batch of circular features has the following attributes: (1) the systematic form error component of every circular feature is very similar; and (2) the random form error component of every circular feature is also very similar. This means that the statistical parameters (mean μ_r and standard deviation σ_r) of the fitted residuals of every circular feature in the batch should be very close to each other.

Once the deterministic profile of a circular feature in the batch is determined, the standard deviation σ_r of the fitted residuals is available. Then, an inner and an outer profile can be obtained by shifting the deterministic profile $3\sigma_r$ at both sides (Fig. 1). The probability for every point on the deterministic profile to lie in between the inner and outer profiles is then 0.9974. The probability for every point on the deterministic profile to lie within the tolerance zone, which is bounded by the two concentric circles in the figure, will be larger than 0.9974. The gap between the two concentric circles that bound the inner and outer profiles thus represents the worst possible roundness value for the batch of circular features.

2.2 Evaluation

According to the above discussion, separating circular form errors into two components, systematic and random, is the key to obtain the deterministic profile from the discrete measurement data for batch roundness evaluation. In this paper, the objective function for characterizing the deterministic profile is defined as

$$\underset{x_0, y_0}{\text{Min}}\left(\sum_{i=1}^{N} e_i^2\right) \tag{1}$$

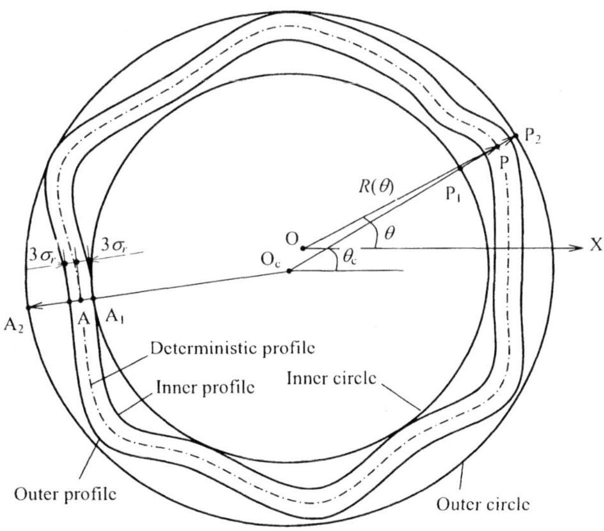

Figure 1 – Batch roundness evaluation.

where N is the number of the discrete data points, e_i the fitted residual of the ith data point ($e_i = R(\theta_i) - R_i$). R_i is the distance between the reference center $O(x_0, y_0)$ and the ith data point (x_i, y_i) and it can be expressed as

$$R_i = \sqrt{(x_i - x_0)^2 + (y_i - y_0)^2} \tag{2}$$

$R(\theta_i)$ is the function value of the deterministic profile $R(\theta)$ at θ_i, and $R(\theta)$ is expressed as

$$R(\theta) = R + SFC \tag{3}$$

where $R = R_0 + dR$, R_0 denotes the nominal radius of the circular feature and dR the change in its radius. dR is a constant and SFC (sum of Fourier components) represents the sinusoidal fluctuation of the deterministic profile in the radial direction. SFC can be expressed as

$$SFC = \sum_{k=1}^{N_d} C_{l_k} \cos(l_k \theta + \phi_{l_k}) \tag{4}$$

where N_d is the number of Fourier components included in the deterministic profile; C_{l_k}, l_k and ϕ_{l_k} are respectively the amplitude, frequency and phase angle of the kth weighted Fourier component (Desta et al., 2003). To guarantee the correct characterization of the deterministic profile, a large sample size of 180 is adopted in the case studies in Sections 3 and 4 using both simulated and experimental data.

To calculate the Fourier components, it is necessary to determine the reference center. The initial coordinates of the reference center are

$$\begin{cases} x_0 = \dfrac{1}{N}\sum_{i=1}^{N} x_i \\ y_0 = \dfrac{1}{N}\sum_{i=1}^{N} y_i \end{cases} \tag{5}$$

Once the reference center is obtained, the Fourier components are calculated by Fast Fourier Transform (FFT). The procedure for characterizing the deterministic profile is an iterative process of solving the optimization problem expressed in Eq. (1). In this process, the Powell's method is applied as the optimization method.

After the deterministic profile is obtained, it can be shifted $3\sigma_r$ at both sides to form an inner and an outer profile. Then, two concentric circles that contain the inner and outer profiles can be determined by the following optimization problem

$$\underset{x_c, y_c}{Min}\left(R_{\max} - R_{\min}\right) \tag{6}$$

where $R_{\max}$ and $R_{\min}$ are respectively the longest and shortest distances from the center $O_c(x_c, y_c)$ of the two concentric circles to the outer and inner profiles. The batch roundness is then given by the gap between the determined concentric circles.

2.3 Profile Confidence Level

Profile confidence level P_{cl} indicates the probability for the characterized deterministic profile to lie in the region bounded by the two concentric circles. It is regarded as the average confidence level of every point on the deterministic profile.

Once the center $O_c(x_c, y_c)$ of the two concentric circles and their radii R_{in} and R_{out} are determined, the profile confidence level can be calculated. For a point P on the deterministic profile, the distances from P to P_1 and P_2 can be determined and let them be d_1 and d_2, respectively (Fig. 1). Then the confidence level of point P is

$$\begin{aligned} p &= \int_{-d_1'}^{d_2'} \varphi(x)dx = \int_{-d_1'}^{0} \varphi(x)dx + \int_{0}^{d_2'} \varphi(x)dx \\ &= \sum_{i=0}^{n_1} \frac{1}{2n_1}\left[\varphi(x_i) + \varphi(x_{i+1})\right] + \sum_{j=0}^{n_2} \frac{1}{2n_2}\left[\varphi(x_j) + \varphi(x_{j+1})\right] \end{aligned} \tag{7}$$

where $d_1'=d_1/\sigma_r$, $d_2'=d_2/\sigma_r$, σ_r is the standard deviation of the fitted residuals, and $\varphi(x) = e^{-x^2/2}/\sqrt{2\pi}$. The profile confidence level can then be calculated as

$$P_{cl} = \frac{1}{2\pi}\int_0^{2\pi} p\, d\theta = \frac{1}{2\pi}\sum_{i=0}^{n} \frac{1}{2n}\left(p_i + p_{i+1}\right) \tag{8}$$

3. SIMULATION STUDY

The general equation for simulating a machined circular profile is given by

$$\begin{cases} x_i = (R + SFC + \varepsilon_i)\cos\theta_i \\ y_i = (R + SFC + \varepsilon_i)\sin\theta_i \end{cases} \quad (i = 1, 2, \Lambda, S) \tag{9}$$

where R denotes the nominal radius, SFC denotes the sum of Fourier components of the deterministic profile (which determines the systematic form error S_e), θ_i is the phase angle of the ith data point, S the sample size, and ε_i a value from a set of normal deviations generated to simulate the random form error. Forty-five samples were produced to represent a batch of 45 circular features. The nominal radius in Eq. (9) was set to 15 mm and SFC consisted of the following five components

$$SFC = 0.005\cos\theta - 0.007\sin 2\theta + 0.015\cos 4\theta - 0.02\sin 7\theta - 0.012\cos 10\theta \tag{10}$$

To investigate the effect of random errors, five cases have been analyzed by gradually increasing the random ratio. The results are listed in Table 1. In the table, $3\sigma/S_e$ denotes the random ratio, $\bar{r}_d$ the average individual roundness of the 45 simulated circular features, V_{r_d} the roundness variation of the 45 individual roundness values, $\bar{R}_d$ the average of the batch roundness values evaluated from the 45 circular features (each circular feature gives an estimated batch roundness value), V_{r_d} the batch roundness variation of the 45 batch roundness values, and $\bar{P}_{cl}$ the average profile confidence level.

When there is no random error ε_i, the resulting batch roundness variation is 0. In this case, the characterized deterministic profiles of the 45 simulated circular features are exactly the same. This results in the systematic form error S_e to be consistently 0.0824 mm, which is also the batch roundness. The profile confidence level is 100%. For the random ratio of 0.25, the typical characterized deterministic profile of the simulated circular features is

$$\begin{aligned} R(\theta) = 15.000011 &+ 0.039654\cos(\theta + 0.564780) + 0.018833\cos(7\theta + 1.562637) \\ &+ 0.015077\cos(4\theta - 0.025804) - 0.010998\cos(10\theta - 0.079321) \\ &+ 0.007483\cos(2\theta + 1.418556) + 0.002648\cos(47\theta + 0.424704) \end{aligned} \tag{11}$$

The calculation results show that the evaluated batch roundness fluctuates around 0.1242 mm with a variation of 0.0109 mm. The profile confidence level is about 99.995% that is larger than 99.74%. Comparatively, the individual roundness fluctuates intensively around 0.0969 mm with a large variation of 0.0201 mm.

As the random ratio further increases, both individual roundness and batch roundness increase. Their variation and the difference between the two roundness values also increase, but the profile confidence level decreases very slightly. However, compared to individual roundness, the evaluated batch roundness values are more consistent which confirms that the batch roundness can be estimated from just one data set sampled from any individual circular feature in the batch.

Table 1 – Results using simulated data.

$3\sigma/S_e$	$\bar{r}_d$ (mm)	V_{rd} (mm)	$\bar{R}_d$ (mm)	V_{Rd} (mm)	$\bar{P}_{cl}$ (%)	$\bar{R}_d - \bar{r}_d$ (mm)
0	0.0822	0.0004	0.0824	0	100	0.0002
0.25	0.0969	0.0201	0.1242	0.0109	99.995	0.0273
0.5	0.1210	0.0342	0.1653	0.0145	99.994	0.0443
0.75	0.1429	0.0411	0.2133	0.0212	99.993	0.0704
1	0.1682	0.0585	0.2593	0.0272	99.992	0.0911

4. EXPERIMENTAL STUDY

To demonstrate the presented concept and evaluation algorithm, 45 bosses with the same dimensions, 20 mm in diameter and 19 mm in height, were machined on one plate by the same end milling operation on a Fadal VMC 4020 vertical machining center. The workpiece material was 6061-T6 aluminum. The cutting tool was M42 HSS end mill with 6.35 mm cutting diameter, two right-handed flutes and 31.75 mm flute length. After the bosses were machined, a DEA *Swift* DCC coordinate measuring machine was used to measure these bosses in a single set-up. For each boss, 180 equidistant data points were taken at the depth of 10 mm.

Although much effort was made to keep the manufacturing conditions the same during machining, it is impossible to make the manufacturing conditions exactly the same all the time. Besides, the measuring conditions cannot be kept exactly the same all the time, either. Therefore, both the systematic form errors and the random form errors will be different from boss to boss.

To judge whether a boss belongs to the same batch, a two-sigma criterion is introduced. For the systematic form errors of all 45 bosses, the mean m_1 is 0.0401 mm and the standard deviation σ_1 is 0.0047 mm. The two-sigma criterion sets a range between 0.0307 mm and 0.0495 mm. Since the obtained systematic form errors for bosses No. 20 and No. 31 (0.0528 mm and 0.0302 mm respectively) fall outside the range, they are regarded as odd bosses and removed from the batch. Similarly, bosses No. 2 and No. 42 were removed due to their large standard deviations of the fitted residuals.

Figure 2 shows the evaluation results for the rest of 41 bosses. It can be seen that the batch roundness fluctuates around 0.0773 mm with a variation of 0.0151 mm. Compared to the individual roundness, which fluctuates intensively around 0.0615

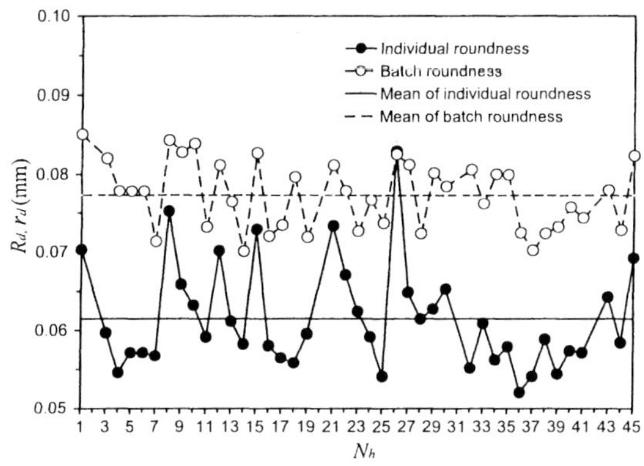

Figure 2 – Batch roundness evaluation results.

mm with a variation of 0.0308 mm, the evaluated batch roundness values are very consistent. Note that the individual roundness of boss No. 26 is a little larger than the corresponding batch roundness, 0.0825 mm. This shows that forming the tolerance region by shifting $\pm 3\sigma_r$ does not guarantee that this tolerance region covers all cases with the same statistical attributes. However, the profile confidence level shows that the average profile confidence level of the rest 40 bosses is 99.986%. This value is indeed high enough to be considered satisfactory in practice.

5. CONCLUSIONS

For mechanical parts with circular features, the concept of batch roundness, which describes the worst possible case of roundness in a batch, helps indicate the overall quality status of the batch of mechanical parts. The simulation study shows that the evaluated batch roundness values are larger than the corresponding individual roundness values. The experimental work shows that in some cases the individual roundness may exceed the corresponding evaluated batch roundness. In general, however, the probability for the individual roundness to exceed the batch roundness is very small as the calculated profile confidence level is satisfactorily high.

Although the concept and the evaluation algorithm presented in this paper is used for evaluating the overall quality status for a batch of circular features, the basic principle is of general significance and can be applied to other types of geometric features. The reason is that every manufactured feature should have a deterministic profile. Once the deterministic profile of the interested feature is characterized, its batch tolerance region can be evaluated similarly.

6. ACKNOWLEDGMENTS

This research was in part supported by Natural Sciences and Engineering Research Council and The AUTO21 Network of Centres of Excellence of Canada.

7. REFERENCES

1. Choi W, Kurfess TR. Dimensional measurement data analysis, part 2: minimum zone evaluation. ASME Journal of Manufacturing Science and Engineering 1999; 121: 246-250.
2. Chou SY, Lin SW, Chen CH. Assessing circularity in three dimensions. ASME Journal of Manufacturing Science and Engineering 2001; 123: 128-134.
3. Desta MT, Feng HY, OuYang D. Characterization of general systematic form errors for circular features. International Journal of Machine Tools and Manufacture 2003; 43(11): 1069-1078.
4. Dowling MM, Griffin PM, Tsui KL, Zhou C. Statistical issues in geometric feature inspection using coordinate measuring machines. Technometrics 1997; 39(1): 3-24.
5. Endrias DH, Feng HY. Minimum-zone form tolerance evaluation using rigid-body coordinate transformation. ASME Journal of Computing and Information Science in Engineering 2003; 3(1): 31-38.
6. Garavelli AC. Performance analysis of a batch production system with limited flexibility, International Journal of Production Economics 2001; 69(1): 39-48.
7. Gou JB, Chu YX, Li ZX. A geometric theory of form, profile, and orientation tolerances. Precision Engineering 1999; 23: 79-93.
8. Heywood MI, Chan M-C, Chatwin CR. Application of stochastic real-valued reinforcement neural networks to batch production rescheduling. Journal of Engineering Manufacture 1997; 211(8): 591-603.
9. Huang J. A new strategy for circularity problems. Precision Engineering 2001; 25: 301-308.
10. Huang J. An exact solution for the roundness evaluation problems. Precision Engineering 1999; 23: 2-8.
11. Kaiser MJ. Generalized zone separation functions for convex perfect forms and incomplete data sets. International Journal of Machine Tools and Manufacture 1998; 38(4): 375-404.
12. Koziarski A, Krolikowski KJ. The calculation of costs of grinding on conventional and CNC grinding machines in small-batch shaft production. International Journal of Machine Tools and Manufacture 1998; 28(2): 123-130.
13. Lai JY, Chen IH. Minimum zone evaluation of circles and cylinders. International Journal of Machine Tools and Manufacture 1996; 36(4): 435-451.
14. Lee R, Leonard R. Technical and financial considerations of AGV use within multi-product and batch manufacture. Computer Integrated Manufacturing Systems 1992, 5(2): 97-104.
15. Lin SS, Varghese P, Zhang C, Wang HPB. A comparative analysis of CMM form-fitting algorithms. ASME Manufacturing Review 1995; 8(1): 47-58.
16. Murthy TSR. A comparison of different algorithms for circularity evaluation. Precision Engineering 1986; 8(1): 19-23.
17. Novaski O, Barczak ALC. Utilization of Voronoi diagrams for circularity algorithms. Precision Engineering 1997; 20(3): 188-195.
18. Primrose PL. Economics of batch manufacture. Computer Integrated Manufacturing Systems 1993; 6(2): 109-116.
19. Rajagopal K, Anand S. Assessment of circularity error using a selective data partition approach. International Journal of Production Research 1999; 37(17): 3959-3979.
20. Roy U, Zhang X. Establishment of a pair of concentric circles with the minimum radial separation for assessing roundness error. Computer-Aided Design 1992; 24(3): 161-168.
21. Samuel GL, Shunmugam MS. Evaluation of circularity from coordinate and form data using computational geometric techniques. Precision Engineering 2000; 24: 251-263.
22. Sharma R, Rajagopal K, Anand S. A genetic algorithm based approach for robust evaluation of form tolerances. Journal of Manufacturing Systems 2000; 19(1): 46-57.
23. Shiu Y-C. An economic batch production quantity model with learning curve-dependent effects: a technical note. International Journal of Production Economics 1991; 25: 35-38.
24. Shunmugam MS. Criteria for computer-aided form evaluation. ASME Journal of Engineering for Industry 1991; 113: 233-238.
25. Suen DS, Chang CN. Application of neural network interval regression method for minimum zone circle roundness. Measurement Science and Technology 1997; 8(3): 245-252.
26. Weber T, Motavalli S, Fallahi B, Cheraghi SH. A unified approach to form error evaluation. Precision Engineering 2002; 26(3): 269-278.
27. Yau HT, Menq CH. A unified least-squares approach to the evaluation of geometric errors using discrete measurement data. International Journal of Machine Tools and Manufacture 1996; 36(11): 1269-1290.

49

A POINT CLOUD SIMPLIFICATION ALGORITHM FOR MECHANICAL PART INSPECTION

Hao Song, Hsi-Yung Feng
Department of Mechanical and Materials Engineering
The University of Western Ontario
London, Ontario, Canada N6A 5B9
hsong2@uwo.ca, sfeng@eng.uwo.ca

A point cloud data set, a set of massive and dense coordinate data points sampled from the surface of a physical object, is emerging as a new representation format of 3D shapes. This is mostly attributed to recent advances in the range finding technology of high-speed 3D laser scanning. A typical laser scanned data set often contains millions of data points and this leads to significant computational challenges in processing the point cloud data for practical applications such as the high-speed laser inspection of mechanical parts. To reduce the number of the massive data points to facilitate geometric computation, this paper presents a simplification algorithm for the laser scanned point cloud data from manufactured mechanical parts, whose boundary surfaces include sharp edges. Due to the distinct feature represented by the points located on or near the sharp edges, these points are first identified and retained. The algorithm then repeatedly removes the least important point from the remaining data points until the specified data reduction ratio is reached. Quantification of a point's importance is based on points in its neighborhood and it indicates the point's contribution to the representation of the local surface geometry. The effectiveness of the proposed algorithm is shown through the simplification results of two practical point cloud data sets.

1. INTRODUCTION

In many disciplines, there is a need to faithfully represent the shape of a physical object in computer. In mechanical engineering, for example, it is often necessary to capture the shape of a manufactured part to evaluate the manufacturing error with respect to the original design. The need of capturing complex 3D shapes is obvious in many applications of computer graphics such as simulation and computer animation. In addition, some medical applications, such as radiation therapy and surgical planning (Lorensen and Cline, 1987), need the 3D surface representation of the anatomy.

Transferring a physical object into computer has been made possible by the development of 3D data acquisition technologies. Modern 3D data acquisition

Please use the following format when citing this chapter:

Song, H., Feng, H.-Y., 2006, in IFIP International Federation for Information Processing, Volume 220, Information Technology for Balanced Manufacturing Systems, ed. Shen, W., (Boston: Springer), pp. 461–468.

devices, such as laser scanner, can quickly capture points from the surface of the object to produce a dense point cloud data set. A typical laser scanned data set often contains millions of data points and this leads to significant computational challenges in processing the point cloud data for practical applications such as the high-speed laser inspection of mechanical parts. Two reasons may cause the data acquisition process to produce too many points. One is the high measurement precision and speed of modern 3D data acquisition devices. The other reason is that, when making measurement, the operator never knows how dense the points should be in order to achieve the description of the shape at a certain accuracy level. Frequently, far more points are captured. In practice, a simplification or decimation of the acquired point set is often necessary for the subsequent applications to operate at a reasonable computational cost.

As the early stage of point cloud processing, simplification may serve to facilitate different subsequent applications, and it is desirable to have some particular characteristics present in the simplified point set for the particular application at hand. In mechanical engineering, it is very common to have sharp edges and sharp corners in manufactured parts. In the point cloud representation of theses parts, the points representing the sharp edges and sharp corners are so unique that the local geometry cannot be correctly reflected without them. Therefore, for the situations that the sharp edge information is critical for the subsequent application, it is desirable to retain those points close to sharp edges or sharp corners in the simplified point set.

In the inspection of mechanical parts, for example, a segmentation process, which determines feature lines of the surface and separates the point set by this lines, is often needed in order to border the correct portion on which the manufacturing errors will be checked and to cut off points in unwanted regions. The feature lines are extracted from the points close to sharp edges or sharp corners and are interpreted as trimming curves in the segmentation process.

This paper presents a simplification algorithm for the point cloud data scanned from mechanical parts, whose boundary surfaces include sharp edges. The algorithm selects a subset of the input point cloud based on points' importance in terms of defining their local geometries. The quantification of a point's importance is derived from its neighborhood – the point itself and a set of neighboring points chosen from the point cloud. The idea is to see whether or not the local geometry can be reliably implied from its neighboring points. If the local geometry cannot be reliably reflected by the neighboring points and their associated properties, the point is considered important in defining the geometry. The simplification is achieved by progressively removing points from the input cloud that are less important than others. Due to the discontinuity occurred at points close to sharp edges, the local geometry at these points cannot be reliably implied from the neighboring points. These points, considered as defining the important feature lines, are explicitly identified and retained in the simplified point set.

2. PREVIOUS WORK

Extensive research work has been reported on simplification of polygonal meshes, a closely related area to point cloud simplification. Actually, many of the reported methods can be adapted to simplification of point clouds (Pauly et al., 2002). Luebke (2001) and Cignoni et al. (1998) presented good reviews on the algorithms of mesh simplification. Luebke (2001) described a useful taxonomy and discussed typical algorithms against some basic items of his classification such as simplification mechanism, the way of dealing with topology, and whether they are static, dynamic, or view-dependent simplifications. Cignoni et al. (1998) took an empirical approach. They tested and compared six representative algorithms by taking into account the computational cost and the quality of the resulting output meshes.

Compared to mesh simplification, direct simplification of point clouds is a new research area. Techniques are usually borrowed from related areas such as image processing and mesh simplification. Pauly and Gross (2001) proposed a spectral framework for processing point-sampled objects by introducing a concept of local frequencies on geometry. They split the model surface to a set of patches that can be represented as scalar height fields. The frequency spectrum of each patch was then obtained by the discrete Fourier transform (DFT). By directly applying signal processing theory, they re-sampled the patches individually and blended the patch boundaries. The simplification obtained by this method, however, unnecessarily depends on the specific layout of the patches.

Moenning and Dodgson (2004) applied the farthest point sampling principle, initially introduced by Eldar et al. (1997) in the context of image sampling, and offered control of density and feature sensitivity in their simplification procedure. The method takes a progressive approach which starts from a randomly picked sample point and repeatedly places the next sample point at the middle of the least-known sampling domain with the help of a pre-computed approximation of the geodesic Voronoi diagram of the input point cloud. Its density control is achieved by a user provided number indicating the upper bound of the geodesic distance from the next farthest point to the obtained (simplified) point set. The control of feature sensitivity, meaning that higher point density in highly curved region and lower density in flat region, is achieved by assigning weights to points according to the estimated local changes such as the mean curvature. Unfortunately, the relations among point set size, density condition, and feature sensitivity were not discussed, leaving their determination a difficult task to the user.

Pauly et al. (2002) adapted and generalized some popular methods in mesh simplification to the point cloud context. The methods of clustering, both incremental region growing and hierarchical clustering, iterative simplification, and particle simulation were tested against the criteria of accuracy, sampling distribution, and computational efficiency. By introducing an approximation of point-to-surface distance, where the surface is represented by a point cloud, they were able to quantitatively evaluate the error of the surface represented by the simplified point set to the surface represented by the original point cloud. Based on their comparison of the adapted methods, they gave useful hints to help choose the most appropriate method for different situations. However, none of their methods were originally designed to devote to any of the error criteria they used.

Lee et al. (2001) presented a 3D grid method for direct point data reduction. The method constructs connected 3D cells in the space to enclose the data points. The cells are generated by an octree-based spatial decomposition procedure and the criterion of the decomposition is defined as the standard deviation of the estimated normal vectors of points within each cell. The simplification is then obtained by choosing a representative point from each cell. Since the number of cells, equivalently data points in the simplification, is related to the number of initial cells and the user-defined tolerance for the decomposition, this method does not produce a simplification with its size controlled in an intuitive way.

A straightforward approach is to judge the importance of each point in the point cloud. The importance is quantified by a measure which indicates either the amount of information contributed to the description of the shape or the redundancy of the point. The simplification is then achieved by removing points that contain the least amount of information or the largest redundancy. Linsen (2001) defined the information content of a point as a weighted sum of factors such as distances to its neighbors, non-planarity, and change of normal vectors evaluated in the selected neighborhood. Alexa et al. (2001) presented a measure which involved the Moving Least-Squares (MLS) surface (Levin, 2004). It is computed as the distance from a point to its projection onto the MLS surface derived from all the other points of the point cloud. They iteratively removed the point with the smallest distance and, after each removal, updated the distances of the affected points. Kalaiah and Varshney (2003) measured the redundancy of each individual point and iteratively removed the point with the largest redundancy. The evaluation of the redundancy is based on the local geometric properties derived from the point's neighborhood.

All the existing simplification methods are based on some smoothness condition of the underlying surface. Manufactured parts, however, frequently show sharp edges and sharp corners which naturally separate the surface into different patches. Although existing methods are able to work on surfaces with sharp features and small blends and produce simplification with relative dense points in the sharp feature regions, they do not explicitly identify the points on or close to the sharp edges and pass the information to subsequent applications.

3. THE SIMPLIFICATION ALGORITHM

The proposed simplification algorithm has two steps. The first step identifies points located on or close to sharp edges (referred as boundary points thereafter). These points are retained in the point cloud and will not be processed by the second step. The second step performs the simplification by progressively removing non-boundary points from the cloud. The removing is based on the quantification of each point's importance. Each time, the least important point is removed and the importance values of points affected by the removal are updated. The process continues until the specified number of points for the simplification is reached.

3.1 Identifying Boundary Points

Different methods have been proposed to identify points located on or close to sharp edges. These methods all involve the quantification of the smoothness of the underlying surface at a point's vicinity and the determination of a threshold. The method employed in this paper establishes a measure derived from the deviation of the normal vectors in a point's vicinity (Song and Feng). It attempts to compensate for the deviation caused by surface curvature and focuses on the component caused by tangential discontinuity. The benefit of using such a measure is that the threshold can be automatically determined by identifying outlying values from the general trend shown by the values calculated at points in smooth regions. As a result, it minimizes user involvement in the process of boundary point detection.

The proposed method needs the information of the unit normal vector of the underlying surface at each data point. The estimation of the normal vectors from the input data cloud is done by the algorithm suggested by OuYang and Feng (2005). The algorithm features a novel definition of neighborhood (a set of neighboring points) for each data point, which is called the Local Voronoi Mesh Neighbors. Establishment of the neighborhood is based on the Voronoi diagram of the input point cloud P. Let $DN_{\mathbf{p}}$ denote the desired neighborhood for point $\mathbf{p}$. If $VN_{\mathbf{p}}$ is the set of points whose corresponding Voronoi cells share a facet with the Voronoi cell of $\mathbf{p}$ in the Voronoi diagram of P, then the algorithm chooses $DN_{\mathbf{p}}$ as a subset of $VN_{\mathbf{p}}$: $DN_{\mathbf{p}} \subseteq VN_{\mathbf{p}}$. The idea is to choose those in $VN_{\mathbf{p}}$ who are close to $\mathbf{p}$ and contribute to the construction of a local mesh centered at $\mathbf{p}$ to reflect the local geometry. In this paper, $VN_{\mathbf{p}}$ is also called the *Voronoi neighbors* of $\mathbf{p}$.

3.2 Evaluating Points' Importance

For a point $\mathbf{p} \in P$ in the smooth region of S, where P is the input point cloud and S is the surface represented by P, suppose it has K neighbors $\mathbf{q}_1$, $\mathbf{q}_2$, ... $\mathbf{q}_K$, and they are all in the smooth region of S (non-boundary points). Let the unit normal vectors at $\mathbf{q}_1$, $\mathbf{q}_2$, ... $\mathbf{q}_K$ be $\mathbf{n}_1$, $\mathbf{n}_2$, ... $\mathbf{n}_K$, respectively, which are estimated from P by the algorithm of OuYang and Feng (2005). The importance of $\mathbf{p}$ is calculated as

$$i(\mathbf{p}) = \sum_{l=1}^{K} \left| (\mathbf{p} - \mathbf{q}_l) \cdot \mathbf{n}_l \right| \Big/ K \tag{1}$$

Geometrically, Eq. (1) calculates the average of the Euclidean distances from $\mathbf{p}$ to the estimated tangent planes at $\mathbf{q}_1$, $\mathbf{q}_2$, ... $\mathbf{q}_K$, respectively. The value reflects the deviation from $\mathbf{p}$ to the underlying surface represented by its neighbors. A large value indicates that $\mathbf{p}$'s position cannot be reliably implied by its neighboring points. Therefore, $\mathbf{p}$ plays an important role in defining the local shape.

The evaluation takes advantage of the established neighborhood and the normal vector information obtained in the previous step for each data point. In the case that some of $\mathbf{p}$'s neighbors are boundary points, they are simply ignored in the evaluation as their associated normal vectors are not reliable.

3.3 Removing Points

The simplification is achieved by progressively removing points from P. In each iteration, it removes one point in the current configuration of P that has the smallest importance value. Suppose $\mathbf{p}$ is chosen to be removed. Those points that previously have $\mathbf{p}$ as one of their neighbors will change their neighborhood configurations. So the normal vectors at these points need to be re-estimated. Since the neighborhood for $\mathbf{p} \in P$ is a subset of $\mathbf{p}$'s Voronoi neighbors, the point that has $\mathbf{p}$ as one of its neighbors can be easily found by exploring $\mathbf{p}$'s Voronoi neighbors. Let $\mathbf{q}$ be such a point that $\mathbf{p}$ is one of $\mathbf{q}$'s neighbors, that is, $\mathbf{p} \in DN_{\mathbf{q}}$. Because $DN_{\mathbf{q}} \subseteq VN_{\mathbf{q}}$, then

$\mathbf{p} \in VN_{\mathbf{q}}$. Due to the characteristic of Voronoi diagram, it is easy to see that $\mathbf{q} \in VN_{\mathbf{p}}$. Therefore, searching in $VN_{\mathbf{p}}$ will cover all the points that previously have **p** as one of their neighbors. By updating the Voronoi diagram of P and re-establishing the neighborhoods, their associated normal vectors can be easily updated.

The impact on the importance values is even bigger. Figure 1 shows the example of the affected points due to the removal of **p**. Suppose **p** is a neighbor of $\mathbf{q}_1$ (it is also a neighbor of $\mathbf{q}_2$, $\mathbf{q}_3$, $\mathbf{q}_4$, and $\mathbf{q}_5$). Due to the removal of **p**, the estimated normal vector at $\mathbf{q}_1$ changes. The changed normal vector, equivalently the tangent plane at $\mathbf{q}_1$, will in turn have impact on the importance values of the points that have $\mathbf{q}_1$ as one of their neighbors (this type of points are marked as triangles in Fig.1). Therefore, if the affected normal vectors are restricted to the immediate ring of points $\mathbf{q}_1$, $\mathbf{q}_2$, ..., $\mathbf{q}_5$ around **p**, the affected importance values will spread into the second ring of points: $\mathbf{t}_1$, $\mathbf{t}_2$, ..., $\mathbf{t}_9$. As a result, finding these points needs to explore not only in $VN_{\mathbf{p}}$ but also in $VN_{\mathbf{q}1}$, $VN_{\mathbf{q}2}$, ..., $VN_{\mathbf{q}5}$.

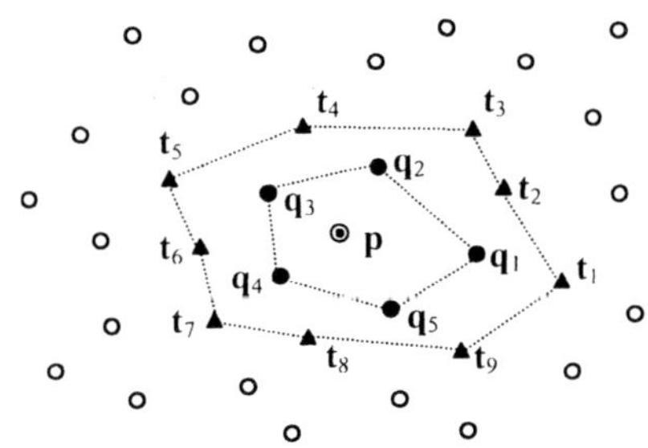

Figure 1 – The affected points (marked as solid dots and triangles) due to the removal of **p** from P.

4. IMPLEMENTATION AND RESULTS

The proposed algorithm relies on the maintenance of the Voronoi diagram of P when P is changing. Actually, it is the spatial structure established by the Voronoi diagram that helps explore the points whose normal vectors and importance values are affected by the removal. Both the construction of the Voronoi diagram and its dynamic maintenance are implemented by using the 3D Delaunay triangulation programs in CGAL (Computational Geometry Algorithms Library) (Boissonnat et al., 2002).

In addition, due to the spatial structure set by the Voronoi diagram, the order in which the points are evaluated does not affect the importance values. In the current implementation, the points are presented as an array and each element of the array records a point's position, associated normal vector, and indices of its neighbors.

The simplification algorithm has been tested using practical data sets. Figures 2 and 3 show the simplification of the point clouds of the Fandisk (16,475 points) and the Bunny (35,947 points). For each example, the identified boundary points are also highlighted to visualize the results obtained from the first step of the algorithm. The running times for identification of boundary points are 190 and 373 seconds for

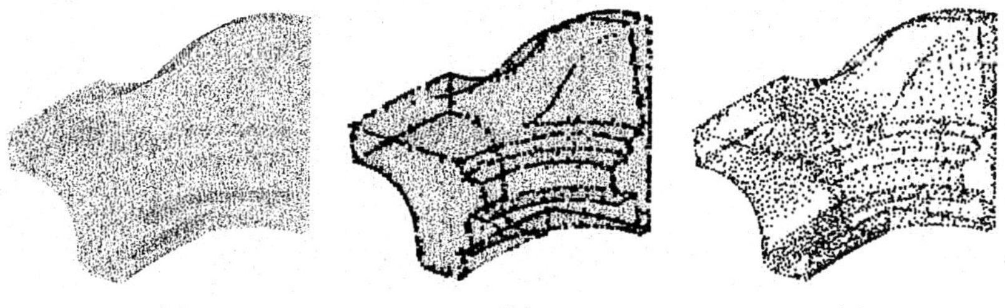

(a) (b) (c)

Figure 2 – Fandisk: (a) the original point set; (b) the highlighted boundary points; and (c) the point set simplified to 20% of its original size.

(a) (b) (c)

Figure 3 – Bunny: (a) the original point set; (b) the highlighted boundary points; and (c) the point set simplified to 10% of its original size.

Fandisk and Bunny, respectively, on a Pentium 4, 3.0 GHz personal computer. The removing process took 1,253 seconds for Fandisk and 1,720 seconds for Bunny.

For the model of Fandisk which has clear boundaries (sharp edges) between different surface patches, the simplified point sets show clear sharp edge information as expected. In smooth regions, there is also a clear trend that points in regions of higher curvatures are denser than those in relatively flat regions. Result of the Bunny, featuring free form surfaces, also show varied point density in regions of different curvatures, which is a favorable characteristic for the simplified point set to achieve both detailed description of the shape and reduction of the data size.

5. CONCLUSIONS

A simplification algorithm for point cloud data has been presented. The obtained simplification completely preserves the sharp edge information and preserves the shape of smooth regions at a reasonable level of detail by progressively removing points in these regions. The removing process is based on the quantification of points' importance calculated from each point's neighborhood. The quantification indicates the point's contribution to the representation of the local geometry.

The algorithm needs the establishment of a neighborhood and the normal vector information for each point. Due to the particular way of establishing the neighborhood, which is the basis for both normal vector estimation and the evaluation of the point importance, finding the points that need to update their associated normal vectors and importance values after a removal becomes efficient. This is achieved by the dynamic maintenance of the Voronoi diagram of the remaining data points.

6. ACKNOWLEDGMENTS

This work was funded in part by the Natural Sciences and Engineering Research Council of Canada.

The authors would like to thank the team of CGAL for making their programs publicly available. The point set of Fandisk was obtained from the web page of Dr. Hugues Hoppe at Microsoft Research. The point set of Bunny was obtained from the Large Geometric Models Archive at Georgia Institute of Technology, which was originally created at the Stanford Computer Graphics Laboratory.

7. REFERENCES

1. Alexa M, Behr J, Cohen Or D, Fleishman S, Levin D, Silva CT. Point set surfaces. in: Proceedings of the IEEE Conference on Visualization (VIS'01), 2001, pp. 21-28.
2. Boissonnat J-D, Devillers O, Pion S, Teillaud M, Yvinec M. Triangulations in CGAL. Computational Geometry: Theory and Applications 2002; 22(1-3): 5-19.
3. Cignoni P, Montani C, Scopigno R. A comparison of mesh simplification algorithms. Computers & Graphics 1998; 22(1): 37-54.
4. Eldar Y, Lindenbaum M, Porat M, Zeevi YY. The farthest point strategy for progressive image sampling. IEEE Transactions on Image Processing 1997; 6(9): 1305-1315.
5. Kalaiah A, Varshney A. Modeling and rendering of points with local geometry. IEEE Transactions on Visualization and Computer Graphics 2003; 9(1): 30-42.
6. Lee KH, Woo H, Suk T. Point data reduction using 3D grids. International Journal of Advanced Manufacturing Technology 2001; 18(3): 201-210.
7. Levin D. "Mesh-independent surface interpolation". in: Geometric Modeling for Scientific Visualization, Brunnett G, Hamann B, Muller H, Linsen L. eds., Springer-Verlag, 2004, pp. 37-49.
8. Linsen L. Point cloud representation. Technical Report, Faculty of Informatics, University of Karlsruhe, Germany, 2001.
9. Lorensen WE, Cline HE. Marching cubes: a high resolution 3D surface construction algorithm. in: Proceedings of the 14th annual conference on Computer graphics and interactive techniques (SIGGRAPH'87), 1987, pp. 163-169.
10. Luebke DP. A developer's survey of polygonal simplification algorithms. IEEE Computer Graphics and Applications 2001; 21(3): 24-35.
11. Moenning C, Dodgson NA. Intrinsic point cloud simplification. in: Proceedings of the 14th International Conference on Computer Graphics and Vision (GraphiCon), Moscow, Russia, September 2004.
12. Pauly M, Gross M. Spectral processing of point-sampled geometry. in: Proceedings of the 28th Annual Conference on Computer Graphics and Interactive Techniques (SIGGRAPH'01), 2001, pp. 379-386.
13. Pauly M, Gross M, Kobbelt LP. Efficient simplification of point-sampled surfaces. in: Proceedings of the 13th IEEE Visualization Conference, Boston, MA, October 2002, pp. 163-170.
14. OuYang D, Feng H-Y. On the normal vector estimation for point cloud data from smooth surfaces. Computer-Aided Design 2005; 37(10): 1071-1079.
15. Song H, Feng H-Y. Automatic detection of tangential discontinuities in point cloud data. ASME Journal of Computing and Information Science in Engineering, in revision.

50

DEGENERATION METHODS IN INTELLIGENT BUILDING CONTROL SYSTEM DESIGN

Julio Arakaki
Pontifícia Universidade Católica de São Paulo/Unifieo – jarakaki@pucsp.br

Paulo Eigi Miyagi
Escola Politécnica da USP – pemiyagi@usp.br

This paper presents a set of special requirements to development of intelligent building control software. It includes the degeneration technique (gradual reduction of the building service level). The method involves an organized activities sequence that results in artifacts such as models and control software specifications. These specifications assure the desired dynamic behavior of system and also include degeneration features. It also presents a specific example related with the control in Intelligent Building which has been adopted as case study and illustrates the application of the proposed method.

1. INTRODUCTION

The productive systems and its respective control systems are organized in components with high cooperation degree. These control systems are becoming highly complex, distributed and its modules are strongly interacted.

In general, the control systems have some degree of robustness and security, in such way that they assure the system functionality in the environments for which they had been projected. However, the complexity of the systems and its control software increase the occurrence probability of unexpected events. In this manner, the implementation of new productive systems or the maintenance of already existing systems, the inclusion of degeneration techniques is desirable because, in fault situations, a gradual reduction of the services level in a system is necessary (Arakaki, 2004).

This paper considers a method (with the degeneration technique) for the control software design in productive systems (Intelligent Building). This method has activities organized in steps. Each step describes the techniques and/or artifacts generated through of models and specifications. The last step results in technical specifications of the control software, with the degeneration requirements incorporated.

The degeneration method considers the ISO/IEC 9126 standard (ISO/IEC 9126, 1998) for the development the control software. This software engineering standardization defines quality of software attributes for the control software

Please use the following format when citing this chapter:

Arakaki, J., Miyagi, P. E., 2006, in IFIP International Federation for Information Processing, Volume 220, Information Technology for Balanced Manufacturing Systems, ed. Shen, W., (Boston: Springer), pp. 469–478.

specifications. The Figure 1 presents the use of control requirements (with degeneration) and ISO/IEC 9126 standard to generate the specifications of control software architecture.

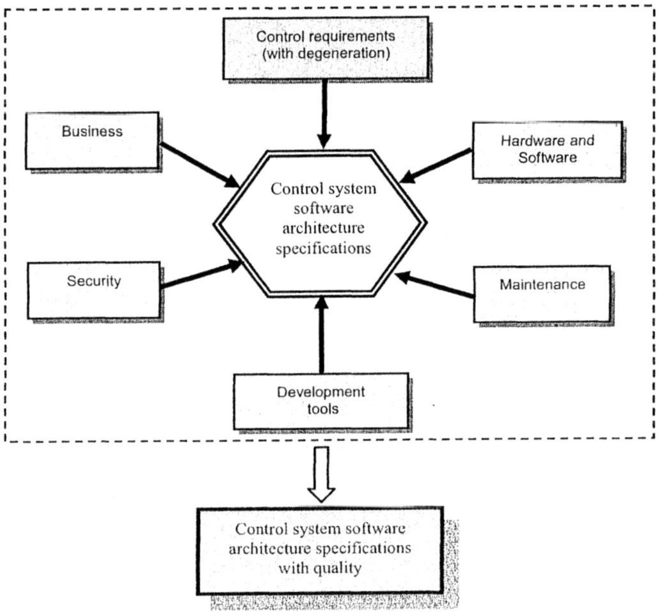

Figure 1 – ISO/IEC 9126 standard and the control requirements (with degeneration) included.

A typical architecture of the discrete event system control was presented in Miyagi (1988), Miyagi (1996) and Hasegawa (1998). In this case, to control a productive system[1], the actuation is function of the command, detection signals and internal state (see Figure 2).

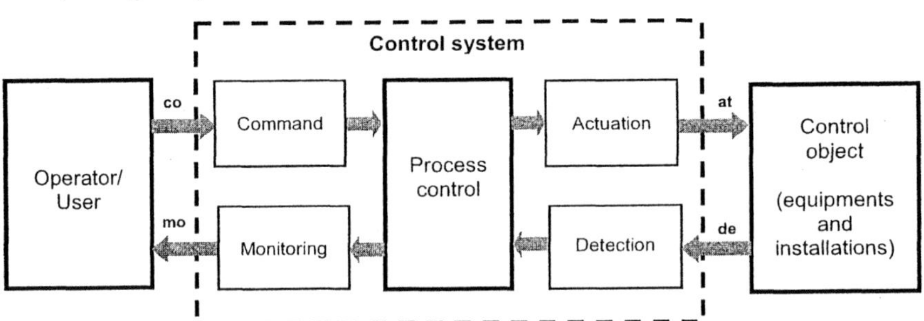

Where: *co* - command *de* - detection
mo - monitoring *at* - actuation

Figure 2 - Control system architecture.

[1] Productive System is based on the event occurrence (Discrete Event System).

The inclusion of the degeneration requirements in a discrete event system (DES) control is explained in the following item.

2. CONTROL SYSTEM WITH DEGENERATION

The new control system architecture is based on the typical architecture of DES, with the inclusion of the degeneration module.

The Figure 3 presents the control of discrete event system with degeneration module included. In this case, the module 'Degeneration' involves the following activities:

- Supervision of critical points;
- Re-configuration.

The actuation with degeneration (at_d) is function (f_d) of: command with degeneration (co_d), detection with degeneration (de_d) and internal state with degeneration (ei_{d_i}):

$$at_d = f_d(co_d, de_d, ei_{d\_i}).$$

The next internal state with degeneration (ei_{d_i+1}) is function of the re-configuration (rec) and of the current internal state:

$$ei_{d\_i+1} = g_d(rec, ei_{d\_i})$$

The reconfiguration with degeneration (rec) is function (h_d) of the critical points (pc) and of the supervision of these critical points [2](su_{pc}):

$$rec = h\ (pc, su_{pc})$$

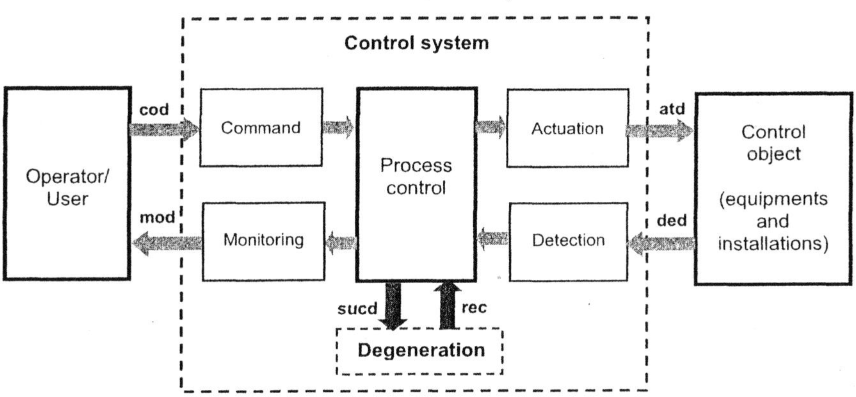

Where: *cod* - command with degeneration; *mod* - monitoring with degeneration; *atd* - actuation with degeneration; *ded* - detection with degeneration; *rec* - reconfiguration to degeneration; *sucd* - critical point to degeneration

Figure 3 - Control system architecture with degeneration module.

[2] Component or configuration with high probability of occurrence of abnormal situations

The degeneration in a control system has the following functionality: initially an unexpected state is detected (through the identification and monitoring of the critical points in the system). After that, the context is evaluated, according to priority criteria for these critical points. Finally, the automatic mechanisms or with human intervention are set to reconfigure the control system to operate in a degraded way as the established programming. The abnormal situation is verified constantly by the components for the treatment of abnormalities (inserted in the control system) and by the components for the degeneration, until the regeneration condition is reestablished (return to the normal functionality, if possible).

3. CASE STUDY: INTELLIGENT BUILDING

In a Intelligent Building (Arkin, 1995; Flax, 1991), each subsystem is managed by a control system that integrates the devices and the operations of that subsystem. These control subsystems interact with other control systems.

In this context, by applying the proposed method, the control software of the Intelligent Building is generated and allows increasing the efficiency and the functionalities of the control systems in unexpected situations.

The Intelligent Buildings had been chosen as study of case for the application of this method because they can be characterized as a productive system. They are based on the discrete events and distributed system and have subsystems with respective controls. These characteristics allow to verify and to analyze the considered method.

Normally, an intelligent building has:

- **Transport/movement** - elements that facilitate and organize the transport and the movement of people and objects. Example: the elevators, the rolling stairs, the rolling mats, and others;
- **Patrimonial and personal security and control of access** – allow the access and exit of people and objects in the interior of the building. The main devices are: electronic ratchets, cameras and internal TV circuit, bar codes readers, fingerprint reader, iris readers, special doors and alarms systems;
- **Energy management** - the energy supply for controls systems and for the building subsystems;
- **Fire** – fire identification system, alarm and fire occurrence control (with smoke detectors, automatic cut-fire doors and automatic devices to eliminate the fire);
- **Illumination** – system to interact with the elements as light bulbs, curtains and auxiliary batteries, to get an excellent illumination in all environments;
- **Ventilation and conditioned air (HVAC)** - systems for keeping comfortable environments (with ventilation, temperature and humidity systems).

Different types of information flow through the subsystems. One subsystem interacts with another. This interaction occurs through communication nets and protocols standards. Each subsystem must have its intelligent, autonomous and cooperative control. These are characteristics of distributed systems.

The inclusion of the requirements of degeneration in the subsystem of Intelligent Building control is illustrated of general form in Figure 4.

In this case, each subsystem has a degeneration component associate with a control component. All the subsystem has an interaction with the global degeneration module that changes the control requirements in a critical situation (not resolved in fault safe system). These changes allow that the control system acts with degeneration (keeping active the necessary operations) or with regeneration, returning to the normal functionality.

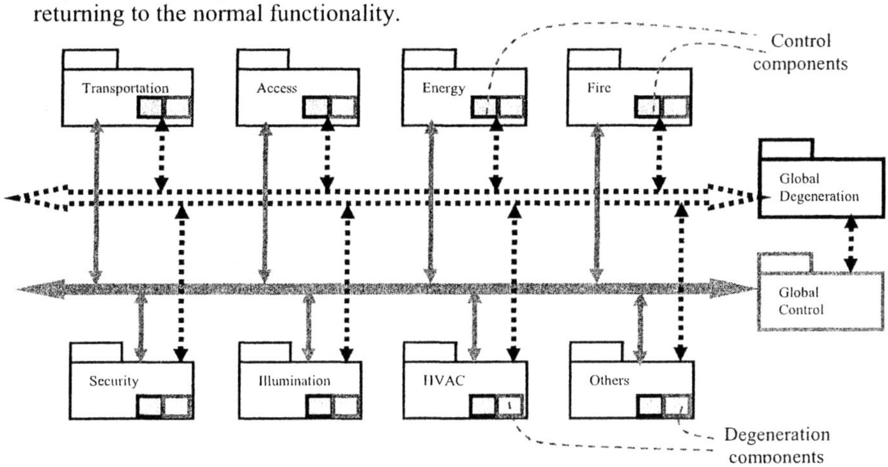

Figure 4. Subsystems in productive system (Intelligent Building with degeneration).

The considered method was applied for the subsystem of Transport and Heating Ventilation and Air Conditioning subsystem (HVAC) in a productive system (Intelligent Building). The modeling, analysis and control of the Intelligent Systems were explained in Bastidas Gustin (2002), Bastidas Gustin (2003) and Miyagi (2002).

4. APPLYING THE DEGENERATION METHOD IN INTELLIGENT BUILDING

To demonstrate this method, subsystems of transport and HVAC (Villani, 2004; Villani, 2005) are used. In sequence, the necessary steps to generate the artifacts (control software specifications) are presented:

Step 1 – System critical points identification

In this step the critical points of the intelligent building are identified. The table 1 presents, as example, two critical points with its respective degenerations.

Table 1. Critical point samples.

Subsystem	Critical point	Degeneration
HVAC	(1) Conditioning of rooms	Reduction of the conditioning in less priority environments
Transportation	(2) Access to the specific floor	Allocation of another elevator

In the HVAC system (Figure 5), the low temperature required in the rooms is very important (ex: hospitals building). However, in days where the exterior use of the building and weather conditions are particularly critical, or still, in the case of imperfections in equipment, the water cold production system (used to set comfortable environments) can not be capable to remove all the necessary thermal load of the building. In this in case, the less priority areas, as reception, corridors, etc. are not conditioned, it must be restricted only to the natural ventilation (doors and windows available).

Step 2 - Modeling of the degeneration component

In this step, the models represented by class diagram are developed to compose this productive system.

Static modeling:

It must be incorporate (to connect) the degeneration procedures to the control system model. To include the degeneration characteristics is necessary to associate, for each control component, a degeneration component.

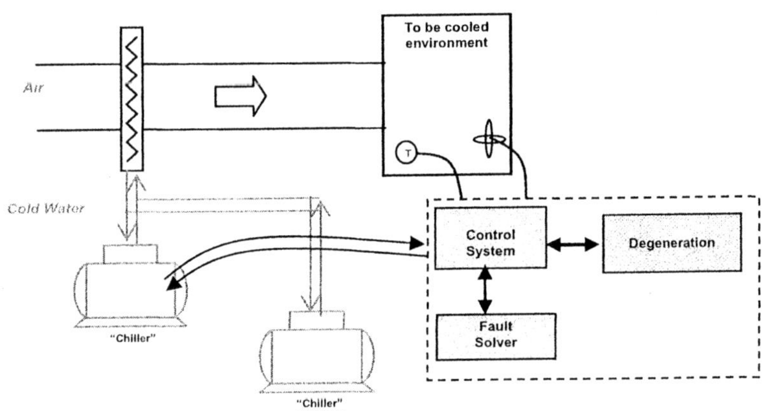

Figure 5 - HVAC sub-system with control and degeneration module.

Figure 6 illustrates a class diagram detailing the association of the control component and degeneration component for the HVAC sub-system. For example, the class structures for the Transport and Access sub-systems are similar, indicating in this way, a high probability of software reuse.

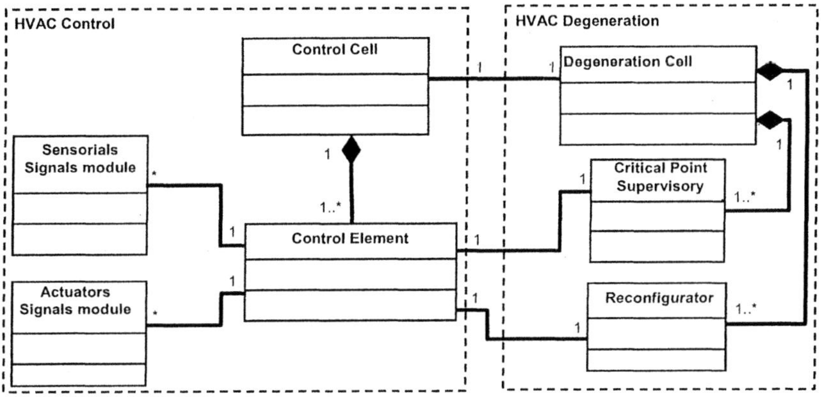

Figure 6 – HVAC with control component and degeneration component.

Dynamic modeling:

The dynamic modeling described here, are related with the critical points 1 and 2 presented in Table 1 (Step 1: identification of the critical points in the intelligent building systems). The model development for the other critical points is analogous.

Figure 7 illustrates a MFG (Mark Flow Graph) model (based on Petri Nets) (Hasegawa, 1998; Miyagi, 1988; Miyagi, 1996) that specifies a procedure to control the temperature, keeping it in a desired value. In this case, the desired temperature is function of sensors signals that detect the temperatures ('Cold', 'Hot') and the heating activity or the cooling activity (linked with the corresponding equipment).

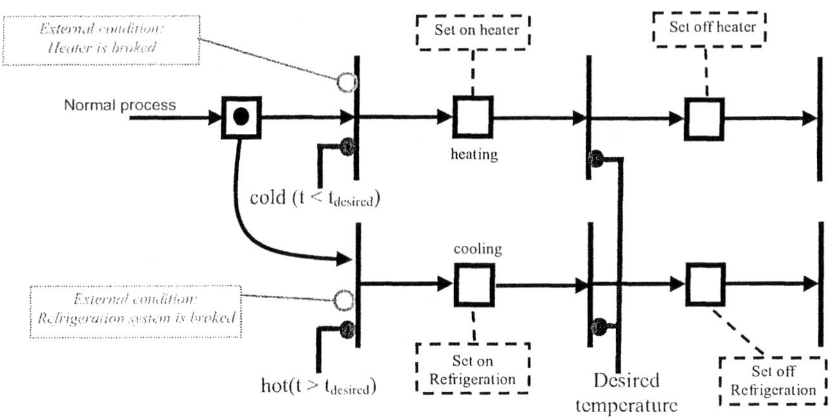

Figure 7. MFG model to keep the temperature in a desired value.

The Figure 8 presents the dynamics for the degeneration in HVAC subsystem, if an unexpected (refrigeration system broke) state happens, and then the fans are turned on until the environments are empty or the temperature reaches the desired value, when the fans are turned off. Thus, the system continues in functioning even in the occurrence of unexpected faults.

Step 3 – Technical specification of degeneration control software

The attributes and the services/operations for the degeneration components are detailed in this step. The technical specifications for the subsystems HVAC and Transport are presented: supervision of critical points components and reconfiguration components.

Component definition:

The control and degeneration components are similar for all the subsystems in analysis. The models of the degeneration components and control components for other subsystems of the intelligent building are similar to the models illustrated in figure 6.

Technical specification of critical point supervisory component

This component is associated with the control component. Thus, it accesses the same information offered for the 'sensor' components in the control component. In Table 2 the specifications for this component are presented (HVAC and Transportation).

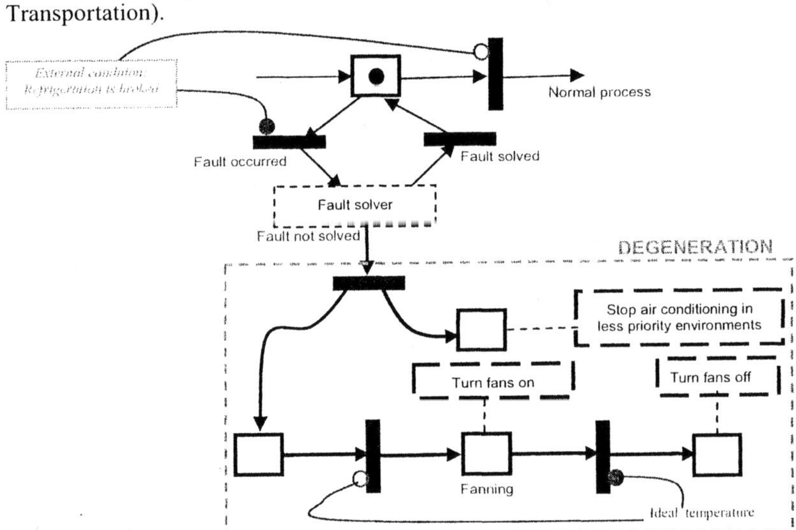

Figure 8. MFG model with degeneration in the refrigeration system.

Table 2. Technical specification of critical point supervisory.

Subsystem	Atributes	Services/Operations
HVAC	_*CriticalBombState: boolean* – indicate the necessity of the degeneration or not. _*CriticalAirState: boolean* – indicate the necessity of the degeneration or not.	*boolean getCriticalBombState()* – get the state of specific sensor for a critical point. *boolean getCriticalAirState()* – get the state of specific sensor for a critical point.
Transportation	_*CriticalEnergyTranspState: boolean* – indicate the necessity of the degeneration or not	*boolean getCriticalEnergyTranspState()* – get the state of specific sensor for a critical point.

Table 3. Technical specification of reconfiguration.

Subsystem	Atributes	Services/Operations
HVAC	*_idAirSegments[]: integer* – store the identificators of ar-condicioning segments. *_idFans[]:integer* – store the identificators of fans.	*disableAirSegments(idSegmentsAr[]: integer)* – send signals to disable less priority segments. *enableFans((idFans[]: integer)* – send signal to enable fans.
ransportation	*_idElevators[]: integer* – store the identificator of elevators. *_idStairsRolling[]: integer* – store the identificators of stairs rolling. *_idStairsDoors[]: integer* – store the identificators of stairs doors.	*disableElevators(idElevators[]: integer)* – send signal to disable elevators. *enableStairsDoors((idStairsDoors[]: integer)* – send signal to enable stairs doors.

Technical specification of reconfiguration component

This component is associated with the control component. Thus, it sends information to the 'actuator' component in the control component. In this way, the corresponding degeneration is carried through activating the corresponding performance components (see Table 3).

5. CONCLUSIONS

Initially was presented the use of ISO/IEC 9126 standards in the control software design with degeneration requirements and the control system archictecture for the discrete event systems with degeneration module included.

The degeneration method was presented through a sequence of tasks necessary to include the degeneration in productive systems. The results was the generation of software artifacts (static and dinamic models and technical specification of components) that will assist the software design for control system in intelligent building. Thus, the use of degeneration requirements in the proposed method permit to increase the system faults tolerance degree.

This paper received financial support from the following entities: FAPESP, CNPq and CAPES. The authors thank specially the Program TIDIA/Kyatera support.

6. REFERENCES

1. Arkin, H.; Paciuk, M. "Service System Integration in Inteligent Buildings". Proc. of IB/IC Intelligent Buildings Congress. Tel-Aviv, Israel. pp. 19-30, 1995.
2. Arakaki, J. "Técnicas de Degeneração no projeto do controle de Sistemas Produtivos". Doctoral thesis, University of São Paulo. São Paulo-SP, 2004.
3. Bastidas Gustin, G. D.; et all. "Modelling integration of systems in intelligent buildings through mechatronics approach". In: 8th Mechatronics Forum International Conference, 2002, Ensc. Mechatronics 2002 : proceedings. Enschede: Drebbel Institute for Mechatronics, 2002.
4. Bastidas Gustin, G.D.; et all. "Open distributed supervisory system design using Petri nets". In: IEEE International Symposium on Industriasl Electronics, Rio de Janeiro, 2003. ISIE'2003. Piscataway : IEEE. v. 2, p. 712-717, 2003.
5. Flax, B. M. "Intelligent Buildings". IEEE Communications Magazine, p. 24-27, 1991.

6. Hasegawa, K.; et all. "Enhanced production flow schema for modeling the complex sharing system". In: 3rd International Conference on Information Technology for Balanced Automation Systems in Manufacturing, Prague, 1998. Inteligent systems for manufacturing: multi-agent systems and virtual and organizations: proceedings of the BASYS'98. Boston : Kluwer Academic, p. 335-344, 1998.
7. ISO/IEC 9126. "Information Technology – Software Product Quality". 1998.
8. Miyagi, P. E.; Hasegawa, K.; Takahashi, K. "A Programming Language for Discrete Event Production Systems Based on Production Flow Schema and Mark Flow Graph". Transactions of The Society of Instrument and Control Engineers, Tokyo, v. 24, n. 2, p. 183-190, 1988.
9. Miyagi, P. E. "Controle Programável – fundamentos do controle de sistema a eventos discretos". Editora Edgard Blücher, São Paulo, 1996.
10. Miyagi, P. E.; et all. "Petri Net Approach for Modelling System Integration in Intelligent Buildings". Journal of The Brazilian Society of Mechanical Sciences, Rio de Janeiro, v. 23, n. 4, p. 341-350, 2002.
11. Villani, E.; Miyagi, P. E. "A Hybrid Petri Net Modeling Approach for HVAC Systems in Intelligent Buildings". Controle & Automação, a Journal of SBA, Campinas, v. 15, n. 2, p. 151-163, 2004.
12. Villani, E.; et all. "A Petri net-based object-oriented approach for the modelling of hybrid productive systems". Nonlinear Analysis, v. 62, n. 8, p. 1394-1418, 2005.

51

A NEW METHOD FOR THE HIERARCHICAL MODELING OF PRODUCTIVE SYSTEMS

Fabrício Junqueira, Paulo E. Miyagi
Escola Politécnica, University of São Paulo, Brazil
fabri@usp.br, pemiyagi@usp.br

Industry reorganization and the increase in the automation level of productive processes results in the augmentation of complexity of the interactions among enterprise subsystems related to monitoring and control. Analysis techniques are used to deal with the complexity, design of new productive systems, and improve the performance of existing systems. In this context there is currently a special focus in distributed simulation. It deals with the execution of simulation in physically dispersed computers connected through a network. In order to explore the potential of distributed simulation, this paper proposes a new method for the hierarchical modeling of productive systems proper for distributed environment. The application of the method is illustrated through an example. This method has been successfully applied to a number of case studies in order to confirm its effectiveness.

1. INTRODUCTION

Markets are becoming global and independent of geographic barriers. More manufacture industries have been established in a distributed and dispersed way. They take advantage of the growth in the communication networks and information technology (IT), which provides the means for a larger transnational cooperation. A change in the way people considers productive systems have also been observed. New technologies and approaches provide managers and engineers with an integrated and dynamic view of productive systems. The enterprise is not considered anymore an isolated entity, but the part of a cooperative consortium of enterprises (Shi and Gregory, 1998).

The incorporation of IT in productive processes has enabled the integration of enterprise heterogeneous systems (Sanz and Alonso, 2001). As an example, an industrial supervisory system interacts with a heterogeneous set of hardware and software (workstations, remote units, programmable controllers, etc.) in order to monitor and control an industrial plant. The integration of heterogeneous systems results in an increase of system complexity.

The increase of complexity in the productive systems demands the development of new analyses solutions. Among them, the use of distributed simulation deserves particular attention. The distributed simulation deals with the execution of

Please use the following format when citing this chapter:

Junqueira, F., Miyagi, P. E., 2006, in IFIP International Federation for Information Processing, Volume 220, Information Technology for Balanced Manufacturing Systems, ed. Shen, W., (Boston: Springer), pp. 479–488.

simulation in geographically dispersed computers connected through a network, which results in a virtual super computer (Fujimoto, 1999; Banks et al., 2000).

In this context the purpose of this paper is to present a new modeling approach suitable for distributed simulation. This new modeling approach is based on the use of object-oriented concepts with Petri nets and progressive refinement techniques such as PFS (Production Flow Schema). The models generated by the proposed approach can be integrated and simulated concurrently with other models in a distributed and geographically dispersed environment.

Some reasons to distribute the simulation among multiple computers are (Fujimoto, 1999): (1) reduction of the execution time through the subdivision of a huge simulation model among processors; (2) integration of simulators, combining simulations that are executed in different manufacturers machines; and (3) fault tolerance, i.e., if a processor fails, other processors continue the simulation.

Kachitvichyanukul (2001) and Banks (2000) emphasize the employment of reusable models based on components. A component can be selected from a repository and used alone, or be combined with others to generate a new one.

An inherent distributed simulation problem is the partition of the model among processors. Nevison (1990) *apud* Fujimoto (1990) use previous knowledge about the modeled system to optimize the simulation. However, the optimization of the simulation becomes impracticable when working with flexible modeling tools. In this case, the modeling tool does not restrict the kind of productive systems under study. It is therefore impossible to use optimization strategies based on the previous knowledge about the system as well as the number of processors being used in the simulation.

On the other hand, models based on Discrete Event Dynamic Systems (DEDS) are intensively used to describe, analyze and control processes in productive systems. These systems are characterized by the occurrence of instantaneous events, which govern their dynamics. The evolution of state of these systems is based on rules that define the conditions for the occurrence of events as well as the new state reached after an event (Cassandras and Strickland, 1992).

Petri net is a graphical and mathematical modeling technique originally proposed by Carl Adam Petri in 1960 to characterize concurrence in computer operations (Murata, 1989; Moore and Brennan, 1996), which is a typical DEDS. Since then, it has been used for modeling dynamic systems in a large range of areas (Srinivasan and Venkatasubramanian, 1998), such as (Elkoutbi and Keller, 1998): communication protocols, distributed algorithms, computers architectures, man-machine interaction. Daum and Sargent (1999), and Kachitvichyanukul (2001), among others, stand out the hierarchical modeling as a way to deal with complex systems. The modeler can divide a complex system in subsystems (and consequently sub-models) that are better managed. Models can be generated in different abstraction levels, helping the verification and validation process.

Besides the system analysis and design considerations, the way that a productive system is modeled is based both on its inherent characteristics, such as system complexity, as well as on personal factors, such as project team experience and the intended abstraction level. In any case, the project team must be able to visualize the productive system (or the main parts under study) as a whole, its parts and behavior, and the relationship among them (their interfaces).

Based on the above consideration, in the following section, a new method for hierarchical modeling of productive system is presented. This method is proper for distributed simulation in the sense that the resulting models and its interactions are explicitly and dearly specified, exploiting the potential of distributed systems (Junqueira et al., 2005a, 2005b).

2. MODELING APPROACH

Figure 1 shows the steps of the proposed approach, which are discussed as follows.

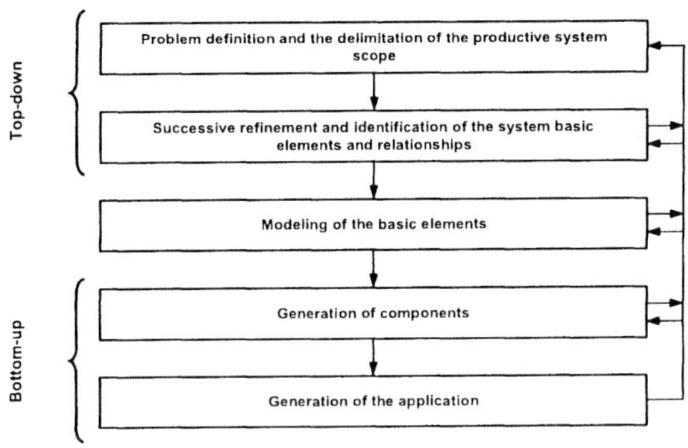

Figure 1 – Diagram of the proposed method for the modeling of productive systems.

Step 1 – Problem definition and the delimitation of the productive system scope

The modeler must delimit the scope of the productive system under study, i.e. what are the departments, equipment (tools) and people (both considered as resources) of the productive system, and what are the characteristics and processes to be modeled and analyzed.

Step 2 - Successive refinement and identification of the system basic elements and relationships

A top-down approach is adopted in this step. The use of modeling techniques such as PFS (Production Flow Schema) helps on the productive system modeling process (Hasegawa et al., 1998; Miyagi, Santos Filho, Arata, 2000). The PFS is a type of high level Petri net composed by *activities*, *distribution elements*, and *arcs*. It is a conceptual model applied in the initial phase of the system modeling process that is gradually translated into a Petri net model.

The modeling process starts at the "top" with a productive system conceptual model that is detailed until the desired level of abstraction. At the end of this step, a set of basic elements that constitutes the productive system had been identified, as well as the relationships among them, i.e., their interfaces and message exchanged formats.

Step 3 – Modeling of the basic elements

At this step, the basic elements functionalities are modeled using Petri net. Each model is called "class" (Figure 2 (a)). Similar to object-oriented program languages, a class describes a set of objects that shares the same attributes, operations, relationships, and semantics.

The model of each basic element can be analyzed isolated from the model of the remaining system, facilitating validation before its use to compose new and more complex models.

Step 4 – Generation of components

Once the basic elements (classes) have been defined, they can be combined to form more complex ones. This step may be arranged in four sub-steps: (4.1) definition of objects; (4.2) encapsulation of objects into components; (4.3) connection of object interfaces; and (4.4) mapping the remained objects interfaces as components interfaces.

The componentization process starts (4.1) using each class as a template to generate one or more objects. A bottom-up approach is adopted, and objects that share some common features, or need to work together to perform a task, are grouped (4.2) to constitute a component (Figure 2 (b)). Then, (4.3) object interfaces are connected (black arrows in Figure 2 (b)). For the proposed method, in a Petri net model, the interfaces are modeled as transitions and the relationship among models are done through transitions fusion (Figure 3) (Sibertin-Blanc, 1993).

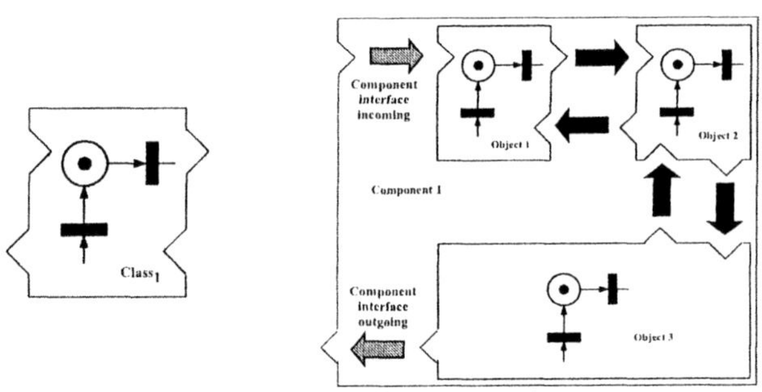

(a) (b)

Figure 2 - (a) a class modeled as a Petri net; and (b) a component constituted by three objects.

To conclude the component model, it is necessary (4.4) to map the remaining object interfaces as the component interfaces. The downward diagonal arrows in Figure 2 (b) are examples of this mapping.

Step 5 – Generation of the application

To generate an application, two or more components are grouped and their interfaces are connected (Figure 4). This step is similar to the previous one. The difference is that applications do not have external interfaces. In other words, making analogy with software elements, a stand-alone component does not execute anything and may be used in different contexts, while an application has all the necessary elements to work alone and has a well-defined purpose.

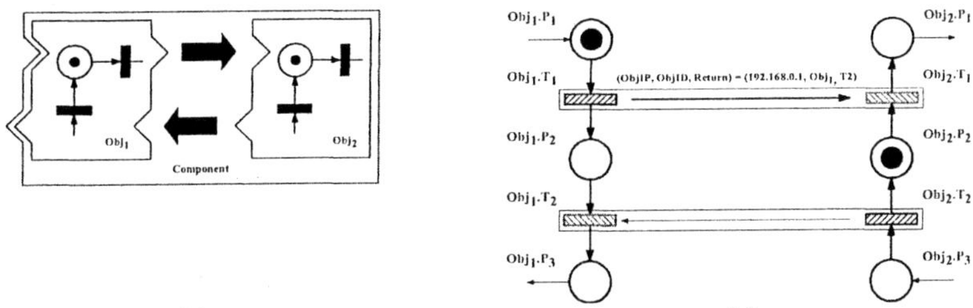

(a) (b)

Figure 3 – Example of object interface: (a) schematic representation; and (b) Petri net representation with transitions fusion.

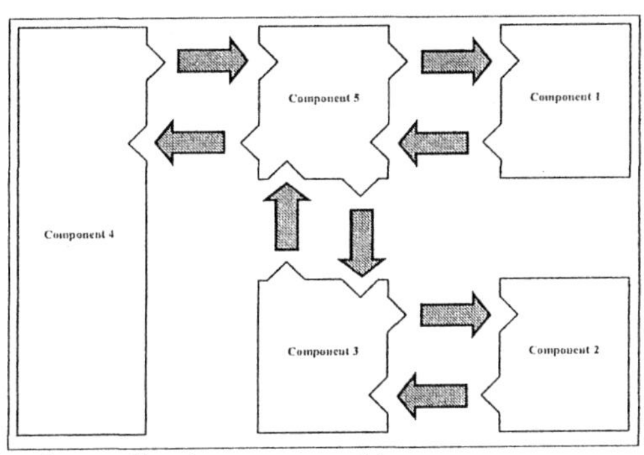

Figure 4 – An application composed by two or more components.

3. A MATERIAL TRANSPORTATION SYSTEM EXAMPLE

To illustrate the application of the method, a practical example is presented. The complete models are in Junqueira (2006); in this paper it is presented a sample of some significative aspects of the method.

Step 1 – Problem definition and the delimitation of the productive system scope

The material transportation system is composed by four stations. Each station one produces or consumes goods. Station A produces goods for Station B, and Station C produces goods for Station D. A vehicle, with unitary capacity, is used for the transportation of goods as showed in Figure 5.

Each station has a processing time. Stations A and C delivery a good (and Stations B and D get a new good) when they finish their processing time. The mechanism to transfer goods from a vehicle to a station and vice-versa is not considered in this example. However, the transfer time is modeled as a vehicle characteristic, i.e., the vehicle will remain stopped until the conclusion of transfer.

Step 2 - Successive refinement and identification of the system basic elements and relationships

Based on the described system (Figure 5(a)), the PFS is used to model the main activities as well as to detail its relationships. Figure 5(b) is the transcription of Figure 5(a) system schema to PFS. It represents the vehicle circular path and the vehicle passage through each station. Some activities of this model may be highlighted: (1) [Transport between stations]; (2) [Station stop] (where goods are (un)loaded); (3) [Vehicle]; and (4) [Station]. Moreover, square dot arrows represent the information exchanged between [Station stop] and [Station], and the long dash arrows, the information between [Station stop] and [Vehicle].

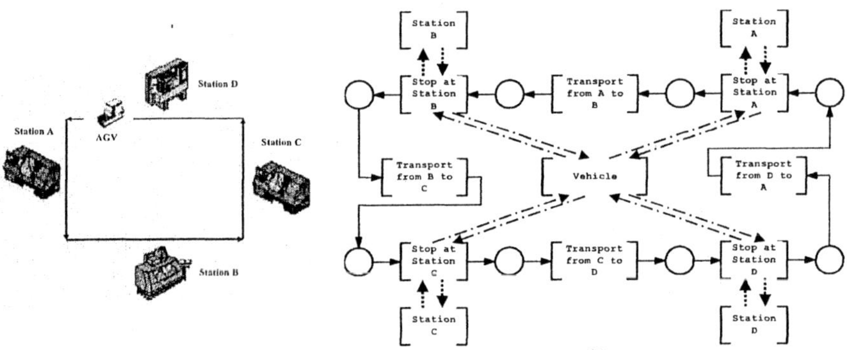

(a) (b)

Figure 5 – (a) vehicle circular path, passing through the Stations; and (b) material transportation system PFS.

Step 3 – Modeling of the basic elements

A Petri net model of the activity [Transport between stations], where the stations are generically identified by X (departure) and Y (arrive) is presented in Figure 6. Place P1 is a pre-condition of this operation. Transition T1 considers the time needed to go from X to Y. Place P2 is the pos-condition of the operation.

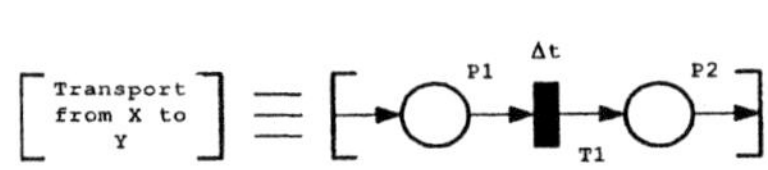

Figure 6 – Petri net model of [Transport between two stations (from X to Y)].

Figure 7 presents the stop at each station (generically called Station Z). Place P1 signs the arrival of the vehicle to the station. The operation beginning is represented by transition T1. Place P2 represents the (un)load process. T2 signs the end of the operation, and place P3 represents the finished state of the operation.

The activity [Vehicle/Station interface] is detailed in Figure 8. Place P1 represents the initial state of the (un)load operation. P2 signs the (un)load operation, and P3, its end. Transitions T1 and T2 are, respectively, the beginning and end of the (un)load operation.

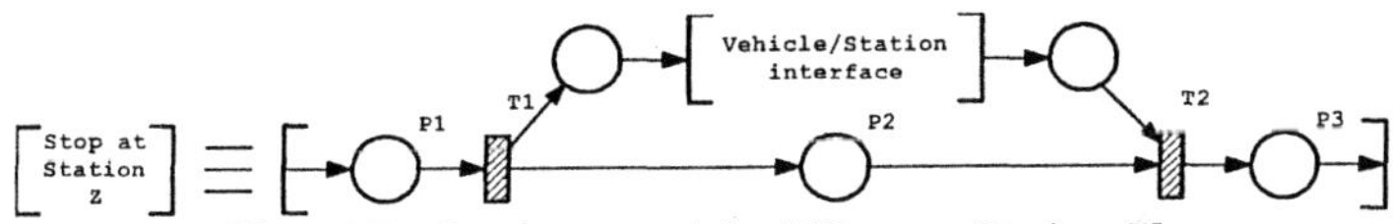

Figure 7 – Petri net model of [Stop at Station Z].

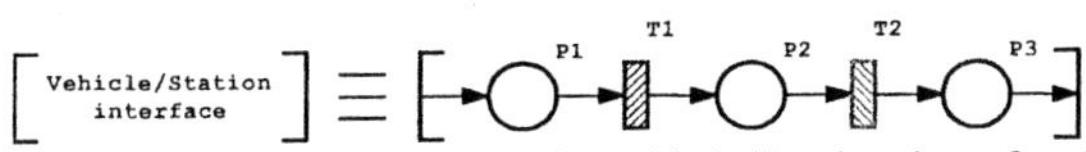

Figure 8 – Petri net model of [Vehicle/Station interface].

Figures 9 (a) presents the Petri net model of [Station Z]. Transitions T1 and T2 are the [Station Z] model interfaces. T1 represents the beginning of the (un)load operation while T2 represents its end. Transition T3 represents the (un)load operation duration. Place P1 represents the station waiting for the end of the (un)load operation. Places P3 and P2 are, respectively, the beginning and end of good processing operation.

The [Vehicle] Petri net model is showed in Figure 9 (b). Place P1 represents the traveling state of the [Vehicle], while P2 and P3 are respectively the beginning and end state of the (un)load operation. Transition T1 represents the vehicle stop and the (un)load operation beginning. T3 represents the (un)load operation duration, and T2, the (un)load operation end. T1 and T2 are also the interfaces of [Vehicle] model.

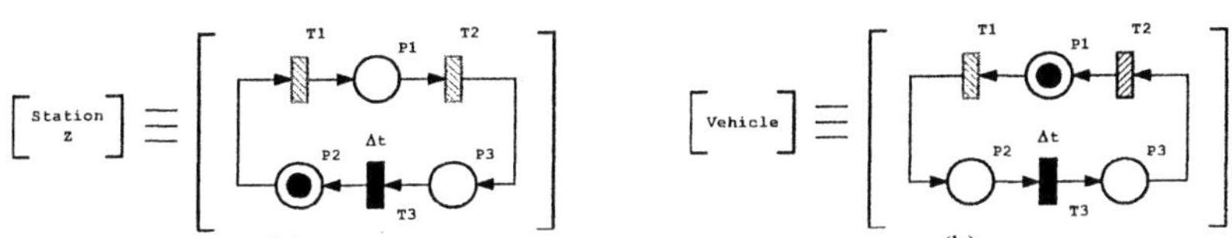

(a) (b)

Figure 9 – (a) Petri net model of [Station Z]; and (b) Petri net model of [Vehicle].

The relationships between [Stop at station Z] and [Station Z] models are made through the fusion of transitions (Figure 10). Transitions T1 of both models are merged, working as a single one. The same is done for T2.

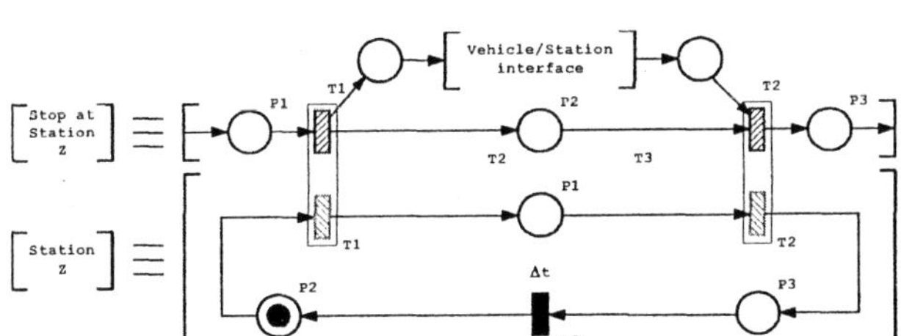

Figure 10 – The relationship between the activities [Stop at station Z] and [Station Z] through transition fusion.

Figure 11 illustrates [Vehicle] and [Vehicle/Station interface] transitions fusion. Transitions T1 of both models behave as an only one. The same occurs with their transitions T2.

Analyzing Figures 5 (b), 6, and 7, a simplification can be applied to part of the model detached on Figure 12 (a). In this example, place P2 of [Stopped in C] model and P1 of [Transport from C to D] model (Figure 12 (b)), for example, can be merged, resulting in the place P1 of the Figure 12 (c) model.

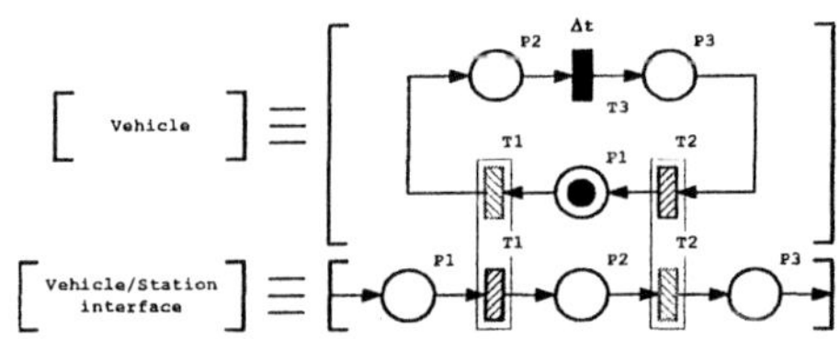

Figure 11 – The relationship between the activities [Vehicle] and [Vehicle/Station interface] through transition fusion.

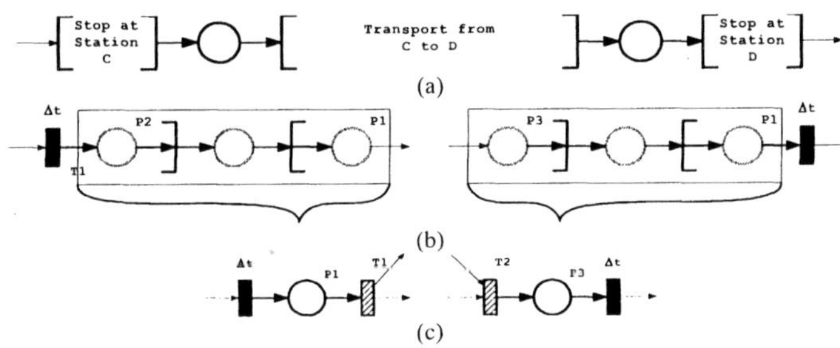

Figure 12 – Application of techniques to reduce the model.

The simplification technique is applied to the model of Figure 5 (b), which results in the [Path] model (Figure 13). Another simplification adopted in this example is the use of only one pair of transitions (T13 and T14) to interface with the [Vehicle] model. Therefore, conflicts for resources become evident in the model and the component generation (step 4) is simplified.

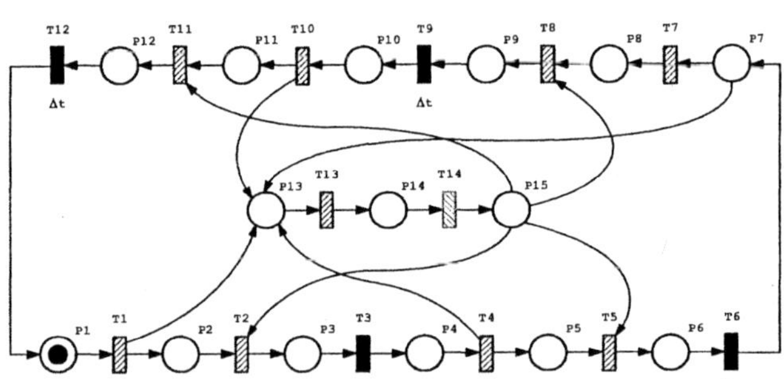

Figure 13 Petri net functional model of Figure 5 (b).

Step 4 – Generation of components

Figures 14 (a), (b) and (c) show, respectively, the AGV objects (based on the class Vehicle), Station A (B, C e D) (based on the class Station), and Path network (based on the class Path).

Then these objects are grouped to compose the Manufacture cell component (Figure 14 (d)). The UML component interface notation is used on Figure 14 object models to show the exchange of information among them.

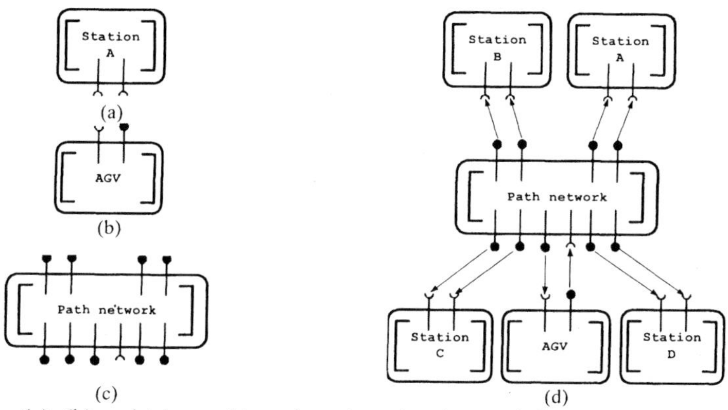

Figure 14 – (a), (b) and (c) are objects based on the classes defined at the step 3; and (d) manufacture cell component.

Step 5 – Generation of the application

For this simplified example, the application model is the same as the component one.

4. CONCLUSIONS

As observed for Daum and Sargent (1999), Kachitvichyanukul et al. (2001) and Banks (2000), researches in the modeling and simulation area are necessary, especially in the direction of developing new hierarchical modeling techniques, as well as the reuse of models.

The proposed modeling technique makes possible the progressive productive system refinement and understating, through successive steps. This approach allows a better characterization of the system elements and the relationships among them. Elements that possess common characteristics can be represented by a single model, guaranteeing its reusability. Concurrently, a library of models may be implemented. In this level, the properties and functionalities of each element can be verified. Then, from the composition of these elements, more complex elements, such as subsystems, can be created, and its functionalities, validated. The whole system model is obtained through successive compositions.

Moreover, this method is proper for distributed simulation, once resulting models and its interactions are explicitly and dearly specified. Thus, distributed simulation can be optimized since models can be better distributed among processors.

5. ACKNOWLEDGMENTS

The authors would like to thank the partial financial support of the Brazilian governmental agencies CNPq, CAPES, and FAPESP, specially the TIDIA/KyaTera committee.

6. REFERENCES

1. Banks, J. (Chair) Simulation in the future. In. Proceedings of the 2000 Winter Simulation Conference, 2000; 1568-1576.
2. Cassandras, CG, Strickland, SG. Sample Path Properties of timed Discrete Event Systems in Discrete Event Dynamic Systems – Analyzing Complexity and Performance in the Modern World. New York: IEEE Press, 1992.
3. Daum, T, Sargent, RG. Scaling, hierarchical modeling, and reuse in an object-oriented modeling and simulation system. In. Proceedings of the 1999 Winter Simulation Conference, 1999; 1470-1477.
4. Elkoutbi, M., Keller, RK. Modeling Interactive Systems with Hierarchical Colored Petri Nets. In: Proceedings of the 1998 Advanced Simulation Technologies Conference, 1998; 432-437.
5. Fujimoto, R. Parallel Discrete Event Simulation. Communications of the ACM, 1990; Vol. 33, No. 10, 30-53.
6. Fujimoto, RM. Parallel and distributed simulation. In. Proceedings of the 1999 Winter Simulation Conference, 1999; 122-131.
7. Hasegawa, K et al. Enhanced production flow schema for modeling the complex resource sharing system. In: BASYS'98 IEEE/IFIP Int. Conf. on Information Technology for Balanced Automation Systems in Manufacturing, Praha, 1998; 335-354.
8. Junqueira, F., Villani, E., Miyagi, PE. A Platform for Distributed Modeling and Simulation of Productive Systems based on Petri Nets and Object-Oriented Paradigm. In: ETFA'05 10th IEEE. International Conference on Emerging Technologies and Factory Automation, Proceedings, IEEE, Catania, Italy, 2005a; 907-914.
9. Junqueira, F., Villani, E., Miyagi, PE. A petri net based platform for distributed modeling and simulation of productive systems. In: COBEM2005 International Congress of Mechanical Engineering, Proceedings. Ouro Preto, Minas Gerais, Brazil; 2005b.
10. Junqueira, F. Modelagem e simulação distribuída de sistemas produtivos. Doctor Thesis, Escola Politécnica da Universidade de São Paulo. Departamento de Engenharia Mecatrônica e de Sistemas Mecânicos, São Paulo, 2006.
11. Kachitvichyanukul, V. (Chair) Simulation Environment for the new millennium. In. Proceedings of the 2001 Winter Simulation Conference, 2001; 541-547.
12. Miyagi, PE, Santos Filho, DJ, Arata, WM. Design of deadlock avoidance compensators for anthropocentric production systems. In: BASYS'2000 IEEE/IFIP Int. Conf. on Information Technology for Balanced Automation Systems in Production and Transportation, Berlin, 2000; 287-294.
13. Moore, KE, Brennan, JE, Alpha/SIM Simulation Software Tutorial. In: Proceedings of the 1996 Winter Simulation Conference, 1996; 632-639.
14. Murata, T. Petri Nets - Properties, Analysis and Applications. In: Proceedings of the IEEE, 1989; Vol. 77, No. 4.
15. Nevison, C. Parallel simulation of manufacturing systems: Structural factors. In Proceedings of the SCS Multiconference on Distributed Simulation 22, 1990; Vol. 1, 17-19.
16. Sanz, R, Alonso, M. CORBA for control systems. In: Annual Reviews in Control, 2001; No. 25, 169-181.
17. Shi, Y, Gregory, M. International manufacturing networks – to develop global competitive capabilities. Journal of Operations Management, 1998; No. 16, 195-214.
18. Sibertin-Blanc, C. A Client-Server Protocol for the Composition of Petri Nets. In: Proceedings 14th International Conference on Application and Theory of Petri Nets. Chicago, Illinois, USA, 1993; 377-396.
19. Srinivasan, R, Venkatasubramanian, V. Automating HAZOP analysis of batch chemical plant: Part I. The knowledge representation framework. Computers Chem. Engineering, Great Britain, 1998; Vol. 22, No. 9, 1345–1355.

52

MODELING AND ON-LINE MONITORING OF MACHINED SURFACE IN TURNING OPERATIONS

Avisekh Banerjee
University of Western Ontario
London, Ontario, Canada
abanerj@uwo.ca
Evgueni V. Bordatchev
National Research Council of Canada,
Integrated Manufacturing Technologies Institute
London, Ontario, Canada
evgueni.bordatchev@nrc.gc.ca
Sounak Kumar Choudhury
Indian Institute of Technology, Kanpur, India
choudhry@iitk.ac.in

Machined surface profile and roughness are important parameters in evaluating the quality of a machining operation. They are resulted from the transformation of the complex tool-workpiece displacements involving the dynamics of the machine tool mechanical system, cutting process, and cutting motions. The focus of this study is the fundamental understanding of the surface profile formation during turning and development of regression and neural network (NN) models of surface roughness incorporating the effects of cutting parameters and tool-workpiece displacements. Also, a bifurcated opto-electrical transducer was developed for on-line monitoring of surface roughness based on the scattering of laser beams from machined surface. The feasibility of on-line monitoring was studied by comparing with actual roughness as well as the prediction results of the regression and NN models.

1. INTRODUCTION

The geometric quality of the machined part with respect to its accuracy, precision and surface finish, is one of the most important measures of the machine tool performance, functionality and final acceptability of the product. Therefore, advanced CNC-based machining technologies will require reliable sensor and control technologies to monitor and control cutting process parameters, and thereby measure and predict the actual surface finish. To achieve un-manned machining, surface roughness, needs to be monitored on-line during the cutting process, so that in case of deviation from the desired limits, control action could be taken.

The actual surface roughness profile can be measured by on-line monitoring the

Please use the following format when citing this chapter:

Banerjee, A., Bordatchev, E. V., Choudhury, S. K., 2006, in IFIP International Federation for Information Processing, Volume 220, Information Technology for Balanced Manufacturing Systems, ed. Shen, W., (Boston: Springer), pp. 489–498.

relative distance of the surface irregularities from a reference plane parallel to the surface with the help of laser beam (Shriashi, 1981, Lee, 1987). An indirect on-line roughness measuring technique based on cutting vibration (Jang, 1996) has also been proposed, as a convenient alternative to direct measurement.

Due to the technical difficulties in on-line monitoring of surface roughness, mathematical modeling of surface roughness based on cutting parameters, for example, first and second order surface regression models (Choudhury, 1997), has been proposed for turning operations. The relative radial displacement between tool and the workpiece (Sata, 1985), cutting vibration (Chen, 2001) and machine tool vibration (Abouelatta, 2001) play an important role in the generation of the surface roughness and had been included as an input to the regression models (Lin, 1998). Apart from regression models, evolutionary computational tool, e.g., neural networks (NN) (Rangwala, 1981, Dimla, 1997, Risbood, 2003), has also been proposed owing to the non-linearity of the cutting process, making the prediction of the surface roughness difficult.

In this paper, both the modeling of surface roughness and its on-line measurements with bifurcated opto-electrical transducer are considered for the purpose of on-line monitoring of machined surface in turning operations. The feasibility of the proposed approach is studied by analyzing the predicted error obtained from regression and NN modeling and measurement error with respect to the actual surface roughness.

2. MODELING

2.1 Systematic Approach

The only task of any machine tool is the production of a workpiece with desired geometric quality. Therefore, modeling of the machining performance should be focused on analysis and prediction of the actual surface roughness. Figure 1 shows a systematic representation of the surface profile formation as a dynamic process involving the dynamics of the machine tool mechanical system and the cutting process. The machine tool mechanical system has three principal components "A", "B", and "C" representing the spindle, carriage, and the tailstock respectively.

Based on the systematic representation and making use of the differential operator $p = d/dt$, dynamics of the machining process with respect to surface formation can be expressed by the following system:

$$\begin{cases} \mathbf{x}(t) - \mathbf{W}(p)\mathbf{f}(t) = 0 \\ \mathbf{f}(t) = \mathbf{f}^*(t) - \mathbf{G}(p)\mathbf{x}(t) \\ \mathbf{r}(t) = \mathbf{Q}(p)(\mathbf{x}(t) + \mathbf{u}(t) + \mathbf{u}^*(t)) \\ \mathbf{z}(t) = \Sigma(\mathbf{r}(t)) \end{cases} \quad (1)$$

where $\mathbf{W}(p)$ is the 3×3 differential matrix operator of the machine-tool mechanical system, which transforms the cutting force vector $\mathbf{f}(t)$ into the vector of workpiece-tool relative displacements $\mathbf{x}(t) = \{x_1(t), x_2(t), x_3(t)\}$; $\mathbf{G}(p)$ is the 3×3 differential matrix operator of the cutting process and transforms the $\mathbf{x}(t)$ vector into additional components of the $\mathbf{f}(t)$ vector; $\mathbf{f}^*(t)$ is the force noise vector of the cutting process, and represents the non-linear properties of the cutting process; $\mathbf{Q}(p)$ is the 3×1 differential matrix operator, which represents the surface profile formation as a dynamic process; $\mathbf{r}(t)$ is the radius vector of the machined surface profile, and results from the control vector $\mathbf{u}(t)$ and applying $\mathbf{Q}(p)$ to $\mathbf{x}(t)$; $\mathbf{z}(t)$ is the product quality vector, its components are ovality, cylindricity, roughness, etc., and is finally formed from the workpiece surface profile, represented by $\mathbf{r}(t)$, and by some statistical operator Σ.

This systematic representation shows that all mathematical models of the surface profile formation should link together along with the actual workpiece-tool displacements and cutting parameters. The representation is used for further development of regression and NN models presented below.

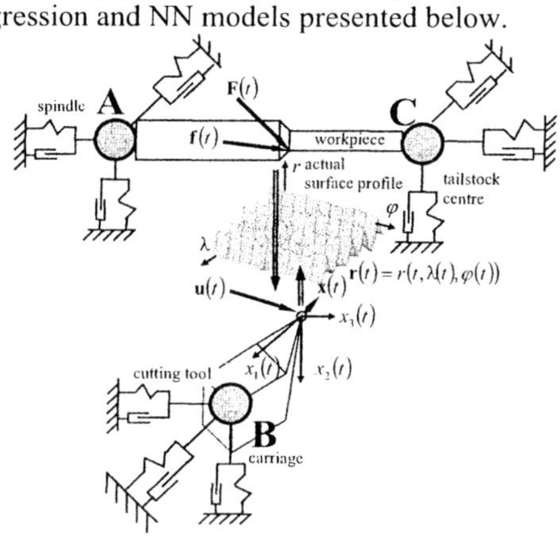

Figure 1 – Formation of surface profile as a dynamic process.

2.2 Regression Analysis

Multivariable regression analysis approach is proposed to determine a functional relationship between cutting process parameters (input variables) and surface finish (output parameter), to predict surface roughness. Both log-linear and second order response models have been implemented such that the effect of different independent variables as well as the effect of their interactions with each other on surface roughness can be modeled. The generic models for predicting the surface roughness $R_a^{predicted} = K \cdot v^{\beta_1} \cdot f^{\beta_2} \cdot \alpha^{\beta_3} \cdot \bar{a}_{disp}^{\ \beta_4}$ (Jang, 1996) based on the measured variables of speed (v), feed (f), rake angle (α), and the averaged acceleration of workpiece-tool displacement ($\bar{a}_{disp}$) can be linearized by taking the logarithm and expressed in terms of a linear regression model as follows:

1) Three regressor model:
   a) Multiple log-linear regression model:
   $$\ln(R_a^{predicted}) = \beta_0 + \beta_1 \ln(v) + \beta_2 \ln(f) + \beta_3 \ln(\alpha) + \varepsilon \qquad (2)$$
   b) Second order surface response model:
   $$R_a^{predicted} = \beta_0 + \beta_1 v + \beta_2 f + \beta_3 \alpha + \beta_4 v^2 + \beta_5 f^2 + \beta_6 \alpha^2 + \beta_7 vf + \beta_8 v\alpha + \beta_9 f\alpha + \varepsilon \qquad (3)$$
2) Four regressor model:

a) Multiple log-linear regression model:

$$\ln(R_a^{predicted}) = \beta_0 + \beta_1 \ln(v) + \beta_2 \ln(f) + \beta_3 \ln(\alpha) + \beta_4 \ln(\bar{a}_{disp}) + \varepsilon \quad (4)$$

b) Second order surface response model:

$$R_a^{predicted} = \beta_0 + \beta_1 v + \beta_2 f + \beta_3 \alpha + \beta_4 v^2 + \beta_5 f^2 + \beta_6 \alpha^2 + \beta_7 \bar{a}_{disp}^{\ 2} + \beta_8 vf + \beta_9 v\alpha + \beta_{10} f\alpha + \beta_{11} v\bar{a}_{disp} + \beta_{12} f\bar{a}_{disp} + \beta_{13} \alpha\bar{a}_{disp} + \varepsilon \quad (5)$$

Regression coefficients (β_i) in above models were identified from experimentally obtained data using statistical package *MINITAB*™.

2.3 Neural Network

Neural Network (NN) consists of massive interconnections of simple processing units or neurons. Each connection is associated with a weight, which is adjusted based on previous input-output mapping, thus loosely simulating a human brain. The network can incorporate past experience and subsequently be used to map the input-output of any generic problem. The inherent non-linearity of the cutting process even for turning has made NN well accepted for predicting surface roughness (Dimla, 1997, Risbood, 2003). An output response is generated from a simple processing unit or perceptrons when the sum of the weighted input signals (w_{ii}) exceeds a certain threshold. A bias value (b_i) is also associated with activation threshold for each processing units as shown in Figure 2(a). Multiple layer perceptrons are used in a feed forward manner such that the signal flows from the input layer to the output layer as shown in Figure 2(b). During the training of the network, the weights are adjusted such that the error between the predicted and targeted output is minimized. For this purpose, an extensively popular back propagation algorithm is used which back propagates the error from the output layer to the input layer along with the adjustments in the weights.

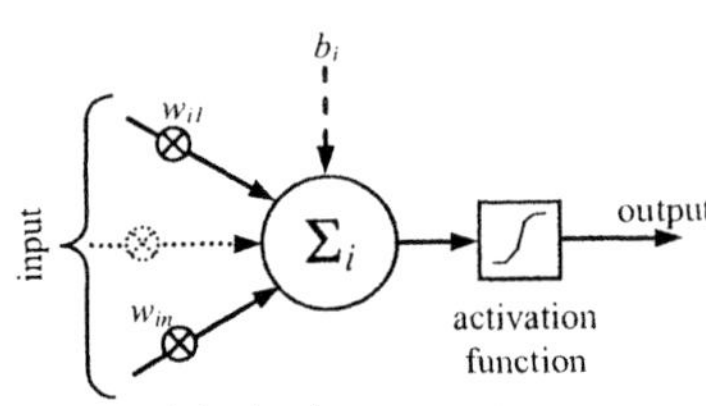

(a) single perceptron

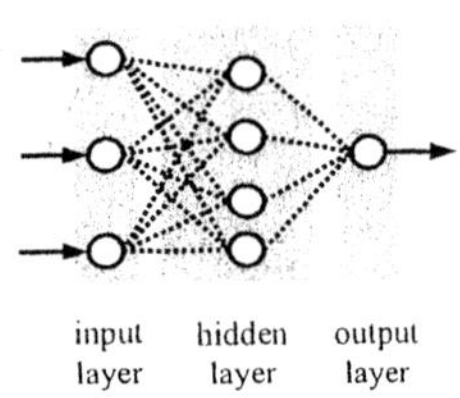

(b) multi-layer perceptron

Figure 2 – Perceptrons

2.3.1 Architecture

The architecture of the NN, i.e. the number of input-hidden-output layers, is intimately linked with the number of data used to train the network and plays an important role in predicting the output. Four different network architectures with a sigmoidal activation function have been developed in this work. The architectures is represented by the input-hidden-output nodes, with the inputs of v, f, α, and $\bar{a}_{disp}$, where as the outputs are $R_a^{predicted}$ and $\bar{a}_{disp}$. The number of perceptrons in the individual layers is selected based on the knowledge of the cutting process as well as the number of training data. For example, for the three input (v, f, α,) model, the output may be only $R_a^{predicted}$ with single hidden layer [3-4-1] and multiple hidden layer [3-2-2-1], but may also have both $R_a^{predicted}$ and $\overline{a}_{disp}$ as the outputs [3-3-2]. Again if $\overline{a}_{disp}$ is considered as an input to the surface generation (Jang, 1996) a [4-3-1] NN model is required.

2.3.2 Training Functions

The NN is trained, during which the weights corresponding to the connections get iteratively modified for minimizing the error between the predicted and target output values. The number of training data should be more than the total number of unknown parameters, i.e. the weights and the biases. A number of training functions have been used with the help of *MATLAB*^{TM} with different algorithms to update the weight and bias values.

2.3.3 Implementation

The 27 experimental data was separated into two sets of 21 and 6 for training and validation respectively such that the training set included the extreme values of the input parameters. Pre-processing of the input data was performed which included removal of the mean value and including unit deviation. A feed-forward multi layer perceptron network was used with different architectures and training functions. After initiating the weight matrix, the training of the network was carried on such that the mean square error was minimized and also repeated for consistent error values. Once the training is completed, the network was used to predict the output with the validation input data. The percentage prediction error (ε_{NN}^{p}) given by the following equation is calculated for all the validation data, averaged and compared to select the most accurate network architecture as well as training functions.

$$\varepsilon_{NN}^{p} = \frac{|Actual - Predicted|}{Actual} \times 100 \tag{5}$$

3. BIFURCATED OPTO-ELECTRICAL TRANSDUCER

3.1 Design and operating principle

The bifurcated opto-electrical transducer shown in Figure 3, is a non-contact type displacement transducer, operating on the principle of scattering of laser beam, and had been used to measure tool wear (Choudhury, 1995) by monitoring the dimensional deviation during the turning operation. It is believed that the transducer can be used for monitoring on-line surface roughness which is superimposed on the dimensional profile, if tool wear, the primary cause of the dimensional deviation is minimized. The transducer consists of bundles of optical fibers with a diameter of 50μm, a 10mW continuous He-Ne Uniphase laser and SI 100s photodiode. The single end section of the transducer is shown in Figure 3, where the inner bundles supply the laser beam which gets scattered from the rough surface and the reflected beams are sent to the photodiode through the outer annular bundles.

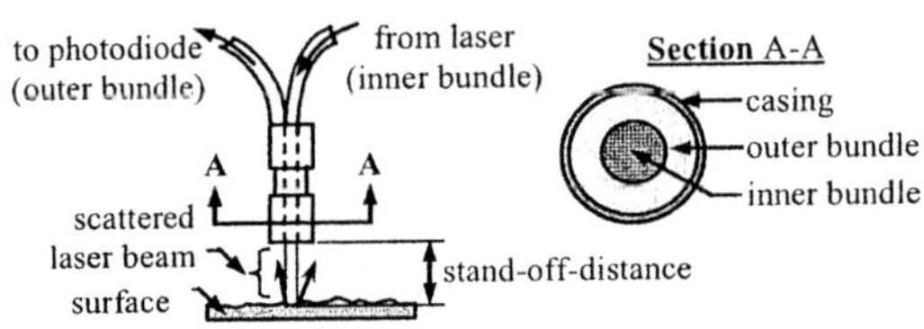

Figure 3 – Bifurcated opto-electrical transducer

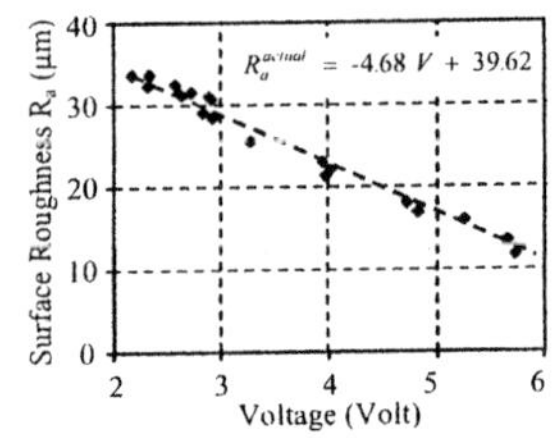

Figure 4 – Calibration chart

3.2 Calibration

Calibration of the bifurcated opto-electrical transducer is required to determine the output voltage (V) in terms of the actual surface roughness (R_a^{actual}) measured off-line. The calibration chart, shown in Figure 4, was obtained as a V - R_a function by measuring pre-machined surfaces with a Taylor-Hobson Surtronic digital surface profiler. The transducer was axially moved over the surface in the axial direction without cutting with a reference stand-off-distance (SoD) of 5.6 mm. Measurements were taken at five different positions, and the equation of the best-fit line was calculated.

4. EXPERIMENTAL SET-UP AND PROCEDURE

The experimental set-up used for on-line measurement and monitoring of actual surface finish is shown in Figure 5. A cylindrical EN24 steel workpiece (80 mm *dia*, 750 mm length) was divided into nine sections of equal lengths and was machined with a manual lathe (HMT LB25) using high-speed steel (10% Cobalt) cutting tool. After turning each section with different feed, the tool was ground to minimize the effects of tool wear and rake angle. A Bruël Kjær 820361 accelerometer was mounted on the tool carriage to measure the accelerations of the tool-workpiece displacements.

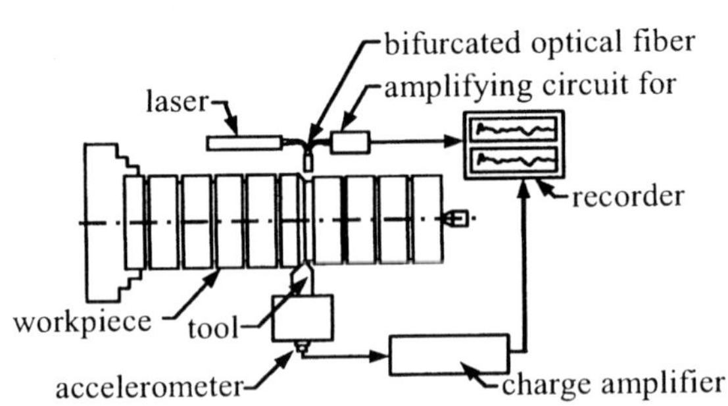

Figure 5 – Schematic diagram of experimental set-up

A 3k full factorial design experiment (Montgomery, 1997) was carried out for three inputs (v , f , α) with three levels (low, intermediate, high) having a total of 27 experiments. The ranges of the input factors (v : 18.6 – 22.6 m/min; f : 0.50 – 0.88 mm/rev; α : -10° – +10°) were selected from handbook (Oberg, 2000) and the machine capabilities such that R_a^{actual} was higher to facilitate the on-line measurement. The depth of cut was kept constant at 0.3 mm, as it has the least effect on the surface roughness (Beauchamp, 1996).

5. COMPARATIVE STUDY

5.1 Correlation between $R_a^{predicted}$ and R_a^{actual}

5.1.1 Regression Models

The minimum least square method incorporated within statistical package *MINITAB*™ was applied to calculate the coefficients (β_i) of the models from 21 experimental data minimizing the prediction error (ε_{reg}^{p}) between the off-line measured R_a^{actual} and calculated $R_a^{predicted}$. The other 6 data corresponding to the intermediate levels of the inputs were used for calculating relative prediction error ($\varepsilon_{reg}^{p} = \{|R_a^{prediction} - R_a^{actual}| / R_a^{actual}\} \times 100\%$) and shown in Table 1.

Table 1 – Comparison of R_a^{actual} and $R_a^{predicted}$ for different regression models

Model		ε_{reg}^{p} (%)
Three Regressor	Multiple Log-Linear	20.71
	Second Order Surface Response	16.73
Four Regressor	Multiple Log-Linear	16.94
	Second Order Surface Response	13.82

Analysis of Variance method and software *ANOVA*™ were used to compare the effect of each input parameter along with their levels on the output variable. Results shown that v had the most significant effect on $R_a^{predicted}$ followed by $\bar{a}_{disp}$, f and α and verified from literature (Escalona, 1998, Chen, 2001).

5.1.2 Neural Network Models

Similar to the regression analysis, the same 21 data were used to train the neural network and the 6 intermediate data to determine the prediction error (ε_{NN}^{p}) between R_a^{actual} and $R_a^{predicted}$. All the above training functions were applied with a maximum number of training cycles of 10,000 to minimize the steady-state mean square error. The architecture, training function as well as ε_{NN}^{p} of the most accurate models with steady state error are given in Table2.

The above table shows that, NN is capable of predicting the actual surface roughness extremely accurately. As expected, more complex architectures or networks with higher number of hidden layers ([3-2-2-1]) give more accurate prediction, with the Levenberg-Marquardt training function working better for such architectures. It can also be observed that the prediction error for $\overline{a}_{disp}$ is much higher as well as when the same is included as an input deteriorates the prediction capability for the actual surface roughness.

Table 2 – Comparison of R_a^{actual} and $R_a^{predicted}$ for different NN models

Architecture	Training Function	ε_{NN}^{p} (%)
[3-4-1]	Gradient Descent	1.141
[3-3-2]	Levenberg-Marquardt	22.97
[3-2-2-1]	Levenberg-Marquardt	0.662
[4-3-1]	Gradient Descent	1.849

5.2 Correlation between $R_a^{on-line}$ and R_a^{actual}

The comparison between $R_a^{on-line}$ and R_a^{actual} obtained from the 27 experiments is shown in Figure 6, which had a correlation of 0.891. If consider R_a^{actual} as the reference, the relative error ($\varepsilon^m = \{|R_a^{on-line} - R_a^{actual}| / R_a^{actual}\} \times 100\%$) was calculated as 14.4%. A main source of this relative error may be relative tool-workpiece displacements and workpiece that change SoD and scattering conditions of the all laser beams from static during off-line measurements to dynamic during actual turning process.

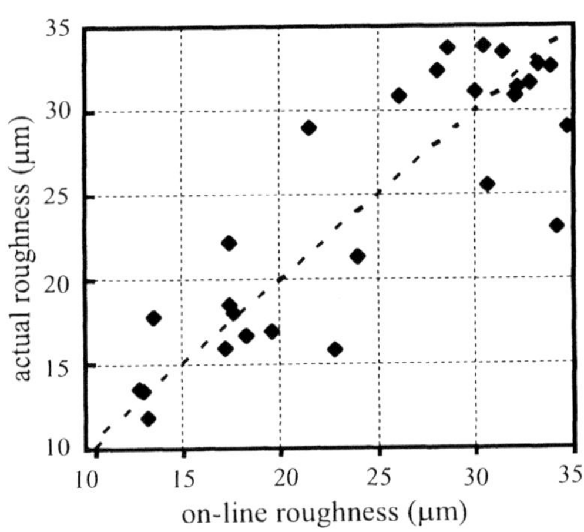

Figure 6 – Correlation between $R_a^{on-line}$ and R_a^{actual}

To summarize the above results, the comparison between R_a^{actual}, $R_a^{on-line}$ (ε^m) and $R_a^{predicted}$ (ε_{reg}^{p}, ε_{NN}^{p}) for the most accurate regression and NN model, with individual errors along with a 15% error ($\varepsilon_{15\%}$) range on R_a^{actual} is shown in Figure 7.

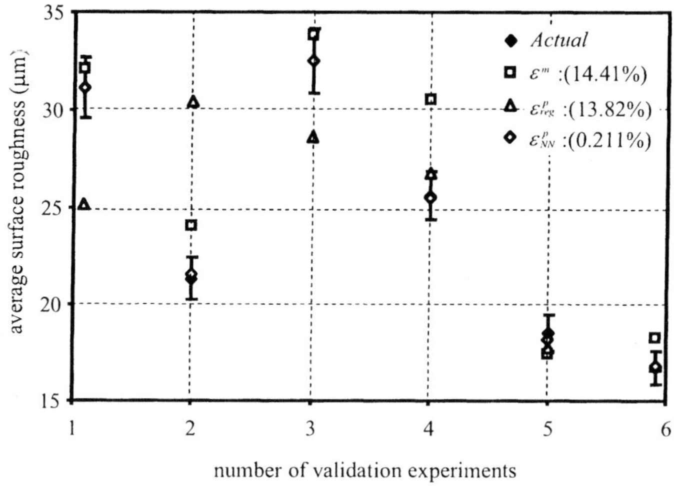

Figure 7 – Comparison of R_a^{actual} , $R_a^{predicted}$ and $R_a^{on-line}$ with 15% error bound

6. CONCLUSIONS

The comparative study of different mathematical models with respect to the actual average surface roughness and the predicted roughness showed that the prediction errors for regression models are higher than NN models for the same validation data, with a minimum error of 13.82% compared to 0.662% for NN. However, the initiation of the weight matrix plays a vital role in the training of the NN and the subsequent selection of the optimally trained network needs much expertise.

The on-line surface roughness monitored using the developed bifurcated opto-electrical transducer closely follows the actual roughness with a correlation of 0.891 with a relative inaccuracy of 14.4% arising between them is primarily due to the relative tool-workpiece displacements and dynamic measurement conditions during the turning. Thus the transducer can be used for on-line monitoring of surface roughness in turning for practical applications within a tolerance of 15%.

Results on comparison of the prediction errors of the modelling and the relative inaccuracy of the on-line monitoring indicate that transducer has better accuracy than most of the regression models. The ability of NN models to predict roughness beyond the input ranges will limit its application significantly. Moreover, the transducer has advanced potential for industrial applications where the entire surface profile is required rather than roughness measured at discrete locations. In such situations, the modeling would become substantially complicated for example for on-line control of the surface roughness in turning where identification of the cutting process dynamics is required.

7. ACKNOWLEDGEMENTS

This study was partially supported by NSERC Discovery Grant R3440A01. The authors will also like to thank the technicians in the Manufacturing Science Lab, Indian Institute of Technology, Kanpur for their help to carry out the experiments.

8. REFERENCES

1. Abouelatta OB, Mádl J. Surface Roughness prediction based on cutting parameters and tool vibrations in turning operations. Journal of Material Processing Technology 2001; 118;269-77.
2. Beauchamp TM, Youssef AY, Masounave J. Effect of tool vibrations on surface roughness during lathe dry turning process. Computers and Industrial Engineering 1996;31(3-4);637-44.
3. Chen JC, Savage MD. Multiple regression-based multilevel in-process surface roughness recognition system in milling operations. The Journal of Technology Studies, Virginia Technical University 2001;28-34.
4. Choudhury IA, Baradie MA El. Surface roughness prediction in turning of high strength steel by factorial design of experiments. Journal of Material Processing Technology 1997;67;55-61.
5. Choudhury SK, Ramesh S. On-line tool wear sensing and compensation in turning. Journal of Material Processing Technology 1995;49(3-4);247-54.
6. Dimla Jr DE, Lister PM, Leighton NJ. Neural network solution to the tool condition monitoring problem in metal cutting- a critical review of methods. International Journal of Machine Tool and Manufacture 1997;37(9);1219-41.
7. Escalona PM, Cassier Z. Influence of critical speed on the surface of finish of turned steel. Wear 1998;218;103-9.
8. Jang DY, Choi YG, Kim HG, Hasiao A. Study of correlation between surface roughness and cutting vibrations to develop an on-line surface roughness measuring technique in hard turning. International Journal of Machine Tool and Manufacture 1996;36 (4);453-63.
9. Lee CS, Kim SW, Yim DY. An in-process measurement technique using a laser for non-contact monitoring of surface roughness and form accuracy of ground surface. Annals of CIRP1987;425-8.
10. Lin SC, Chang MF. A study on the effects of vibrations on surface finish using a surface topography simulation model for turning. International Journal of Machine Tool and Manufacture 1998;38(3);763-82.
11. Montgomery DC. Design and Analysis of Experiments. New York: John Wiley & Sons, 1997.
12. Oberg E, Jones FD, Horton HL, Ryffell HH. Machinery's Hand Book, 26th Edition, Industrial Press, (2000).
13. Rangwala S, Dornfeld D. Sensor integration using neural networks for intelligent tool condition monitoring. Transaction of ASME: Journal of Engineering for Industry. 1981;112;219-28.
14. Risbood KA, Dixit US, Sahasrabudhe AD. Prediction of surface roughness and dimensional deviation by measuring cutting forces and vibrations in turning process. Journal of Material Processing Technology 2003; 132;203-14.
15. Sata T, Li M, Takata S, Hiraoka H, Li CQ, Xiao XG, Xing XZ. Analysis of surface roughness in turning operation and its application. Annals of CIRP 1985;34(1);473-76.
16. Shriashi M. In-process measurement of surface roughness in turning by laser beams. Transaction of ASME: Journal of Engineering for Industry 1981;103;203-9.

AUTHOR INDEX